Urban Hydroinformatics

Data, Models and Decision Support for Integrated Urban Water Management

Urban Hydroinformatics Series

Urban Hydroinformatics

Data, Models and Decision Support for Integrated Urban Water Management

Roland K. Price and Zoran Vojinović

IWA Publishing
London • New York

Published by **IWA Publishing**
Alliance House
12 Caxton Street
London SW1H 0QS, UK
Telephone: +44 (0)20 7654 5500
Fax: +44 (0)20 7654 5555
Email: publications@iwap.co.uk
Web: www.iwapublishing.com

First published in Paperback 2011, Reprinted in Hardback 2012, Paperback in 2017

Cover design by www.designforpublishing.co.uk
Cover image by Jana Vojinović
Typeset by MPS Limited, a Macmillan Company, Chennai, India, www.macmillansolutions.com

British Library Cataloguing in Publication Data
A CIP catalogue record for this book is available from the British Library

Library of Congress Cataloging-in-Publication Data
A catalog record for this book is available from the Library of Congress

ISBN: 9781780409344 (Paperback)

ISBN: 9781780400945 (eBook)

Contents

Acronyms

ADI	Alternating Direction
ADWF	Average Dry Weather Flow
ALS	Airborne Radar Scanner
AMI	Average Mutual Information
AMP	Asset Management Planning
ANN	Artificial Neural Network
AOB	Ammonia Oxidizing Bacteria
ARMA	AutoRegressive Moving Average
ASCE	American Society of Civil Engineers
ASCII	American Standard Code for Information Interchange
BOD	Biological Oxygen Demand
BMP	Best Management Practice
CAD	Computer Aided Graphics
CCTV	Closed Circuit TeleVision
CFD	Computational Fluid Dynamics
COD	Chemical Oxygen Demand
CSO	Combined Sewer Overflow
CSTR	Completely Stirred Tank Reactor
DAP	Drainage Area Plan
DEM	Digital Elevation Model
DEN	DENitrifiers
DHI	Danish Hydraulic Institute
DO	Dissolved Oxygen
DOS	Disk Operating System
DSS	Decision Support System
DTM	Digital Terrain Model

DWF	Dry Weather Flow
EC	European Commission
EM	ElectroMagnetic
EMC	Event Mean Concentrations
EPS	Extended Period Simulation
ESS	Effluent Suspended Solids
EU	European Union
FD	Flood Directive
FIR	Finite Impulse Response
FOSM	First Order First Component
FSR	Flood Studies Report produced by NERC UK in 1975
GDP	Gross Domestic Product
GIS	Geographic Information Systems
GLUE	Generalised Likelihood Uncertainty Estimation
GPS	Geophysical Positioning System
ICT	Information and Communication Technology
ID	Identification
IDW	Interpolation Distance Weighted
IR	InfraRed
IPCC	Intergovernmental Panel on Climate Change
ISO	International Standards Organisation
LIDAR	LIght Detection And Ranging
MBR	Membrane BioReactors
MCS	Monte Carlo Simulation
MDG	Millennium Development Goal
MIMO	Multiple Input Multiple Output
MISO	Multiple Input Single Output
MLP	Multi Layer Perceptron
MT	Model Trees
MW	MicroWave
NIR	Near InfraRed
NOB	Nitrite Oxidizing Bacteria
NPDES	National Pollutant Discharge Elimination Scheme
NPV	Net Present Value
OFWAT	Office of Water
PAC	Powdered Activated Carbon
PAO	Phosphorous Accumulating Organisms
PDE	Partial Differential Equation
PE	Person Equivalent
PHA	Poly Hydroxyl Alkanoates
PL	Person Load
PSIR	Pressure-State-Impact-Response
RAM	Random Access Memory
RBF	Radial Basis Function
RS	Remote Sensing
RTC	Real Time Control

SAR	Synthetic Aperture Radar
SBH	Sludge Blanket Height
SCADA	Supervisory Control And Data Acquisition
SERPES	Sewerage Rehabilitation Planning Expert System
SRM	Sewerage Rehabilitation Manual
SS	Suspended Solids
STW	Sewage Treatment Works
SVM	Support Vector Machine
SWIR	Short Wave InfraRed
SWMM	Storm Water Management Model
SWO	Storm Water Overflow
SWQO	Storm Water Quality Objective
TIN	Triangulated Irregular Network
TIR	Thermal InfraRed
TIS	Tanks-In-Series
TKN	Total Kjeldahl-Nitrogen
TN	Total Nitrogen
TMP	TransMembrane Pressure
TOC	Total Organic Carbon
TP	Total Phosphorus
TSR	Time Series Rainfall
TSS	Total Suspended Solids
UPM	Urban Pollution Management (Manual)
UNEEC	UNcertainty Estimation based on Local Errors and Clustering
USEPA	United States Environmental Protection Agency
USGS	United States Geological Survey
UV	UltraViolet
UWWTD	Urban Waste Water Treatment Directive
VFA	Volatile Fatty Acids
VIS	VISible light
VNIR	Visible Near InfraRed
WADNES	Water Distribution Network Expert System
WAP	Wireless Application Protocol
WaPUG	Wallingford Procedure User Group/Wastewater Planning User Group
WFD	Water Framework Directive
WRc	Water Research Centre
WS	Wallingford Software
WWF	Wet Weather Flow
1D	One Dimensional Space
2D	Two Dimensional Space
3D	Three Dimensional Space

Preface

Ever since Mike Abbott coined the word in 1991, 'hydroinformatics' has emerged as a serious field of study in academia and has been embraced by a number of consultants and operators in the water industry. UNESCO-IHE Institute for Water Education, where Abbott was Professor of Hydroinformatics, became the home of the subject, and therefore, from the beginning hydroinformatics was identified with the management of water in developing as well as developed countries. This distinction between the economic development of countries was seen as a moral and ethical imperative to encourage post graduate students at the Institute, almost all of whom came from developing countries, to take advantage of the particular informatics viewpoint of hydroinformatics in managing water. The MSc course on hydroinformatics took an overview of how information can be used to manage water better, covering topics that ranged from monitoring the physical world, to managing and analysing the resulting data, to creating a virtual world through physical based and data driven modelling, to providing support for decision making, to involving all stakeholders in the different key areas of water management, to appreciating the ethical consequences of interventions in the physical world of water, and to engaging in social justice. As a result the course came to be structured in such a way that the underlying theory of hydroinformatics was addressed first, followed by the application of the theory to particular areas of water management. In the early days the major areas of application were to the management of rivers, estuaries and coastal waters. When one of us (RKP) arrived in 1997 as the successor of Mike Abbott who had just retired, professional experience in urban drainage was contributed to the course, and a greater emphasis was put on urban hydroinformatics. However, it was not until the arrival at UNESCO-IHE of the other of us (ZV) in 2004 that research and education in this branch of hydroinformatics was initiated seriously. Since then the MSc course modules on urban hydroinformatics have grown considerably, with particular emphasis on urban flood risk management.

This book therefore is our first response to the rising significance of urban water management world-wide, an initiative to promote a specifically hydroinformatics viewpoint of how urban water systems can be managed better, and a support to the development of post-graduate face-to-face and on-line courses on urban hydroinformatics. As such, the book is structured in such a way that the particular urban water systems are identified first in terms of the service they provide, the possible failures of the service, and ways of correcting the failures. Our understanding of the nature of hydroinformatics is developed in Chapter 3, followed by a sequence of chapters that explore the outworking of hydroinformatics theory to urban water systems: data acquisition and management, modelling paradigms, decision support systems, involvement of society, and asset management. There follows another sequence of chapters focussed on water distribution, wastewater and stormwater collection, wastewater treatment, receiving water impact, urban flood management, urban water management in developing countries, and the water industry and the institutional framework. The book concludes with a chapter on the future of urban hydroinformatics.

Our knowledge and expertise in urban water management is limited primarily to urban drainage and urban flooding. Therefore we have not presumed to write in detail about other aspects of urban water management; instead we have asked leading professionals and academics to contribute chapters on water distribution, wastewater treatment and social and management aspects. We are grateful to them for their contributions to this book. We have also included in the book a number of specific case studies on different topics written by colleagues from around the world.

The book provides an introduction to urban hydroinformatics, and is therefore the first in a series with this general title. It is written primarily for post-graduate students wanting to have an overview of urban hydroinformatics and it supports two UNESCO-IHE regular/on-line modules: 'Urban Water Systems Modelling' and 'Urban Flood Modelling and Disaster Risk Management'. We believe that the material in the book provides support for the training of professional hydroinformaticians specialising in urban water management, and for those students wanting to embark on a research career. The book forms the basis for post-graduate courses, and is accompanied by a DVD consisting of lectures by the authors and leading hydroinformaticians, software and modelling data for exercises in connection with the UNESCO-IHE courses. There is also a website (www.urbanhydroinformatics.com) where we intend to provide access to relevant information and to maintain a conversation on the subject of urban hydroinformatics. In particular, we welcome critically constructive comments on the material in this book.

We are aware that we have only touched on the basic aspects of the subject – there is so much more that can be written. Therefore we have had to be selective about the material we include. We trust that this book will do much to raise the awareness of hydroinformatics as a whole, and improve the debate on what it is so that we have a better platform for communicating ideas and concepts in the future.

Whereas we have involved other authors in writing specific chapters in this book, we are responsible for the views expressed. We hope to have the opportunity to correct any errors or serious omissions in future editions of this book. We sincerely hope that the book will challenge and encourage students all over the world to advance their knowledge and practice of urban hydroinformatics.

Roland K. Price and Zoran Vojinović
The Hague, May, 2010

Corresponding author e-mail: z.vojinovic@unesco-ihe.org

Acknowledgements

We are grateful to colleagues around the world for their contributions to this book. First we express our appreciation of those who have contributed chapters to the book and their comments: Prof dr. Dragan Savić and dr. Zoran Kapelan (Water Distribution, Chapter 9), Prof dr. Peter Vanrolleghen and Prof dr. Ingmar Nopens (Wastewater Treatment, Chapter 11), and Prof dr. Meine Pieter van Dijk (contribution to Chapter 7). The second group of contributors include those who have allowed us to include specific case studies for which they were responsible: John Hateley; dr. Mukand S. Babel, Md. S. Islam, and Aldrin A. Rivas; dr. Sebastian C.F. Meijer and Prof dr. Damir Brdjanovic; Tjitte Nauta, Kok Meng Tan and Hans Eikaas; dr. Ole Mark and dr. Sutat Weesakul. The third group includes our colleagues at UNESCO-IHE and elsewhere who have supported this initiative: Jan Luijendijk and our colleagues in the HIKM department, Prof dr. Mike Abbott, Prof dr. Andras Szollosi-Nagy, Jetze Heune, Richard Kellagher, Andrew Walker, David McConnell, Prof dr. Nilo Nascimento, Hans Emeis, present and former PhD students: Wilmer Barretto, dr. Shreedhar Maskey, dr. Emmanuel Tumwesigye, Solomon Seyoum, dr. Durgalal Shrestha, Arlex Sanchez Torres and Ahmad Fikri Abdullah and the many UNESCO-IHE Masters students who have stimulated us by their enthusiasm for the subject and their many questions and insights which have caused us to look afresh at the issues involved. We are especially grateful to Prof dr. Mike Abbott for valuable comments on sections of the text. We are indebted to Maggie Smith and the production team at IWA for helping us to realise our dream of writing this book. Finally, we could not have even begun to contemplate writing without the encouragement of Thea and Jana; thank you for your patience and loving support.

Roland K. Price and Zoran Vojinović
The Hague, May, 2010

Foreword

It is usual in any work of this kind that it should be supplied with a foreword by one who is associated with its subject and its aims, while at the same time outlining the environment within which the work itself is situated. I shall do my best to fulfill these requirements.

The subject of Urban Hydroinformatics is one of the most important of all the immediate practical applications of hydroinformatics. It covers a wide field of applications that have major social and economic consequences for the lives and fortunes of those living and working in urban areas and it is correspondingly a major consumer of human and financial resources, whether these are of private or public origin. This is, as a result, the biggest area of application of hydroinformatics in money terms at the present time.

It is then necessary to ask and to answer the question of why this should be the case: why does urban hydroinformatics occupy such a leading position, such as can be measured in money terms? This question is clearly inseparable from the question of why so many persons who use the productions of hydroinformatics should come together in urban areas, and then very specifically in cities. The answer is of course the same as that of every bank robber: because that is where the money is!

This rather facile observation leads however to another question again, of why so much money is concentrated in urban areas and most predominantly in cities. This can only be answered by posing a further question of what money does, or at least has the capacity to do. For the engineer – although not necessarily for the banker! – money is an imitation of what in thermodynamics is called negentropy, as that which is capable of creating order. Thus, all the orderings of materials incorporated into buildings, roads, metros, etc in the city are products of the money that has been put into their construction. None of this can be justified economically, however, if the resulting buildings, roads, metros etc. do not produce more money, directly or indirectly, than was put into their construction. This repayment money is itself a product of the negentropy that is incorporated in the information production that the investment generates as most of this is further structured into human information and some of this is refined in turn into human knowledge and some of this knowledge is structured yet again into human understanding – and all of these ever more complex attributes of hand and mind together become encapsulated in tradable, physical, goods and services. It is this process that is the conditio sine qua non for the existence and further development of every urban conglomeration, of every urban area.

Of course in strictly thermodynamic terms, the negentropy incorporated in the buildings, roads, metros, etc, together with that created in people and encapsulated in their productions will always be less than the entropy produced in their construction and their employment and this difference can only be balanced by the supply of further negentropy, expressed in terms of energy, as provided by such agencies as the hydrologic cycle, vegetation, sunlight, fossil fuels and suitably enriched uranium. Money is that which facilitates these processes in the world of commerce by providing a common denominator for all of these negentropic human productions and energy sources that create order, so that they can be more easily traded. Money also appears of course in charges for waste removal and disposal, while this kind of money evaluation has recently been extended further into entropic productions in the form of taxes on carbon dioxide emissions, together with carbon trading, although these last forms still remain secondary, and even controversial matters.

Thus an urban area, with ever more mind-based order being necessarily produced for it to function, is one with a correspondingly higher entropy production than a rural area, where less ordering of this kind is required. This entropy production can take many forms, such as heat, waste water, waste paper and plastics, automobile gases, scrapped household appliances among many, many others, and in almost every case these must be removed from the working areas of the urban conglomeration in order that these working areas can function effectively. Since so much of this waste contaminates water, there is a corresponding need for waste water collection and processing alongside fresh water supply and distribution and both of these again become more concentrated, and so more intensive, in urban areas.

This process can only be justified, however, if the urban area provides more, and more relevant information, and with this knowledge, and even understanding – including the understanding that information, expressed in terms of bits, has a direct and indeed quantifiable relation to negentropy, such as is well established, and especially in its limiting case, in classical thermodynamics. Indeed the whole raison d'être for the existence of an urban area is the capacity of its populace to produce negentropy, in certain desired forms, faster than is the case in a more rural area. The function of integrated water management is to facilitate this process by reducing entropy production in the water sector through the exercise of information, knowledge and understanding – and hydroinformatics is the principle means for realising this function.

We should correspondingly expect that with the ever more intensive traffic over the World-Wide Web, there should be a much lesser need for cities, and especially megacities, some of which are becoming 'metacities', in a post-modern condition of society, understood as one in which there occurs a transmutation of a society of knowers into a society of consumers of knowledge. It is however the most general observation that this is not in fact happening: the part of the population living in cities continues to increase and the megacities are still becoming ever more 'mega', to become 'meta'. Of course many persons working at the outer limits of technology now live in the countryside and maintain contact with their fellows primarily over the Web, but these remain for the most part exceptions.

To this the observation should however be added that in China, which accounts for about a half of all the world's new infrastructure, the so-called first and second tier cities, the mega-cities-cum-meta-cities, have recently been growing more slowly, at least partly because the costs of accommodation in these cities have grown at an apparently unsustainable rate, while prices in third, fourth and fifth tier cities have remained much lower and their populations have increased more rapidly. Since, correspondingly, China is potentially by far the largest market for hydroinformatics' products and services, this change is singularly important.

These two tendencies together – the retreat to the countryside by those who are 'web-connected' and the related tendency to move to smaller cities that the web makes possible even to those who are otherwise not so 'well connected' – leads to a situation whereby the future of hydroinformatics, and especially in China, depends to an increasing degree upon the migration of hydroinformatics onto the web. This is however a very difficult and complicated process and one which makes new demands both on technological advances and on changes in their closely related social environments.

It was this difficulty and complication, alongside the apparent total indifference of the existing players in the hydroinformatics market to this need, that led the present writer to form the so-called Water-Knowledge Initiative, that brought together seven persons at its core. One of its major achievements, albeit so-far little reported in publications, was its success in placing a proprietary computational engine on a server in such a way that it could be accessed and used over the web while using a standard desktop computer. This was only possible, however, when a specific new user interface, and preferably another kind of viewer also, were made available. The most extensive presentation of this technology, from business model, through software implementation to practical application, was provided in the 2009 doctoral thesis of Emmanuel Tumwesigye. This was carried out at the National University of Ireland in Cork under the supervision of Professor Philip J. O'Kane and a convincing practical application of this system was made to the upgrading of the urban drainage system of Cobh, a suburb of Cork. An invaluable support was provided by Stanislav Vanecek and his staff of *DHI Hydroinform* in Prague while four university colleagues in Cork, although not a part of our core group, also contributed to the success of this demonstrator.

There is currently something of a split however within the hydroinformatics community that in turn reflects a range of different splits between the functions of server and client. Hamish Harvey, another member of the WaterKnowledge team, has pursued what are for most of us the more esoteric advances in software tooling – HTML5, Flash Player, Silverlight, etc – that are needed in building the specifically hydroinformatics tools that will make possible sustainable business operations over the web, and especially those involving active stakeholder participation in projects and issue-based, participative-democratic procedures generally. Abbott and Vojinović (2009) have estimated, by extrapolation, that these developments will increase the number of users of hydroinformatics tools and services by two orders of magnitude by reducing client costs by one order of magnitude and increasing the size of the tool-producing and developing business operations by a further order of magnitude. However, as Hamish Harvey has emphasised within our group, none of the presently possible approaches allow us to avoid a significant software development. Outside of our group, others again have been pursuing the possibilities presented by cloud computing technologies and tools and this provides another split again in the enabling technologies. All of this, however, remains work in progress.

This must then suffice to provide a contemporary sociotechnical environment for the present book, as one that goes into the hydroinformatics' background and applications at a hitherto unprecedented detail and depth. I congratulate the authors on their excellent work.

M. B. Abbott
Brussels, May, 2010

Abbott M. B. and Vojinović Z. (2009). Applications of numerical modelling in hydroinformatics. *J. Hydroinf.*, **11**(3–4), 308–319.

Introduction

Urbanization is an unstoppable process. Very recently we have crossed a psychologically very important threshold: half of humanity lives now in urban communities. This trend of urban concentration is to continue. In a few decades time two third of the global population will leave in urban and peri-urban areas.

With a nine billion people in the world within four decades this phenomenon already poses unprecedented challenges, such as urban water supply and sanitation, urban risk management including a multitude of complex, and occasionally multiple, urban disasters, whether one considers urban flash floods or rainfall induced landslides. All these point towards the strong need of developing and using effective tools that could be applied to manage urban water systems in a sustainable fashion. Sustainability requires the appropriate consideration of all the relevant issues in a complex urban setting, ranging from hydrological data collection up to the challenges of social inclusion and the service to the poor. For it is perhaps the uncontrolled growth of bidon villes and favellas in the peri-urban areas in the developing countries where social exclusion is strongest and the needs for social justice and solidarity are the highest. It is often the poor people who suffer the most from inadequate water supply, the lack of sanitation and the resulting public health hazards as well as urban flash floods and mudflows.

Managing complex urban water systems in a sustainable manner is, therefore, a key for any development in the future. Balancing the social, environmental and economic equations in the urban context is a hard act. It clearly requires a systematic integration of all the relevant elements. Integrated urban water management, therefore, is a must if humanity is to survive.

Where are we now then with integrated urban water (systems) management today?

The past three decades have witnessed an unparalleled development in urban water systems management. Urban hydrology became a subject of its own, largely due to the global research efforts undertaken with the framework of UNESCO's International Hydrological Programme. Urban hydraulics has been given a new impetus by the fast progress in computational hydraulics over the past two decades. Real-time data collection systems became a routinely applied tool – except in the developing countries where the overwhelming majority of the urban water issues are. With the rapid progress in computer and computational technology the computational barrier is practically gone. Automatic on-line urban water control systems are now wide spread – except in the developing countries where the overwhelming majority of the urban water issues are ...

So there was a strong need to integrate available knowledge on various elements of the urban water cycle as well as to develop new approaches and put them together into a system framework that could equally be used in the developing and so-called developed countries. A system was badly needed that connects data, models and decision support to derive sustainable urban water policies for design and management.

The overall system that could offer this is the general hydroinformatics framework that has been elaborated by several years of efforts promoted by Professor Mike Abbott of our Institute. The work that is presented in this volume, however, was focused down by Roland Price and Zoran Vojinović on one particular problematique: the urban water complex. It is indeed the first text book that addresses the integrated management of urban water in a manner that could be applied in both developed and developing countries. Although it is written mainly as a text book for post-graduate students who have a strong interest in hydroinformatics applied to urban water systems, practicing engineers and researchers involved in design, modelling and decision support will find it useful as a reference. The book uses examples from real world case studies to support and demonstrate the key issues explained in the text.

If one is challenged with the question what will have the largest impact on how water will be managed forty years from now the honest answer is: it will be population growth. Climate variability and change will certainly be an add-on as a source of increased uncertainty but it will be the nine billion human beings, more than half of them living in urban conglomerates by then, that will change the hydrological cycle in an unprecedented manner. Urban systems management will indeed be key to human survivability. I believe this book is a major contribution to go down the alley of sustainable urban water systems management. I wish to thank, therefore, my colleagues, Professor Price and Associate Professor Vojinović, for all their efforts that went into this book and for their generosity to share their wisdom with all of us.

András Szöllösi-Nagy
Rector, UNESCO-IHE Institute for Water Education

Delft, May, 2010

Chapter 1

The Imperative for Urban Water Management

1.1 GLOBAL URBAN WATER ISSUES

The need to address urban water issues becomes all the more urgent as migration to urban areas increases the pressure on the infrastructure and services provided. According to UN reports, the global population growth forecasts are sending out an alarming signal to governments and city managers for them to address some of the most essential urban infrastructural aspects in order to avoid catastrophic consequences of urbanisation. During the 20th century the world's *urban* population grew from 220 million to 2.8 billion, and is estimated to double during the first 30 years of the 21st century. In 2030 81% of the world's urban population will live in developing countries. The largest migration pressures will come in the 18 megacities (with populations in excess of 10 million) situated in coastal areas. About 75% of people residing in low-lying areas are in Asia, and the most vulnerable are the poor. Slums (or 'informal' urban areas) are growing at the same rate as urban growth. It is expected that most of the growth will occur in developing countries, particularly in regions already under water stress, and where there is limited access to safe drinking water and adequate sanitation facilities. In 2008 50% of the world's population lived in urban areas; this is estimated to change to about 60% in the year 2030, with 95% of the increase taking place in informal urban areas of developing countries in Africa and Asia.

Over the last 20 years there has been a seed change in the water industry of many countries from unregulated national monopolies to regulated decentralised utilities. But the public utilities in developing countries do not perform well due to low motivation, poor management, inadequate cost recovery and political interference.

Regulation has assumed increasing importance in Europe. For example, the EC Urban Waste Water Treatment Directive (CEC, 1991) concerns the collection, treatment and discharge of urban wastewater and the treatment and discharge of wastewater from certain industrial sectors. Its aim is to protect the environment from any adverse effects due to the discharge of such waters.

884 million people do not have access to safe drinking water and 2.5 billion are without adequate sanitation facilities.

Diarrhoea causes 2.5 million deaths/year.

Up to 30% of fresh water supplies are lost due to leakage in developed countries, and in some major cities losses can be as high as 40% to 70%.

A person living in Sub-Saharan Africa uses about 10–20 litres of water a day; on average, a Canadian uses 326 litres a day.

Agriculture accounts for over 80% of the world's water consumption.

Lack of safe water and sanitation costs sub-Saharan Africa around 5% of its Gross Domestic Product (GDP) each year.

The UN suggests that each person needs 20–50 litres of safe freshwater a day to ensure their basic needs for drinking, cooking and cleaning. The daily drinking water requirement per person is 2–4 litres, but it takes 2,000 to 5,000 litres of water to produce one person's daily food.

Reference: UNESCO, 2009

Similarly, the EC Drinking Water Directive (CEC, 1998) is intended to protect human health by laying down health and purity requirements which must be met by drinking water within the European Community. It applies to all water intended for human consumption apart from natural mineral waters and waters which are medicinal products.

A third directive, the EC Water Framework Directive (CEC, 2000) requires member states to identify all the river basins lying within their national territory and to assign them to individual river basin districts. A competent authority had to be designated for each of the river basin districts by 2004. In addition, member states have to produce a management plan for each river basin including an economic analysis of water use. These directives are discussed further below.

Whereas developed countries are forging ahead with their regulatory regimes to improve the management of their river basins (or watersheds), taking into account urban as well as rural water usage, developing countries are struggling to manage their urban water resources. For them, urban water issues are particularly urgent because the state of urban water assets in these countries is very poor, and available resources are extremely limited. Some of the most vulnerable and essential water-related services impacted by urban population growth are the growing demand for water supply and distribution, the collection and treatment of an increasing amount of wastewater, the drainage of excess stormwater runoff, the maintenance of the quality of surface waters adjacent to and downstream of urban areas, and the management of groundwater beneath urban areas.

Better management of urban water systems is of major concern in helping to meet the Millennium Development Goals (MDGs) (United Nations, 2009), especially in providing an improved urban environment affecting Goal 1: Eradicating extreme poverty and hunger, Goal 4: Reducing child mortality, and Goal 7: Ensuring environmental sustainability (especially targets 9, 10 and 11). Almost one billion people are at risk because they do not have access to safe drinking water. This limits their health and exacerbates their poverty. In addition two billion people are without effective sanitation, which again constrains their development and leads to an ongoing degradation of their environment. Furthermore, climate change appears to be generating a greater frequency of damaging floods and droughts in different parts of the world. Whereas in many cases these disasters are basin wide, particular urban areas have been subject to torrential rain and to inundation from rivers and major urban streams.

GOAL 7: ENSURE ENVIRONMENTAL SUSTAINABILITY

Target 3: Halve, by 2015, the proportion of the population without sustainable access to safe drinking water and basic sanitation

- Almost half of the world's population face a scarcity of water
- More people are using improved sanitation facilities, but meeting the target will require a redoubling of efforts
- In developing regions, nearly one in four uses no form of sanitation
- Though access to improved drinking water has expanded, nearly one billion people do without
- Women shoulder the largest burden in collecting water

Reference: United Nations, 2009

1.2 WATER ISSUES IN SOME MAJOR CITIES

In order to appreciate some of the problems faced by some of the major cities consider the following:

1.2.1 Bangkok

Bangkok has a metropolitan area of 1,568.7 km^2 and a population in excess of 12 million. It is situated in the flat deltaic plain of the Chao Phraya River. The city is at an average height of about 2 m above sea level, and is vulnerable to intense rainfalls during the rainy season between May and November; daily rainfalls in excess of 400 mm are not unknown. Flooding in the city is caused by rainfall directly on the urban area or by the overtopping of embankments along the Chao Phraya River due to high water levels from flood flows upstream or high surges originating from storms in the Gulf of Siam. The duration of annual flooding has been reduced to a matter of days rather than weeks through the installation of huge pumping stations that discharge flood water from the city to the river. Flood management in the city has been exacerbated by over-abstraction of water from aquifers below the city for water supply purposes which has lead to extensive and severe land subsidence within the city boundaries. This was as much as 15 cm per year in some parts of the city. Fortunately this has been remedied by the construction of a large pipe line conveying water to the city centre from the Chao Phraya to the north. Further difficulties with flooding are generated through the large demand for land for development, which has lead to many natural drainage channels being filled in. Another issue concerns the difficulties in flushing the remaining klongs (or canals) to reduce the pollution caused by wastewater discharges from housing and industry, and in removing the refuse that is regularly dumped in them. See Figure 1.1.

1.2.2 Jakarta

Jakarta, the capital city of Indonesia, is another case in point. With a population in excess of 8.5 million (23 million in Greater Jakarta) and an area of 665 km^2, Jakarta is situated on the north-west coast of Java on a low lying plain with about 40% of the city below high tide level. There are a number of small rivers flowing through the city from hills to the east. These rivers are prone to flash floods generated by tropical storms on the hills to the east where rapid development has lead to deforestation. Flooding in the city is exacerbated by high tides that inundate the coastal districts. Major flooding occurred in 1996 and 2007 when 70% of the city was inundated with water up to 4 m deep. Like Bangkok the drainage channels are used for garbage disposal and squatters line the banks; see Figure 1.2. Added to this there is a growing sedimentation problem in the rivers. Less than a quarter of the population is properly served by water supply and drainage facilities: most rely on rivers, lakes and private water vendors. Wastewater treatment is negligible, and there are difficulties in flushing and maintaining the drainage channels and canals. Raising funds through property and land taxes has proved difficult because of problems in identifying owners of property.

Figure 1.1 Pumping station outfalls, Bangkok (Photograph: Sutat Weesakul)

Figure 1.2 Flooding in Jakarta (Photograph: Hendra Murtidjaja)

1.2.3 Dhaka

Dhaka is the most densely populated city in the world, with more than 12 million people in the Metropolitan area of 1463 km^2. It is the only city in Bangladesh with a water-borne sewage system, but this serves only 25% of the population while another 30% are served with septic tanks. Only two-thirds of households in Dhaka are served by the city water supply system, which was introduced originally in 1874. More than 9.7 million tonnes of solid waste are produced in Dhaka city each year. The city is flat and close to sea level, which makes it susceptible to flooding during the monsoon season due to heavy rainfall and cyclones; see Figure 1.3.

Figure 1.3 Dhaka flooding (Photograph: DHI, Water & Environment)

1.2.4 London

The capital of the United Kingdom, Greater London has a population in excess of 12 million, and covers an area of approximately 1570 km^2. Water supply is from the River Thames, the River Lea, and some underground resources. There is a ring main with a number of contributing potable water treatment works. The pipes are largely cast iron and are deteriorating significantly, which contributes to a reasonably high leakage rate.

During the 19th century the British parliament authorized the development of a modern, combined sewerage system for London to counter the obnoxious smell from wastewater and to alleviate numerous cholera outbreaks. Intercepting sewers were constructed along the lines of the rivers draining into the River Thames. The flow was mainly under gravity, but pumping stations were built to raise the sewage from the north of the Thames to a major treatment works at Beckton, (now serving 3.4 million people), and similarly from the south of the Thames to Crossness (receiving sewage from about 2 million people); see Figure 1.4. In addition, the Mogden treatment works now serves 1.8 million people. Further major improvements were made both to the sewerage systems and the treatment works in the 20th century. Yet the regular expansion of the population living in London and redevelopment has placed increased pressure on the system, which has limited capacity. Flooding therefore does occur with an unacceptable frequency in some areas. The plan now is to build one or more tunnels for storage and transfer of sewage, with a single tunnel being 35 km long and up to 9 m in diameter. Called the Thames Tideway scheme, it forms one of the city's most ambitious wastewater projects since its first sanitation system in the mid-19th century. 32 million cubic metres of untreated sewage is discharged annually into the river; so the tunnel will improve the capacity of London's sewerage system, help prevent sewage discharge into the River Thames, and allow more wastewater to be treated to a higher standard.

Figure 1.4 Crossness treatment works, London (Source: Thames Water; reprinted with permission)

1.2.5 Chicago

Chicago is the third largest city in the USA with more than 9.5 million people in the Metropolitan area of 18,684.2 km^2. The city is built on marsh land adjacent to Lake Michigan; see Figure 1.5. With over 1000 mm of rainfall and more than 1000 mm of snowfall, the urban area is prone to flooding. Initially, the wastewater drainage in the city was designed to discharge directly to the Chicago River, which turned out to be ineffective in transporting sewage away from the city. In the late 19th century the flow in the Chicago River was reversed to take wastewater away from Lake Michigan. But matters became worse as the urban area expanded and there were regular major flood events in the metropolitan area. In the last half of the 20th century a major deep tunnel system 176 km long that had been planned as early as 1834 was constructed. The tunnel and surface storage lakes were intended to hold the raw sewage for treatment and diversion to the Chicago River. Drilling for the tunnel was completed in 2005. Fish have returned to the Chicago River though pollution levels are still high. This tunnel has effectively limited the discharge of raw sewage into the Chicago River and enabled its restoration.

1.2.6 Shanghai

Shanghai has over 20 million people in an area of 6,218 km^2. A highly advanced and modern city it is constructed near the coast and has a sizeable lake above it which drains through the city to the sea. The low lying nature of the urban area has lead to the construction of polder areas with associated pumping to keep water levels low, especially during high rainfall events. Flooding generally occurs with rainfall in excess of 35 mm/hr.

1.2.7 Belo Horizonte

Belo Horizonte (BH) is the capital of the State of Minas Gerais (Brazil), which in economic terms (gross product) is the third among the 26 Brazilian states. The city lies at 20° South latitude and 44° West longitude and has an altitude of 750 to 1,300 metres. It is located in a mountainous region of tropical soils that originated from the decomposition of metamorphic rock. Tropical highland weather predominates in this area, with an average yearly rainfall of 1,500 mm and an average yearly temperature of 21°C. BH has 2,227,400 inhabitants with a population density of 6,900 inhabitants/km^2.

Figure 1.5 Chicago (Photograph: Tzvetan Stantchev)

Figure 1.6 Shanghai (Photograph: Laura Kwak)

In Belo Horizonte as well as throughout the BH metropolitan area, a separated sewerage system has been adopted, although illicit inter-connections between the wastewater and stormwater networks prevail, resulting in heavily polluted receiving bodies in the urban area and in the Velhas River downstream of the city. Another source of water pollution by wastewater is the lack of interceptor pipelines (Figure 1.7) as part of the main sewerage system.

Figure 1.7 Pollution of receiving water bodies by wastewater due to lack of interceptors (Photograph: Nilo Nascimento)

1.3 PROBLEMS FACED BY MEGA CITIES

In general, the problems with water that are faced by mega cities all over the world are similar. Many cities are situated on the coast or on a major river or lake. They can be highly prone to flooding both through rainfall directly on the urban area, which may generate large amounts of excess runoff that exceeds the capacity of the existing drainage system, and through overbank flows from rivers or spillage from an adjacent lake, or inundation from coastal surges combined with high astronomic tides. Besides the resulting flooding of parts of the urban area, there can be contamination of the flood water due to the flushing of wastewater drainage systems, damage to water distribution systems, mud slides generated on unstable hill slopes, rising groundwater levels due to recharge, and so on. The city authorities have therefore to identify the risks their city faces from flooding, to take action to reduce these risks, to provide adequate forecasting and warning facilities of flood events, and to plan for the restoration of the social and economic life of the community following a disaster.

Each city has the need to sustain its growing population through the provision of adequate fresh water supplies that are normally distributed through pipe networks operating under pressure. The sources for water supply are best located outside the city, preferably from groundwater aquifers, or from rivers upstream of the city. Reservoirs are often used to provide temporary storage of water. In each case some treatment may be required before the potable water is distributed. Correspondingly, there needs to be a means of collecting the wastewater that is produced by domestic, public and industrial use of the potable water. Such wastewater collection is often combined with the drainage of stormwater from the urban area, though there is an increasing trend to keep the collection of wastewater and stormwater separate. What is of primary concern to many cities is to provide some treatment at least of the wastewater component in order to remove harmful pollutants, such as solids, nitrogen and phosphorus compounds, heavy metals and particular bacteria, including e-coli, before returning the effluent to the environment.

With the rapid growth of urban areas in many parts of the world, water engineers face ongoing challenges to extend the services they provide, in particular, water supply and distribution, wastewater collection and storm drainage, in such a way that they do not prejudice future developments. What is more, the artificial nature and construction of the distribution and collection networks implies that they deteriorate in time and therefore need both regular maintenance and eventual renewal or rehabilitation. In particular, the rehabilitation of existing water and wastewater treatment plants and distribution, wastewater and storm drainage networks today should be designed in such a way that expected rehabilitation works needed to cater for urbanization in the future are not made more difficult than necessary. In particular, the discounted costs of investment in and maintenance of urban water services over a period of time into the future should be minimized.

Consideration of such an issue by many cities in developing countries is secondary as they face an ongoing daily struggle to maintain and improve their existing urban water systems. Good potable water supply sources from groundwater are often limited, and a city is then dependent on withdrawing polluted water from adjacent rivers. There is a greater need for treatment, which can be expensive, and an excess of demand over supply of water and the uncertainty in electricity power supplies can make both the treatment of potable water and its distribution through the piped network intermittent. Besides the contamination of the supplied water, the decrease in pressure in the pipe network can lead to infiltration from groundwater which contaminates further the distributed water. Such infiltration is symptomatic of the leakage of water from the pipe network when the network is operating under pressure. Many cities experience high rates of leakage from their distribution networks due to deficiencies in the joints between pipes and cracks induced by poor construction, adverse surface loading and human interference. Together with infiltration illegal connections contribute to much of the distributed water being 'unaccounted for'. Securing payment for water supplied to end users is also an ongoing issue for many water authorities. A number of cities are turning to water meters installed at the demand points in order to charge more appropriately for the volume consumed by individual users.

If the provision of potable water is problematic for some cities in developing countries, the implementation of adequate wastewater collection and storm drainage can be even more difficult. Part of the problem is that many cities have inadequate measures to collect the tonnes of garbage (or solid waste) that are produced daily. Local urban streams are obvious places for dumping solid waste and their banks attract squatters, thereby restricting the capacity of the channels to convey flood flows. There is continual pressure to fill in drains for industrial development or housing. Where there are storm drains and conduits these silt up with detritus. Wastewater is discharged directly to the drainage system without treatment. Floating garbage contributes to visual pollution and clogs flow structures including pumping stations. Generally there is a lack of pumping capacity to evacuate flood flows from the drainage channels in the city, or to provide adequate flushing of the system. This leads to serious pollution throughout a city, giving rise to water related diseases and adverse effects on groundwater. Water sources for supply are often the aquifers directly below the urban area, which are tapped by hotels and industry. This has serious consequences for land subsidence in the city, which tends to lower the terrain, making flooding more serious. Fortunately, many cities have become aware of the problem and have put in place alternative sources of supply, from nearby surface waters, reservoirs or groundwater aquifers some distance from the city.

1.4 SOLUTIONS TO PROBLEMS OF CITIES IN DEVELOPING COUNTRIES

There are various solutions for the problems of cities in developing countries. A significant approach for many cities situated in low lying delta areas is to construct polders in the areas most difficult to drain. Coupled with appropriate pumping resources these polders can be drained continuously with discharges to higher level canals or rivers. This approach facilitates flushing to reduce pollution in the drainage channels. At the same time there has to be adequate provision made for solid waste collection and the maintenance of the channel conveyance. Where possible wastewater should be collected separately and receive at least primary treatment and preferably secondary treatment. Water abstraction from below urban areas for supply purposes should be stopped where such abstraction causes land subsidence, though abstraction to limit the rise of groundwater in adversely affecting the foundations of buildings and other urban infrastructures such as tunnels will need to continue.

There are a number of institutional and financial issues faced by cities in developing countries in regard to water. First, they need to evaluate the value of water to the community. For society to thrive it has a wide range of needs, not least water supply, wastewater collection and stormwater drainage. These services require considerable financial investment. This means that there has to be adequate economic and fiscal arrangements in place. In particular there should be facilities for tax collection in order to pay for the capital investment in constructing the supply, collection and treatment facilities, and to maintain them in an efficient and effective manner. The various systems and networks are valuable assets that needs appropriate treatment and respect. Whereas it is comparatively easy to justify financial charges for the amount of potable water delivered (as metered, say), it is considerably more difficult to arrange corresponding charges for wastewater collected, or for the treatment of such wastewater before being disposed back into the environment. This is because everyone is aware of the value of freshwater to their life and wellbeing, even if they are also aware that such water is in one sense freely available as rainfall: the actual collection, storage, treatment, distribution and availability of potable water on demand is nevertheless an important service directly benefiting the individual. In contrast, the collection of wastewater and stormwater and preferably the treatment of the wastewater component are services that are assumed as a benefit more to the community than to the individual, and therefore they are perceived as being of less importance. What is more, when the services are successful, the fact that they are 'out of sight' supports the contention that they are 'out of mind'. We know of course that this is a dangerous mind set that fails to appreciate the contributions that the collection and treatment services make to guarding public health, protecting life and property from flooding, preserving an aesthetic urban environment, and sustaining the local ecology. The reversal of the mind set comes about through the public impact of any failures of the services provided and the ongoing awareness of the population that the services have to be funded by them from one source or another.

Because the primary beneficiary of urban water services is the local community it is important that it is involved not only in paying for them but also in requiring their efficient and effective design, operation and management through specialist service organisations. For example, it is well known that considerable amounts of potable water are lost from distribution networks through leakage, as much as 70% in extreme cases. Leakage can be reduced by identifying significant leaks and repairing the corresponding pipes. Alternatively the network can be renewed every 50 years, say. Better quality control over construction of the network can also improve reliability and reduce leakage. Another aspect of water distribution is the assessment and forecasting of demand. As the population becomes more affluent there is an increase in the individual's demand for and use of potable water. In most cases the increase in demand occurs without the individual being aware. On the other hand, if individuals were aware then they could take action to reduce their use of potable water, and to contribute to a reduction in demand, which would make the service more efficient. Correspondingly, any reduction in water consumption should lead to a corresponding reduction in the generation of wastewater. The separation and collection of human urine and feces independently of other wastewater would contribute significantly to a reduction in the 'end of pipe' treatment and have the benefit of using the urine and faeces for the production of fertilizers. Additionally, wastewater could receive on-site preliminary treatment, enabling it to be used as 'grey' water for purposes not associated with human consumption.

1.5 URBAN WATER MANAGEMENT IN DEVELOPED COUNTRIES

The need to collect both wastewater and excess rainfall-runoff (or stormwater) and the fact that originally they were collected together in what are now called combined sewers has raised the ongoing issue of whether it is better to collect the wastewater and stormwater together in a single network of conduits, or to do so separately in two networks. The advantages of having a single network in which the wastewater and stormwater are collected together are that only one network needs to be designed and constructed, and the periodically irregular stormwater flows can help scour any sediment deposits from the wastewater flows, formed during quiescent periods. The disadvantages include the difficulty of separating out the wastewater component for treatment from the stormwater, the need to avoid excessive flows going to treatment during largish storms by diverting flows through overflows to receiving waters, and the inevitability of surface flooding with dilute sewage for some frequency at various points throughout the catchment. On the other hand, having two separate networks with treatment only at the end of the wastewater network appears to increase the overall capital cost and the operational costs while improving the efficiency of the treatment works and reducing the risk of pollution impact on the environment. There is an even more urgent need to reduce inflows and infiltration to a wastewater network, and to exclude wastewater connections to the storm sewers. The capacity of the wastewater network is determined by the population served, the agreed discharges from public buildings, industry and commerce and the accepted safety factor to cover unforeseen events and urban development. Alternatively, the capacity of a stormwater network has to provide protection in different areas of the catchment against flooding for prescribed frequencies: there will always be rarer floods that cause damage in the catchment. The additional issue, which is not always acknowledged, is the need to design the urban surface facilities to manage the overland flood flows for these rarer events and to minimize the damage caused. Urban flooding is not only generated by excess rainfall-runoff on the urban area; it can also be due to urban rivers overflowing their banks or limiting the discharge of storm flows from storm sewers into the receiving streams due to high water levels in the streams. The level of protection provided by a storm sewer system to surface flooding can be increased by minimizing the amount of stormwater entering the drainage network through source control, that is, by diverting rainfall-runoff to infiltration facilities or detaining excess flows. Generally source control is most effective for the more frequent events and becomes less effective the smaller the frequency of event. What can be done in detaining flows at source can also be done through storage in the system. The use of storage tanks and diversions becomes an important alternative to increasing pipe sizes and in reducing the discharge (though not necessarily volume of flow) to treatment and flows to overflows. In combined systems, frequently operating storm overflows can cause significant short term and long term damage to the receiving streams and the associated environment. Storage facilities with the overflows can help reduce the frequency of operation of the overflows and reduce the sediment and pollutant concentrations in the overflow discharges. Real time control of the storage tanks can greatly improve their efficiency in managing flows and pollution.

It is apparent that the sophistication encountered in designing, constructing, maintaining, operating, expanding and sustaining the water distribution, wastewater and stormwater collection, and wastewater treatment systems implies that the systems are extremely valuable assets for the community. They should be developed and operated efficiently and effectively to the benefit of the whole local community.

It has to be recognized however, that the provision of services requires considerable capital investment and maintenance expenditure. The sophisticated water supply and drainage systems of the developed world cannot be replicated necessarily by the developing world. There is a need therefore to find institutional and cost effective means of providing the relevant services that are consistent with the funding available and which can be upgraded appropriately as the capital and revenue income increase. What is important is that the systems involved should be sustainable by future generations without having to undo the infrastructure developed previously.

Proper management of the urban water system therefore requires formal attention to the engineered and natural subsystems, which are subject to both deterministic and stochastic phenomena, while meeting the demands and expectations of stakeholders and customers. This is, in a limited sense, an optimization problem, dependent on having a good understanding of the performance of each subsystem, the interactions between them, and the integration of the various management tools and aspects. In turn, this requires the ability to conceptualize computer models for each subsystem, to link the outputs from and the inputs to the models, and to apply properly the management tools and constraints.

1.6 URBAN WATER MANAGEMENT IN THE USA

Water management in a given country is in part a function of its culture. So, for example, urban areas in the USA have developed extensive suburban districts with low density housing. It is therefore no surprise that the governing principle is that each river system and its watershed must be managed as an integrated system for purposes such as energy production, flood control, irrigation, water supply, recreation and navigation. Water quality must be maintained and improved through proper treatment of wastewater. As early as 1956 the Federal Water Pollution Control Act funded publically owned treatment works. By 1965 the Water Quality Act promoted the development of water quality standards by the States. Increased funding for restoring and maintaining the nation's waters was made available following the 1972 Clean Water Act in which the National Pollutant Discharge Elimination System (NPDES) permit program was introduced. Basin plans were required to deal with point source pollution. The 1987 amendments to the Clean Water Act expanded attention to deal with toxins, nonpoint sources, wetlands and water quality standards. Considerable emphasis is given to cooperation between agencies and local citizens and organizations for watershed stewardship.

About 80% of the USA population lives in so called urban areas. Therefore many small streams in a watershed are affected by urban development and as a consequence are degrading through polluted runoff and deliberate dumping. It is estimated that urban areas contribute more than 11% of the non-point source pollution into the nation's waters: construction alone contributes 3%. The health of urban watersheds is of major concern. Source control, in its widest sense, and education are seen as important ways of managing these watersheds.

1.7 URBAN WATER MANAGEMENT IN EUROPE

Specific legislation on water in Europe has already been introduced above. Such legislation focuses on the protection of natural water courses and resources, and achieving a uniform standard throughout the European Union. For example, in 1976 the Bathing Water Quality Directive (European Commission, 1976) regulated the bacteriological quality of coastal bathing waters. A revision in 2002 identified two indicators: Intestinal Enterococci and Escherichia coli. Thresholds have been set for these indicators at 200 IE cfu/100 ml and 500 EC cfu/100 ml with 95% compliance.

Subsequently, different directives have been introduced to improve the environment and quality of life of citizens. Key directives included the Urban Waste Water Treatment Directive (UWWTD) (CEC, 1991), which addresses emissions from waste water treatment plants (with less attention given to discharges from collection systems), the Water Framework Directive (WFD) (CEC, 2000), which focuses on river basins achieving standards of previous Directives and new ecological quality standards, and the Flood Directive (FD) (CEC, 2007), which is concerned with action to reduce flood risk through structural and non-structural facilities. The UWWTD insists that each community with more than 2000 population equivalent has a collection system, and that due consideration is taken for polluted discharges from overflows. Meanwhile the aim of WFD is to achieve or maintain a good status of ground and surface waters by setting standards that are legally binding on the river basin authorities. Finally the FD is designed to reduce flood risk throughout the EU.

So far as drinking water is concerned the Council issued the Drinking Water Directive in 1998 (CEC, 1998). The objective is to protect the health of EU consumers and to make sure the water is 'wholesome and clean'.

1.8 INTEGRATED URBAN WATER MANAGEMENT

The different aspects of water management mean that there are many factors to be considered in providing communities with safe water and hygienic living conditions. In particular it is important that the local communities are involved in decision making concerning the use of water resources for food production and directly for human consumption, and to conserve the quantity and quality of freshwater and terrestrial ecosystems on which all depend. Audits are needed to appraise the water resource base for economic, political and social decision making. Demand should be identified with stakeholder participation. This implies that there should be an integrated management of urban water resources, and this must be pursued through the concept of the watershed or catchment. The challenge as always is to maintain an adequate supply of usable water, and to limit damage done to the environment through poor collection and drainage systems, and from flooding. In pursuit of the integrated management of urban water it is vital that the best use is made of information about the urban systems and knowledge about how to design, operate and maintain them. The keywords here are 'information' and 'knowledge'. Inevitably, this raises the need for the appropriate use of informatics and knowledge management, terms that become the focus for hydroinformatics as introduced formally in Chapter 3.

1.9 THIS BOOK

The book covers aspects such as artificial systems for water distribution and collection in urban areas, the interaction with and impact on natural systems, including rivers, lakes, coastal waters, groundwater, the types and levels of service provided, design of assets, assessment of hydraulic, structural and environmental performance, failures of the services provided, identifying and managing risk, impact on water resources, security of supply, safe disposal of wastewater, stormwater drainage, urban catchments nested in rural/natural catchments, flooding from rivers flowing through urban areas, flooding from intense rainfall on urban areas, water quality and public health issues, source control, etc. It also addresses the topics of asset management, rehabilitation/adaptation, sustainability and involvement of stakeholders in the decision making process.

The book is structured so that it explores first the flow of information from the physical world in the form of data, to the virtual world of mathematical and computer based modelling, to the organizational world where decisions on how best to manage the urban water systems are taken, and then to society which is both the beneficiary of the urban water systems and ultimately responsible for interventions in the physical world. The remaining chapters explore the implications of this analysis for the different urban water systems. Above all, the book addresses urban water management from a hydroinformatics point of view. Therefore this book is not intended primarily for design engineers. Instead, the focus is on the management of information in connection with the urban water systems that most of us are dependent on.

1.10 KEYWORDS

By the end of the chapter you should have encountered the following terms; ensure that you are familiar with them!

Adequate sanitation facilities
Demand*
Distribution*
Drainage*
Flooding*
Groundwater abstraction
Integrated management*
Leakage*
Legislation*
Maintenance and rehabilitation*
Millennium development goals
Pollution*
Safe drinking water
Solid waste
Subsidence
Urbanisation

The items with an asterisk are dealt with in more detail in later chapters.

1.11 QUESTIONS

What are the key issues facing mega cities around the world?
What particular issues face megacities in developing countries, and in developed countries?
What progress is being made to meet the Millennium Development Goals concerning water?
How is legislation being used to manage water in the (urban) environment?

1.12 REFERENCES

CEC (1991). *Directive Concerning Urban Waste Water Treatment*. 91/271/EEC.
CEC (1998). *Directive Concerning Drinking Water*. 91/83/EEC.
CEC (2000). *Directive Concerning Water Framework – Integrated River Basin Management for Europe*. 2000/60/EEC.
CEC (2007). *Directive Concerning EU Floods*. 2007/60/EEC.
UNESCO (2009). *The 3rd United Nations World Water Development Report: Water in a Changing World (WWDR-3)*. UNESCO Publishing/Earthscan, Paris, London.
United Nations (2009). *Strengthening the Global Partnership for Development in a Time of Crisis*. MDG Task Force Report.

Chapter 2
Urban Water Systems

2.1 A BRIEF HISTORY OF URBAN WATER SYSTEMS

Access to water has been critical in determining the progress of civilization. The first significant civilizations occurred along major rivers, such as the Nile, Euphrates, Ganges and Yellow River, where agriculture could support growing populations. As people congregated together trade developed and the first conurbations were formed. Walled cities provided some defence against marauders, and their location required a secure water supply. So from the beginning water determined the location and sustainable size of cities.

As societies became more sophisticated they were expert at developing irrigation and water supply systems with dams and dykes, largely through the construction of open channel networks. One of the oldest dams was built between 2650–2150 BC to harvest rainwater from the Wadi Garawi in Egypt. Whereas the primary need was to sustain agriculture, canals also supplied water to urban areas. Cities emerged to facilitate government over a region. They were centres for administration, religion, services, defence and trade. The largest city population in 650 BC was 120,000. The Romans even constructed the first aquaduct in 312 BC to bring water in a tunnel 16 km long, providing 73 million litres of water per day. They used pipes to distribute water underground to their villas, public buildings, baths and fountains. They also were aware of the need to collect wastewater and to convey it to an adjacent water course where it could be disposed of safely. Laws and penalties governed the use and maintenance of aquaducts. Every farmer had responsibilities to care for the water in their area, and there was collective responsibility for the apportionment of water and policing irrigation arrangements. The basic principles of urban water supply, distribution and wastewater collection and disposal were already worked out. But it would be some centuries before the next major advances would be made in urban water systems. Interestingly, the climate during the Roman period was warmer than it is today. It was undergoing what is termed the Roman Warming. In the middle of the sixth century the climate became colder, only for it to warm again during the period 900 to 1300. Subsequently, during the period 1300 to 1850

Europe in particular underwent the 'little ice age'. It was a time of large fluctuations in temperature from year to year. Many people starved as crops failed and disease was rampant in overcrowded cities.

Islam spread advanced irrigation and water technology across the Mediterranean, focusing on dams, qanats (networks of horizontal tunnels to transport and distribute underground water to the surface by gravity), the noria (a large wheel driven by flowing water and achieving a high hydraulic lift), and the undershot and overshot water mills. Bridges across major rivers attracted urban settlements (see Hassan, 2004). Cities on major rivers were vulnerable to climate change, such as prehistoric farmers, the Hohokam in Arizona on the Gila river, who succumbed to floods, and the Mayan civilization in Central America which was devastated by droughts.

In about 1850 the climate generally began to become warmer. A primary impetus for reviewing and improving the supply and drainage of water in urban areas was the industrial revolution when there was a major movement of people from rural areas to the cities that accommodated them as workers in the factories. Row upon row of houses was constructed by the factory owners for their workers. Many of the factories were situated adjacent to good water supplies, and the workers obtained their water from the same sources and local springs. The wastewater at first flowed down the middle of the streets where solid waste was also dumped. Runoff from heavy rainfall helped to flush away stagnant water and particulate solids into the nearest water courses. But conditions were generally grim, and disease and ill health affected many, restricting life expectancy to 35 to 40 years. Beer was consumed, as much because it was safer to drink than raw water due to the alcohol it contained than for the sense of light-headedness and well-being it induced, and once tea had been brought from warmer climates it rapidly became a preferred beverage due to the fact that it was made with boiling water, in which bacteria had been destroyed. The awareness that cholera and typhoid were enhanced by poor water supplies and the lax disposal of wastewater prompted initiatives to think in terms of treating the water before distributing it through piped networks; wastewater was be collected also by a separate network of pipes and disposed of to the nearest water course. Provision would also be made to collect excess rainfall-runoff in the same pipes as the wastewater, such that advantage could be taken of the periodic higher flows to flush any sedimentary deposits. Water distribution pipes were made from the new materials being processed by the factories, namely iron, copper and lead, whereas the storm and wastewater conduits were made out of clay or brick. In the larger cities the underground sewers constructed in the Victorian era were of a size that men could walk through. In London many kilometers of such sewers were built, conveying sewage directly to the Thames. Typically the main trunk sewers were generously over-sized leaving a considerable benefit to later generations as the urban areas continued to expand and put additional loads on the existing facilities. The industrial revolution benefited not only the manufacture of pipes but also provided a range of ancillary equipment such as pumps, operated first by burning coal, then diesel, and subsequently by electricity. Screens were developed to filter out solids of different sizes, and tanks were devised so that undesirable bacteria could be destroyed through treating the water with chlorine.

If water supply and distribution and sewerage made the urban environment more acceptable, the waters such as streams and lakes receiving the raw sewage became heavily polluted. The problem of wastewater disposal had only been partially solved. The receiving streams possessed some treatment ability but as the urban areas expanded and grew so the capacity of the streams to accommodate the wastewater was unacceptably reduced in many cases. The only sustainable solution was to treat as much of the wastewater before disposing it back into the environment. Consequently, wastewater treatment facilities were devised, and were situated just before the water was discharged to the receiving stream, generally on the flood plain of the stream or river. The first concern was to remove gross and particulate solids through screens, filters and sedimentation tanks. Subsequently biological processes were developed to enhance the removal of fine solids and a range of chemicals such as nitrates and phosphorus. As concern for other pollutants has grown, so a range of different processes has been refined to remove further chemicals such as heavy metals and bacteria such as E-coli.

Countries such as Britain, The Netherlands, Germany and France, who were in the vanguard of the industrial revolution, were the first to develop modern forms of water distribution and storm and wastewater collection (or drainage) systems. But if they received the early benefit from these systems they also came to realize that these systems required periodic renewal as well as regular operation and maintenance. Unlike the natural streams and rivers or coastal waters where urban water effluents are discharged, the artificial water supply and drainage networks are subject to decay: pipes or constructed conduits age and break due to ground movements and vibrations, through differential stresses and strains affecting the integrity of the structures, and through surrounding cavities formed by infiltration or exfiltration. They are also prone to accidental interference by developers and contractors. The performance of the drainage systems in particular is very dependent on the capacity of the systems to limit or prevent the accumulation of sediment at any point in the networks. Sediment deposits or solidified fats from restaurants can severely restrict the capacity of a conduit to convey the sewage. Such a restriction can have adverse effects on water levels upstream, leading to surcharging and even surface flooding with an unacceptable frequency. Regular maintenance can do much to remove deposits and to repair the structural integrity of the conduits. But such maintenance only preserves the performance of the system as it was designed or constructed: it is not intended to improve the performance for changed loadings on the system due to expansion of the network to serve developments of the urban area, or to meet more restrictive standards on the operation of overflows or the frequency of surface flooding. There is a need for periodic assessment of the system in order to rehabilitate its performance to meet the prevailing standards. Note that this is not rehabilitation in the sense that the system is being restored to its original design, but to a level of performance that is acceptable to the current or future needs of the community.

The strategic importance of urban water systems has generated considerable engineering ingenuity, and as the science of water and its management has progressed so engineers have developed mathematical tools to assess demand for water, design water treatment works, size conduits for water distribution networks under pressure, design drainage networks for wastewater and stormwater flows, explore the operation of storm overflows, design wastewater treatment works and examine the impact on receiving waters. Such tools have been considerably enhanced by the availability of computers, and the management of urban water assets has received considerable impetus from the ICT revolution. This book explores some of the benefits that have accrued to the water industry through ICT-based tools.

In the remainder of this chapter, we examine each of the distinct phases of the urban water cycle, namely

- Water resources for supply
- Potable water treatment
- Potable water distribution
- Wastewater collection
- Stormwater and combined sewerage
- Wastewater treatment
- Impact on receiving waters
- Urban flood management
- Groundwater management in urban areas (See Figure 2.1)

These phases are discussed from the point of view that each phase involves a system that may be essentially natural or manmade, and which in the context of the urban area provides a particular service to the community and to the environment. The interest in each case is to define the nature of and reasons for the service(s) provided, the nature of and reasons for the possible failures of the service(s) and the corresponding risks, and the means of investigating the associated issues. Note that the word 'cycle' in the phrase 'urban water cycle' is strictly inappropriate in that water does not circulate within the urban systems but within the general hydrological cycle. Nevertheless, 'urban water cycle' is commonly used to describe the phases above. The chapter concludes with introductions to some related topics including the need for an integrated understanding and the management and control of water in the whole urban water cycle.

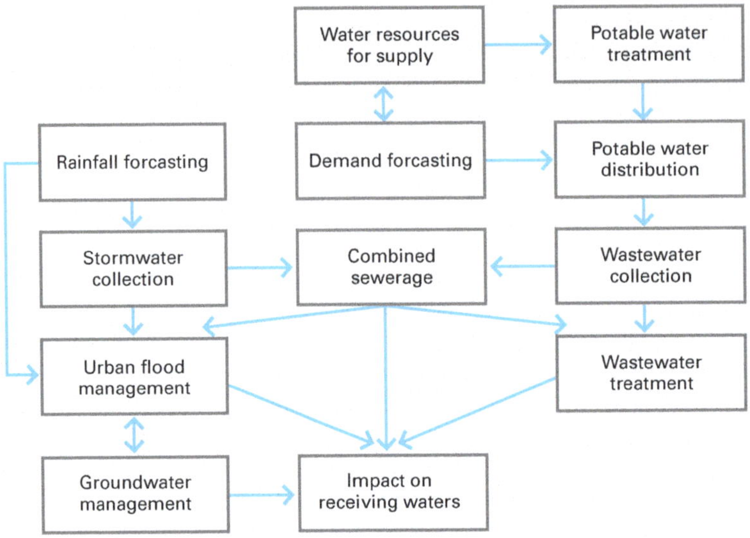

Figure 2.1 Urban water phases

2.2 WATER RESOURCES FOR SUPPLY

Originally water in the natural environment was treated as a resource in its natural state. Wells were dug in order to tap the water below ground. Usually the wells reached down to water residing in aquifers or saturated zones determined by the strata of rock and sedimentary layers. The water in the aquifers could originate from some distance away through infiltration from surface rainfall or water courses, and could therefore be of considerable age. The filtration through the geological medium would tend to remove pollutants, though residual chemicals such as arsenic could be dissolved from the strata and be in solution. Aquifers near coasts could also be contaminated with sea water and therefore salt.

Wells and springs are often limited in the quantity of water they generate. Agriculture consumes large quantities of fresh water, mostly in the form of rainfall. But in some countries, such as Egypt, agriculture is predominantly dependent on irrigation. The large quantities of water required can only be provided through sizeable canals that convey significant quantities of water directly from the natural rivers through to the fields. The need for secure guaranteed supplies of water to meet the demand from urban areas has lead to the impounding of water at various scales. Harvesting of intermittent rainfall at the house level has been traditionally adopted in many parts of Asia. A key aspect is keeping the stored water uncontaminated over time.

At larger scales, water in streams is impounded off-line in reservoirs that can be replenished when flows in the supply stream are large enough. At still larger scales, flows in major rivers can be retained in reservoirs on-line behind a dam that traverses the river valley at some constriction. Such dams were first built successfully by the Romans in about 100AD, and subsequently their benefit was recognized around the world. Dams offer a range of services beyond that of water supply, and including flood control, hydroelectric power generation, recreation, maintenance of navigation depths downstream. They also have the disadvantages of limiting sediment transport on downstream, affecting the morphology of the river, and detaining water in a stationary state where adverse chemical and biological processes find more favourable conditions to multiply algae, some of which can be toxic to humans and biota. In addition, dams and reservoirs take up large areas of land, and often small, and even larger, communities are displaced. Nevertheless, the benefits of dams and there impounded waters have been sufficiently attractive to outweigh other objections.

Very often a city will have a range of sources for its potable water. This promotes the security of the supply. What is important is that cities in developing countries in particular, have come to rigorously prevent water abstraction through boreholes tapping the aquifer below the urban area. This is because such abstraction can lead to permanent subsidence of the geological strata with adverse effects on the ground surface above. Because many cities are situated in delta or coastal areas they can ill afford the subsidence that occurs and which can make the provision of basic water services and an acceptable risk of flooding difficult to maintain. Correspondingly, it may be necessary to control groundwater levels in urban areas to prevent them becoming too high through recharge during flood conditions. Some infiltration from groundwater to collection and drainage systems can therefore be appropriate.

Table 2.1 Water supply

Service provided:	The provision of adequate water to storage facilities such as tanks or reservoirs in terms of security of supply, of sufficient quantity and quality (that can be treated economically to a standard fit for human consumption) from one or more sources, which may include local storage of water harvested from rainfall, off-line storage from intermittent streams, major on-line reservoirs, groundwater aquifers, or recycled drainage water. Typical sources of fresh water are: • Surface water • River • Groundwater • Desalination
Reasons for providing the service:	To meet the needs and demands of domestic, public and industrial users for potable water
Failures of the service occur when:	There is an intermittent supply, or an inadequate quantity of water, of a quality that is too expensive to treat to standards fit for human consumption
Reasons for a failure of the service include:	Inadequate volume of water for supply, breakdown of pumping facilities, breakdown of a desalination plant, short-term pollution of a resource, such as eutrophication with toxic algae in a reservoir, increase in long term pollution of a resource, such as increasing salinity in a coastal aquifer or growing arsenic concentrations in an inland aquifer.
Some actions to correct failures of the service:	• Effective monitoring and control of systems • Regular maintenance and inspection
Stability of a water supply system is affected by:	• Climate variability • Geology • Land use • Dam structure • Environmental and social impacts
Modelling software products:	• MODFLOW (USGS) • MIKESHE (DHI)

2.3 POTABLE WATER TREATMENT

The supply of water from a range of different sources, whether ground or surface water, can involve differences in quality. In a number of cases the quality will not be adequate to meet standards for human consumption. For example, there may be the need to remove particulate solids, nitrates and phosphorus from water that has received, say, agricultural runoff. In addition, there is often the requirement to destroy pathogens through dosing the water with chlorine or filtering it with ozone.

Table 2.2 Potable water treatment

Service provided:	The production of potable water fit for human and or industrial consumption sufficient to meet demand with minimal disruption for delivery through a water distribution system using chemical treatment processes
Reasons for providing the service:	To produce potable water fit for human consumption, washing, food processing, domestic waste disposal, hospital care, industrial processing
Failures of the service occur when:	The water issuing from the treatment works is inadequate in terms of quantity or not fit for human consumption
Reasons for a failure of the service include:	The dosing of the water with chemicals such as chlorine has been incorrect or the residence time of the water under treatment has been too short
Some actions to correct failures of the service:	• Implementation of monitoring and predictive control measures • Regular inspection and maintenance • Close monitoring of processes • Review dosing or increase residence time
Stability of a potable water treatment plant is affected by:	• Storage failures • Pumping failures • Screening failures • Processing failures
System structure:	A potable water treatment plant consists of a number of distinct processes connected in series or in parallel. After screening floating or suspended debris, fine particles are removed by coagulation and flocculation. Larger particles can be removed by sedimentation or flotation. Subsequently, the water is passed through filters of sand or anthracite before being disinfected with chlorine or ozone. More advanced membrane filtration may be used, and fluoride may be added.
Modelling software products:	OTTER (WRc) Watpro (Hydromantis) WTPmodel (USEPA)

2.4 POTABLE WATER DISTRIBUTION

Human beings need a regular daily supply of water, preferably on demand. Although water can be transported in containers, the volume required makes this form of supply expensive. The alternative is to distribute potable water in pipes from a reliable source to domestic, public and industrial demand points, and/or to reduce the volume supplied by on-site reuse or recovery of waste water. Because we do not yet have the treatment technology at a sufficiently inexpensive cost the former option remains the one preferred in most countries. A distribution network in a large city may consist of a ring main circulating much of the urban area with separate sub networks derived from the ring main. In each case the distribution networks are designed to operate under pressure. This means that the variation in the topography of the urban area does not normally restrict the layout of the network. To facilitate fluctuations in demand across the network, the layout is looped and the flow is regulated by valves that are manually or remotely controlled. The ability to separate sections of the network is of considerable value when diagnosing problems to do with leakage or uncounted for water. The structural integrity of the network can be at risk when leakage leads to the burst of a particular pipe. Repairs to the pipe can introduce polluted water to the network, which has to be flushed through the system before the water is safe to use. Contamination can also occur due to infiltration at points where leakage normally occurs due to a significant drop in pressure because of a pump failure. Pressure waves in the pipe network induced by sudden changes in the pumped discharge or closing valves or outlets at demand points can be dangerous for the operation and structural integrity of the network. This water hammer phenomenon is minimized by installing air vessels at strategic positions in the network to dampen down the oscillations.

Table 2.3 Potable water distribution

Service provided:	Potable water fit for human and/or industrial consumption sufficient to meet demand with minimal disruption is delivered to individual domestic, industrial and public service consumers (including schools, shops, offices, hospitals, fire hydrants and standpipes) in a prescribed urban area through artificial conduits normally operating under pressure, and preferably under gravity from storage reservoirs (such as water towers).
Reasons for providing the service:	To deliver potable water of sufficient quantity and quality to demand points
Failures of the service occur when:	There is inadequate supply of potable water to demand points due to contamination, reduction in flow or violently fluctuating flows
Reasons for a failure of the service include:	• Excess chlorine or other residual treatment chemical is present in the water • Polluted groundwater infiltrates to the network due to reduction in pressure caused by pump or valve failure or repairs to the network • Biological films formed on the inside of conduits in the network are dislodged and convected with the flow • There is unacceptable growth of biological entities in water detained for some time in part of the network before delivery • Excess leakage (possibly through a pipe burst or accidental damage to a pipe) reduces the water available, including unaccounted for water (such as illegal connections) • Ancillary structure failure (such as a pump or valve) leads to potential infiltration • Demand exceeds supply (such as due to fire fighting or an unusual public event) • Urban development puts too much strain on the expansion of the existing system • Excess pressures due to water hammer cause unacceptable stresses in pipes and structures that can lead to incremental failure of the system (e.g. increase in leakage)
Some actions to correct failures of the service:	• Prolong residence time in chlorine contact tanks to remove chlorine residuals. • Flush different parts of the network by adjusting valves to remove pollutants and aging water • Clean interior walls of pipes for biological films • Identify leaks through a mixture of acoustic technology and flow measurements made in prescribed sections of the network at times of low demand (eg at night). • Rehabilitate pipes that leak through relining or replacement • Bypass pipes that leak excessively (eg pipe burst) and repair • Maintain base pressure in the network to reduce the risk of infiltration • Carry out more frequent demand analysis • Install additional air chambers to absorb water hammer shocks

(*Continued*)

Table 2.3 Potable water distribution (*Continued*)

Stability of a distribution system is affected by:	• Extensions to include new customers • Maintenance involving cleaning pipes • Repairs involving structural failure (burst pipes) • Renovation of pipes and ancillary structures • Rehabilitation of distribution networks
System structure:	A water distribution system consists of a network of pipes conveying potable water from the treatment works by pumping to elevated storage reservoirs from which water can flow under pressure and under gravity to the consumers through a network of pipes. The pipes are generally arranged as a looped network of water mains in which the flow can be directed flexibly via different routes using adjustable valves. The pumps and valves may be controlled remotely and adjusted to deliver appropriate pressures at demand points. Air chambers are positioned in the network to absorb any water hammer events that may be triggered by a sudden pump or valve failure.
Modelling software products:	Physically based simulation modelling software for water distribution networks is an invaluable tool for a range of different tasks associated with managing a distribution network. Key products available in 2010 include: • EPA-NET (USEPA) • H_2ONET (MWH Soft) • InfoWorks WS (MWH Soft) • MIKE URBAN (DHI) • WaterCAD/GEMS (Bentley) Most products can perform time dependent hydraulic and water quality simulations. Prescriptive sizing of pipes in a predetermined network layout can be performed by software that minimises the energy requirements to meet the pressures required at demand nodes.

2.5 DRAINAGE OF WATER FROM URBAN AREAS

Human beings generate wastewater through their consumption of supplied water including washing, food preparation, urine, conveyance of faeces and other waste materials, industrial processes, and so on. In rural areas wastewater can be disposed of through septic tanks and soakaways to groundwater. This is not so convenient in urban areas due to the high volumes of wastewater and the pollution of the groundwater resource. Instead communities have provided conduits to convey wastewater away from houses and public buildings either to the nearest water course or to treatment facilities. Since the industrial revolution and the manufacture of pipes or brick-lined underground tunnels, wastewater has been conveyed together with excess rainfall-runoff (stormwater) for disposal in receiving streams, with the wastewater component increasingly treated by removing solids and various chemical components such as nutrients to maintain the water quality of the receiving streams. There are on-going debates about the relative merits of collecting wastewater and stormwater separately or together in one system. Also, in developing countries the performance of underground pipe networks deteriorates due to poor maintenance, and therefore a range of simpler alternatives are adopted. These include separating human faeces and urine from wastewater generated by washing, food preparation, etc which can be conveyed reasonably safely by open drainage to a suitable collection or disposal point, and having a completely separate storm drainage system made up of open drains.

Often wastewater management is dealt with on its own, though it is apparent that it should be done in conjunction with solid waster management. Many developing countries are overwhelmed by the accumulation of solid waste, much of which is dumped in drainage channels, restricting conveyance of flood flows and increasing the pollution levels in the water. Ideas about collecting urine and faeces apart from wastewater have to include notions such as source treatment and solid waste collection.

Table 2.4 Collection of wastewater and stormwater

Service provided:	The collection of wastewater, originating largely from a water distribution system, from housing and industry and delivery for treatment and subsequent discharge to receiving waters or for discharge directly to receiving waters, and the collection of excess rainfall-runoff from the urban area and the discharge of it to an adjacent water course, or the collection of both wastewater and stormwater in the same conduits and delivery up to a fixed factor of the cumulative wastewater flow to treatment and the discharge of any excess flow through combined sewer overflows to an adjacent water course the conveyance of stormwater safely in natural channels passing through urban areas
Reasons for providing the service:	• To convey safely waste material carried in water (sewage) from housing and industry to treatment for the removal of the waste and the discharge of the effluent to the receiving waters • To minimise the exposure of human beings to dangerous bacteria in wastewater causing infectious diseases • To establish and maintain safe living environments • To establish and maintain acceptable water quality in rivers and waterways receiving wastewater and stormwater • To improve and maintain the drainage of low-lying swampy areas • To limit the frequency of flooding and inundation in cities • To secure water resources through the improvement of water quality • To reutilize treated water • To reutilize sludge produced through treatment of wastewater
Failures of the service occur when:	• Wastewater does not drain away from the generating source • There are excessive stormwater flows into a separate wastewater network • There are excessive wastewater flows into a separate stormwater network • There is excessive infiltration to a separate wastewater or combined sewer network • There is too frequent surface flooding or polluted water on the urban surface leading to a deterioration in the acceptability of the environment • There is a deterioration of public health due to exposure to bacteria in wastewater • There is too frequent or there are excessively polluted discharges to receiving waters from combined sewer overflows or from overflows in a separate wastewater network leading to a general deterioration in water quality of short term fish kills
Reasons for a failure of the service include:	• Backup in house connections due to blockage due to sediment or congealed fats in conduits or inlets • Illegal connections of stormwater conduits to the separate wastewater network • Illegal connections of wastewater conduits to the separate stormwater network • Structural collapse of conduits or inlets • Too infrequent street cleaning • Significant infiltration into conduits through cracks or deteriorating joints • Inadequate hydraulic capacity of particular conduits in the network • Blocked outfalls • An overflow setting which is too low • Ancillary structure failure (pumps, gates, flap valves) • Failure of flood control system

(*Continued*)

Table 2.4 Collection of wastewater and stormwater (*Continued*)

Some actions to correct failures of the service:	• Remove blockages • Identify and close illegal connections • Assess structural integrity of conduit, and repair conduits that are vulnerable to collapse • Increase frequency of street cleaning • Increase infiltration to the ground through source control • Reduce infiltration by focussing on the repair of leaking conduits • Identify conduits with significant infiltration and repair the leaks through such methods as grouting, relining etc • Remove physical blockages at outfalls • Revise the setting of overflows • Provide back-up systems
Stability of a drainage system is affected by:	• Frequency of the surface flooding is greater than acceptable • Blockages in drainage conduits • Collapsed sewers which can be catastrophic in some circumstances • Inadequate capacity of inlets limits the flow into the below ground conduits • Sediment deposition reduces the capacity of the conduit • Infiltration dilutes wastewater going to treatment and can lead to more frequent operation of overflows • Extraneous inflows due either to infiltration or illegal connections • Inadequate hydraulic capacity, possibly due to changes in demand • Too frequent polluted discharges through CSOs thereby putting a heavier pollution load on the receiving stream • Extensions to include new customers • Maintenance involving clearing sediment depositions and rubbish removal • Repairs involving structural failure (collapsed sewers) • Renovation of pipes and ancillary structures • Rehabilitation of sewerage systems
System structure:	A drainage network consists of a network of conduits, often closed and underground (though many parts of the world still retain open surface channels for drainage), with uniform cross-section and gradient, which collect wastewater from housing and industry and excess stormwater to convey them safely under gravity and usually with free surface flow to treatment or an adjacent water course. The network is usually dendritic (having a tree-like structure) with the conduits joining at nodes called manholes, which are large enough to provide some degree of access for inspection purposes. As such they also act as (dynamic) storage facilities for sewage. Because wastewater flows that can be accepted for treatment are limited, diversion structures called overflows exist to convey excess flows to an adjacent water course. Outfalls normally have flaps to limit the ingress of water from the receiving stream and screens to remove gross solids. Flows in the network may also be controlled through the use of designated storage facilities, which may be separately operated in real time. Pumps are used to raise sewage for treatment and may also be used to raise sewage in networks in flat terrain. Severe surface flooding may be deliberately channelled along streets or other designated flow paths. This major system provides above-ground drainage of the urban area, and is designed to convey flood waters to an adjacent water course.

(*Continued*)

Table 2.4 Collection of wastewater and stormwater (*Continued*)

Modelling software products:	• CivilStorm (Bentley Systems) • InfoWorks CS (MWH Soft) • MOUSE (included in MIKE URBAN) (DHI) • SewerCAD/GEMS (Bentley Systems) • SIMPOL (WRc) • SOBEK-Urban (Deltaris) • SWMM (USEPA) • Win*DES*/Win*DAP* (Micro Drainage)

2.6 WASTEWATER TREATMENT

The wastewater resulting from domestic, public and industrial use of potable water supplied to urban areas is heavily polluted. The wastewater largely derives from the potable water conveyed to demand points through the distribution network. The domestic uses of potable water for washing, food preparation, disposal of sanitary waste, together with the range of industrial uses introduces significant contaminants and materials into the flow. Discharging untreated wastewater back into the natural environment can lead to serious deterioration of the receiving waters. It is preferable therefore to treat the wastewater, removing the solids and unacceptable concentrations of chemicals such as nutrients, heavy metals and harmful bacteria. This is normally done as an end-of-the-pipe process, though some treatment of industrial wastewater takes place at the site of origin. The treatment consists of a range of processes linked in series or in parallel. The raw sewage is normally lifted from the drainage network so that it can flow under gravity through the works. After screening for grit and floatables the sewage passes through sedimentation tanks and on to one or more activated sludge processes. Tertiary treatment to remove heavy metals, other toxic chemicals and bacteria are addressed using sand filters, membranes and disinfection.

Table 2.5 Wastewater treatment

Service provided:	Treatment of wastewater collected from housing and industry by removing solids, nutrients, pathogens and possibly heavy metals before discharging the effluent to the environment in accordance with national or international standards
Reasons for providing the service:	• To limit the discharge of particular pollutants to the natural environment • To make wastewater safely available for reuse • To limit the danger to human and environmental health of disposed wastewater • To meet national and/or international standards
Failures of the service occur when:	Polluted water with a quality that fails to make the required standard is released too frequently to the natural environment
Reasons for a failure of the service include:	• A breakdown in a specific treatment process • Accidental release of untreated wastewater
Some actions to correct failures of the service:	• Implementation of monitoring and predictive control measures to control potential hazards • Identify and close illegal discharges • Introduce screens to remove unwanted litter • Regular inspection and maintenance • Close monitoring of processes

(*Continued*)

Table 2.5 Wastewater treatment (*Continued*)

Stability of a treatment plant system is affected by:	• Illegal discharges • Blockages • Various mechanical failures • Sludge scraper failure • Blowers failure • Chlorination failure
Plant structure:	It includes elements that can undertake physical, chemical, and biological processes to remove physical, chemical and biological contaminants. The following levels of treatment can be undertaken: • Primary: mechanical separation of solids • Secondary: chemical and biological processes to remove fine solids, nitrogen, nutrients, ecoli. • Tertiary: refined chemical and biological processes to remove further pollutants including heavy metals
Modelling software products:	• AQUASIM (EAWAG) • BioWin (EnviroSim Associates) • GPS-X (Hydromantis) • SIMBA (ifak e. V. – Institut für Automation und Kommunikation) • STOAT (WRc) • WEST ++ (University of Ghent, Belgium)

2.7 RECEIVING WATERS IMPACT

Wastewater, treated or untreated, and stormwater are of such quantities such that they have to be disposed of in the natural environment. Normally the effluents will be discharged to a receiving water body, such as a stream (or river) or canal, lake or coastal waters. These receiving waters have a limited treatment capacity of their own, especially the flowing rivers and the coastal waters. For example, many cities have in the past discharged raw sewage directly to these receiving waters. Unfortunately, their treatment capacity is limited and in many cases they experience unacceptable pollution, either in the short term due to dramatic decreases in dissolved oxygen, or the long term due to the build up of toxic chemicals. In addition, one of the most serious consequences of discharging to the receiving waters is the failure to remove floatables from stormwater, such as plastic bags and sanitary towels, which get trapped in the vegetation lining the river banks or are washed up on beaches near to sea outfalls.

The relationship of the urban area to the streams flowing through it is complex. Urbanisation augments the runoff from the original natural catchment by increasing the peak flows (due to the reduction in infiltration) and decreasing the response time of the runoff (due to the higher overland and pipe flow velocities). These factors affect the morphology of the streams. A more serious effect however, is the potentially high concentrations of various pollutants in the effluents, with untreated effluents from wastewater causing serious deterioration of the receiving waters, supported by dilute sewage from combined sewer overflows, let alone accidental releases of untreated sewage. The consequences of these events on the water quality of the receiving waters means that improvements in the quality of these waters can only be achieved if the drainage system, consisting of collection network, treatment works and receiving waters, is regarded as an integrated system. Finally, urban rivers or coastal waters adjacent to an urban area, carry with them their own risk of extreme events, whether flooding or pollution, which may originate from regions remote to the urban area.

Table 2.6 Impact on receiving waters

Service provided:	• To receive (treated) wastewater from a treatment works • To receive stormwater from separate stormwater outfalls or from combined sewer overflows with a frequency no greater than a prescribed standard • To limit the frequency of above-bank flows from natural streams flowing through the urban area
Reasons for providing the service:	• To dispose wastewater safely and efficiently in the environment, and particularly to rivers and coastal waters.
Failures of the service occur when:	Polluted water with a quality that fails to make the required standard is released too frequently to the natural environment
Reasons for a failure of the service include:	• The effluent being discharged to the receiving water damages the ecology in either the short or long term, or affects the health of users of the receiving water as a resource • Failure of particular processes in the treatment works. • Failure to screen stormwater outfalls • The frequency of operation of the overflows is too large • The background pollution is too high • The receiving water discharge is too low • The receiving water level is too high
Some actions to correct failures of the service:	• Improving quality control and assurance procedures in the treatment works • Increasing storage within the drainage network to limit the frequency of operation of the overflows • Monitoring the background pollution, water levels and discharges in the receiving stream and using real time control to detain polluted discharges from the drainage network to the receiving streams
Stability of a receiving water is affected by:	• Illegal and uncontrolled wastewater discharges • Point and non-point source pollution • Failure of a wastewater treatment plan
Receiving water system:	Urban streams can originate outside the urban area and can contain significant levels of pollutants depending on the nature of the catchment, whether there is non-point source pollution from advanced agriculture (for example, sugar cane production), or point source pollution from village/urban communities with untreated wastewater, including waste disposal sites. Sometimes major urban conurbations are situated near lakes. These waters are particularly fragile and sensitive to pollutant loading from societal activities around the lake, including agriculture, fisheries, runoff from urban catchments, and even discharges of raw sewage. The danger to these lakes is the unacceptable growth of toxic algae and bacteria that are harmful to human health and the ecology of the water body. This is especially true where the water from the lake is used for water supply. A third category of receiving water is the waters around our coasts. Many large conurbations are situated on the coast for ease of communication, transport, fisheries, recreation and so on. It is very tempting to regard the ocean as a natural treatment works, and therefore cities around the world either discharge wastewater directly to the coastal water at the coastline, or further out to sea through submerged outfalls. Although the sea water contains salts that are beneficial in treating wastewater and the coastal currents disperse high concentrations of pollutants, there can be long term build-up of particular pollutants such as pathogens, which linger in the waters along beaches making bathing dangerous

(*Continued*)

Table 2.6 Impact on receiving waters (*Continued*)

Modelling software products:	• Rivers: MIKE11 (DHI), SOBEK (Deltaris), InfoWorks RS (MWH Soft), SIMPOL (WRc) • Lakes: DELFT3D (in both 2D and 3D mode) (Deltaris), MIKE21 (DHI), MIKE3 (DHI) • Coastal waters: DELFT3D (Deltaris), MIKE21 (DHI), MIKE3 (DHI)

2.8 URBAN FLOOD MANAGEMENT

Society's need for housing and roads leads to the paving of the ground surface, thereby limiting the infiltration of rainfall and increasing the runoff over the surface. Such runoff accumulates according to the local topography, and can quickly become unacceptable during a storm event. Special provision therefore is made to dispose of the excess rainfall-runoff by diverting it into artificial conduits, whether as open channels or underground pipes, which convey the stormwater to an appropriate water course. The potentially large flows that can be generated dictate the size of the conduits. In many cases the same conduits are used to convey wastewater away from housing and industry, such that the periodic stormwater flows can be used to flush away sedimentary deposits. The high cost of constructing the storm drainage network means that the conduits are normally designed to convey a certain frequency of discharge. For smaller frequencies of flow the conduits will flow full, leading to overtopping of open channels or the surcharging of underground pipes. In both cases, for a sufficiently small frequency of discharge surface flooding will occur, either because flow is discharged from the conduits onto the surface or runoff over the surface is prevented from entering the conduits because they are already full. It is important to note that the storm drainage network will always fail for a storm event of sufficiently low frequency (or large enough return period).

But surface flooding does not only occur due to the lack of capacity in the constructed storm drainage network. The urban area was originally (part of) a natural catchment that consisted of a surface network of natural streams. When the urban area was developed many of these natural streams were filled in or formed part of the constructed drainage network. Some of the larger streams may have been left as open channels because they drain catchments extending beyond the urban area. These streams are also prone to flooding for storm events of sufficiently small frequency.

We can conclude that urban flooding is a complex process that is generated normally by lack of capacity in the natural and constructed drainage networks and which results in overland flow through the built up environment. Such overland flow is dictated as much by artificial structures such as buildings, walls and streets or roads as it is by the underlying ground surface topography. This has lead some countries, such as Australia and Canada, to make special provision in their drainage systems for below-ground piped drainage to deal with the more frequent storm events (the minor system) and above-ground flow paths to cope with the more severe, less frequent storms (the major system).

Table 2.7 Urban flooding

Service provided:	• To drain excess rainfall runoff over an urban area safely and efficiently to natural streams of adequate capacity in order to limit the frequency of surface flooding, and to restrict the flooding from the urban streams to an acceptable frequency • To identify areas at risk of flooding • To provide adequate warning of impending flooding so that people who may be affected can secure possessions and resort to safety • To restore damaged infrastructure and services and to provide assistance to the community for its recovery

(*Continued*)

Table 2.7 Urban flooding (*Continued*)

Reasons for providing the service:	• To limit damages due to flooding, including the weakening of walls and foundations, the soaking of household contents, plaster and decorations, and the malfunction of electrical and other services • To avoid flood flows in the urban area that delay traffic, increase the risk of physical and mental harm to individuals, and lead to erosion and deposition of sediment
Failures of the service occur when:	• The frequency of surface flooding in an urban area is unacceptable, whether generated by the inadequate capacity of the constructed drainage system, or the overtopping of the embankments adjacent to urban streams • The identified areas at risk of flooding are inaccurate • Inadequate warning is provided to the community about the impending flooding • Little or no resources are provided to assist recovery after a flood event
Reasons for a failure of the service include:	• Excess rainfall-runoff • Inadequate drainage network capacity • Overtopping of flood embankments from flood flows in urban streams, too high water levels in urban lakes, and storm surges in coastal waters • Inadequate capacity of the constructed storm drainage network or the urban streams for the required design frequencies of discharge, due to the deterioration in the performance of the network or the streams as urban development has taken place upstream, or the design capacity has been reduced by sedimentation or other restrictions on the flow; alternatively overland flow paths may have been interfered with in unacceptable ways such that the major drainage system is inadequate • The mapping of areas at risk made too many assumptions and used inadequate models • The warning system did not provide sufficient lead time for the community to respond, or there was a breakdown of confidence in the system, or people were not trained on what they should do, or there was a failure in the communication of the warnings • The regional or national authority is not geared up to provide the necessary assistance
Some actions to correct failures of the service:	• Introduce better maintenance procedures to limit the build up of sediment deposits in urban streams or local encroachment on the stream capacity • Rehabilitate the storm drainage network by enlarging conduits, diverting flows or introducing (additional) storage facilities, including tanks or wet/dry ponds • Improve hazard mapping for floods in the urban area • Improve (or if necessary, introduce) the flood warning system • Set aside funds and resources for post disaster recovery
Some problems with flooding in urban areas:	There are several ways in which urban flooding may occur, and the diagnosis of what is the cause of flooding can be difficult. More frequent flooding may be due to: • Climate change • Urbanization • Blockages in the channel or pipe systems • Inadequate capacity • Rare storms • Flash flooding originating in the rural catchments of urban streams
Urban flooding environment:	It is important to realize that urban flooding can be generated by a number of sources, including inadequate capacity of the below ground urban network to accommodate the excess rainfall-runoff from the urban area, poor designation of overland flow paths to convey above ground rainfall-runoff safely to adjacent receiving waters, inadequate capacity of receiving waters to accept drainage network outfall or overflow discharges

(*Continued*)

Table 2.7 Urban flooding (*Continued*)

	and/or above ground flood flows, overtopping of embankments along urban streams due to flood flows in those streams and the subsequent inundation of the urban area, and coastal storm surges (or even astronomically high spring tides) leading to overtopping of embankments through the mean sea level rise and/or wave overtopping
Modelling software products:	There is a number of modeling software products for the different phases above: • Urban drainage: InfoWorks CS (WS), MOUSE (DHI) • Above ground urban flood flows: SOBEK2D (Deltaris), MIKE21(DHI), InfoWorks CS (MWH Soft) • Groundwater: MODFLOW (USGS) • Urban rivers: MIKE11 (DHI), SOBEK1D (Deltaris), InfoWorks RS (MWH Soft) • Coastal waters: MIKE3 (DHI), MIKE21 (DHI), DELFT3D (Deltaris), SOBEK2D (Deltaris)
The features of a model for urban flooding:	There may be the need to integrate several distinct models for the analysis of urban flooding, including: • Lumped rainfall-runoff from urban catchments and flood discharges in 1D below-ground drainage networks • 2D overland flood flows in urban areas with flows originating from rainfall-runoff, discharges from below ground drainage networks, and overtopping of embankments from urban streams, lakes and coastal waters • Lumped rainfall-runoff from rural catchments, 2D flows in urban lakes and 1D flood flows in urban streams, and 2D storm surge and astronomic tides with wave overtopping in coastal waters • 3D groundwater flows with links to exfiltration/infiltration to 1D below-ground drainage networks, and leakage from 1D water distribution networks

2.9 GROUNDWATER MANAGEMENT IN URBAN AREAS

Even though the paved surfaces in an urban area limit the infiltration of rainfall groundwater continues to flow under the urban area. The nature of the groundwater is dependent on a number of factors, not least the geological structure. Interaction of surface and groundwater occurs with urban streams, rainfall infiltration through pervious surfaces, sewerage networks, and flows originating from beyond the urban area. With the growing trend for source control in which infiltration in an urban area is deliberately increased through porous pavements, and groundwater recharge through soakaways and infiltration trenches, there is an attempt to limit the excess rainfall-runoff into the drainage system and to move back towards the state of the groundwater before the urban area existed. Of course that is not always desirable, especially if the area was originally poorly drained. In such cases there would be the need to deliberately abstract water to maintain groundwater levels below a certain value in order to limit uplift on buildings and other infrastructure. In addition there is the interaction of groundwater with water supply and drainage networks. Normally water distribution networks operate under pressure so that where there are cracks in the pipes or weak joints there is leakage from the pipes to the groundwater. Obviously such leakage can lead to a further weakening of the pipe structure due to movement of fine material in the backfill round the pipe being removed by the exfiltrating water, though the impact on groundwater is simply to increase the amount available. If the pressure in the distribution pipe drops for any reason, such as a failure of a pump, or the shutting down of the system due to lack of water, then the pressure of the groundwater around a pipe can be sufficient to cause infiltration to the pipe. Such infiltrating water can be polluted, in which case the situation for the water distribution network is serious: it has to be flushed before it can be used for its proper function. Infiltration to a drainage network, whether a separate wastewater or combined network, can lead to overloading of an end-of-line treatment works and the consequent operation of overflows discharging polluted effluent to receiving waters. In general there is a need to limit infiltration to the drainage network, unless it is deliberately designed to drain not only surface and/or wastewater but also groundwater. In comparison to surface flooding and corresponding discharges in the drainage networks the response time of groundwater to change is rather slow, but rising groundwater levels can damage buildings and lead to standing water on the surface, especially where there are depressions in the ground surface. The other serious consequence for the groundwater is its water quality. Seepage from wastewater networks can lead to growing pollution in the groundwater. Managing such pollution can be very difficult.

The management of groundwater below urban areas is therefore an important issue in that it affects all of the services provided.

Table 2.8 Groundwater below urban areas

Service provided:	The control of abstraction from groundwater aquifers below urban areas such that surface subsidence is negligible, the uplift on buildings with deep foundations and below ground storage is minimized, and infiltration to distribution and collection systems is acceptable
Failures of the service occur when:	• Abstraction from groundwater aquifers below the urban area for water supply is too large such that the reduction in aquifer levels leads to irreversible geological subsidence, which results in subsidence of the ground surface, or • Abstraction of groundwater is too small such that the uplift pressures on buildings and structures leads to unacceptable stresses in the structures, and infiltration to wastewater sewers in particular means that sewage is diluted, or infiltration to distribution systems in which the pressures are reduced due to some failure of the systems leads to pollution of the potable water
Reasons for a failure of the service include:	• Failure of groundwater pumping • Unacceptable subsidence leading to increased flood risk • Too small or too large an infiltration to the drainage network • Groundwater recharge from lateral areas is too large • Unacceptable climate change
Some actions to correct failures of the service:	• Install back-up pumping capacity • Rehabilitate the urban drainage network • Introduce boreholes to control groundwater flows from recharge areas • Reassess frequencies of events due to climate change and carry out remedial works
Some problems with groundwater in urban areas:	• Too low or too high groundwater levels affecting subsidence and integrity of structures • Pollution from surface infiltration and/or wastewater component in urban drainage networks • Possible infiltration of polluted groundwater to water distribution networks with low pressures
Modelling software products:	• MODFLOW (USGS) • MIKESHE (DHI)

2.10 INTEGRATED URBAN WATER MANAGEMENT

Urban water systems have evolved in order to provide a balance between ecological needs, human well-being, technical expertise, and the availability of natural resources. Since the primary focus of this book is on the sustainable management of urban water cycle, all individual systems which form this cycle (i.e., water supply, treatment and distribution, stormwater and wastewater collection systems, wastewater treatment and impacts on the receiving environment including groundwater and surface waters) will be covered. Some of the issues elaborated include:

- continuous supply of safe and acceptable drinking water,
- public health hazards due to wastewater,
- flooding,
- wastewater treatment and reuse,
- legislation, regulations and institutional arrangements for managing water-conflicts and opportunities,
- stakeholder awareness and understanding of urban water systems and involvement in their management,
- need to manage urban water in an integrated way according to urban water cycle principles.

Historically, the five main components of urban water cycle (potable water supply, treatment and distribution, drainage/sewerage collection systems, WWTP, receiving waters and groundwater) have been addressed separately, and their interactions were to a certain extent disregarded. As a consequence, cost-effective solutions were limited to a single component. The importance of understanding the degree of interdependence and interaction between these components is essential for those concerned with urban water management and therefore special attention is given to this aspect. The purpose here is to present an integrated overview of the different components, and to review the current state-of-the-art in modeling urban water systems in an integrated manner with the objective of generating cost-effective solutions while minimizing impacts on the receiving environment, mitigating adverse effects and implementing management programmes. Note that each component is addressed more thoroughly in separate chapters below.

An important consideration of the different components of the urban water cycle is the time scale associated with each phase. Rainfall-runoff from an urban catchment is comparatively fast due to the paved surfaces and the comparatively small size of the catchment. Typically the response time of an urban catchment to rainfall is one hour. This includes the time for runoff and flow through the drainage network to the outfall. Water distribution networks respond immediately to changes in the boundary conditions, so the time scale for these networks is a direct function of the time scale for the boundary conditions, namely the supply of water from the treatment works and the demand points. Normally, it is the demand points that determine the time scale, and this, in a typical distribution network serving 50,000 people, would also be the order of an hour.

The time scale for urban streams depends on the nature of the catchments for those streams. If the streams originate outside the urban area then it is reasonable to suppose that the response time of the catchments to rainfall is a matter of several hours. The larger the stream is, the longer the response time.

Where coastal waters are involved, the time scale of tides is twelve hours, whereas a coastal surge may have setup due to wind stress directed towards the coast, or a Coriolis affected surge originating from wind stress and pressure drop over the open ocean. In the first case the time scale is comparatively short, namely about an hour, whereas in the second case the response time could be ten hours or more.

The response of groundwater aquifers is comparatively slow due to the small velocities through the porous media. Unless the groundwater levels are close to the surface, such as in a number of cities in delta areas or close to the coast, the time scale associated with significant groundwater changes may be several days. For delta cities however, groundwater may make a significant contribution to surface flooding or to uplift on structures within a day.

In general therefore, the linking of groundwater to urban drainage or water distribution networks can be done on a time scale longer than the duration of a storm event, and should take into account the cumulative effect of a succession of storm events defined by a time series rainfall. This points to the groundwater model running in the background with the models for the urban drainage system, wastewater treatment works and the (dynamic) receiving waters (rivers and coastal waters), each of which has a comparatively short response time, integrated in a feed-forward mode (that is, data flowing from the drainage network to the treatment works and to the receiving waters), and possibly in a feed-back mode (for example, with high water levels in the receiving waters blocking flows from the urban drainage network). In each case the interaction with the groundwater will be determined on an event-averaged basis. The remaining phase is water distribution. Here the model for the water distribution network has an interaction with the groundwater through leakage (or more rarely, infiltration). Again an ongoing background and event average should be derived and act as an input to the groundwater model.

2.10.1 Asset management

The urban water systems consist largely of artificial networks or designated above ground flow paths, or in the case of urban streams, highly canalised channels, often with designated storage areas or ponds. As such the systems should be regarded as highly valuable assets belonging to the city community or other responsible organisation. Like any such asset they should be subject to proper management in terms of planning, demand and financial forecasting, optimum design and construction, efficient operation and maintenance, rehabilitation and renewal, appropriate investment, revenue collection, proper control of costs, and so on. The performance of the assets should meet their service level requirements, regulatory standards, customer needs and expectations, and maintenance needs in response to reported defects in their hydraulic, structural and environmental status. This obviously requires a high level of knowledge about the assets, and should include a register of the assets, demand and loading analyses, an assessment of their structural condition, information on their hydraulic performance and environmental impact, the history of each component, and forecasts of their useful life, with all data stored and accessible in a GIS. Such levels of asset management are barely achievable in highly affluent societies, but they provide a target to aim at, whatever the economic level of a given city. Asset management is important no matter what type of organisational structure is implemented, namely public, private or a public-private partnership.

The basic philosophy of asset management is to meet a required level of service in the most cost-effective manner through the creation, acquisition, maintenance, operation, rehabilitation and disposal of assets to provide for present and future customers. Over the past years, the practice of asset management in the urban water sector has evolved through different stages. Initially, the focus was on condition- and performance-based asset management practice, and more recently, there has been a shift towards the service level and risk-based asset management. Service level and risk-based asset management seeks to accomplish the practice through consideration of the quality of service delivery and associated risks.

The role of hydroinformatics in asset management is in the provision of tools and techniques that can enhance our capabilities in capturing, storing and assessing various aspects of urban water assets. This refers to the development and use of modelling systems, information and communication technologies, decision support systems and expert systems. Some of the key benefits of applying hydroinformatics technologies within the overall asset management process can be summarised to the following:

- To assess the performance and condition of assets
- To asses impacts on asset management due to external factors and trends
- To evaluate different scenarios

- To compare lifetime costs against compliance with standards of performance
- To optimise operational and capital expenditure
- To provide effective means of communication to stakeholders (intuitive GIS maps, 2D and 3D animations, etc.)

An important aspect of asset management is to determine the process of collecting and inserting the data into the system that ensures cost-effective data storage and retrieval. This is usually done within a data management framework which links various types of databases with GIS. Within such a framework, users can better understand the way in which the condition and performance of the asset change over time and what is their relation with the provision of services to customers.

2.10.2 Sewage and drinking water system renovation and rehabilitation

Sewerage systems, just like water distribution systems, are valuable assets for the local communities. They are expensive to both construct and maintain. What is more, their performance deteriorates in time due to the expansion of the urban area, renewal of the original area, and the natural decay of the drainage structure. Also, as society develops standards, such as levels of service, increase and generally become more stringent. All of these factors imply that there is a need periodically to review whether the drainage system performance has deteriorated unacceptably and a major overhaul of the system is needed to meet the latest standards and expectations. This process has come to be termed the rehabilitation of the system. The word 'rehabilitation' carries with it the notion of restoring something to its original state; here the word is used in the sense that the drainage system is being renewed such that it not only has a similar if not improved performance in terms of the original system, but that the system also serves a larger and developed area. Necessarily, a system needs to be rehabilitated when the levels of service, especially for surface flooding and for pollution in receiving waters, are unacceptable or the system structure is at risk. Procedures for rehabilitating sewerage and water distribution systems have been developed and refined in various developed countries in recent years. These procedures focus on the three major aspects of hydraulic performance, environmental impact and structural integrity. Hydraulic performance refers to the primary service provided by the networks (supplying potable water to demand points in urban areas, and draining wastewater and excess rainfall runoff away from urban areas), whereas environmental impact refers to the necessary interaction with the environment, whether the urban area itself or the receiving waters locally and downstream, and the possible adverse effects. The structural integrity acknowledges that the artificial, constructed nature of the systems is subject to decay over time, and therefore there is the need to renew or strengthen the structure periodically.

The major nature of a rehabilitation project indicates that considerable care should be taken over its formulation and implementation. It is recommended that the people living in the area served by the networks are best placed to identify deficiencies in the network performance, and they should be consulted through formalized questionnaires. These questionnaires can be computerized and made available over Internet or mobile phone networks. In addition, network inspections by trained engineers can be reported over mobile phones directly to a database. These sorts of facilities enable decision making to be more efficient, and can be regarded as hydroinformatics components of a decision support system.

2.10.3 Urban pollution management

The activities of the urban society lead to the production of waste materials, many of which are discharged directly to the environment, such as car exhaust fumes or light and sound pollution. A poor society may also discharge wastewater containing effluent from washing, food preparation, backstreet manufacturing, and latrines directly to the local environment such as the street, possibly with rudimentary drainage channels. Because of health and aesthetic reasons, more developed societies install underground pipe networks to convey wastewater more safely away from the urban area, and with treatment to dispose of it to the appropriate receiving waters. There are limitations on the size of the treatment works, and it follows that because rainfall in particular is stochastic there will be storm events that a treatment works in a combined or even separate drainage system will be unable to accommodate, and overflows operate to discharge excess polluted water directly to the receiving waters. Obviously this consequence can be reduced but there will always be pollution events that need to be managed. The discharge of polluted water to a receiving stream normally has both a short and long term effect. In the short term, the drop in dissolved oxygen in the stream due to the activity of the biological material in the wastewater can be serious for fish and other ecology. If the dissolved oxygen falls below 4 mg/l for any considerable time fish struggle to survive and can die. Similarly, a frequent discharge of pollutants to the stream can lead to a build up of pollutants attached to sediments which are deposited on the bed of the stream. These sediments prolong the effect of a particular pollutant on the water quality of the stream, especially when the sediment is eroded and brought into suspension due to higher velocities in the stream.

A particular issue with receiving streams concerns the trapping of any floatables originating from the urban area, such as plastic bags and condoms. Such gross pollution can be minimized by installing screens on all outflows from the drainage system, with the need to maintain such screens and to keep them clear so that they do not become blocked.

Obviously there is a need to manage the water quality in the receiving streams in order to improve or maintain the health of the streams, and to enhance the aesthetic appearance of the waters.

2.10.4 Emergency flood warning

Some cities are prone to extreme flooding on a regular basis, whether due to monsoon or hurricane type rainfall or flash flooding originating in mountains above the city. Flash flooding is particularly difficult to anticipate, and the short response time for the runoff to rainfall means that there is comparatively little time after the rainfall has started for the authorities to take action to protect the urban community. For this reason certain cities bordering the Mediterranean sea, such as Genova and Athens, have put in place emergency flood warning systems that depend on having rainfall forecasts, through weather radar or through simulation, using local area models coupled with global circulation of the atmosphere models. The idea is to gain sufficient lead time so that the community can take evasive action as necessary; see Abebe and Price, 2005.

Conceptually, emergency flood warning systems consist of three distinctive units: data acquisition and processing, model simulations and results mapping, and information dissemination. In the first part, telemetry (i.e., data acquisition) and pre-processing are concerned with the use of meteorological data and simulation of a hydrodynamic (HD) model to predict an upcoming flood event given the required lead time. In the second part, either the model simulations are undertaken and results are obtained or the pre-prepared library of model results is searched for appropriate information (e.g., risk maps, safe areas notification, evacuation routes, traffic control etc.). Finally, the third part is concerned with information dissemination. At this stage, all necessary real-time information applicable not only to flood managers but also to the public is crucial for the success of such system.

Recent developments in weather radar techniques have made emergency flood warning systems to become an even more valuable tool in urban flood management. Weather radar gives between three to six hours advance warning: longer time thresholds have to be sought from simulation modelling of the atmosphere. Because of the chaotic structure of the equations describing the processes in the atmosphere (that is, the simulations are very sensitive to the initial conditions) the models are normally run a number of times with slight variations in the initial conditions, giving an ensemble of forecasts, each of which is deemed to be equally likely.

The actual forecasts need to be made taking into account not only the simulation ensemble but also all other relevant information, such as synoptic data from satellites and radar and on the ground rainfall data.

2.10.5 Real time control

The networks for water distribution, wastewater collection and storm drainage, and the treatment plants for potable and waste water lend themselves to real time control (RTC). Such control enables the capacities of the networks and plants to be better utilised during fluctuations in demand or loading. In particular, RTC can provide improved management of undesirable consequences of extreme events, such as a pipe burst or collapse in a distribution or collection system, overloading of part of a storm drainage network, the operation of storage tanks, the failure of a treatment process, the breakdown of a pump, or too frequent an operation of a combined sewer overflow. The RTC may be effected through the dynamic operation of ancillary structure facilities, such as a valve, orifice, gate or pump.

An RTC device consists of regulators, which are physical structures to modify flows and levels; sensors to detect particular variables that describe the performance of the system, including rainfall, water levels, flows (velocities) and water quality determinands; controllers to instruct how the regulators should be operated based on data received from the sensors; and a communication system. There are different levels of control, including local control, regional/unit process control, and central control. Similarly, there are alternative modes of operation, such as manual and supervisory control. What RTC does is give added flexibility to optimise the use of the existing capacity of a given asset. Interestingly, RTC can be an integral component of the design for a treatment works or a water supply and distribution system, but is usually introduced to a collection or drainage network only after it has been constructed. In the future, there may be benefits in incorporating RTC in the design of new drainage networks.

2.11 KEYWORDS

By the end of the chapter you should have encountered the following terms; ensure that you are familiar with them!

Abbott*
Asset management*
Climate
Civilization
Dams
Databases
Failure of service
Flash flooding
GIS*
Groundwater*
Irrigation
Industrial revolution
Infiltration*
Integrated management*
Leakage*
Potable water treatment
Pipe burst/collapse*
Public health
Real time control*
Receiving waters*
Rehabilitation*
Remote sensing*
Service provided*
Stormwater drainage*
Wastewater drainage*
Wastewater treatment*
Water supply
Water distribution*
Wastewater treatment*
Urban flooding*
Urban pollution*

The items with an asterisk are dealt with in more detail in later chapters.

2.12 QUESTIONS

What was the influence of climate change on the development of civilization and cities?
What are the arguments for and against combined or separate collection systems for wastewater and stormwater collection?

What issues of public health can be addressed by good water supply and drainage systems in urban areas?
What are the advantages and disadvantages of including real time control in the design and operation of water distribution and collection systems?
How should the integrated urban drainage system including the receiving waters be managed?

2.13 REFERENCES

Abebe A. and Price R. (2005). Decision support system for urban flood management. *J. Hydroinf.*, **7**(1), 3–15.
Hassan F. A. (2004). Water and ethics: a historical perspective. In: Series on Water and Ethics. UNESCO, Paris.

Chapter 3

Hydroinformatics

3.1 ON ORIGINS

Besides the technical aspects of operating a water treatment plant, or rehabilitating a sewerage system or designing a flood prevention structure, there are significant social aspects. These include the design and implementation of social, legal, policy, economic and institutional arrangements (human-to-human relations), in contrast to the engineered tools and systems that transform the material world to provide benefits to societies (human-to-material relations). In this respect, the applications of hydroinformatics technologies to the management of urban water systems are like the applications of almost every technology nowadays in that they can be viewed as effecting changes in the way that knowledge is produced and employed within society. Abbott and Vojinović (2009) describe these changes as transformations, or even mutations, from *a modern* to a *post-modern* condition of society, that is from a society of *knowers* to a society of *consumers of knowledge*. Such changes can be related to the distinctive generations of computational hydraulics and subsequently hydroinformatics (see below).

Hydroinformatics emerged when Abbott (1991) identified a synergetic relation between numerical modelling and data collection and processing at the end of the 1980s. Since then, hydroinformatics has incorporated the social dimension by being concerned with the study of processes in which social aspects of water infrastructure cannot be separated from technical aspects. Hydroinformatics therefore integrates knowledges from the social and technical domains to create so-called *conjunctive knowledge*. It follows that the study and practice of sociotechnology is an essential aspect of hydroinformatics in order to understand how technical interventions have social consequences and how the resulting social changes in turn generate new technical developments and so on. The sociotechnological nature of problems associated with the effective management of urban water systems arises from the large number of stakeholders and the diversity in their interests and attitudes. Furthermore, an awareness of the writings of philosophers such as Locke, Hume and

Heidegger leads to the acknowledgment of the severe limitations of a technocratic mode of thinking. This must be countered by developing the means of involving stakeholders in infrastructure development together with the corresponding technical aspects. As stated by Abbott (2007) and described in much more detail by Abbott and Vojinović (2010) the role of the hydroinformatician in this process is to create sociotechnical environments in which the transmutations necessary to provide states of social justice can be catalysed through the creation of appropriate technologies. These technologies are those that can transform data and technical information into a form of knowledge and understanding that can be assimilated by the non-technical stakeholders so that they can contribute meaningfully and responsibly to the decision-making process. Therefore, one of the aims of hydroinformatics is a completely transparent decision making process involving all stakeholders in a water infrastructure project along with the relevant professionals directed to realisations of social justice, which are understood as transcendental states of social being (Abbott and Vojinović, 2010).

From one point of view, hydroinformatics has been practiced ever since human beings found they could exercise some form of management over the streams and rivers that they lived near to. Water engineers distilled good and better practices in hydraulics and water resources, and passed on the associated information to each other and to successive generations of engineers. Communities were involved in decision making about the water infrastructure to a greater or lesser extent. In some cases, municipalities and local communities have been able to take advantage of political consultation at the ballot box. However, from another point of view, hydroinformatics has only emerged following the widespread adoption of digital computing and the corresponding information and communication technologies, which throw open many possibilities for generating and communicating appropriate information in forms that any committed stakeholder can assimilate and for establishing stakeholders, professionals and managers in new intentional communities. This suggests that as we move from a modern condition to a post-modern condition of society, technological developments are often found to be strongly conditioned by our social arrangements and the two aspects (technical and social) are inseparable and must be recognised as such.

The emergence of hydorinformatics has occurred at a time when there is a growing worldwide water crisis that is in part due to human society's failure to be aware of the problem and therefore its possible solutions. The crisis is perceived in different ways by different groups. For example, the environmentalists and the agriculturalists see it as a conflict over the availability of water resources for the natural environment and food security. For some groups it is learning how to deal with problems of climate change, whether resulting in drought and possibly accelerated desertification or more frequent and severe flooding in rivers and along coasts. Other groups have to deal with the problem of existing resources being unacceptably polluted by industrial effluents or untreated wastewater from urban areas or by diffuse (non-point) sources of fertilizers and pesticides from agriculture or of eroded sediment arising from deforestation. These concerns and problems appear to be increasing in number and extent, and solutions are no longer simple to identify and implement due to the complex nature of the processes involved, the vested interests of certain stakeholders and growing costs. Increasingly, water managers are realizing that they have to take an integrated view of several distinct sub-systems which traditionally have been treated separately and in isolation from each other. Each sub-system has had its own experts with their particular knowledge and experience. Therefore attempts to address the integrated system have involved bringing these different experts together and requiring them to communicate and collaborate in order to find effective solutions. This is as true in urban water management as it is elsewhere in the water sector. For example, drainage engineers need to talk with treatment engineers and similarly with river engineers in order to resolve problems of rehabilitating urban sewerage systems in a way that not only achieves the objectives for the systems but maximizes the benefits to the society as well as to the river and minimizes the costs of wastewater treatment. The sociotechnical approach is applicable to all urban water problems irrespective of scale, although a larger scale may bring with it a greater need for diversity in the expertise that must be mobilised.

Obviously, a way to explore solutions within an integrated urban drainage system of combined sewerage network, wastewater treatment works and receiving waters is to model each phase separately and then to integrate the three models. What we are unable to do with our minds in terms of the numerical calculations we can do with the computer. The individual models can be surprisingly detailed and can accommodate the behaviour of what appear to be very complex systems. Predictions can be made about the response of the system to changes in the various boundary conditions, such as rainfall, or internal conditions dependent on parameters, or even functional representations of different identified water quality phenomena. Computational hydraulic modelling has become an integral part of the urban water manager's tool kit. We can ask therefore how this has come about, and what contribution it has made to the emergence of hydroinformatics.

In speaking about hydroinformatics as a sociotechnical discipline in which social and technical aspects cannot be separated, those who wish to engage in this kind of work are said to *draw upon technology*. The processes and aspects involved are by no means simple and straightforward to understand, and indeed the whole notion of technology itself needs to be carefully examined and thoroughly understood. Such comprehension is of fundamental importance for every hydroinformatician.

3.2 GENERATIONS OF COMPUTATIONAL HYDRAULIC MODELLING

Hydroinformatics emerged as a serious field of study at the beginning of the 1990s when the attention that had been focused on computational hydraulic modelling and on the computational engines at the heart of the modelling systems was being transferred to the intelligent use of the graphics user interfaces that facilitated access to the engines and thereby links to other management functions such as asset management. As stated above the notion of hydroinformatics was first described by Mike Abbott in his seminal book (Abbott, 1991). He had traced the development of computational hydraulics in terms of particular generations, which can be identified approximately with developments in computer hardware and software.

3.2.1 First generation of computational hydraulic modelling

Early efforts in computational hydraulic modelling during the first half of the 20th century were focused on innovative numerical analysis techniques based on hand calculations to solve the governing equations for gradually varying free surface flow in open channels, namely the Saint Venant equations, or the more simple forms of equation for flood and reservoir routing. For example, extensive hand calculations were done using graphical techniques based on the method of characteristics for the Saint Venant equations, and these were programmed for the first commercial computing machines. Abbott (1988) summarises the first generation as follows:

> *'The first generation was then one in which various methods that were already in use for 'hand calculation' were programmed to run on the machine. No differentiation had then to be made between 'human friendly' and 'computer friendly' methods. In this mode of employment the computer was said to be used only as a 'super slide rule' and it is a measure of how far this period is now removed that most of today's young engineers do not even know what a slide rule is, or was'*

Furthermore, Abbott (1988) summarises the challenges and opportunities inherent in the first generation and the requirements for the second generation as:

> *'It soon became clear that the computational methods most suited to the ways of working of the human mind were rarely if ever those best suited to the ways of working of the digital machine. A number of new 'machine-friendly'-even if 'human-friendly' – procedures were correspondingly developed.'*

3.2.2 Second generation of computational hydraulic modelling

The 1960s saw the growing availability of commercial computers, and were characterized by the writing of software code to solve the Saint Venant equations for each particular water system, such as the Thames estuary (Rossiter *et al.*, 1966). These applications benefited from a large amount of research that had been done on numerical analysis in anticipation of a growing sophistication in digital calculating machines. In particular, the characteristic method lent itself to graphical analysis, and explicit and implicit finite difference schemes had been proposed years before without the authors having access to the computing facilities necessary to capitalize on such developments. The computing hardware was very basic in comparison with what we are used to today. Only mainframe machines were available, allowing multiple users but with limited power. Input of source code and data was confined to paper tape, then to punched cards. It was only towards the end of the decade that electronic monitors became more widespread. Programming languages included FORTRAN and PASCAL. The engineers developing the modelling codes for the applications were the same people who instantiated and applied the models, using the codes for research and commercial purposes.

3.2.3 Third generation of computational hydraulic modelling

The 1970s heralded the third generation of computational hydraulics. Mainframe computers were increasing in power and accessibility, and software codes were being written more for generic use than for each application separately. The structure of the software assumed the input of data in one or more files, a batch run of the engine and the output of results as sets of numbers. Codes were modularized for different functions including data input and validation, modelling distinct physical processes, such as rainfall-runoff and channel flows, calibration and verification, and analysis of results. As such they became software modelling systems, including FLUCOMP from HR Wallingford (Price, 1980) and SYSTEM 21 from DHI (Abbott *et al.*, 1973). Software engineering specialists were graduating from universities, and commercial codes were beginning to be written by such specialists working alongside the research engineers. The modelling systems were being sold commercially and the corresponding marketing and support services were being created.

Abbott (1988) compares the second and third generation by outlining their characteristics, opportunities and challenges:

> *'The needs of engineering practice were such that the second-generation approach proved inadequate: the lead times for model construction and testing were too uncertain and the corresponding costs were too high for most of the applications arising from hydraulic, coastal, and ocean engineering. Accordingly, from about 1970 onwards, a third-generation was developed, which consisted of constructing design systems or, as they are nowadays most commonly called, modelling systems. These are software packages that construct and run any model of a given, wide class when the description of the model is provided to the package in a definite, specified format. The modelling system is, in effect, 'a factory for building models.' Whereas the second-generation approach necessitated the writing and testing of special codes for each model, the third-generation approach necessitated only the coding and testing of the system; thereafter, the models could be constructed and run without recourse to further coding. The introduction of the third-generation approach reduced lead times and costs by at least one order of magnitude and made possible a considerable increase in the range and power of numerical modelling methods in practice. In other areas, such modelling systems are commonly called shells. Where as second-generation modelling established the scientific paradigm in this field, as computational hydraulics, third-generation modelling established new technological paradigm concerning how this science could best be applied within society at large.'*

The third generation was regaled by the appearance of the first personal computers towards the end of the 1970s. During the 1980s rapid advances were made in miniaturizing the main frame such that its power could be put on the desks of individuals. The IBM PC (personal computer) running under DOS (disk operating system) symbolized the beginning of a new era in computing in which number crunching power was accessible to all. By the mid 1980s the mainframe codes had been shoehorned into the more limited RAM (random access memory) of the IBM PC, and the first commercial modelling packages for PCs were being marketed, including WASSP and WALLRUS from Wallingford Software and MOUSE and MIKE11 from Danish Hydraulic Institute. A new, larger range of users was being supported, namely water engineers who did not necessarily have a background in computational hydraulics and the associated mathematics, but for whom computational modelling gave the ability they did not have previously to explore options in managing their water systems. Modelling was becoming a necessary competence in the water industry. New coding languages were becoming available, such as C and C++ whose potential would be fully realized in the next decade. Certainly the writing of code was now in the hands of software professionals. The major modelling software suppliers in Europe were the hydraulic laboratories, including Danish Hydraulic Institute (DHI), Delft Hydraulics (Netherlands) and HR Wallingford (UK). They were becoming increasingly commercial in making their products available to the industry. Indeed, HR Wallingford set up a subsidiary software house, Wallingford Software, through which to release its primary modelling product for urban drainage. In contrast, product development for the simulation modelling of rivers, estuaries, coastal waters, and urban water systems in North America was funded largely by the US Environment Protection Agency and the US Army Corps of Engineers. As a consequence, many of these products were (and continue to be) made available in the public domain, and therefore supplied as source code free of charge. This, in turn, continued to open up more information on infrastructure projects, not only to engineers and managers, but to a growing number of stakeholders who were able to ask intelligent questions about what was being designed and implemented. The scene was set for the emergence of hydroinformatics.

3.2.4 Fourth generation of computational hydraulic modelling

The fourth generation, which is still very much with us, is mostly associated with stand-alone packages that combine large quantities of electronically-encapsulated knowledge. These packages were becoming widely distributed during the period 1985–90, and in the process transformed parts of the established major knowledge-providers from *performers of projects* into *providers of products*. As a consequence there was a more general social transformation taking place from knowledge being provided directly by developers to knowledge being provided in packaged forms for consumption by end users. As noted above, there was a general social transformation *from societies of knowers* to *societies of consumers of knowledge*.

Towards the end of 1980s, the ubiquitous PC with its integral screen had driven the development of interactive graphics in handling a wide range of computing activities. The graphics user interface was to bring about another radical change in the way modelling was carried out. The fourth generation of modelling has changed the perspective of modelling by introducing ‘modelling to non-experts’, as a result of tool proliferation. What the new Windows graphical operating system from Microsoft did was to make the PC more user-friendly, and thereby to open up personal computing to an even wider range of users. The fourth generation of computational hydraulics therefore began to put more emphasis on the interface than on the engine. Modular facilities, such as databases and geographic information systems (GIS), complemented the extensive video-graphics tools that were available in the modelling packages to visualize the networks, ancillary structures and the results of the simulations. Additional tools became available to assist users with model building and calibration. Such tools as optimisation routines, automatic calibration and others were introduced. The philosophy of software development and the use of it have made a shift from software as an aid to experts in their analyses to tools for decision making. Modelling software became increasingly subservient to tools for data handling and asset management, and was integrated in multi-user systems such that teams of engineers could work simultaneously on multiple projects, as in the large UK Water Companies. Again, the user base of the commercial modelling software had grown rapidly, such that users of specific modelling packages and systems were in considerable demand during the 1990s. By the year 2000 most of the developed countries of the world had at least some experience of using sophisticated modelling systems for urban water management, especially water distribution and wastewater and stormwater collection.

So what has characterized the first decade of the 21st century? Something that might have been anticipated to be important, but which turned out to be largely irrelevant is the rule-based expert system. This received vigorous attention in Europe and North America during the mid-1980s in order to counter initiatives in Japan. But though such systems have merit, and they have and are being embedded in some commercial systems, they were seen as being limited to rational logic and unable to reflect the vagaries of human reasoning. Therefore they have not become an important part of the latest modelling systems. What emerged in the 1990s out of the exploration of rule based expert systems are facilities for the support of decision making, and these are incorporated in decision support systems (DSS). The research in new tools such as 'expert-systems' was further pursued with a view to support the users (Price *et al.*, 1994). Such systems capitalize on the availability of a range of tools to address different aspects of the process leading up to the making of a decision, whether it is selecting a preferred option for improving flood defences for an urban area, or for designing a major extension of a water distribution network to deal with a significant urban development, or for rehabilitating a sewerage network whose performance has become inadequate; see Chapter 6.

Data-driven models, fuzzy set systems, genetic programming, Kalman filtering, global optimisation and other sub-symbolic methods have also entered the field. Some of these methods were based upon representing intelligent behaviour by sequences of symbols, construed as rules and facts, so that it constituted a symbolic paradigm. Most of these tools still remain at the experimental level and in research institutions as in-house tools.

3.2.5 Fifth generation of computational hydraulic modelling

The forces behind the fifth generation of modelling are greatly influenced by the change in the nature of the knowledges and by an increasingly active stakeholder participation in projects, policy formulation and on-line management. The need for the fifth generation of modelling was described by Cunge (1989):

> *'…, it should be stressed that there are two important points concerning these new systems. The first is that they cannot be built by an amateur. The widespread idea that since modelling systems run on a micro they can be built or modified by anybody who knows hydraulics and something about MS DOS is completely wrong. Actually the fourth generation micro-based systems accelerate considerably the separation the user and the manufacturer of the software. The second fact is that of fourth generation systems fall short of the (high!) expectations of the average buyer. They are very efficient tools for the experienced user (who, however has no access to the source-software) but there is still an important step needed before the systems can be sold to and used by the average engineer without larger knowledge in hydraulics. The latter observation indicates a need for future systems of a fifth generation which will integrate the knowledge (integration is meant here as opposed to a number of rules simply added as an appendix to a system and pompously called "expert system").'*

It was the introduction and proliferation of the Internet, leading to the World-Wide-Web, that provided the technical means necessary to catalyse and so promote the process of social transformation *from societies of knowers* to *societies of consumers of knowledge*. In particular, these developments have revolutionized the way in which the communication of information is carried out, with large amounts of data and information stored at nodes (servers) and accessible to anybody with a computer or mobile phone connected to the Internet anywhere in the world. Also, individuals can contact each other through electronic mail, streaming video and audio. There is a growing movement to store individuals' data and information centrally in secure data centers. Powerful modelling systems can be run on servers and accessed remotely by intelligent laypeople who have an interest in how particular water systems behave. This way anyone can have access to information about the performance of the systems. In addition, the communication facilities allow stakeholders to interact on-line in making decisions. What with the rapid growth in the use of mobile phone technologies coupled with computing there are a considerable number of options for the way in which developments will occur in the future. What is sure however, is that the systems emerging in the 2010s will be markedly more extensive and integrated than those at present, and will be growing in user friendliness. The systems will open up a more transparent communication of information and knowledge with all stakeholders, and there will be corresponding social changes as a result. As Abbott and Vojinović (2010) state that 'the purpose of stakeholder participation is to induce a change in the built environment that aligns with a positive change in the social environment'. The implications of this are considerable in that Abbott and Vojinović (2010) claim that 'stakeholder participation .... is a means of realising social justice'. It is to be hoped that these social changes will lead to the inclusion of more and more stakeholders in decision making, not in a reactive manner to the proposals and decisions of managers, but in an interactive sharing of transparent understanding, information and knowledge.

The fifth generation of modelling in hydroinformatics (which is assumed to have inherited the first four generations of computational hydraulic modelling) is directed to making electronically encapsulated modelling knowledge available over the Internet while supporting this with human knowledge and understanding. An example of such system was developed by Tumwesigye (2009) who has demonstrated how the instantiation and operation of complete models via GIS interfaces using remotely-installed systems can be realised over the Internet. In this work, a numerical modelling system of the urban drainage network of Cobh, a suburb of Cork, was installed on a server in one place and it was shown how it could be accessed from any other place in the world using a suitable Internet connection. The results obtained from the particular model so employed can then be used to provide such information such as flood predictions for surface-run-off, flood planning and mitigation or other information that may serve to inculcate knowledge and understanding. Such facilities make it possible to devolve the processes and responsibilities of preventive and remedial actions to local authorities, for example, in an efficient and responsible way.

These innovations are nowadays well-known and familiar, but it still happens that some more technocratically-minded persons regard the origins of the fourth generation, and even the fifth generation, as consequences of purely technical innovations, while overlooking their social origins and implications. Indeed, it is largely because the developments have mostly been regarded as technical and not sociotechnical, that further progress has been slow and inefficient; it can even be argued that the development of software for computational hydraulic modelling has proceeded in an inappropriate way as a consequence. It is against this background that the development of the fifth generation of modelling software together with other hydroinformatics tools has been formulated as an essentially sociotechnical task. This task has been integral to what has become known as The WaterKnowledge Initiative (Abbott *et al.*, 2006).

3.2.6 The role of Hydroinformatics in urban water management

At the beginning of the computer modelling era, before the development of intuitive graphical user interfaces, computer models were producing series of numbers which were not easily understandable to the wider audience but only to a few experts (this is the modern condition of a society of *knowers*). In return, the market for computer models was very restricted and physical laboratory models remained viable. As soon as graphical user interfaces were introduced, the market for computer models was opened up, enabling almost anyone to use such facilities in their daily work activities. Nowadays, most graphical interfaces of water modelling systems are able to display patterns of water movement, such as using velocity vectors, and there is a positive trend in extending such facilities as the range of end-users continues to expand and they become a post-modern condition of society of *consumers of knowledge*. Such facilities are clearly characterised by a change in the way that knowledge is produced and employed within society, and paradoxically, they lead to a state of society where individuals tend to know more but understand less. This can be explained by the fact that seamless interfaces for model instantiation are very powerful when it comes to shortening the time for instantiation but they create less opportunities for users to understand what is being computed, why and how. Moved by the easiness of powerful graphical user interfaces many of these users consider it no longer relevant to engage an expert, but rather to rely on their individual knowledge and understanding. This represents a challenge for hydroinformaticians to develop modelling technologies that ensure safe and reliable model instantiation and at the same time provide users with the ability to discriminate between information and disinformation in order that they come up with more reliable decisions.

Perhaps of all its features, the primary role of hydroinformatics within urban water management practice is in the development and installation of sociotechnical arrangements that can truly enable a holistic assessment of urban water assets, starting from planning and design, through construction and operation to dismantling and disposal. Without such arrangements, which necessitate ever greater improvements in computational and communication technologies, urban water utilities will not be able to strengthen their self-organising relations and so adapt themselves to changing environments and continuously evolving standards. The task of developing and installing the necessary sociotechnical arrangements calls in the first place for a *true leadership* which can provide enabling environments for two or more knowledges to operate simultaneously and interactively. Note that it is important to distinguish between leadership and management: leadership is about providing direction while management is about providing order. Furthermore, it is also important to be aware of so-called pseudo-leaders, that is, persons who occupy titles and positions of leaders but have little or no true leadership capabilities and for whom competent people are only a nuisance. For any professional, it is important to be aware of the existence of such individuals in order to prevent them making incorrect decisions that are often made to suit personal interests. Organisations that fall into the hands of pseudo-leaders have a greater likelihood of failing throughout this process. Abbott (2004) has discussed at length different concepts of direct and indirect leadership and their regular and irregular forms and the interested reader should follow up this reference. What is important to realise here is that there are different forms of leadership and that the preferred form is the one that creates and stimulates enabling environments by placing emphasis on human understanding while serving the interests of society, and is governed by ethical norms and ethical behaviour.

Another important role of hydroinformatics in urban water management is in providing technologies for the efficient integration and processing of a large amount of very diverse data and information of a range of water infrastructure facilities and combining the knowledge from them. With such technologies and its emphasis on both social and technical aspects hydroinformatics is best able to realise the principles of *integrated urban water management*. This is a popular term that is widely used but just as often misused. To be properly understood, this principle concerns on the one hand the integration of different structural and non-structural measures (which differs from a traditional technocratic approach to urban water management in which only the structural measures are evaluated, with only minor reference to 'social aspects'), and on the other hand it is concerned with the integration of views by all relevant parties, namely the stakeholders, within the decision making process. As noted above, the general purpose of stakeholder participation is to induce a change in the built environment that aligns with a positive change in the social environment (Abbott, 2007). When this positive social change is not realised, the process of attempting to transform the built environment suffers accordingly.

The purpose of stakeholder participation within the urban water management process itself is to transform the social environment from a reactive one, (in which each stakeholder reacts individually during the course of a purely technical transformation and solely on the basis of the most immediate social consequences for that individual stakeholder) into an interactive one (in which each stakeholder interacts with the other stakeholders within a community, whereby the different stakeholders come to cooperate in the forming and shaping the transformation as a whole, making of it an essentially sociotechnical transformation); see Abbott and Vojinović, 2010.

In terms of the future prospects of urban water management practice and the role of hydroinformatics, our current difficulties are no longer so much associated with numerical-computational software as such, but much more with the way in which the software is to be used: that is, how it integrates into its social context and how it in its turn it reacts to this context. With the changes brought by fifth generation modelling, it is not only that such arrangements can be addressed, but it also becomes possible to emerge with new business modalities in which both parties (*knowledge provider* and *knowledge consumer*) can benefit.

3.2.7 Artificial intelligence and Hydroinformatics

We now return to consider the relation between hydroinformatics and artificial intelligence and the contribution that the latter can make to the former. One way in which it appeared that artificial intelligence techniques were becoming relevant to computational hydraulics was in providing support for decision making through rule based expert systems. As mentioned above, in the mid 1980s there was considerable consternation in North America, and to a lesser extent in Europe, at the huge investment being made by Japan in expert system technologies. Various governments initiated research and development programmes to explore rule based expert systems for various commercial and military applications. Although interest in expert systems declined once their restrictions were recognized, some interesting work was done that laid the foundations for future developments. In particular, in the urban water area a number of expert systems were developed, such as SERPES (Walker, 1988) and WADNES (Walker, 1988) in the UK, which addressed the issues of sewerage rehabilitation and the real time control of water distribution networks respectively, and systems due to Baffaut (1989) for calibrating SWMM and León *et al.* (2000) for water networks management. Although these systems do not appear to have been generally accepted and used they prepared the way for other systems that offered better opportunities for acceptance by practicing engineers.

What the expert systems highlighted was the way in which engineers reasoned when addressing certain recurrent issues. In particular, as in other areas, engineers appreciate the fact that collectively they home in on preferred ways of doing things, whether in designing a structure or a system, or implementing what have come to be termed 'best management practices'. These consist of identified tasks in sequence or parallel, designed to achieve a specific management goal. Such tasks can be treated as objects with a range of attributes, including input and output data or information, the tools that may be needed to complete the task, warnings or errors that might be triggered while performing the task, documentation accompanying the task, sub-tasks, and so on. Given the interpretation of engineering and management procedures in terms of task objects, a range of established, provisional or new procedures can be implemented within decision support systems:

But the term 'artificial intelligence' covers a far wider range of technologies than rule based expert systems. In particular, it refers to modelling techniques that work largely with data, looking for patterns in sets of data and connections between different sets. In other words, these are data-driven modelling techniques which are focused on revealing new information from data as opposed to physically-based computational modelling approaches that were the forerunners of hydroinformatics.

Data-driven techniques have been used by engineers for pragmatic objectives. For example, the well known Saint Venant equations for gradually varying flow in open channels short circuit the theoretical development of the turbulent dissipation of energy in a Newtonian fluid based on the Navier-Stokes equations by assuming that the stress due to the friction on the flow from the boundary of the domain is proportional to the square of the mean velocity integrated over depth (and over the cross-section for one dimensional flow). The 'constant of proportionality' is regarded as a function of the depth and geometry, the nature of the flow ('rough' or 'smooth' turbulence), the properties of the water and the roughness of the boundary texture. Numerous laboratory experiments were done by Prandtl and others to define appropriate functions for different purposes; see Prandtl (1952). Engineers talk about the 'friction slope' in the momentum equations in terms of the Manning, Chezy, Strickler, Darcy or Colebrook-White formulations. In each case, the functional forms are appropriate for a particular range of Reynolds number (Ackers, 1961), and include parameters whose values have to be selected for the particular application. The Colebrook-White formulation includes two separate regression expressions for rough and smooth turbulence conditions that are integrated pragmatically into one expression; see Colebrook and White (1937).

The statistical regression technique has been used widely in hydraulic and hydrologic modelling for urban water systems. For example, the UK model for event-based urban rainfall-runoff includes a regression equation relating the volume of runoff to the percentage paved area in the catchment, the soil type and the antecedent precipitation; see Chapter 5. The rate of runoff is modelled using one or two linear reservoirs in series for which the storage constant is derived as a function of the catchment area, the mean slope and a measure of the rainfall intensity. Again, the function is determined from data using regression. Where design unit hydrographs are used, the parameters for the hydrograph are usually determined from regression relationships based on catchment characteristics.

Regression analysis is appropriate where data is limited. With access to large amounts of data however, data driven modelling techniques become more acceptable. For example, pattern recognition techniques can be used to classify storms, or an artificial neural network (ANN) can relate overflow operation to the generating storm events. Similarly, an ANN or another data driven modelling technique such as model tree, evolutionary programming, fuzzy logic generator or nearest neighbour can replicate the output from a physically based model for a large drainage network that takes a long time to run on a computer. The data driven model will be considerably faster to run, and can be used in the integrated modelling of a drainage-wastewater treatment-receiving water system to explore performance in different scenarios.

A data driven model *induces* new information from existing data. Like the physically based model it relates the (temporal) output to the (temporal) input. However, the parameterized function that connects the output to the input is determined by the nature of the technique and is completely independent of the physics of the situation. The values of the parameters are derived from the data by training. In this sense the data driven model is a 'black box', unlike the physically based model which encapsulates knowledge about the physics of the flow in the form of algorithmic equations. These equations describe principles such as conservation of mass, momentum and energy at a point. The equations are solved numerically for a discretisation of the model space occupied by the flow. In this sense, the physically based model *deduces* new information from existing data and takes into account the geometry of the domain.

Data driven modelling is therefore a modelling paradigm that is distinct from physically based modelling. There is a third modelling paradigm that is becoming recognised and is beginning to make a contribution to urban water management. This third paradigm is intelligent agent modelling. It has become widely used in various fields, especially those involving the behaviour of biological communities such as human beings, animals and plants, and the growth and decay of inanimate objects that are the result of human endeavour; see Wolfram (2002) and Ball (2006). As an example, consider the growth of urban areas. A city develops subject to a large range of factors, including the nature of the terrain, transport communications, economic prosperity, industrial and business development, etc. Given a raster map of the city and its surrounding area, each raster can be regarded as an intelligent agent subject to simplified rules involving the more important factors affecting urbanization (Barredo *et al.*, 2003); see Figure 3.1 which shows the success of the intelligent agent software product

MOLAND in modelling the growth of Dublin over a 50 year period. For the water engineer the interest is in the provision of extensions to existing water supply, wastewater collection and storm drainage services to meet the needs of newly developed areas. Again, intelligent agents can be assigned to the service networks in both the raster cells which are existing urban areas and peripheral cells which are potential subjects of urbanization. Generally, the networks will be aligned with the major roads. However, the extensions to the networks have to take into account the capacity of the existing facilities. This leads to the notion of periodic rehabilitation of the existing networks, which minimizes the discounted investment over a period of time, say a hundred years.

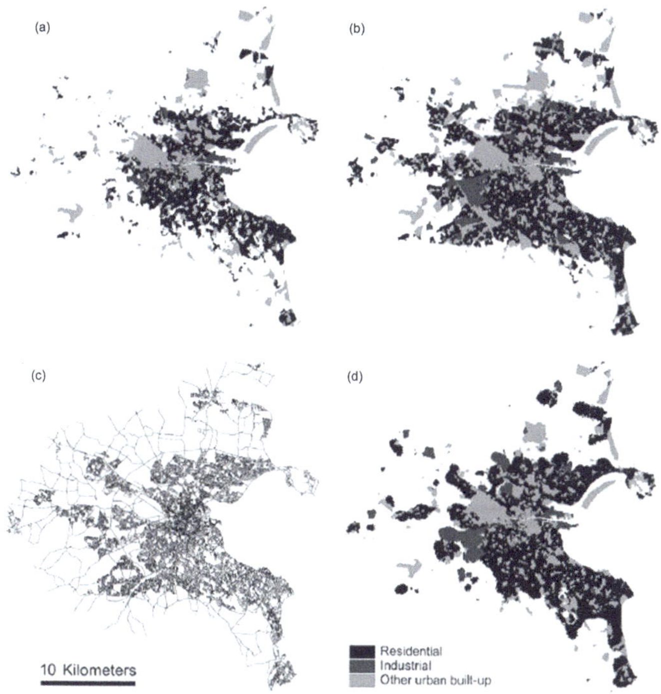

Figure 3.1 Urbanisation of Dublin, Ireland over a 50 year period, from a) to b) (actual), and from c) to d) (modelled with MOLAND) (Barredo *et al.*, 2003)

The modelling paradigms are discussed in more detail in Chapter 5.

3.2.8 Data management and decision support

The successful operation of urban water systems necessitates the installation and operation of a wide range of off-line and on-line measuring instruments. Coordination of on-line data collection systems are realised through SCADA systems, telemetry devices which are radio-accessible and capable of transmitting streams of data on a continuous basis to the appropriate decision-making centres (e.g., flood control centres, etc.) and other, more general, environments (e.g., water utility web sites, etc.). Nowadays, in many cities around the world most of this equipment is already available and installed, but it needs to be extended to increase the level of coverage and resolution across wider geographical areas. Some of this equipment is most conveniently situated at geographically fixed places (e.g., ground-based meteorological stations, GPS stations, etc.)

while other equipment is mobile, such as in the case of weather and water quality measuring equipment. Furthermore, other instrumentation again may be mounted on satellites. Whatever the application, the resulting on-line data streams have to be fed into the running models, both as boundary and other ancillary conditions and as means of correcting modelled descriptions of ongoing events by assimilating this data into the model results themselves. The role of hydroinformatics in this process is to bring together expertise of a social, scientific, technical and economic nature and to open up the possibilities that are being offered by the rapidly advancing technologies of communication.

There has been a great stress in some countries on data acquisition and decision support. This is particularly true in northern Europe, where countries such as UK, Netherlands, Germany, France, Sweden and Denmark have established the commercial nature of the organizations providing the urban water services, whether as completely privatized companies such as in the UK, France and Germany, public-private partnerships as in The Netherlands or public organizations as in Sweden. In the case of the UK, the decision was made in the late 1980s to privatize the then water authorities, placing the water supply, wastewater collection and treatment, and stormwater drainage services in the private water companies and the management of the receiving waters in the UK Environment Agency. One of the objectives in privatizing the water companies was to raise much needed investment capital to renew the aging water supply and sewerage assets. The UK Environment Agency became the 'game-keeper' for the potential water companies, the 'poachers', so far as the quality of the receiving waters are concerned. The UK Environment Agency also became the guardians of the water resources for supply purposes such that the water companies extract supplies from groundwater and surface waters under licence. The UK government set up a separate Office of Water (OFWAT) to regulate the performance of the water companies, requiring asset management plans and setting financial limits on charges to customers. The requirements of asset management planning implied the need for better data on the nature of the assets and their performance. This led to improved asset surveys and extensive modelling to identify the performance of the networks in terms of service levels. The large number of separate distribution and drainage networks (arising from more than 2,500 water treatment works and 9,000 sewage treatment works in the UK, and more than 2,400 waste water treatment plants with more than 50,000 population equivalent in Europe) meant that there was an obvious need to distil best management practices for the modelling and analysis of the performance of existing systems and to find efficient and effective ways of rehabilitating them. As a consequence a number of engineering and management procedures for water distribution and sewerage networks have been developed and implemented in many European countries and elsewhere in the world. The procedures form the basis of decision support systems in urban water management. These systems are being complemented by improving ways of involving the public in the decision making, using Internet and mobile phone technology. With the widespread availability of the World Wide Web, more and more people have access to information and facilities, including tools for analysing data and modelling alternative real world situations. In particular, the computational models that have until now been the province of the modelling expert can be made accessible to stakeholders over the Internet such that they can carry out limited investigations of their own into the consequences of different assumptions and scenarios. This is bringing laypeople more and more into the frame as regards responsible decision making about issues that affect them in the area of water management. It opens up better opportunities for reaching decisions that are agreed by a much larger proportion of society and hopefully are more effective as a consequence.

3.2.9 The nature of knowledge

If we wish to use our knowledge through the application of models to improve urban water services it is important to understand the flow of information which starts from data (which has the capacity to provide information) and continues through information (which has the capacity to provide knowledge), knowledge (which provides a change in one's state of awareness), understanding (which refers to the ability to comprehend something or somebody) and ends up with wisdom (which refers to the state which governs the way in which knowledge and understanding are applied).

Abbott (2005) pointed out that knowledge itself has the capacity to change the behaviour of the person acquiring the knowledge. With technological advancements it is not only our ability to communicate that has increased but also the amount of knowledge available to us. The two main examples of this are the Internet and mobile telephony. Having greater access to knowledge our society has the ability to change our behaviour by changing our states of awareness.

The nature of the knowledge is that it flows from a higher to a lower 'level' and Abbott (2005) has defined this flow in terms of a knowledge gradient. When we speak about using our knowledge through the application of a technology that enables us to intervene in the physical environment or to provide services, we refer to the so-called *encapsulation of knowledge*. In this respect, flood protection structures, whose design require the modelling services of engineering consultants, can be regarded as objects that encapsulate knowledge. Since we live in the post-modern society in which the knowledge is declared to be a central theme, our understanding of the nature and function of knowledge becomes important. Furthermore, dealing with stakeholders and attempting to create environments that can enable their meaningful involvement also calls for an understanding of these environments. Therefore we explore further knowledge management, knowledge sharing, knowledge creation and knowledge transfer.

3.2.10 Knowledge management

With the growth of capitalism and the recognition of the vulnerability of the markets with their periodic fluctuations, commercial companies have increasingly sought competitive advantage in a market place that is moving from a manufacturing to a service based economy by making better use of their knowledge. They aim to secure their advantage both in terms of what is called explicit knowledge, such as the culture and procedures inherent to their businesses, and tacit knowledge, which resides in the minds and experiences of their employees. Knowledge management is therefore as much about people and their social interaction as it is about their use of technological tools to address issues of understanding and managing water-based systems in the natural and urban environments; see Price (2000).

Plato defined knowledge succinctly as 'justified true belief'. This has the value of stressing that knowledge is both personal and context dependent. A person has to 'own' the knowledge for it to 'come to presence'. People generate knowledge in the form of concepts that originate from their 'unconscious' minds. This is a complex process that neuro-biologists and psychologists are only beginning to come to grips with. It depends heavily on past experiences, as well as an openness to allow the mind (whether 'unconscious' and 'conscious') to think laterally. Any insights are verified through the experience that the person has of the world. In this way, knowledge is justified based on relevant and valued information, which in turn is derived from particular structured data. In other words, there is the possibility of personal change in one's actions following the assimilation of new knowledge. This is a basic assumption about the value of education, and is directly relevant to the issue about transferring knowledge about water. The corollary is that knowledge is always dynamic and changing in the light of personal experience. So, for example, knowledge about the operation of a combined sewer overflow (CSO) in a particular combined sewerage system changes (and hopefully improves) after a new discharge event, but from the point of view of the individual the knowledge changes after the personal experiences of that individual. Such experiences might be unrelated to the CSO.

3.2.11 Sharing knowledge

It is all very well for an individual to generate new knowledge, but such knowledge remains the tacit property of the individual unless it is shared. Sharing knowledge is a complex and multi-faceted process that depends on the nature of knowledge itself. Nonaka and Takeuchi (1995) were the first to distinguish between tacit knowledge that resides in the minds, experiences and competencies of individuals, and explicit knowledge, which is tacit knowledge that has been encapsulated in documents, methods, procedures, models, tools and other human artefacts. The process of making knowledge explicit demands communication of that knowledge to other individuals or groups. This is one of the purposes of language through which different forms of communication can be done depending on the shared meanings attributed to the words. Only a small proportion of the tacit knowledge of an individual is made explicit, whether by default or intention, primarily because it is mainly embedded in the 'unconscious' rather than the 'conscious' mind of an individual. Any discussion about knowledge must recognise these forms of knowledge and therefore the (primary) importance of people and their tacit knowledge as well as explicit knowledge.

Nonaka and Takeuchi (1995) provided a simple template for the transfer and accompanying transformations of knowledge; see Figure 3.2. Encoding is the process of making knowledge that resides internal to a person external and formalised such that others can access it independently of the originator. Once knowledge is made explicit, the process of synthesis can lead to the combination of knowledge from different sources to generate new knowledge. Then there is the process of an individual internalising particular explicit knowledge to make that knowledge part of his or her worldview. This assimilation of knowledge is a vital part of the learning process and is a function of both the way in which the knowledge is presented and the capability and background experience of the receiver. Finally, tacit knowledge is transferred between individuals through socialisation that is best done face-to-face in that communication then includes body movements, gestures and facial expressions as well as words and language. None of these four transformations is trivial in that each is dependent on the complex psychological make up of every human being involved.

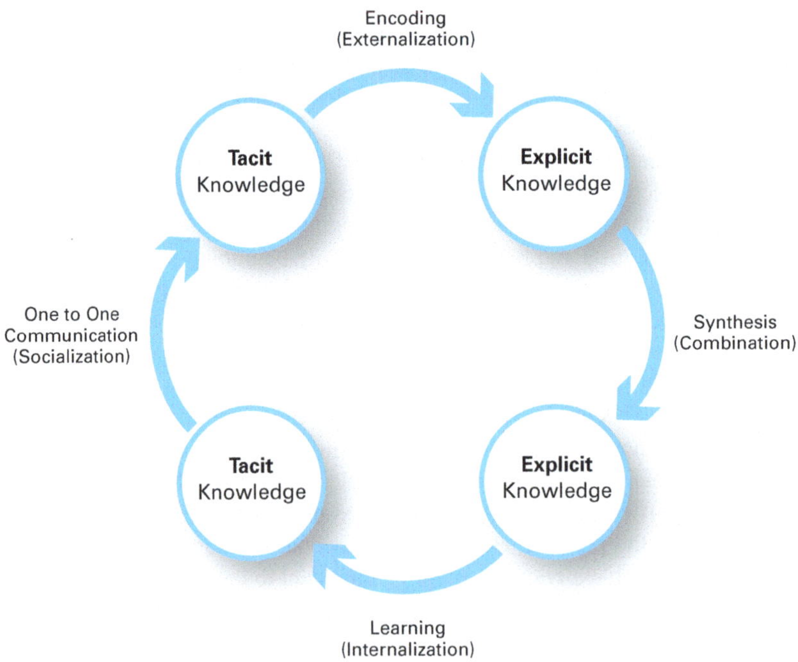

Figure 3.2 Tacit and explicit knowledge transfers (after Nonaka and Takeuchi, 1995)

3.3 HYDROINFORMATICS AND THE FLOW OF INFORMATION

The previous paragraphs have explored briefly the nature of knowledge and its management in human society and organizations. Why do we introduce this subject here? It is because hydroinformatics is concerned with how the flow of information and the exchange and transformation of knowledge within a sociotechnical context can be facilitated to best implement the management of our integrated water resources. In other words, we cannot appreciate what hydroinformatics is without having some appreciation of how information and knowledge about water management are generated and propagated in human society and organizations. What we are talking about is the flow of information from the physical world, which we perceive with our five senses and has its own physical and social realities, to the virtual world, which is conceived through the mathematical mental activity of human beings, to the organizational world that is at the heart of good management, and then to the stakeholders in society who make decisions or are directly affected by the decisions to intervene physically in the physical world. Necessarily, information flows between each 'world' in the opposite directions as well.

To appreciate this further, consider first three domains: the physical, the mental and the mathematical; see Penrose (1994) and Figure 3.3. Each of these domains is associated with profound mysteries or questions that have challenged, and continue to challenge, generations of scientists and philosophers. For example, how have we, as physical human beings, emerged within the physical world? And how have we evolved our level of consciousness such that we can inhabit our mental world in which we conceptualise and form links between our concepts, apparently independently of our physical nature? How is it that the mathematics that we so enthusiastically develop and embrace appears to undergird the physics of the universe independently of our mental world?

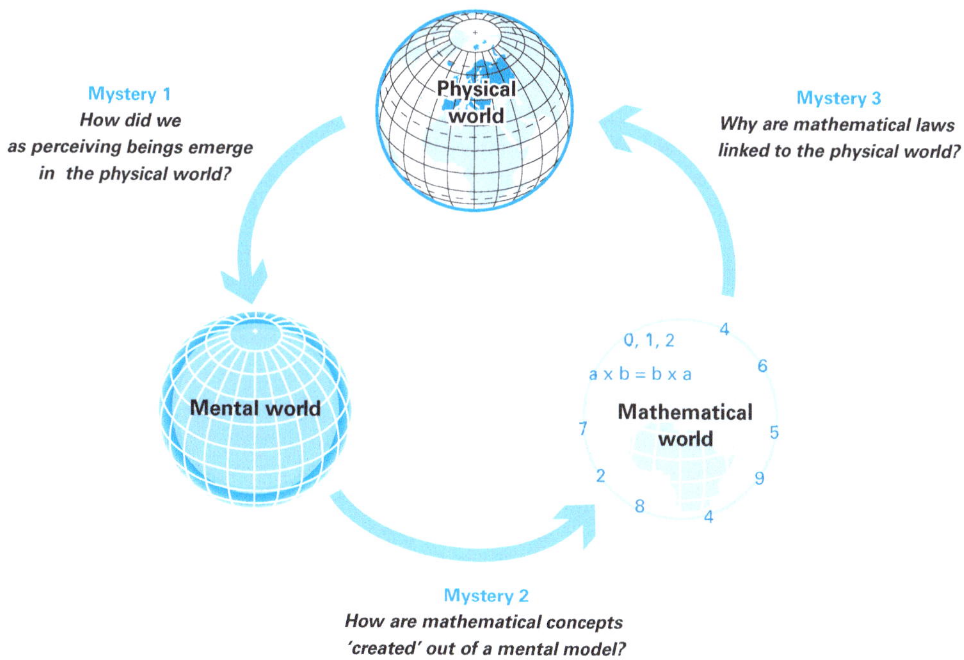

Figure 3.3 Three mysteries (after Penrose, 1994)

Whatever the resolution of these mysteries, it is apparent that mathematics is essential to our appreciation of our physical world, especially in our management of water in the natural rural and artificial urban environments. As human beings we are part of the biotic system that inhabits our planet, and increasingly we recognise that what is good for the whole of life on the earth is good for us as an animal species. Survival of the human race is consistent with sustaining all life and resources. As a species endowed with consciousness and our mental capabilities, we have the potential to dominate all other species and to abuse the earth's resources. Our destruction of tropical forests, our pollution of the oceans, our rape of the earth's fossil fuels, our emission with abandon of greenhouse gases, and our drive to industrialise, appear to result in unforeseen and unwelcome climate change. Sea levels are set to rise as the glaciers and polar caps, and especially the Antarctic, melt, threatening densely populated coastal regions around the world. Weather events appear to become more extreme, posing greater flood risks in many areas already familiar with severe flooding and more devastating droughts in regions already under water stress. There will be greater pressure for human migration from areas at risk to regions of lower risk, which will inevitably put strain on national boundaries and create a growing refugee population. Better management of water is therefore crucial to maintain national and social stability. It follows that it is necessary for the integrated management of water resources to take every advantage of the opportunities and benefits that the technologies available for manipulating and communicating information have between tools and people. This is where hydroinformatics comes into its own.

3.3.1 The physical and societal domains

We return now to the physical and societal domains in the physical world. By 'physical' we mean everything that pertains to our world that we can identify with our five senses. Water exists in this physical domain and interacts in complex ways with other physical entities and constructs. We need to intervene safely in the physical domain or make reliable and safe predictions in order to manage water better, and for the benefit of society and the sustainability of the planet. This means that in the 'societal' domain, which is the accumulated mental and social abilities of human individuals and societies in general, we need to formulate decisions about how to improve the performance of systems involving water in the physical domain, and to take appropriate action to do so. As Abbott points out this is only half of the story: we must also expect, indeed work for, a concomitant change in the social environment (Abbott, 2007).

3.3.2 The virtual world and the scientific interface with the physical world

Down the centuries we have achieved better management of our resources such as water using our mental ability to deconstruct our physical world into objects and their associated attributes that enable us then to deduce the underlying relationships between them.

The scientific method of observation, analysis, deduction and/or induction, has lead to the formulation of laws pertaining to the physical world and our ability to construct with these a virtual world in parallel to the physical world and replicating a part of it. This virtual world is a mental construct assisted increasingly by the latest information and communication technologies that enable us to visualize in growing detail the relevant part of the original physical world system or situation. In particular, the digital solution of the underlying mathematical equations, coupled with the digital data acquired from the physical world and made possible by the rapidly increasing power of computers, enables us through the numbers that are generated to replicate the particular aspect of the physical world. Indeed, through the calculations we can describe in far more detail what is happening in the physical world than we can measure. In this sense it can be argued that our involvement with computers is a commitment to virtual reality. Rivers and flood plains with flowing water computed using a computational hydraulic model can be portrayed on a computer screen in 3D scenic realism, something that makers of commercial films have long had the capacity to do. Alternatively, storm and wastewater in an urban drainage system can be visualized in underground pipes, whether under free surface or pressurized flow, as though it were being videoed in reality; see Figure 3.4.

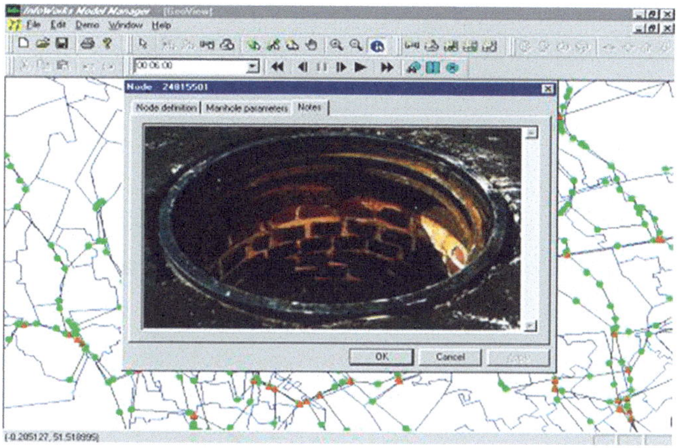

Figure 3.4 Network visualization (MWH Soft)

Again, the flow is calculated with a computational hydraulic model and the digital results are presented as a graphic picture. What the models, equations and relationships provide is a virtual world in which people can carry out experiments that are impossible in the physical world. They have full control over what happens in the virtual world, and they can explore a range of options and scenarios affecting the performance of the targeted physical world system. This assumes of course that the virtual counterpart of the physical world system is complete, consistent and accurate within known and acceptable levels of uncertainty. Completeness refers to the inclusion of all necessary structures, phenomena and processes, while consistency concerns the similarity between the detailed performance of the virtual system and the physical world system. Accuracy is about the magnitude of any divergence between the monitored physical world system and its virtual counterpart; see Figure 3.5.

The virtual world created by the computer is based on numbers. We can envisage what the numbers represent by generating graphics which put the flow of water in the context of our images of the physical world counterpart. But our experience of the virtual world is an enhancement of that of the physical world only in so far as we have a much more detailed measurement of what is happening in the model and solution spaces: we still do not necessarily understand more about the interaction of the basic physical processes governing the flow of water other than the assurance that the original algorithmic equations may or may not be sufficient to generate an adequate representation of the phenomena observed in reality.

The experience of virtual reality taunts us with the prospect of being 'in the picture' ourselves. This is in contrast to our traditional view where there is a rigorous separation between society and nature. For us nature has been regarded as a 'pool of resources' for us to master and dominate. We do this by compartmentalizing our knowledge and using its power to reign over and manage these resources. Now however, we need to see ourselves increasingly as part of the system. Knowledge as we have explained above is personal and contextual; it is itself a resource, and therefore subject to the management process. There are producers and consumers of knowledge. Abbott and Jonoski (2001) point out that knowledge can be 'sold, exchanged and renewed indefinitely, without depleting the original store of knowledge'. So it is not like other scarce resources. Also, access to knowledge is increasingly measured by access to electronic networks. 'Knowledge circulation processes in the third world have positive potential for fighting poverty …. (through) new kinds of sociotechnical arrangements' (Jonoski, 2002).

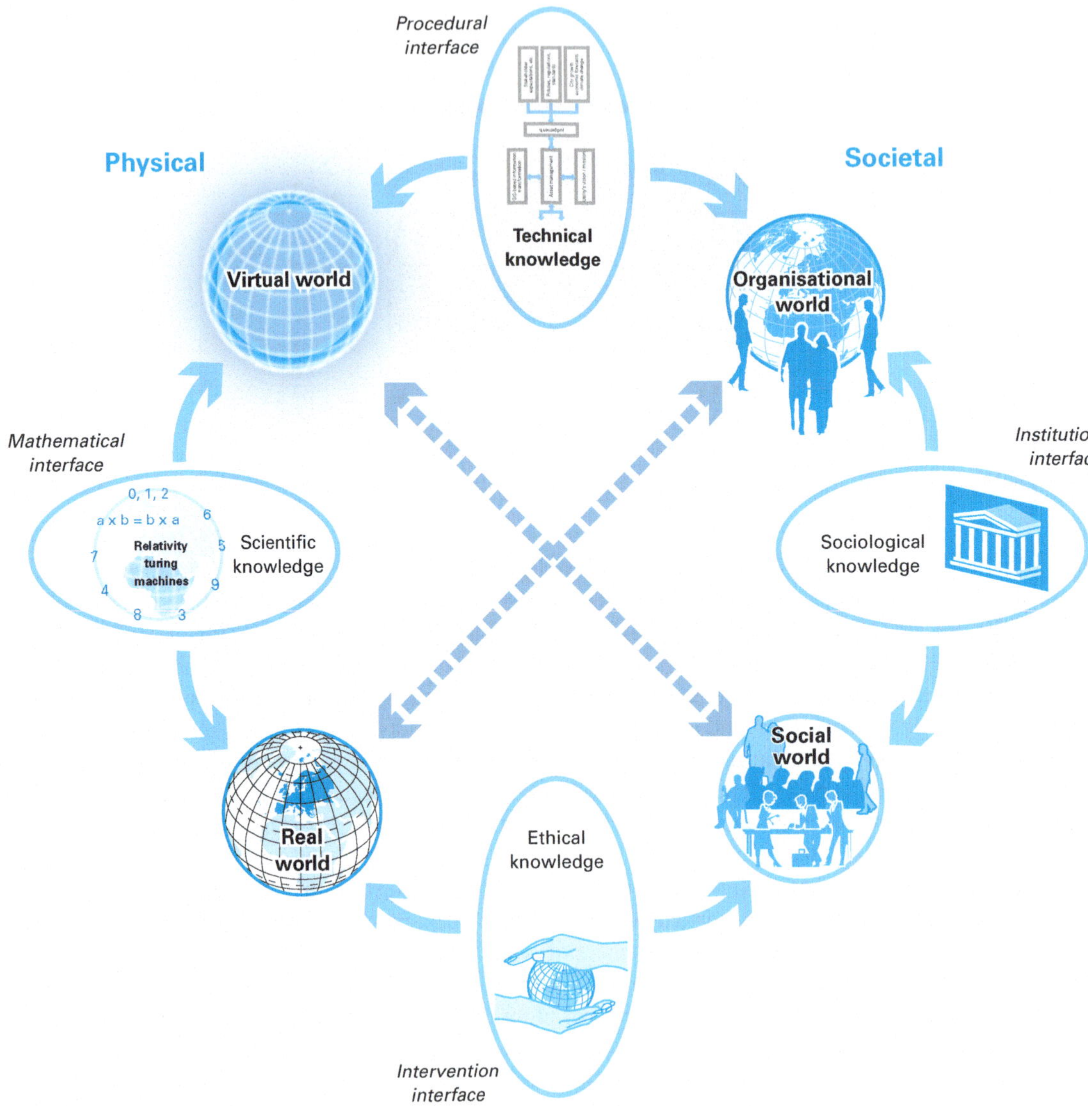

Figure 3.5 Physical and Societal domains

This raises the possibility of modelling not only the physical world but the social world. Indeed, modelling social behaviour has become commonplace through techniques such as cellular automata and intelligent agents. For example, models exist of social response to common events such as people queuing to get through a door, or individuals making investment decisions that are inter-related; see Ball (2006). The danger of this social modelling approach however, is that it can be regarded as a replacement of the actual participation of stakeholders in decision making. This would continue to fall into the technocratic trap that philosophers such as Heidigger and Ellul have warned us against.

3.3.3 The organizational world and the procedural interface with the virtual world

It is apparent that the virtual world is facilitated by technology and built on sound science. Yet we interact with it through knowledge. The building of the virtual world in terms of its tools and models is an expert task, which depends on certain professionals. These professionals are normally employed in certain organizations that have been delegated responsibility by society to manage the water resources on its behalf. Such organizations may not only be responsible for the physical design and operation of one or more urban (and natural) water systems, but they also develop and apply the large resource of knowledge associated with the systems. They are therefore continually developing their capacity to manage both the water and the corresponding knowledge. It is no surprise therefore that hydroinformatics is concerned as we have discussed above not only with managing water but also knowledge and thereby the capacity of organizations and society to make best use of their water as a resource.

The management of urban water systems is continually evolving as new research results become available and changes occur in society and the environment. But the management of urban systems includes much that is routine, and procedures for good management are well established in many organizations. With the hundreds of thousands of similar systems world-wide there is considerable opportunity for using such procedures as a means of exchanging information and transferring knowledge. For example, the model builders and users follow acknowledged procedures in routinely building and operating models in the water sector. Hydraulic models of urban drainage or water distribution systems are now commonplace, and considerable experience has been accumulated in their construction and application. As a result the accumulated experience has been formulated in best management practices (BMPs), including data acquisition, model building and calibration or training, design guidelines and standards, rehabilitation of systems, implementation of real time control, forecasting and warning systems, and so on. Such practices consist of recommended sequences of tasks and sub-tasks as objects with specific attributes. The tasks may even be supported by rule based expert systems. The BMPs and similar procedures are all intended and designed to provide support for decision making at different levels. The procedural approach to building the virtual counterpart to the physical world and operating the models and associated tools to explore different scenarios for managing the physical world system is so critical that it is viewed here as a vital interface between the societal and physical domains, and in particular between the organizational and virtual worlds. In other words, society assigns responsibility to groups of professionals to design, operate and manage particular physical world systems to the benefit of society according to its economic, social, political and environmental requirements. The people involved in the management of the physical world system make extensive use of appropriate tools for modelling and decision support. These tools have made possible the conversion from 'symbols' (in terms of the underlying mathematics) to 'signs' (the models that point to the specific results that come from them); see Abbott (1991) (1993; 1994). Individual managers used to be 'knowers' of the knowledge associated with their systems, that is, they were heavily dependent on their tacit knowledge. Now however they are primarily 'consumers' of knowledge, that is, they know where to find and to utilize the explicit knowledge they need at any time: their primary task is to use that knowledge effectively in getting the best value out of the systems they manage and operate; see Abbott (2002). The professionals manage the physical world system, such as the river basin or the urban drainage system, according to the guidelines laid down and monitored by society, and with the involvement of affected stakeholders who in better circumstances have access to the virtual world tools and models to ensure transparency of the decision making process. The access given to the virtual world tools and models by stakeholders is increasing as developments in ICT are adopted by the managers in the procedural world. Big changes are occurring, with opportunities to integrate modelling software from different suppliers (Open Modelling Systems; see Blind and Gregersen (2005)) to run modelling software remotely on servers, to obtain on-line assistance and advice on a commercial basis from experts over the Internet on building and operating models, and to provide appropriate interfaces for lay people (stakeholders) to run instantiated models. This opens up a range of possibilities for consulting the public over their observations of the failures of a given system, their worries and concerns, and demands and expectations. This recognizes that the indigenous knowledge of people 'on the ground' is important in contributing to understanding and improving the performance of a given system.

3.3.4 The social world and the institutional interface with the organizational world

An intervention on the physical world such as the construction of a wastewater treatment works, or the rehabilitation of water distribution or sewerage networks, is aimed at benefiting the particular local community, and therefore the social environment. The involvement of the public and stakeholders in a collaborative decision making process in the water sector is normally through facilities provided through local government. Where privatization of the water utilities is in place, institutional safeguards are normally provided by the national or regional governments to ensure that the stakeholders continue to be involved and that the company managing the water system is sensitive to local needs and ambitions. So, for example, before undertaking a major rehabilitation scheme for sewerage, the company should consult the local community to help in the diagnosis of the problems in the system, and to set appropriate goals and targets for the performance of the system. Again, as Abbott points out, the company may well be surprised at the level of interest and contributions that will emerge as such consultation goes ahead (Abbott, 2007). The public may well be interested in the levels of service provided in different areas of the city, the frequency of surface flooding, the limitation of sewer collapses, the maintenance of the network to limit blockages, the disruption to traffic of remedial works, the enhancement of the environment, particularly of urban streams, the use of park areas as temporary flood storage, and so on. Questionnaires can be used to collect public concerns and expectations. Mobile telephony can facilitate the reporting of problems and issues. Access to models of virtual situations such as areas sensitive to surface flooding, or what happens to surface flood flows in extreme events, or the impact of a discharging overflow on the pollution distribution in the receiving waters, can help generate new insights and ideas about how to improve the performance of the system. But whereas those delegated with responsibility for urban water management by society have in the past been accustomed to take the initiative (or not) in approaching their customers and seeking their involvement in the decision making process, the latest communication technologies, such as web-sites, blogs and twitters, are opening up new possibilities of members of the local community (including customers) to take the initiative instead. This is producing a paradigm change in the involvement of society in decisions that affect its environment. Note that individuals are strictly unpredictable in their behaviour and involvement in decision making, whereas it is assumed that corporately they have some predictability, though even that may be tested by unforeseen events.

3.3.5 The physical world and the intervention interface with the social world

The result of societal involvement and collaboration with the water organizations should be safe and reliable intervention in the physical world. The intervention has the support of the community such that they take responsibility for it, whether in terms of its construction or operation.

A well-known example of the outworking of this approach in a major construction project is in the design and implementation of the tunnel and bridge connection between Denmark and Sweden. The latest modelling tools were used to predict currents in connection with remote sensing and monitoring of water levels and velocities. The information was needed to minimize the dispersion of dredged material excavated from the trench used to position the tunnel sections. The material was brought into suspension and could easily be convected with the current to sites where there were vulnerable grasses and fish breeding grounds. A particular communication system was devised to convey the data and model results to a group of interested and active stakeholders in the construction process. In this way the vulnerable eco-systems were preserved and costs reduced; see Thorkilsen and Dynesen (2001). These authors observe that: 'A common feature of all the different phases of the project was the widespread uses of ICT solutions, not only functioning in the role of technical problem solvers, but also as facilitators for communicating complex decision problems to all interested persons and parties'. And again, 'The task of informing the politicians, the responsible authorities and the public as a whole in a truthful and realistic way about the consequences of the construction and inseparably from this, the empowering of these to act as genuine stakeholders in the management of the natural environment, was a task of hydroinformaticians over and above the tasks within the physical project itself'.

Ideally society and the water organizations should follow principles of social justice in which each intervention should not prejudice the rights of others, whether downstream or in the future, of access to safe and adequate water supplies, and of their ability to sustain their (aquatic) environment.

3.3.6 Hydroinformatics summarised

Hydroinformatics therefore is concerned with the provision, integration and operation of information management tools for water in the environment covering the four major domains of the physical world, the virtual world, the organizational world and the social world, with the appropriate interfaces between these worlds, and with bringing about concomitant changes in both the built and social environments. In other words, hydroinformatics provides the sociotechnological tools and facilities for water professionals and stakeholders to implement integrated water resources management in general, and to facilitate urban water systems in particular. The remainder of this book considers urban water systems management from this point of view.

A careful consideration of the nature of hydroinformatics as presented here leads to the acknowledgement that hydroinformatics is more than just a set of techniques to manage information and knowledge concerning water in the environment: it is concerned with change in both the physical built and social environments. Although we can talk generally about the principles on which hydroinformatics is based, the subject only comes into being as it is substantiated in application. This book focuses on the application to urban water systems management. We could also apply hydroinformatics to flood or drought management in rivers or coastal areas, or to groundwater or water resources management. In each case the databases will be different as will the models and the procedures; but the underlying principles are the same.

Finally, we would point out that hydroinformatics as a sociotechnology effecting interactive changes in the social as well as the technical infrastructure environments is still in its infancy: the demonstrations of its power to effect change in urban water management practice are few. Part of the difficulty lies in the need for leadership that is truly ethical. We trust that such leadership will emerge from present and future hydroinformaticians.

3.4 KEYWORDS

By the end of the chapter you should have encountered the following terms; ensure that you are familiar with them!

Abbott
Artificial intelligence
Communities of practice
Computational engine*
Computational hydraulic modelling*
Data driven modelling*
Digital computing
Hydroinformatics*
Information management*
Intelligent agent modelling*
Internet*
Knowledge management
Nonaka and Takeuchi
Organizational world
Physical built environment
Physical world
Physically based modelling*
Procedural interface
Procedures*
Rule based expert systems
Social environment
Social world
Tacit and explicit knowledge
User interface*
Virtual world

The items with an asterisk are dealt with in more detail in later chapters.

3.5 QUESTIONS

What are the advantages and disadvantages of physically based and data driven modelling?
How did digital computing contribute to the emergence of hydroinformatics?
What is the relationship between information and knowledge management?
What are the benefits and dangers of adopting a procedural approach to modelling?

3.6 REFERENCES

Abbott M. B. (1988). Contributions of computational hydraulics to the foundation of computational hydrology. In: *Recent Advances in the Modelling of Hydrologic Systems*, Vol. 345. NATO ASI Series C. Kluwer Academic Pub, Sintra, Portugal, pp. 509–537.

Abbott M. B. (1991). *Hydroinformatics: Information Technology and the Aquatic Environment*. Avebury Technical Brookfield, USA.

Abbott M. B. (1993). The electronic encapsulation of knowledge in hydraulics, hydrology and water resources. *Adv. Water Resour.*, **16**(1), 21–39.

Abbott M. B. (1994). Hydroinformatics: a Copernican revolution in hydraulics. *J. Hydraul. Res.*, **32**, 3–13.

Abbott M. B. (2002). On definitions. *J. Hydroinf.*, **4**(2), i–xxvii.

Abbott M. B. (2004). European Industrial Leadership, Knowledge Engineering and European Institute for Industrial Leadership, Brussels, www.eiil.net

Abbott M. B. (2007). Managing the inner world of infrastructure. *Proc. ICE Civ. Eng.*, **160**, 26–30.

Abbott M. B., Chambers J. and Hing C. C. (2005). *From a Knowledge Society to an Understanding Society*. European Institute for Industrial Leadership, Knowledge Engineering BVBA, Brussels.

Abbott M. B., Damsgaard A. and Rodenhuis G. S. (1973). System 21,"Jupiter": a design system for two-dimensional nearly-horizontal flows. *J. Hydraul. Res.*, **11**(1), 1–28.

Abbott M. B. and Jonoski A. (2001). The democratisation of decision-making processes in the water sector II. *J. Hydroinf.*, **3**(1), 35–48.

Abbott M. B., Tumwesigye B. M. and Vojinović Z. (2006). The fifth generation of modelling in hydroinformatics. In: *7th International Conference on Hydroinformatics*, September, Acropolis, Nice, France, 4–7 September.

Abbott M. B. and Vojinović Z. (2010). Realising social justice in the water sector: 1. *J. Hydroinf.*, **12**(1), 110–130.

Abbott M. B. and Vojinović Z. (2010). Realising social justice in the water sector: 2. *J. Hydroinf.*, **12**(2), 225–239.

Ackers P. (1961). The hydraulic resistance of drainage conduits. *Proc. Inst. Civ. Eng.*, **19**, 26525.

Baffaut C. (1989). Expert system for calibrating SWMM. *J. Water Resour. Plann. Manage.*, **115**, 278.

Ball P. (2006). *Critical Mass: How One Thing Leads to Another.* Farrar Straus & Giroux, New York.

Bandler R., Grinder J. and Andreas S. (1982). *Neurolinguistic Programming and Transformation of Meaning*. Real People Press, Moab, Utah.
Barredo J. I., Kasanko M., McCormick N. and Lavalle, C. (2003). Modelling dynamic spatial processes: simulation of urban future scenarios through cellular automata. *Landscape Urban Plann.*, **64**, 145–160.
Baudrillard J. (1970). *La Société de Consommation.* Denoël, Paris. [The Consumer Society].
Blind M. and Gregersen J. (2005). Towards an Open Modelling Interface (OpenMI) the HarmonIT project. *Adv. Geosci.*, **4**, 69–74.
Colebrook C. and White C. (1937). Experiments with fluid friction in roughened pipes. *Proc. Roy. Soc. Lond. Ser. A Math. Phys. Sci.*, **161**(906), 367–381.
Cunge J. A. (1989). Review of recent developments in river modelling. In: *Proceedings of International Conference on Hydraulic and Environmental Modelling of Coastal, Estuarine, and River Waters*, Aldershot, 3–19. Bredford: Gower Technical/Aldershot, UK, 19–21 September.
Jonoski A. (2002). *Hydroinformatics as Sociotechnology.* Swets & Zeitlinger BV, Lisse.
Krogh G. v., Ichijo I. and Nonaka I. (2000). *Enabling Knowledge Creation*. Oxford University Press, New York.
León C., Martín S., Elena, J. M. and Luque, J. (2000). EXPLORE – hybrid expert system for water networks management. *J. Water Resour. Plann. Manage.*, **126**(2), 65–74.
Lyotard J.-C. (1986). *La Condition Postmoderne, Rapport sur le Savoir* (*The Postmodern Condition: A Report on Knowledge*, trans. Bennington G. and Massumi B., Manchester University, Manchester). Minuit, Paris, 1979.
Nonaka I. and Takeuchi H. (1995). *The Knowledge-Creating Company: How Japanese Companies Create the Dynamics of Innovation*. Oxford University Press, USA.
Penrose R. (1994). *Shadows of the Mind.* Oxford University Press, Oxford.
Prandtl L. (1952). *Essentials of Fluid Dynamics*. Blackie, London.
Price R. and Samuels P. (1980). A computational hydraulic model for rivers. *Proc. Inst. Civ. Eng. Part 2* (69): 87–96.
Price R. K. (2000). *Hydroinformatics for River Flood Management*. Kluwer, Amsterdam.
Price R. K., Samuganathan K. and Powel K. (1994). Hydroinformatics and the management of water-based assets. *J. Hydraul. Res. Hydroinf.*, **32**(Extra Issue), 65–82.
Rossiter J. R., Lennon G. W., Lacey G., Mcdowell D. M., Barr D. I. H., Odd N. V. M., Robson C. L. and Andrews A. L. (1966). Computation of tidal conditions in the Thames by the initial value method. *Proc. Inst. Civ. Eng.*, **34**(1), 47–62.
Thorkilsen M. and Dynesen C. (2001). An owner's view of hydroinformatics: its role in realising the bridge and tunnel connection between Denmark and Sweden. *J. Hydroinf.*, **3**(2), 105–135.
Tumwesigye E. (2009). *The WaterWebFrame Modelling Framework and Business Context in Development towards 5th Generation of Hydroinformatics*. PhD Thesis, University College Cork, Ireland.
Vuijst J. D. (1999). In: *Kennismenagement in de Publieke Sector.* Elsevier, Knowledge management demands external drivers. The Hague. [Kennismanagement vergt externe trekkers].
Walker R. (1988). *Water Industry Expert Systems Club Final Report*. WRc Engineering, Swindon.
Wenger E. and Snyder W. (2000). Communities of practice: the organizational frontier. *Harvard Bus. Rev.*, **78**(1), 139–146.
Wolfram S. (2002). *A New Kind of Science*. Wolfram Media Inc., Champaign, IL.

Chapter 4
Data Management

4.1 INTRODUCTION

The purpose of data collection in urban water management is to provide practitioners and decision makers with information necessary to make better decisions. In general terms, the data is needed for various aspects of asset management and planning, operation and maintenance activities. With such data it possible to understand and predict the behaviour of an individual urban water system and its interactions with other parts of the urban water cycle. Typically, an urban water system database represents a collection of spatial, temporal and other data (e.g., water distribution or sewer pipe network data, water levels, flow rates, public perception of utility's levels of service, etc.). Collection of such data is of the utmost importance for making cost effective investment and operational/maintenance decisions. This is true whether for water distribution or drainage networks. In particular, for drainage networks engineers have developed the strategy of carrying out short term flow surveys in which as many as 50 flow meters may be installed in a (large) drainage network to collect flow data over a period of several weeks. Such data is invaluable for calibrating and verifying a physically based model and for improving the understanding of the structure and performance of the network.

There are a number of important aspects that need careful consideration and planning to achieve a successful data collection campaign. Obviously, the first aspect is to clearly define the needs and objectives. Together with this, consideration needs to be given to the minimum degree of accuracy in order to select the right equipment and measurement techniques. It is very important that the survey contractor is in direct contact with the person who has made the plan for the campaign as the position of survey sites is likely to be a compromise between where the data is required and where it can be effectively and safely collected. This is particularly important for temporal data surveys. In addition to ground survey technologies, the use of Remote Sensing (RS) technologies also plays an important role in the collection of data for management of urban water systems. Remote Sensing can be regarded as the process of acquiring, processing and interpreting data or information about objects without being in contact with them.

Since the data collection is a very costly and time consuming task, it is important to have a systematic and coordinated approach during the implementation of all related activities. Once collected, the data needs to be validated and processed before it is stored and used. This is usually done within a data management framework, which links various types of databases with Geographic Information System (GIS). There is a considerable body of literature which describes various aspects of data collection in urban water management (e.g., Fletcher and Deletic, 2007) and this chapter aims to provide an overview on some of the main principles and issues that should be considered when designing a data collection programme.

4.2 DEFINING OBJECTIVES OF DATA COLLECTION

Since the data collection can be very costly and demanding task, defining a set of clear objectives is the most important aspect. Collecting the data without clear purpose and justification can result in collecting potentially unnecessary data and/or missing essential data. Some of the questions that need to be asked while defining the set of objectives for data collection are:

- What is the main driver for data collection?
- Is such data already available?
- Who will benefit from this data?
- How often will this data be used?
- What is the minimum spatial and/or temporal resolution required?
- What is the minimum degree of accuracy?
- Do we have the necessary software and hardware to process and store the data?
- What will be the overall cost of data collection and storing?

Once the above questions are addressed the scale of a feasible and affordable data collection campaign could be determined.

4.3 PREPARING FOR A DATA COLLECTION CAMPAIGN

Once the scale of a data collection campaign is defined, details such as existing data gap analysis, asset identification, site and equipment selection, variables selection, budgeting, financial, social and institutional requirements should be considered.

The objective of gap analysis is to identify which of the data needs of the project can be met by the existing data and which data will have to be collected through new data collection activities. Aspects to consider are availability, quality and compatibility of existing data.

Having defined the needs for new data, the next step in designing a data collection programme is to specify a unique identification for each asset for which data is to be collected. The objective of asset identification is to develop a system of unique numbers for each asset (e.g., manholes, pipes, water meters, valves, hydrants, reservoirs, catchments, etc.) in order to minimise sources of error in data collection and data entry. In doing this, it is important to develop a system of numbers which is logical and has meaning on the ground. Furthermore, it has to allow for extension in order to accommodate newly created assets or the development of more sophisticated database.

For temporal data collection it is important to have an understanding of the water system infrastructure within the area under consideration and the characteristics of different site locations. Sometimes it is necessary to undertake a site investigation prior to the data collection programme in order to obtain necessary information and minimise the potential waste of resources and time. For example, a careful selection of a raingauge site is important when planning a flow survey. Data from an inappropriately selected raingauge site will affect the success of any model calibration. Furthermore, the geometry of assets must be measured accurately and measurements from design drawings should not be assumed. In the case of a buried infrastructure, a specific survey may be required.

4.4 SPATIAL DATA COLLECTION

An overview of some key spatial data that is required for management of different urban water systems is given in Table 4.1.

Table 4.1 Overview of spatial data for different urban water systems

Water supply, treatment plant and distribution networks	Storm water and wastewater networks and wastewater treatment plants	Receiving waters (streams, rivers, watercourses, lakes, wetlands, coastal waters)
Landuse data	Landuse data	Network layout of streams, rivers, watercourse, lakes, dams, groundwater aquifer
Network layout of streams, rivers, lakes, dams, groundwater aquifer	Catchment and sub-catchment characteristics boundaries and their aerial coverage	Contributing catchment areas, their characteristics and aerial coverage
Pressure zone characteristics, boundaries and their aerial coverage	Pipe/open channel network layout	Coastal zone layout
Pipe network layout	Manhole data (IDs, X and Y coordinates, elevation)	Coastal bathymetry
Junction data (IDs, X and Y coordinates, elevation)	Channel cross sections (IDs, X and Y coordinates, geometry, chainage)	Channel cross sections (IDs, X and Y coordinates, geometry, chainage)
Tanks and reservoirs (IDs, X and Y coordinates, bottom elevation, tank diameter)	Storages (IDs, X and Y coordinates, bottom elevation, storage area)	Location of flow, level, pressure and water quality measurement devices (ID, X and Y coordinates)
Pipes (IDs, start node, end node, length, diameter, elevations)	Gravity pipes, rising mains and inverted siphons (IDs, start node, end node, length, diameter, inlet and outlet elevations)	Stormwater outlets
Pumping stations (IDs, X and Y coordinates, elevation)	Pumping stations (IDs, X and Y coordinates, wet well geometry and elevation)	Connecting CSOs and SSOs
Valves (IDs, X and Y coordinates)	Catch-pits and gully pot data	
Water treatment plants (IDs, X and Y coordinates)	CSOs and SSOs (IDs, X and Y coordinates, geometry)	
Location of flow, level, pressure and water quality measurement devices (IDs, X and Y coordinates)	Weirs, Orifices (IDs, X and Y coordinates, orifice geometry)	

(Continued)

Table 4.1 Overview of spatial data for different urban water systems (*Continued*)

Water supply, treatment plant and distribution networks	Storm water and wastewater networks and wastewater treatment plants	Receiving waters (streams, rivers, watercourses, lakes, wetlands, coastal waters)
Cadastre and property data	Culverts/bridges (IDs, X and Y coordinates, geometry)	
Population data (per census area unit)	Outlets (IDs, start node, end node, length, diameter, inlet and outlet elevations)	
Groundwater abstraction wells (IDs, X and Y coordinates)	Wastewater treatment plants and its elements (IDs, X and Y coordinates)	
Groundwater recharging wells and ponds (IDs, X and Y coordinates, geometry)	Location of flow, level, pressure and water quality measurement devices (ID, X and Y coordinates)	
Asset condition data	Cadastre and property data	
	Population data (per census area unit)	
	Ponds, storages, tanks, soakaways, infiltration wells and basins, swales (IDs, X and Y coordinates, geometry)	
	Asset condition data	

For example, for drainage network systems, the following details are needed:

- Cover level (or ground level) of the manhole,
- Invert level of the manhole,
- Manhole geometry,
- Incoming pipe length,
- Outgoing pipe length,
- Incoming pipe invert level,
- Outgoing pipe invert level,
- Incoming pipe geometry,
- Outgoing pipe geometry.

A typical drainage manhole structure consisting of two incoming pipes and one outgoing pipe is presented in Figure 4.1.

For open channel networks (artificial or natural), the following details are needed:

- Cross section geometry (distances and elevations); see Figure 4.2,
- Chainages, and
- Long sections.

In addition to the pipe/channel network, details of the ancillary structure geometry are also necessary. Ancillary structures typically include constructed weirs/overflows, bifurcations, pumping stations, storage tanks flow control devices and inlets/outlets; see Figure 4.3.

In addition to the above mentioned spatial data, there are also other data which are needed irrespective to the particular system under consideration. Such data includes digital terrain model, land use characteristics, layout of property boundaries, road network layout, geology and soil distribution, administrative boundaries, layout and boundaries of census areas, etc.

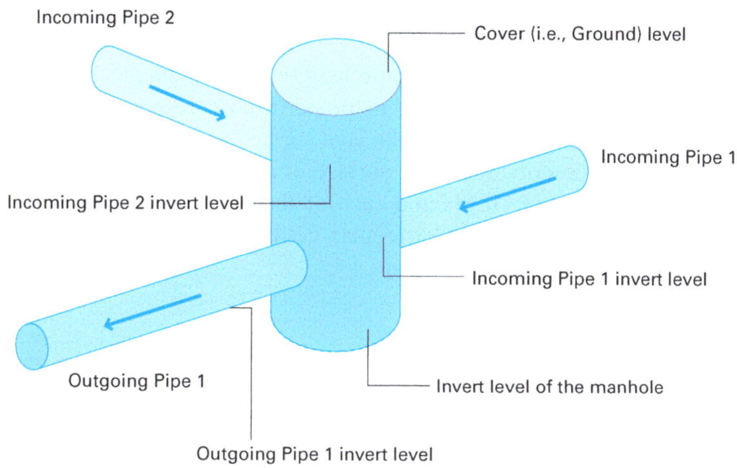

Figure 4.1 Manhole/Pipe details

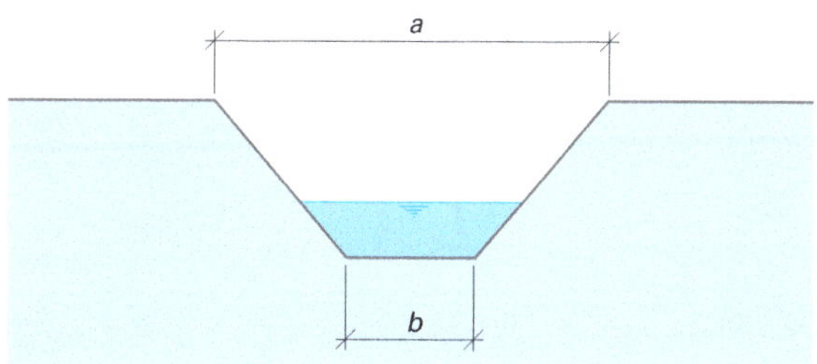

Figure 4.2 Channel cross section

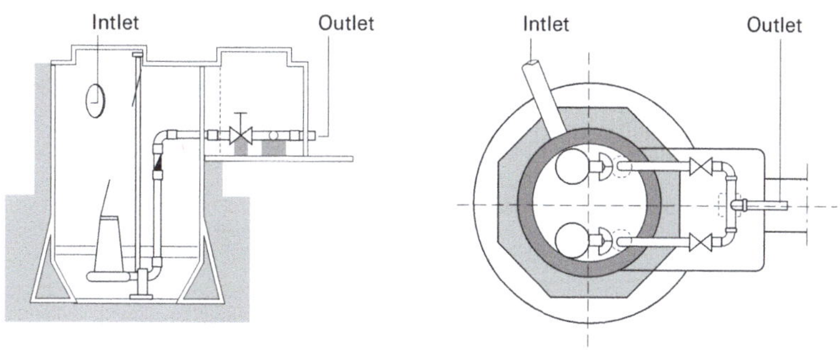

Figure 4.3 Pumping station geometry (cross-sectional view left and top view right)

4.5 DIGITAL TERRAIN DATA COLLECTION

A digital terrain model (DTM) refers to a topographic map which contains terrain elevations, and as such it is used to represent the terrain (or land) surface and its properties. Correspondingly, the term DEM (or digital elevation model), although usually associated with the land surface, refers to the elevations of any surface. For example, a DEM may be used to describe the water surface of a lake or the water elevations in a groundwater aquifer. There are various structures of DEMs and each of them have their own merits and short comings. Typical structures are: line model, triangulated irregular network (TIN) and grid network. The line model describes the terrain elevations by contours; see Figure 4.4.

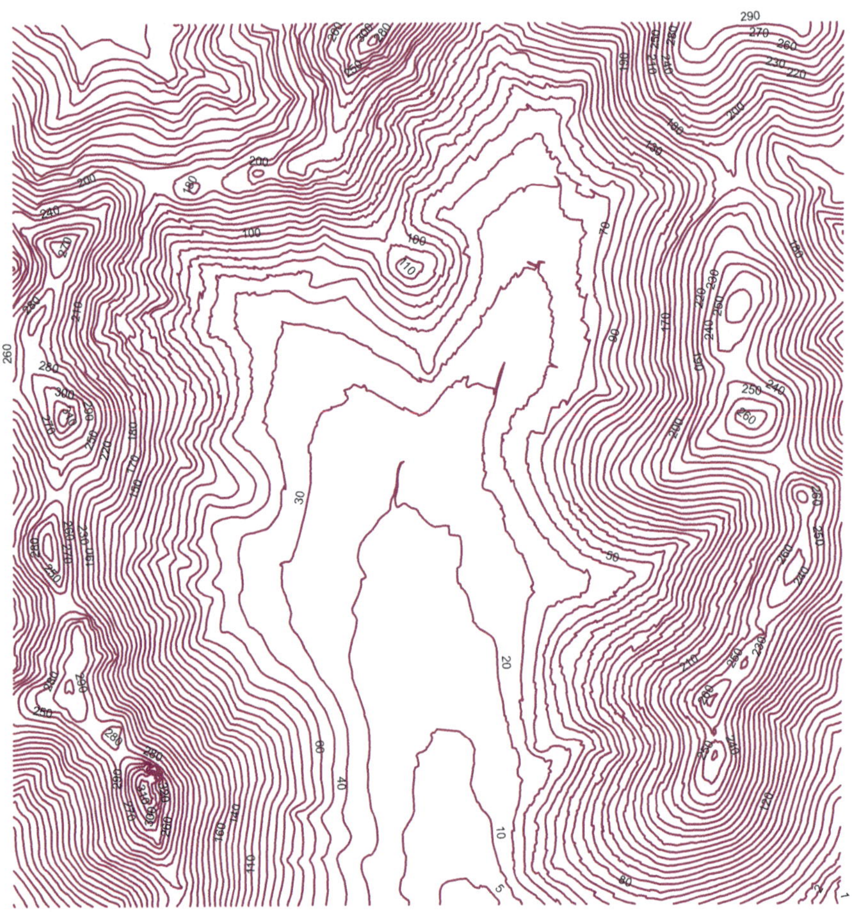

Figure 4.4 Example of a line model – contour map

The TIN model describes the surface as a triangular elemental plane; see Figure 4.5.

The grid structure usually contains regularly spaced square grids. Each grid (or cell) is bounded by three or four adjacent grids (or cells).

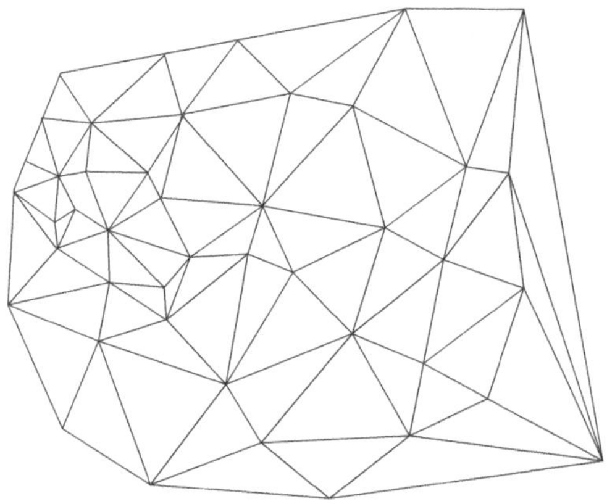

Figure 4.5 Example of a TIN

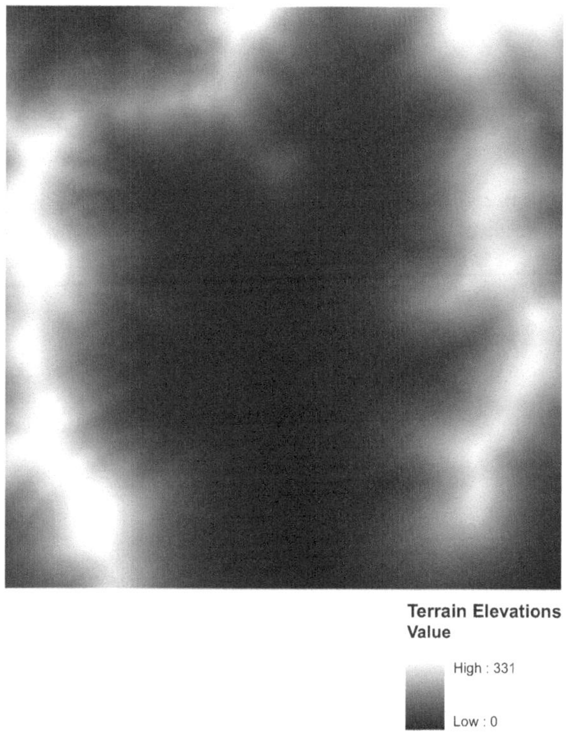

Figure 4.6 Example of a grid structure DTM processed in a raster-based GIS

Typically, DTM data can be obtained from ground surveys (e.g., total stations together with Global Positioning System – GPS; see Figure 4.7), aerial stereo photography, satellite stereo imagery, airborne laser scanning or by digitising points/contours from analogue format, such as a paper map to digital format, so that they can be stored and displayed within a GIS package and then interpolated.

Figure 4.7 Total station

Aerial Stereo Photography is one of the oldest and most versatile forms of remote sensing of the Earth's surface. It is taken from a flying object (airplane, kite, balloon etc.). Most professional Aerial Stereo Photographs are taken from special planes which are equipped with cameras at the bottom pointing straight down to the earth; see Figure 4.8. These (large format) photographs are then used to calculate the elevations of the terrain (and buildings) through photogrammetry.

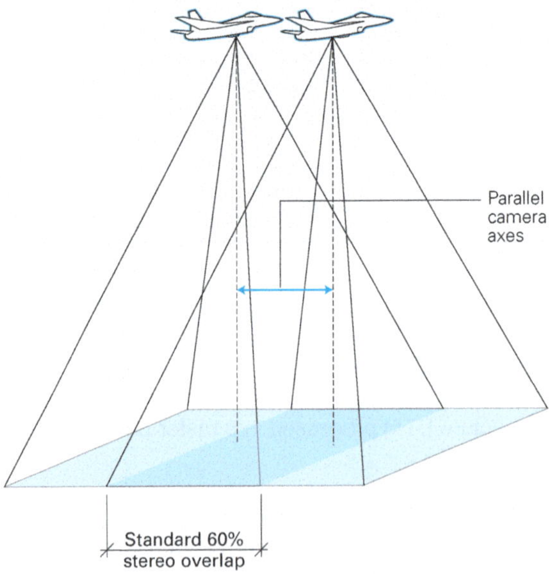

Figure 4.8 Aerial Stereo Photography

Satellite stereo imagery is another remote sensing technique which consists of two satellite images of the same location on Earth, taken from two different perspectives during one orbital pass; see Figure 4.9. The pair of images is collected in-track, or on the same ground path just moments apart, to maintain the tonal consistency between each image, enabling better interpretability. Typically, each stereo pair contains an image collected at a low elevation angle (above 60 degrees) as well as an image collected at a higher elevation angle (above 72 degrees) with 30º–45º convergence (0.54 to 0.83 base-to-height ratio).

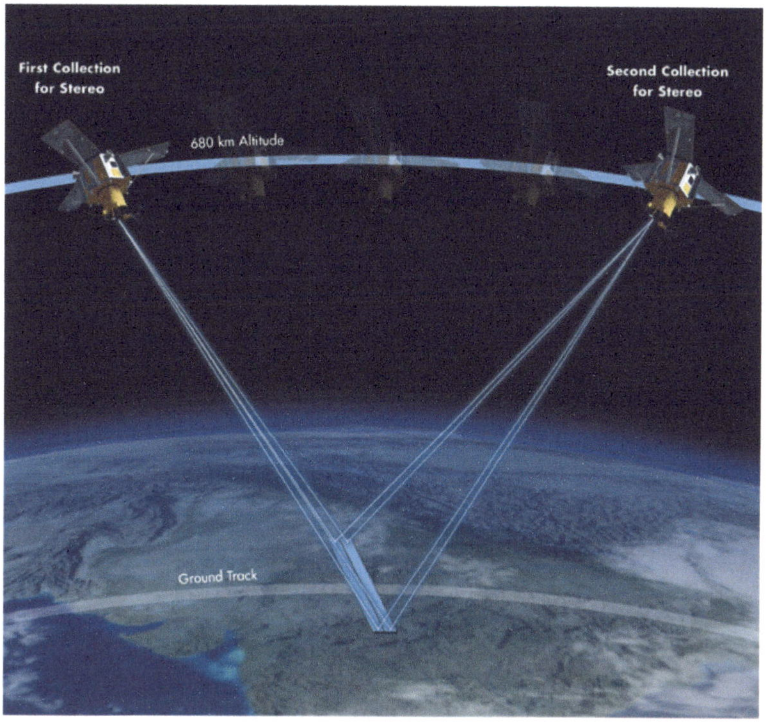

Figure 4.9 Satellite stereo imagery

Airborne Laser Scanning (ALS) or Light Detection and Ranging (LIDAR) is yet another remote sensing technique that is used to measure the elevation of an area in an accurate and economical manner. It can deliver information on terrain levels to a desired resolution. The end result from an ALS survey is a large number of spot elevations spread across the area. Typically, spot heights are provided at a spacing of 0.5 to 5 m with a horizontal accuracy of 0.3 m and vertical accuracy of 0.15 m.

A LIDAR scanner is mounted in an aircraft together with a GPS positioning device, and an inertial navigation system is used to capture the raw data; see Figure 4.10. The laser emits a pulse of light along a vector which is scanned from side to side in the direction of the flight path resulting in a set of zigzag scan lines; see Figure 4.11. Each pulse returns some reflected light to the device from soft objects such as vegetation and eventually a hard object such as the land or a building. The distance is calculated by measuring the time of travel of the pulse, and the spot point determined from the position of the aircraft (by an onboard GPS unit), the distance and the angle of the light vector. Pulse scanners have a range of up to a kilometre, and the rate of data capture which is limited by the physics of light travel and reflectance is between hundreds and thousands of points per second. Depending on flying height and speed the scan width can vary from 200 m to 1 or 2 km with 500 or more points per scan line and greater than 66 scan liners per second. The requirement to obtain ground level data at a higher resolution comes at the expense of large data files which are typically hundreds of thousands of megabytes, often extending to several gigabytes. The complexities associated with manipulating such large data files has created the need to undertake further processing of the raw data and to deliver a manageable set of points. Typically, thinning, filtering and interpolation are part of this process; see Figure 4.12, Figure 4.13 and Figure 4.14. The thinning (or reduction of data points) is usually attained by removing neighbouring points which are within a defined elevation tolerance. Filtering is a process of automatic detection and interpretation of bare earth and objects. To date, there are many filtering algorithms which have been developed. Some of them are: progressive morphological filter, Morph and Morph2D, elevation threshold with expand window (ETEW), maximum local slope, iterative polynomial fitting, polynomial 2-surface filter, adaptive TIN (ATIN), etc. For further elaboration of different filtering algorithms and their application to drainage system modelling see Abdullah *et al.*, 2009. Morphological filtering is done by classifying whether a point is on the ground, on a piece of vegetation (e.g., bush or a tree) or on a building. This ground spot height data is then saved for the creation of a digital elevation model (DEM).

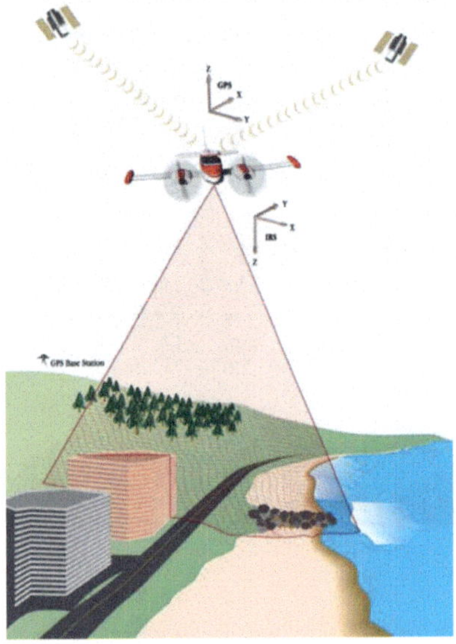

Figure 4.10 Use of LIDAR for collection of digital terrain data

Interpolation refers to the process of constructing new data points within the range of a discrete set of known data points. There are many interpolation methods available. The typical ones used in most GIS packages are: inverse spline, natural neighbour, kriging, distance weighted (IDW), etc.; see Figure 4.15.

In addition to LiDAR data collection technology, the airborne interferometric synthetic aperture radar (IFSAR) for digital mapping is another technology that can be used to obtain DTM data. IFSAR is an aircraft-mounted sensor designed to measure surface elevations. The data collected is used to produce topographic imagery. Radar pulses are aimed at targets on the earth's surface, and the return signals from the ground are received by two antennae that record the elevations at specific ground coordinates. Similar to the LiDAR technology, the ground coordinates are determined by the GPS and inertial measurement unit (IMU) technology. Post-processing of these data produces topographic information in the form of orthorectified radar imagery (ORRI).

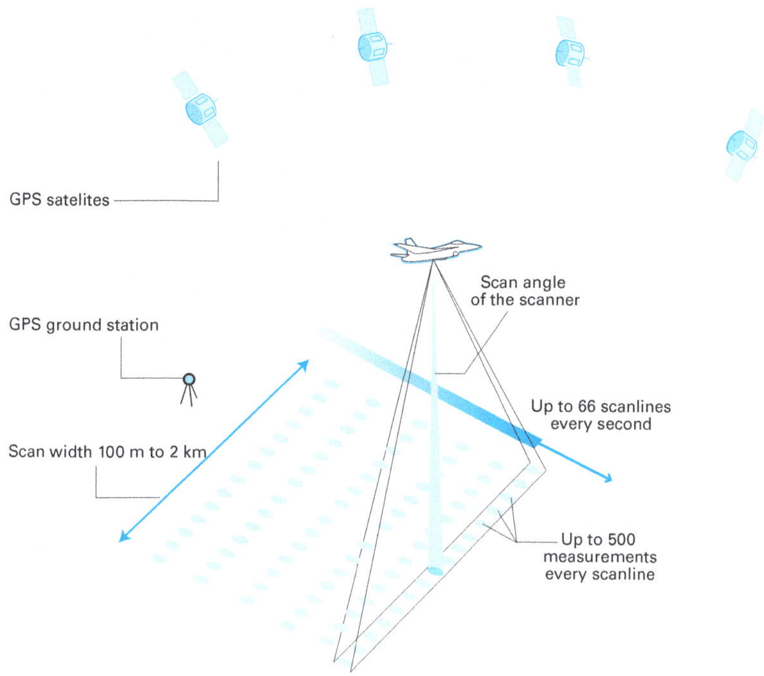

Figure 4.11 LIDAR scanning

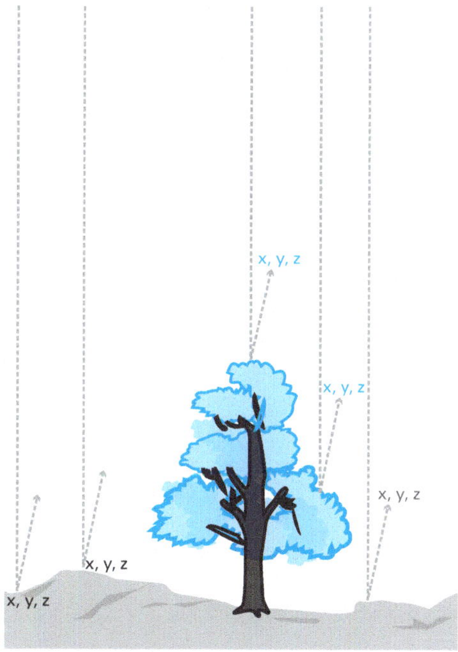

Figure 4.12 Obstruction by morphological objects

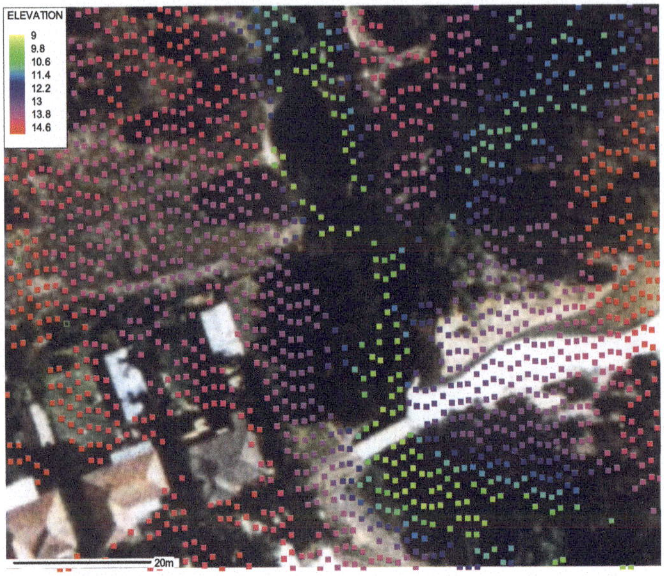

Figure 4.13 Filtering of ALS points

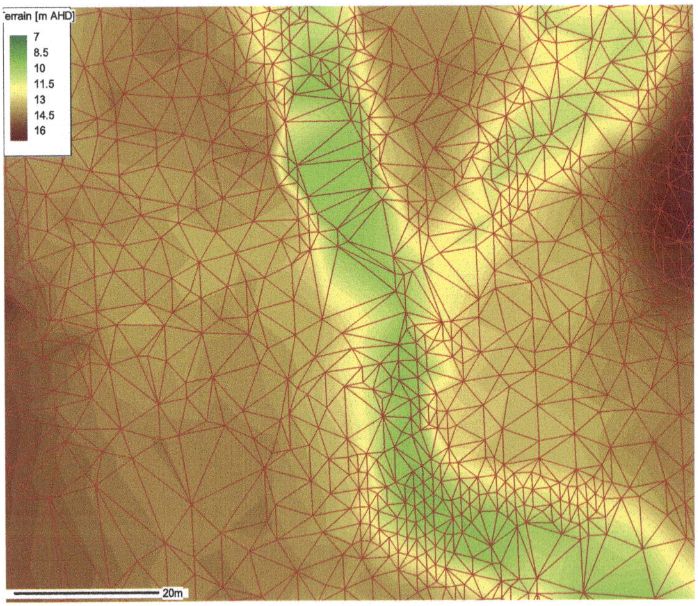

Figure 4.14 Interpolation of filtered data

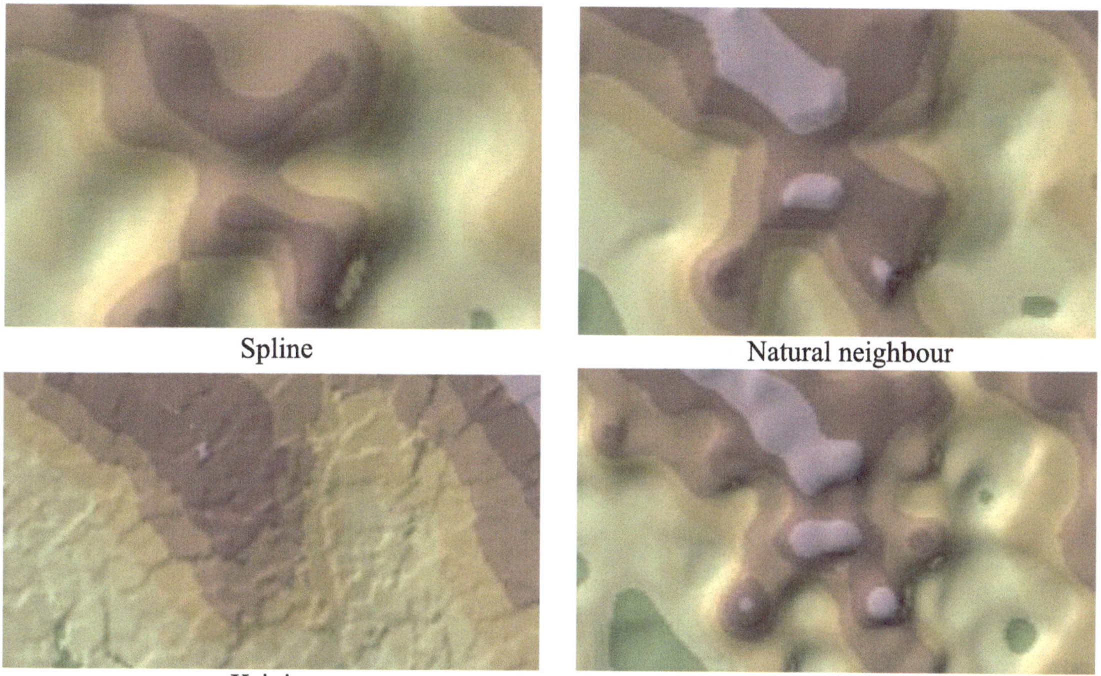

Figure 4.15 Common interpolation methods used in GIS packages to create DTMs

4.6 REMOTE SENSING

As stated in the opening section of this chapter, remote sensing is the science of acquiring, processing and interpreting data or information about objects without being in physical contact with them. The characteristics of a particular object can be identified by analyzing the interaction between the electromagnetic (EM) energy and the object. EM radiation which is reflected or emitted from an object is the usual source of remote sensing data. Each object has a unique characteristic of reflection or emission in different environmental conditions. Remote sensing attempts to identify and understand the object or the environmental condition through the uniqueness of its reflection or emission. Remote sensing involves an interaction between the incident radiation and the targets of interest. A typical remote sensing system (see Figure 4.16) involving imaging systems includes the following elements:

i) Energy Source or Illumination
ii) Radiation and the Atmosphere
iii) Interaction with the Target object
iv) Recording of Energy by the Sensor
v) Transmission, Reception, and Processing
vi) Interpretation and Analysis
vii) Application

A remote sensor is a device used to detect the electromagnetic (EM) radiation reflected or emitted from an object. Cameras or scanners are examples of remote sensors. A carrier of the sensor is called a platform (aircrafts, satellites, etc.). EM radiation is a carrier of electro-magnetic energy by transmitting the oscillation of the electro-magnetic field through space or matter. The range of all possible EM radiation is known as the electromagnetic spectrum. The EM spectrum is classified into different types according to the frequency of the electromagnetic wave. These types include (in order of increasing frequency): radio waves, microwaves (MW), infrared radiation (IR), visible (VIS) light, ultraviolet (UV) radiation, X-rays and gamma rays. This variability of the EM radiation is used in remote sensing to discern different objects or the same object in different contexts.

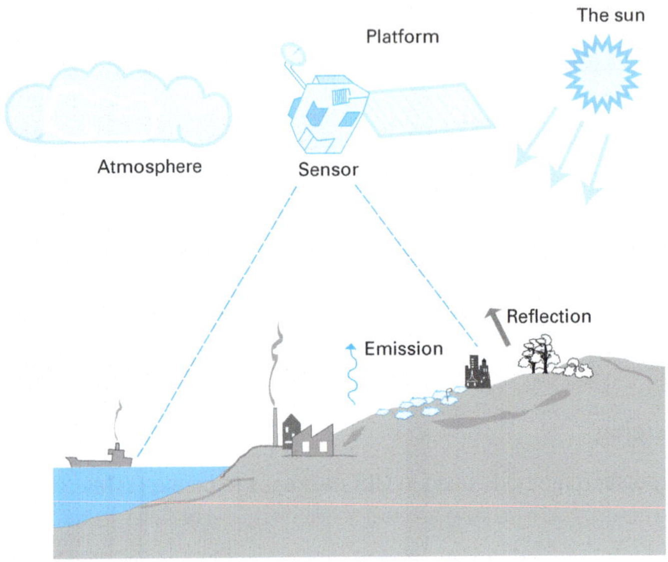

Figure 4.16 Data collection by remote sensing

The EM spectrum regions used in remote sensing are near UV (0.3–0.4 μm wave length), visible light (0.4–0.7 μm), near shortwave and thermal infrared (0.7–14 μm) and micro wave (1 mm–1 m). Visible light corresponds to the spectral colours. They are, in order from the longer wavelengths in the visible region, the so called rainbow colours; red, orange, yellow, green, blue, indigo and violet. Short wave infrared has been used for geological classification of rock types. Thermal infrared (TIR) is primarily used for temperature measurement, while micro wave is utilized for radar and micro wave radiometry.

The unique characteristics of matter are called spectral characteristics. Reflectance is defined as the ratio of the incident flux on a sample surface to the reflected flux from the surface. Reflectance ranges from 0 to 1. Reflectance with respect to wavelength is called the spectral reflectance. A basic assumption in remote sensing is that the spectral reflectance is unique and different from one object to another. Spectral reflectance is assumed to differ with respect to the type of land cover. This is the principle that in many cases allows the identification of land cover using remote sensing by observing the spectral reflectance or spectral radiance from a distance far removed from the surface.

Near infrared (NIR) is very useful for vegetation surveys and mapping because such a steep gradient at 0.7–0.9 mm is produced only by vegetation. Because of the water content in a leaf, there are two absorption bands at about 1.5 μm and 1.9 μm. Rock surveys are mostly done in the short wave infrared range (1.3–3.0 μm).

Remote sensing is classified with respect to the type of energy sources: passive and active remote sensing. Passive remote sensing makes use of sensors on satellites that detect the reflected or emitted EM radiation from natural sources. On the other hand, active remote sensing makes use of sensors such as radar that detect reflected responses from objects that are irradiated from artificially-generated energy sources.

A number of spaceborne satellites have been used to assess natural hazards since 2000. These satellites are able to quantify physical geography phenomena associated with the movements of the earth's surface (earthquakes, mass movements, water (floods, tsunamis, storms), and fire (wildfires). Most of these satellites contain active or passive sensors that can be utilized by the scientific community for the remote sensing of natural hazards over a number of spatial and temporal scales. The most useful satellite imagery for the assessment of earthquake damage comes from high resolution (0.6 m to 1 m pixel size) passive sensors and moderate resolution active sensors that can quantify the vertical and horizontal movement of the earth's surface. High-resolution passive sensors have been used successfully to assess flood damage while passive and active sensors have been used for prediction maps of flood vulnerability areas. Example of how delineation of flood hazard areas can be accomplished from satellite images is given in Chapter 13.

Recent moderate resolution sensors are able to provide near real-time data on fires and provide quantitative data for fire behaviour models. Limitations currently exist due to atmospheric interference, pixel resolution, and revisit times. However, a number of new micro-satellites and constellations of satellites will be launched in the next five years with increased resolution (0.5 m to 1 m pixel resolution for active sensors) and short revisit times (daily <2.3 m resolution images from passive sensors) that will significantly improve our ability to assess and predict natural hazards from space; see Gillespie Chu *et al.*, 2007.

There have been over 73 successful launches of earth observing satellites between 2000 and 2006 (ITC, 2007). Most of these satellites contain passive or active sensors that can be utilized by the scientific community for the remote sensing of natural hazards. Passive sensors record reflected (visible and infrared wavelengths) and emitted energy (thermal wavelengths). Since the year 2000, a number of new high-resolution passive sensors that provide panchromatic, visible, near infrared, and thermal imagery have become operational; see Table 4.1. Panchromatic imagery contains the entire wavelength of visible light and provides a black and white image similar to a black and white photograph. Currently, spaceborne satellites such as IKONOS 2, EROS, QuickBird 2, SPOT 5, and OrbView 3 can provide panchromatic imagery with 0.6 m to 2.5 m pixel size and have repeat times of 3 to 14 days. These satellites can also provide multi-spectral imagery using individual bands in the visible (blue, green, red) and near infrared wavelengths to provide true colour and infrared imagery. The resolution of this multi-spectral imagery ranges from 2.5 m to 10 m pixel size and repeat times are from 3 to 16 days. Shortwave infrared and thermal imagery provide information on temperature, and sensors such as ASTER and MODIS, have a number of bands that collect this information. Most shortwave infrared and thermal imagery have relatively large pixel sizes of 30 m to 1 km but a higher temporal resolution with repeat times of one day. Finally, the Defense Meteorological Satellite Program-Operational Linescan System (DMSP-OLS) and MODIS can provide daily panchromatic night-time light imagery (500 m) and thermal imagery (1000 m) at a global spatial scale.

Active sensors send and receive an energy pulse to create an image. Radar (RAdio Detection And Ranging) is the most commonly used active remote sensing technique and can provide moderate resolution imagery and digital elevation models of the earth's surface. Unlike passive sensors, radar can penetrate cloud cover, providing imagery over a target area both day and night regardless of weather conditions. There are a number of satellites (Table 4.2 and Table 4.3) equipped with radar sensors that have been used in the remote sensing of natural hazards. The pixel size from radar sensors (10 m to 30 m) is significantly larger than data provided by passive sensors.

Radar sensors measure microwave backscatter within discrete wavelengths or bands such as the X, C, and L bands. The shortest X (3 cm) and C (5.6 cm) bands reflect off the top of objects on the Earth's surface while longer wavelengths, such as the L band (24 cm), are able to penetrate vegetation and substrate. Recent satellite missions such as ERS-2, SRTM, and Envisat have provided high-resolution topography imagery using interferometric synthetic aperture radar (SAR) techniques. Interferometric synthetic aperture radar sends out a microwave pulse and records the reflected pulse at two points separated by a baseline distance. The resulting parallax can be used to measure elevation with a high degree of accuracy; see Metternicht *et al.*, 2005. These active systems are able to survey large areas due to their larger pixel and swath size but have long repeat times of between 24 to 46 days.

Table 4.2 An overview of some satellites with passive sensors that provide panchromatic (Pan), Visible and near infrared (VNIR), short wave infrared (SWIR), and thermal data used in remote sensing of natural hazards

Satellite (sensor)	Launch	Pan (m)	VNIR (m)	SWIR & thermal (m)	Spectral bands	Swath size (km)	Repeat time (days)
GeoEye-1	2008	0.5	15		8	15.2	3
GeoEye-2	2008	0.5	15		8	15.2	3
TerraSAR-X	2007	1	15	20	7	10–100	2.5
WorldView-1	2007	0.5	4 x 2		Panchromatic	17.6	1.7
WorldView-2	2007	0.5	8 x 2		Panchromatic	16.4	1.1
IKONOS 2	1999	1	4		4	11	3–5
Terra(ASTER)	1999		15	30.90	14	60	16
Terra(MODIS)	1999		250	500,1000	36	2300	1–2
EROSA	2000	1.9			0	13.5	3–7
Quick Bird 2	2001	0.6	2.5		4	16.5	3
Spot 5	2002	2.5	10	20	5	60	5–26
Aqua(MODIS)	2002		250	500,1000	36	2300	1–2
Orbview 3	2003	1	4		4	8	3
POMOSAT 2	2004	2	8		4	24	14
ALOS	2006	2.5	10		4	35	2–46
KOMPSAT 2	2006	1	4		4	15	28
DMSP(CLS)	2006	560		560	2	2,960	1
EROSB	2006	0.7	2.8		1	7	2

Table 4.3 Recent satellites with active sensors that can be used in remote sensing research of natural hazards

Satellite (sensor)	Launch (year)	Pixel (m)	Bands	Polarization	Swath size (km)	Repeat time (days)
ERS-2	1995	26	C	VV	100	3–35
Radarsat-1	1995	10	C	HH	45–500	24
SRTM	2000	30, 90	C, X	VV	225	0
Envisat(ASAR)	2002	30	C	All	500	35
ALOS(Palsar)	2006	7–100	L	All	40–350	2–46

4.7 TEMPORAL DATA COLLECTION

The measurement of a physical variable over a period of time can be referred to as the process by which a numerical value is assigned by the sensor at specific time instances (or time steps). A sensor is a device that can detect a signal or physical condition. Some of the reasons for placing measuring devices within a pipe (or open channel) network are:

- To gain understanding of the system's performance,
- To identify blockages, leakage (or exfiltration), infiltration and inflow,
- To obtain data for model calibration purposes,
- To operate system in real-time,
- To answer questions from the public.

Typically, different quantities such as temperature, pressure, velocity, level, pH, suspended solids, and total coliforms will require different kinds of sensors. It is important to note that there is a wide range of sensors that can be found on the market and their accuracy (the closeness of the agreement between the result of a measurement and the true value of the measurement) and time of response, range of measurement, repeatability (ability to give the same result when used to measure the same quantity several times in succession) and robustness vary according to the application and environment. There is always a trade off between the accuracy, repeatability and robustness that can be achieved for a given cost.

Selection of the appropriate measurement equipment for a site may not be simple or easy. Difficulties can arise due to presence of vibrations, direct sunlight, high humidity, aggressive environment, high or low temperature, difficult structural conditions, etc. Furthermore, special consideration must to be given to sites which have a high potential for vandalism, in which case appropriate housing arrangements may need to be constructed; see Figure 4.17.

Figure 4.17 Enclosure to prevent vandalism (Photograph: Zoran Vojinović)

The following temporal data are commonly collected from different locations within an urban area:

- Meteorological data (temperature, precipitation, evapotranspiration, humidity and wind speed),
- Flow rates,
- Velocity,
- Depth,
- Pressure,

- Pumping data,
- Contaminant concentrations,
- Temperature,
- Biological and chemical parameters.

In designing a temporal data collection campaign it is important to be aware of potential uncertainties associated with the measuring equipment as well as with the nature of the phenomena under consideration. In this respect, careful definition of the temporal and spatial scale is necessary. The temporal scale refers to two dimensions such as frequency and duration of measurements. The spatial scale also refers to two dimensions, nominally, the size of the area under consideration and the density of the gauges.

To minimise potential uncertainties associated with the measurement process it is necessary to undertake the calibration of any instruments before their installation and use. In this way, we can minimise the error of the measurements, remove the potential bias, identify dominant sources of error, quantify their effects and estimate uncertainties.

4.7.1 Meteorological data

Typical meteorological data that can be collected from urban environments involves temperature, precipitation, (rain, snow and hail), evapotranspiration, humidity and wind speed. Out of these five quantities, precipitation is the one most utilised by urban water engineers and it is often measured not only daily but also at much sorter time scales. Depending on the climate, topography and other local conditions it can be highly variable in both time and space. Typically, such data is collected by raingauges which in effect represent point measurements of rainfall, and in the case of multiple stations, the readings between different stations need to be interpolated in order to obtain some spatial representation of the precipitation (e.g., using Thiessen polygons, IDW, kriging, etc.); see Figure 4.18.

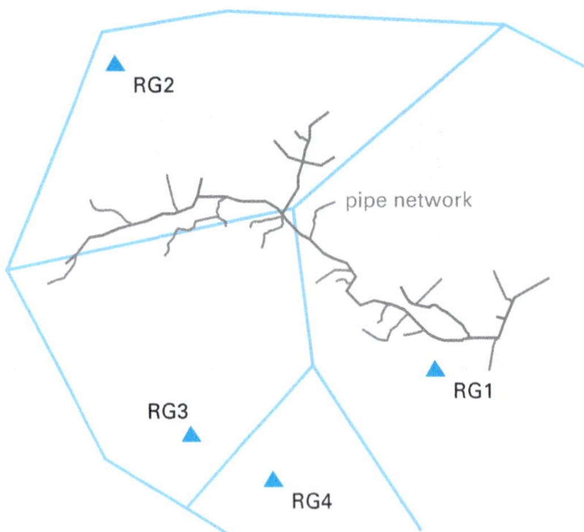

Figure 4.18 Thiessen polygons (RG – raingauge)

Types of rain gauges include: graduated cylinders, weighing gauges, tipping bucket gauges, and simple buried pit collectors. Each type has its advantages and disadvantages for collecting rain data. The most commonly used raingauge is the tipping bucket; see Figure 4.19.

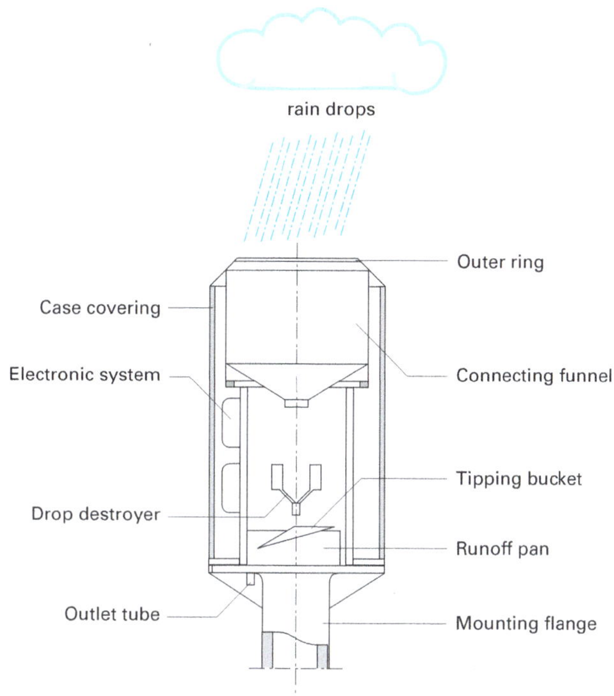

Figure 4.19 Tipping bucket rain gauge

The tipping bucket raingauge consists of a large copper cylinder which is typically set into the ground. At the top of the cylinder is a funnel that collects and channels the precipitation. The precipitation falls onto one of two small buckets or levers which are balanced in the same manner as a scale. After a certain amount of precipitation the lever tips and an electrical signal is sent to the recorder. The recorder consists of a pen mounted on an arm attached to a geared wheel that moves once with each signal sent from the collector.

If the rainfall data is to be used for model calibration purposes it is very important that the sites are selected in such a way that they will provide a good representation of the rainfall across the catchment.

In principle, the raingauges should be placed twice as far from an obstacle as the obstacle is high off the ground. For example, if an object is 3 meters high, the rain gauge should be at least 6 meters from that object. The top of the rain gauge should be at least 0.5 m off the ground in order to minimise the effects of the wind near the ground and to increase the capture of the rainfall.

Apart from precipitation measurements that can be undertaken using raingauges, recent developments in weather radar techniques have made radar a very valuable tool for collecting precipitation data in urban areas; see Figure 4.20. Such techniques have the advantage of measuring rainfall at all points over a catchment rather than at a few locations. This creates a special advantage over the use of raingauges, especially where spatial effects are important.

The method of measurement involves a radar beam being reflected from falling raindrops; see Figure 4.21. The intensity of the reflected beam is related to the intensity of the rainfall. Before use it is necessary to make several corrections to the data:

(1) Reflections from ground based objects such as hills and tall buildings (called ground clutter) must be removed.
(2) High intensity signals caused by ice crystals at the bottom of the cloud layers (bright band) must be corrected for.

(3) Other corrections are required for the size of the raindrops, which can be made using the Doppler frequency shift to measure their velocity.
(4) It is also usual to calibrate the signal against rainfall intensities measured at rain gauges within the catchment. This may involve constructing a rainfall intensity map from the rain gauge results and comparing this to the rainfall intensity map from the radar.

Figure 4.20 Weather radar in Bangkok (Photograph: Zoran Vojinović)

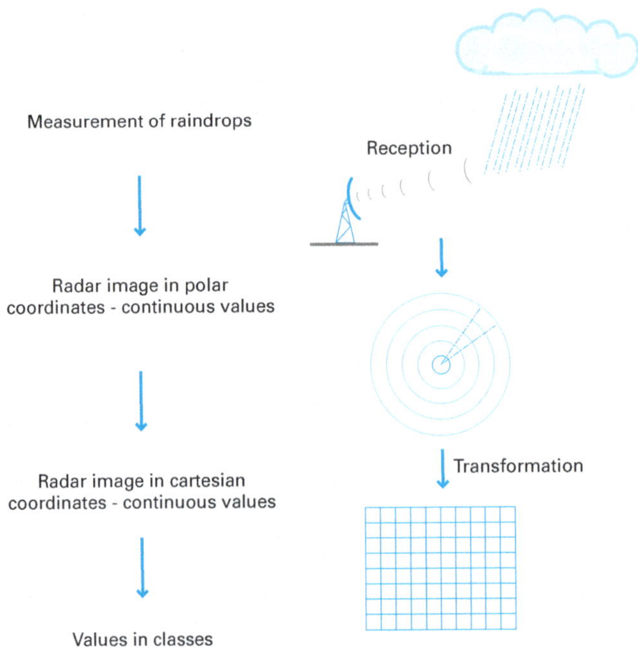

Figure 4.21 Radar measurement processing (after Einfalt *et al.*, 1990)

All of these effects vary during the passage of a storm, and continuous recalculation of the corrections is required.

For urban water management applications, the results are typically produced as a series of rainfall intensities on a rectangular grid with a mesh size of between 0.5 and 1 km; see Figure 4.22. The temporal frequency is usually provided at 15- or even 5-minute intervals. The appropriate mesh point is chosen for each contributing area in the catchment.

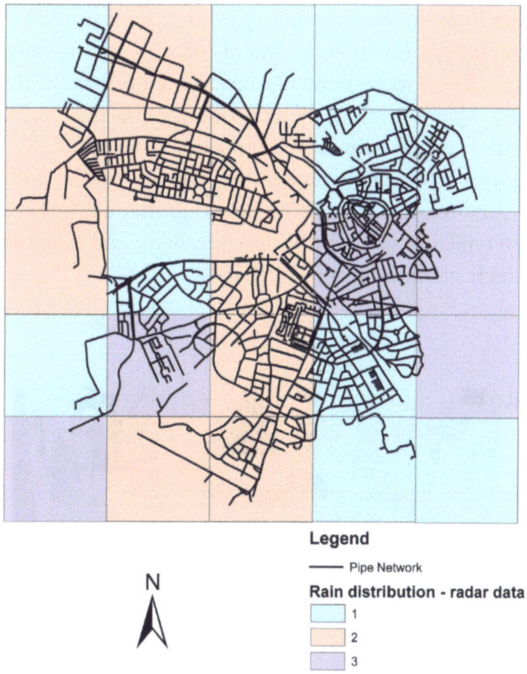

Figure 4.22 Example of a rain distribution over the city of Amersfoort, Netherlands (KNMI radar network with the spatial resolution of 1 km cells – cell areas are represented in different colours)

4.7.2 Water supply, treatment and distribution

Water supply and treatment systems consist of several components: intake points such as natural streams or groundwater, pipelines and pumping stations, which are used to transport untreated water, and treatment plants.

Typical measurements at the intake points are inflows and water levels. In the case of natural streams it is necessary to monitor upstream and downstream water levels and water quality parameters. In the case of lakes or dams, the variables of importance are inflow rates, water level and downstream flow rates. For groundwater extraction, it is important to monitor the groundwater table, and this is normally done using boreholes and piezometric wells. Typically, the groundwater monitoring programme would involve a number of observation wells installed at appropriate locations and depths. The location of such wells must be defined at a sufficient spatial distribution, which can enable reconstruction of the groundwater table or piezometric surface. Also, groundwater quality can be monitored by the use of individual sensors or multiple parameter probes and by taking samples from observation wells.

The water transport system consists of pipelines and pumping stations and it is used to convey the water from the intake point to a treatment plant and from the treatment plant to the distribution network. Particular quantities that need to be monitored within the water transport system are flow rates, pressure, pumping operation and water quality.

In terms of the monitoring within the treatment plant, the variables of interest are: inflow and outflow rates and various chemical parameters.

Water distribution systems typically consist of reservoirs, pipe networks, pumping and buster stations. Reservoirs are used for temporary storage of water and to balance temporal variations in supply and demand. Typical variables of interest for monitoring reservoirs are inflow, outflow, water level, leakage and chlorine residuals, and for monitoring water distribution networks, pressure, flow, turbidity, residual chlorine, valve status and consumer consumptions. Flow meters function on one of three main principles: pressure difference, rotation and magnetic or ultrasonic waves. Flow measurements based on pressure difference are usually done using Ventury meters and Orifice plates. Magnetic flow meters measure the flow velocity by generating a magnetic field and releasing a small electric current which is proportional to the flow velocity; see Figure 4.23 (a). Ultrasonic flow meters utilise ultrasonic waves to sample the velocity profile within the pipe. Inferential meters measure the flow of water by the rotational speed of a vertical or horizontal rotor or vane, which is transferred to the register. Volumetric meters are typically used for billing purposes, and they measure the flow of water by the rotation of the moving element which sits in a measuring chamber.

(a)

(b)

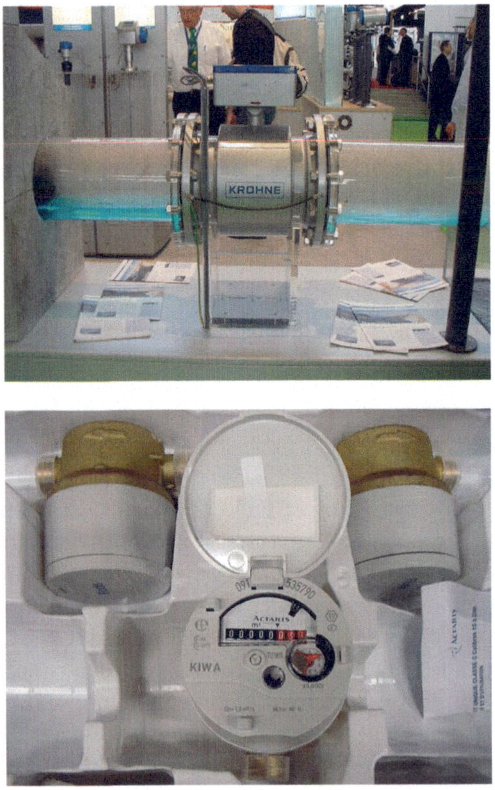

Figure 4.23 Water pressure gauge and water flow rate meters: a) electromagnetic and b) volumetric (Photograph: Nemanja Trifunovic)

Figure 4.24 Water depth meter at the pumping station in Bangkok (Photograph: Zoran Vojinović)

4.7.3 Wastewater and storm water systems and treatment plants

As discussed previously, wastewater and storm water systems typically include pipe and open channel networks, ancillary structures, above-ground or under-ground storages and wastewater treatment plants. The quantities of interest for measurement are flow rates, velocities, levels (or depths), overflow rates, inflow and exfiltration rates and various biological and chemical parameters. The flow monitoring units normally include the depth, pressure and velocity sensors, the data loggers, the mounting bands, cabling, data modems, power supply, meters (or rulers), structures in the pipes (weirs, flumes, etc.), portable devices and other ancillary equipment; see Figure 4.25 and Figure 4.26. Depth measurements can be obtained from meters (or rulers, 'depth of flow stick') readings (Figure 4.23), pressure sensors or ultrasonic sensors. Meters or rulers are effective if the flow is flat, slow, and since they require a person to read them they are only good for spot or single readings. They are not efficient for closed conduits because it is difficult to find a suitable location (without debris) and because by standing in the flow and placing the ruler in the flow an effect of a dam will be created and as such it may misrepresent the actual flow depth.

Another way of measuring flow depths is using a pressure sensor which measures the pressure of the flow on the top of the sensor; see Figure 4.25. As the pressure changes, it records a change in the depth of the flow. The device works with barometric or atmospheric pressure. The sensor must be fully immersed in the flow on order to give accurate readings. Barometric sensors may have a vent tube that has to be kept out of the flow at all times in order for them to function properly. Some of the problems that can be encountered with pressure sensors are: too low a flow for measurement, a blocked vent tube or a loss of power to sensor (low battery of electrical failure). The depth sensor is normally capable of measuring the depth of the flow over a range from 30 mm up to and including 6 m above the invert of the conduit and with a minimum instrument reading resolution of a 2 mm increment. Typical accuracy is in the order of +/−1 – 10% and the accuracy tends to be worst at low flow conditions. To prevent loss of data from damage or fouling of the instrument, multiple ultrasonic level sensors are usually installed to provide redundancy.

Figure 4.25 Pressure sensors (Photograph: Zoran Vojinović; AIT Hydraulic Laboratory, Bangkok)

(a) (b)

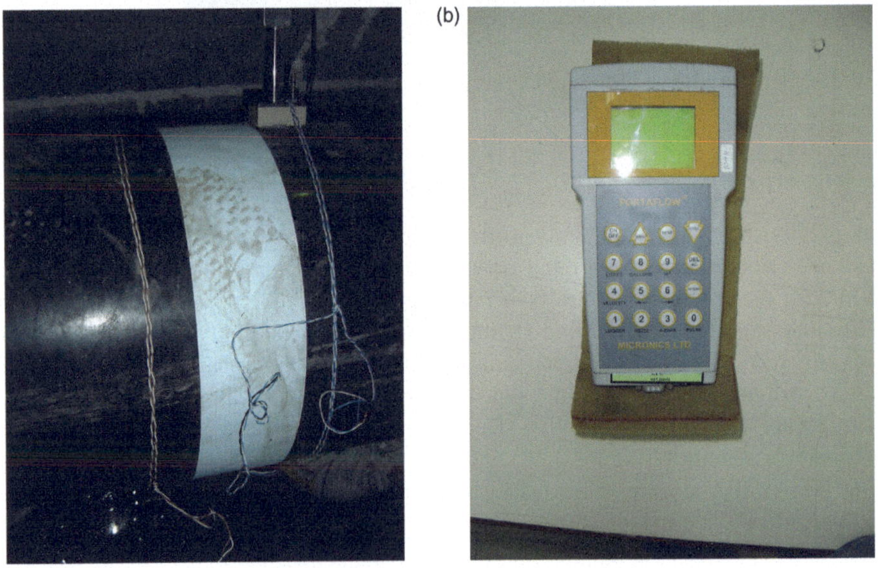

Figure 4.26 Ultrasonic sensor: a) diagonal mode operation and b) handheld logger (Photograph: Zoran Vojinović; AIT Hydraulic Laboratory, Bangkok)

Ultrasonic sensors are also used to measure the depth of flow; see Figure 4.26. They measure sound reflections through the air space above the flow. The sensor must both transmit and receive an ultrasonic pulse. The ultrasonic signal is usually 40 KHz which does not bring any harm to people. Some of the typical problems associated with ultrasonic sensors are: low charge in the batteries that result in weak signals and therefore shallow depth readings, installation difficulties (due to the fact that the sendsors need to be installed level to the surface of the flow), a dead band (which is the minimum distance from the face of the transducer at which the sensor can measure), loss of power, etc.

The velocity sensor is normally capable of measuring the velocity of the flow in a drainage system over a range from −1.0 m/s to +4.5 m/s inclusive, for depths of flow of 30 mm or greater above the invert of the drainage pipe. The typical installation of such sensor is in the upstream pipe of a manhole such that it faces into the flow; see Figure 4.27. It operates using doppler sound waves; see Figure 4.27. The sensor is mounted on a metal ring which is inserted at the downstream end of the pipe; see Figure 4.28 and Figure 4.29.

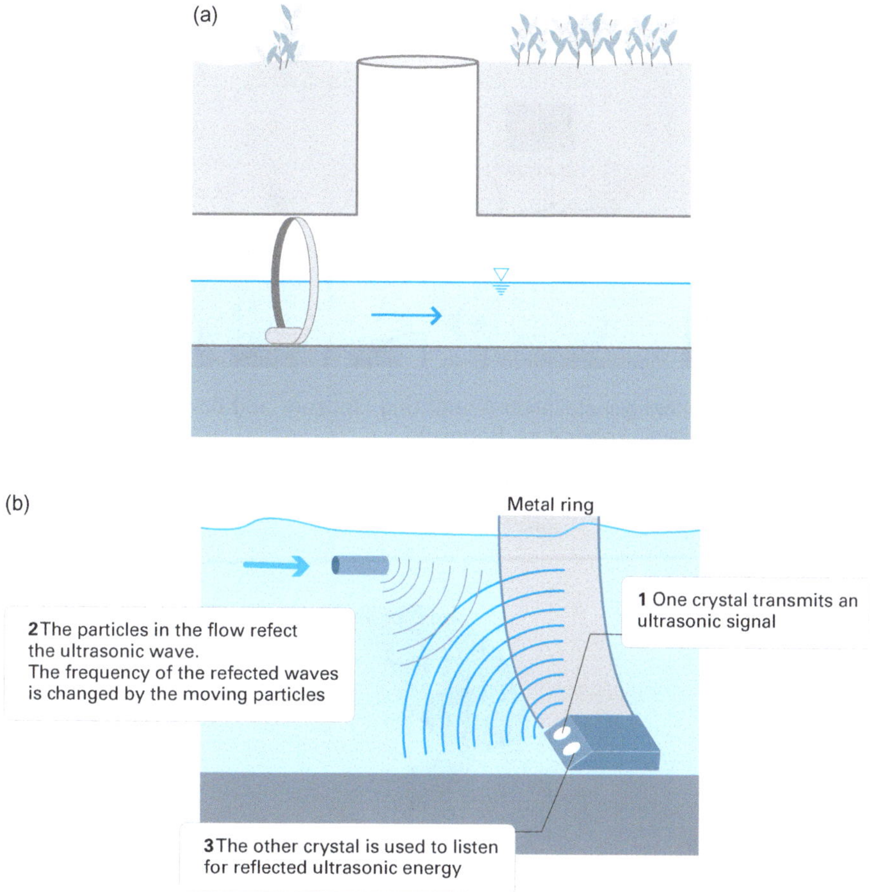

Figure 4.27 Operation of the velocity sensor

The sensor devices incorporate a means of measuring, recording and storing the velocity and depth of flow at the required location and at a time interval specified in increments from 1 minute and larger. They are also capable of and suitable for discrete measurements at regular sampling intervals: 5 minutes for temporary monitoring and 15 minutes for strategic gauging under both open channel and surcharge conditions.

For discrete velocity profiles in sewers a suitable hand held velocity measurement instrument (i.e., pygmy or propeller meter) is normally used for in-situ calibration. The depth of flow measured in the sewer at the deepest point is needed to determine what velocity profile to use.

Field calibration of the measuring instruments is an integral part of the measurement process and confirms the accuracy of the field measurements. An initial calibration is required at the setup of the instrument, and thereafter it is important to ensure a programme of field calibrations which validate sensor readings across a representative range of flows, that is, the minimum, average and peak day flows as well as elevated flow due to wet weather, where practicable.

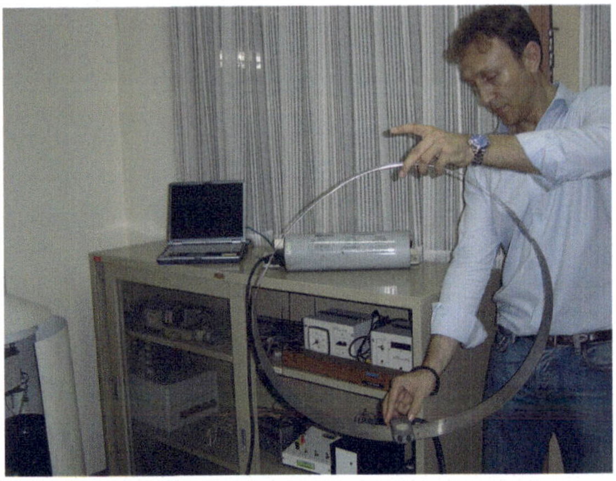

Figure 4.28 A typical flow monitoring installation (metal ring, sensors and flow transmitter connection to the computer – Courtesy of AIT Hydraulic Laboratory, Bangkok)

Figure 4.29 Installation of a velocity sensor in a pipe (Photograph: Sutat Weesakul; AIT Hydraulic Laboratory, Bangkok)

The dimensions of the sewer in which the flow monitor sensors are installed are always measured and recorded at the time the flow monitor is installed and again upon removal or relocation of the sensors.

The major vertical and horizontal dimensions of regular conduits are measured and recorded. Where conduits are of an irregular shape, sufficient measurements are taken and recorded to allow the cross sectional area to be adequately calculated by accepted mathematical means. The industry practice is that the accuracy of measured sewer dimensions is ± 5 mm up to a dimension of 1 m and ± 10 mm for greater dimensions.

In the case of low water depths at a gauging site, a weir structure is usually installed. This structure is normally located perpendicular to the direction of the flow in plan view, with a vertical wall in section view, such that the top of the V-notch weir plate is horizontal. In plan view, the weir is placed in the manhole such that the flow through the calibrated weir plate falls freely onto the manhole invert. The weir pool should incorporate at least two thirds of the manhole area. The notch of the 'V' is positioned such that it is vertically above the centre line of the pipe, and so that it is at least 50 mm higher in level than the soffit of the inlet pipe(s) entering the manhole. The weir structure is normally fixed to the manhole such that the manhole's structural integrity is not affected. The weir is installed and maintained such that it remains in position during dry and wet weather flow conditions without affecting its structural integrity as a measuring device.

The level or pressure sensor is placed in or above the pool created by the weir such that it measures the level of the pool unaffected by draw-down near the V-notch. The level sensing equipment is used to measure accurate levels in the range from the notch of the V-notch weir up to the overtopping level of the weir structure.

Independent level measurement calibrations are usually performed on at least three occasions during the gauging period. Additional calibrations may be required to confirm the accuracy of the recorded data.

In principle, regular audits are usually required, and they include the analysis of the following:

- Scattergraphs of Depth vs. Velocity,
- Verification of the flow using a dual equation calculation of the flow at each site,
- Flow balancing,
- Statistical analysis of the field verification results.

The following deliverables are normally obtained as the final output from the flow gauging process:

- Site geometry and schematics of the incoming and outgoing lines;
- Edited depth, velocity and flow data recorded at a specified time interval in an electronic form,
- Graphical plots of edited depth, flow and velocity data for each monitor to a scale such that individual storms can be easily examined,
- A quality flag of any missing data and an assessment of the accuracy of the data presented, so that basic quality coding of the data may be completed,
- Details of any equipment failure or location problems (e.g. silting, excessive ragging, high, low or layered/variable velocity), and an indication of the 'drift' in the depth measurement for each monitor,
- A summary of the month's results including comments on the performance of the monitor sites,
- Copies of the check sheets of velocity and depth,
- An installation sheet and monitoring summary for each site,
- Project progress reports delivered at the agreed frequency,
- Processed data, quality coded, in ASCII, Access or any other database format,
- Inflow/infiltration related monitoring (if required) would include:
    - Average dry weather hydrograph for each site,
    - Comparison of the theoretical average dry weather flow and the actual dry weather flow,
    - Typical wet weather hydrographs for the rain events,
    - Quantification of the four components of flow – wastewater; high groundwater; rain-related inflow and rain-induced infiltration,
    - Ranking of the subcatchments in terms of severity for each of the components of inflow and infiltration (I/I),
    - Recommended source detection programme.
- A summary of quality control measures including flow balancing; review of scattergraphs and a statistical assessment of the set of calibrations for each site,
- Commentary of any hydraulic or system features apparent from the flow data,
- Final report discussing all the foregoing points.

Overflow weirs are monitored with simple rectangular or V-notch weirs or pressure/ultrasonic level sensors. If the intention is to use the data for model calibration purposes, the placement of flow gauges should:

- avoid downstream backwater effects (e.g., be located at the furthest point upstream of the major confluence),
- choose straight sections, both in plan and grade,
- avoid drop connections;
- choose leaf catchments where possible in order to avoid subtraction errors,
- avoid sections with low velocity as the velocity sensors are less accurate below 0.6 m/s and such sites are prone to siltation.
- avoid steeper section and situations where the likely flow depth is below 40–50 mm.

The monitoring of pumping stations should be accurate in circumstances where a particular station has sufficient wet weather capacity, i.e., peak wet weather inflows are lower than the duty flow rate of the pump(s). If a pumping station floods regularly during wet weather (in which case overflow is suspected) and if there are only one or two major incoming sewers, it may be better to gauge these sewers, being aware of ay drawdown effects from the wet well level fluctuations. An overflow may be deduced from the level data, but it may require a gauge or level sensor at the inlet to the overflow. It should be possible to estimate the overflow potential from the pump start/stop records during selected rainfall events prior to the gauge selection.

In terms of the water quality parameters (such as BOD, COD, DO, TSS, TN, TP, heavy metals, etc.) within the pipe/open channel network, these are measured by sampling, which can be either grab, composite or continuous. Such samples are typically taken at any instance in time (in a dry weather or wet weather) and at any point within the system depending on the objectives of the monitoring programme. Typically, grab samples are used for dry weather flow sampling, whereas, continuous auto-samplers are used for wet weather flow sampling.

The sampling interval can be different for different parameters. Recent developments in gauging equipment have enabled higher sampling frequencies using sensors even though most of the parameters require laboratory analyses. Measurements with higher sampling frequencies are used in situations where the purpose is to construct pollutographs. However, if the purpose is to obtain event mean concentrations (EMC), the same samples can be combined to produce the mean values. Whatever sampling equipment is chosen (e.g., auto-samplers, gravity-fed samplers, probes, etc.) careful consideration needs to be given to the site selection and time interval to ensure that representative samples are collected.

Sediment transport, deposition and re-entrainment through pipe and open channel networks can be characterised over time by regular sampling a few times a year. Such samples are then taken to laboratory for further analyses (e.g., particle size, density, etc.).

In terms of the wastewater treatment plant monitoring, apart from the regular monitoring and sampling activities throughout different treatment processes, monitoring at the inlet (i.e., influent) and the outlet (i.e., effluent) to the treatment plant are essential to assess the removal performance.

Since the flow gauging and water quality sampling is a very expensive and time consuming task, before embarking on such programme it is important to consider whether particular gauge/sampling information provides the necessary information and therefore worth installation. Furthermore, the proposed site should be analysed whether it is appropriate in terms of the slope, drop junctions, depth and potential major access problems.

4.7.4 Receiving waters

Receiving waters in urban areas typically refer to various types of water bodies which often provide a range of goods and services that are highly valued by the community (e.g., rivers, lakes, coastal waters, wetlands, etc.). With an ongoing process of urbanisation these water bodies tend to receive greater impacts from storm water runoff, overflow discharges and effluent. Typically, legislative and regulatory requirements are used as standards for protection, enhancement and restoration of receiving waters. For example, in Europe it is the European Water Framework Directive (CEC, 2000) which prescribes such requirements and in the United States there are different ecosystem monitoring and protection programmes managed by state and regional EPA offices.

Tracer studies are often carried out in order to detect contaminants within or coming from wastewater systems. Tracers are compounds, usually dyes or salts used for measuring, mapping, and monitoring wastewater systems. Fluorescent tracers are often chosen because they are cost-effective and easily and accurately measured on-site with a portable, field ready fluorometer. The ideal tracer is nontoxic, usable in small quantities, cost effective, easy to measure at very low concentrations, and stable during the course of the study. In such studies the travel time and distribution of the tracer is of interest; see Figure 4.30.

Figure 4.30 Tracer equipment (Source: Ole Mark)

To determine the status of receiving water bodies at different spatial and temporal scales, various quantities need to be monitored. For such water bodies, indicators such as temperature, pH, turbidity, carbon and oxygen levels are of particular importance. These are usually measured using different sampling programmes and collected samples are processed in laboratories. Furthermore, bacterial concentrations, unpleasant odour and visual disturbances are also monitored. Sampling from receiving waters can be grab, composite or continuous as for pipe and open channel networks. Typical environmental monitoring systems are multi-parameter water quality measurement systems. They contain various sondes, which are placed in the water to gather water quality data. Sondes may have multiple probes, with one or more sensors that read water quality data. Typical parameters that can be measured are: dissolved oxygen, conductivity, specific conductance, salinity, total dissolved solids, resistivity, temperature, pH, depth, level, flow, turbidity, nitrate-N, ammonia-N, ammonium-N, chloride, chlorophyll, etc. An example of a sonde is shown in Figure 4.31.

4.8 OTHER DATA

In addition to the above data, we may also need to collect other kinds of data such as:

- infrastructure condition data (for further discussion on condition surveys refer Chapter 8),
- geological and soil data,
- customer-related data,
- flooding and surcharge incidents,
- system design and environmental standards,
- economic indicators,
- construction and maintenance cost data,
- etc.

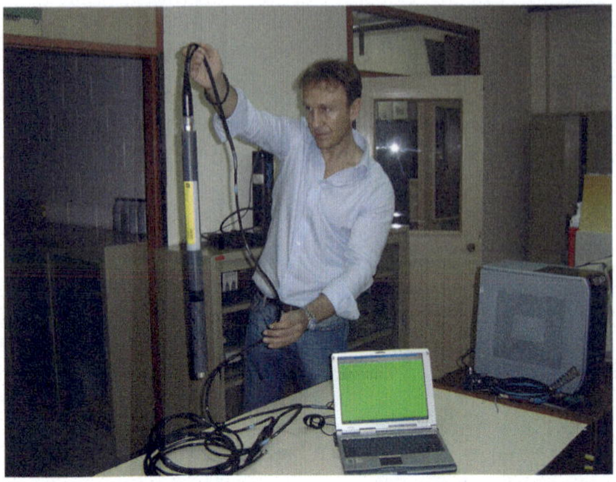

Figure 4.31 Example of an environmental sonde connected to the computer (AIT Hydraulic Laboratory, Bangkok)

4.9 MEASUREMENT UNCERTAINTY

Measurement uncertainty arises due to the stochastic nature of the phenomena that has been measured as well as errors due to the measuring equipment and practice. These errors provide a fundamental limit to the accuracy with which subsequent models can be developed. Even though the science and technology concerning data measurement activities has advanced considerably over the past two decades the associated uncertainty has not been fully eliminated. Therefore, understanding the sources of uncertainty which are inherent within the data measurement process is an important aspect of urban water management. The process of quantifying errors and identifying their dominant sources is often referred to as *uncertainty analysis*. Measurement uncertainty can be defined as the sum of the following uncertainties:

- sampling errors: for example, measurements taken at the point in the sewer pipe which does not give the maximum flow;
- instrument errors: for example, sensor readings are biased consistently below the true value;
- practice error: for example, the rainfall not being measured accurately due to poor placement of a rain gauge.

To date, certain engineering fields have made attempts to categorise different error types based on their effects on measured values (e.g., AIAA, 1995; ANSI/ASME, 1985 and ASME, 1998). They divide the total measurement error into two components: bias (fixed) component and precision (random) component. Some other standards (ISO, 1993) have grouped uncertainties into two categories according to the method used to estimate their numerical values: Type A (which refer to the use of statistical methods and which are founded on frequency distributions) and Type B (which refer to other means and which are founded on a priori distributions). Most of the international standards which are currently available are based on the following assumptions:

- uncertainty sources are normally distributed;
- elemental uncertainty estimates for individual variables are combined using the root sum of squares (RSS) method;
- propagation of the elemental errors to the final measurement results are made using first order Taylor series approximation;

- systematic and random uncertainties are RSS combined and multiplied by a coverage factor to obtain confidence of results;

Since there will be always a degree of uncertainty associated with the collection of data from urban water systems it is necessary to develop a set of QA procedures (concerning choice of the measuring equipment, site selection, instrument calibration, data processing, etc.) which will ensure that the errors arising from the measurement process are minimal.

4.10 DATA VALIDATION, PROCESSING, HANDLING AND STORAGE

Upon the completion of data collection activities all collected data needs to be validated and processed before it is used and stored; otherwise, it should be treated as raw and inaccurate. This will ensure that all outliers in the data which may occur due to a poor sensor reading or disturbances at the gauging site are detected and removed. This also refers to the removal of biased values and other data anomalies. Various statistical methods such as variance and principal component analysis can be applied in this context. Short term missing or invalid data can be replaced by carefully calculated estimates. Certainly, every detail of such validation and processing work must be documented and discussed.

Once the data is collected and processed it is then necessary to be stored adequately in a database so that it can be easily shared among different users. Building up appropriate databases using the latest relational or object-oriented database technology can produce considerable cost savings in terms of planning, design, maintenance and operations. Having the data 'to hand' is also extremely important for job management, customer service, and satisfying regulatory requirements. Furthermore, an appropriate database can facilitate the coordination of all activities required for providing the desired level of service in a cost-effective manner.

A comprehensive urban water system database uses data from a variety of sources which can be spatial and non-spatial. The classes of data needed include the following:

- asset description
- customer related data
- ageing and condition data
- performance data (CSOs, SSOs, flood incidence, blockages, collapses, inflow/infiltration, exfiltration, leakage, intermittent water supply, etc.)
- domestic and industrial discharges,
- condition related data
- historic data
- policy related data
- environment related data
- financial and economic data
- measurements (flow, level, pressure, rainfall, climate, water consumption, effluent, etc.)
- land use
- population
- policy changes on levels of service
- changes in design policy: source control, CSO performance standards
- Etc.

Once the data is validated, processed and stored in appropriate databases, its manipulation and maintenance is usually undertaken within a system that can handle spatial and temporal data.

4.11 GEOGRAPHIC INFORMATION SYSTEMS

An important aspect of urban water systems management is to determine the process of collecting and inserting the data into the system that ensures cost-effective data storage and retrieval. This is usually done within a data management framework which links various types of databases with a Geographic Information System (GIS). Within such a framework, users can better understand the way in which the asset condition and performance change over time and what is their relation with the provision of services to customers.

Geographic Information Systems (GIS) are a versatile and intuitive technology which has a broad application in urban water management in general. The use of a GIS system within the water/wastewater utility can provide benefits at different organisational levels and different users (e.g., managers, planners, scientists, engineers, technicians, operators, etc.). Overall, some of the benefits are:

1. More efficient information sharing between different stakeholders;
2. Ability to offer an improved map service;
3. Improved integration between different departments/organisations;
4. Opportunity to reduce the sets of manual maps held and the associated storage costs;
5. Develop and implement practices for managing land development, including land value assessment, and seek to ensure that such information is readily available;
6. Greater efficiency resulting in increased staff capacity and savings;
7. Faster and more extensive access to geographic information throughout the department or authority;
8. The ability to track and monitor growth and development over time;
9. Improved analysis e.g., the extent of flood prone areas, emergency response planning, underground and above-ground infrastructure information, housing issues, use of public facilities, etc;
10. Improved productivity in providing public information;
11. Better communication of information between the public and decision-makers;
12. Better decision-making;
13. Improved ability to aggregate data for specific sub-areas;
14. Improved quality of services by each stakeholder;
15. Better targeting and coordination of services;
16. Ability to perform and display different types of professional analysis; and
17. Improved policy formulation.

Geographic stands for spatial data (i.e., the data that can be referenced in space, or georeferenced), and displayed in the form of a map. *Information* may be viewed as a communication of value-added information. *System* may refer to the integration between hardware, software and users. Particularly, such spatial technology is invaluable in analysing spatial data, in producing thematic maps as well as in presenting numerical model results.

GIS tools have been developed to geo-reference various data sets into a spatial framework (or to a grid which includes a model of the terrain). With the use of GIS technology, asset data can be linked to the cadastre and property databases and as such used very efficiently in the overall urban planning process.

Geographic data can be characterised by three major aspects: *position*, *attributes* and *time*. Position usually refers to the location within the coordinate system, attributes refer to the characteristic of an entity contained in the database (pipe diameter, condition, material, etc.) and time refers to the point (or period) of time when the geographic data were collected. There are four geographic data types: *points*, *lines*, *polygons* and *continuous surfaces*. Their spatial distribution is represented in two basic types of spatial model: *vector* and *raster* (or *tessellation*) models. The approach used in the vector model is to specify precisely the position of points, lines and polygons with either one or a series of X-Y location pointers. For raster models, the approach used is based on a grid (or cellular) organisation of the spatial data.

Table 4.4 Different users of GIS

Kind of user	Kind of data	Kind of analysis/tools
Data providers	Raw data, qualitative, quantitative	Flexible, large range of tools
Researchers	Raw data, pre-processed, quantitative	Exploratory, explanatory models
Policy preparers	Pre-processed, quantitative, qualitative	Decision support
Policy makers	Strategic information, quantitative, qualitative	Decision support, communication
Public	Information, quantitative, qualitative	Communication

Most of the water/wastewater assets are geographically referenced (i.e., georeferenced). Therefore, it makes sense to store all these data along with their geographical locations within a spatial information system. This is an obvious step for the networks and the ancillary structures, the above ground catchments, rainfall gauge locations, and so on. The value of the GIS comes when the data can be interrogated from a relationship perspective. So, for example, a query can quickly determine the houses served by a particular water main, or identify the nearest sewers to a factory. The value of the GIS comes however, in relating other services to each other, along a main road for project and job management, or quickly determining what part of a water distribution network needs to be shut down for servicing.

Overall applications of GIS systems can be summarised to the following:

- utility networks and cadrastal mapping
- topographic mapping and thematic mapping
- surveying and photogrammetry
- remote sensing and image processing
- civil engineering
- mining engineering
- geography
- soil sciences
- military
- public services
- tourism (etc)
- land use planning and management
- mineral exploration
- environmental impact studies
- management of water resources
- natural hazard mapping
- forestry and wildlife management
- soil degradation studies
- monitoring desertification

There are four basic functions of GIS:

- data input and pre-processing,
- data management (storage and retrieval),
- manipulation and analysis,
- data output (reports and publications).

Some of the common applications of GIS in urban water management are:

- Combination of different maps with different scales,
- Classification of features,

- Creation of digital elevation models (DEMs),
- DEM interpolation,
- Combination of different factors so that new spatial contexts for different urban water management measures can be visualised,
- Manual and automatic creation of thematic maps,
- Discrimination of various types of pervious and impervious areas,
- Analysis of the DEM to identify possible flow patterns and routes,
- Cross-referencing the underground infrastructure against the surface terrain features,
- Delineation of the catchment into subcatchments and assignment of land features,
- Flood risk and flood hazard mapping,
- Processing of radar rainfall data,
- Creation of model input files,
- Presentation of model results.

It is still questionable however, how much can be done by interacting with a GIS and how much is better left to the judgment of an engineer. Certainly, the identification or delineation of catchment boundaries and streamlines is sometimes done best by the engineer on the ground. However, for most cases, a combination of a GIS-based delineation followed by a manual adjustment is likely to be the most efficient approach; see Figure 4.32 and Figure 4.33.

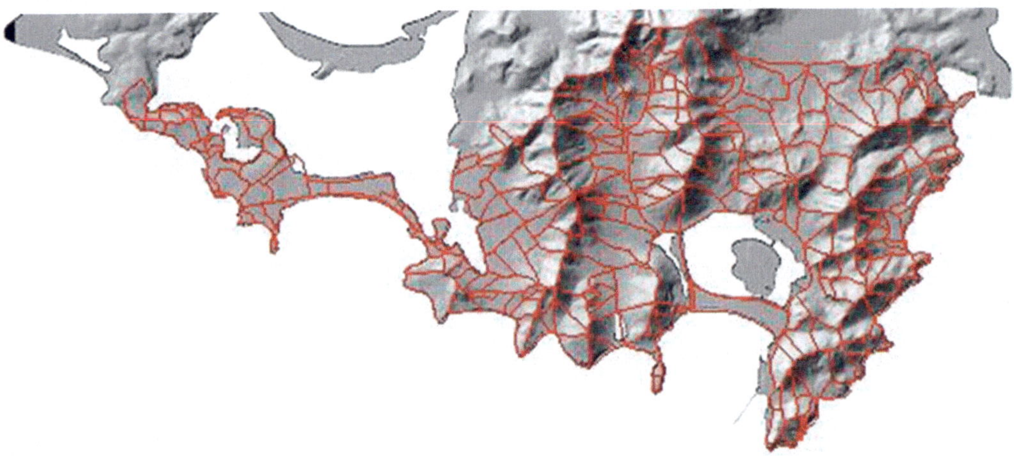

Figure 4.32 Example of catchments/sub-catchments delineation initially pre-processed in GIS and then manually post-processed

A Geographic database refers to the collection of spatially referenced data that acts as a model of reality. Two main components of geographic data can be defined:

- attributes or properties
- geographic position

Each entity usually has a number of attributes (properties). Attributes may be:

- nominal (identity e.g. manhole, pipe, maize, granite, lake)
- ordinal (ranking e.g. class1, class2 etc)
- scalar (value e.g. ground level, invert level,water depth, elevation, BOD)

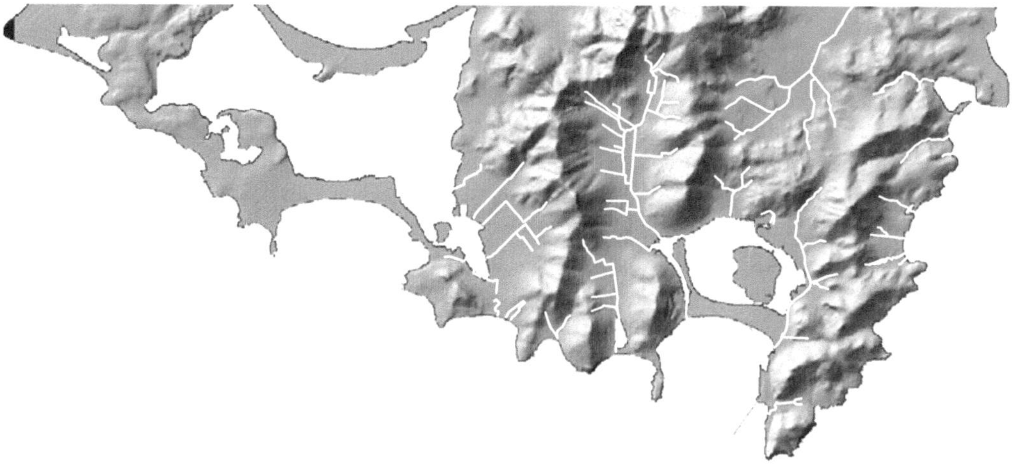

Figure 4.33 Example of streamline definition initially pre-processed in GIS and then manually post-processed

There are four common categories of analysis procedures:

- retrieval, (re)classification and measurement operations
- overlay operations
- neighbourhood operations
- connectivity operations

Neighbourhood operations include:

- retrieval functions
- interpolation functions
- topographic functions

Connectivity operations include:

- contiguity functions: connected areas that share common characteristic(s)
- proximity functions: connected areas having the same distance to a given point, line or area
- spread functions: connected areas that result from spreading, dilution or accumulation of phenomena from a given point, line or area.

Nowadays, access to spatial data as well as to advanced mapping and spatial analysis over the Internet is becoming more common. In the past, an individual would have to buy an expensive software package to use and manipulate the data needed for GIS. With the advent of Java based programming, software applications for web-based GIS work are now available. Some of these programs require the user to buy some additional software, and others require plug-ins to be added to web browsers, but some programs require no special additional software at all. With fast Internet connections the amount of information that can be transferred over the Internet is immense. A typical web-based system is a computerised system that delivers information to the user using a 'thin-client' Web browser like Internet Explorer or Netscape Navigator. The computer server that hosts the system application is linked to the user's computer by a network with the Transmissions Control Protocol/Internet Protocol (TCP/IP) protocol. Example of a web-based decision support system is given in Choi *et al.*, 2005. The same system can be applied to the management of urban water systems where a web-GIS system can be utilised for catchment delineation, map interfaces, data preparation routines, model instantiation and web communication for Internet-based system operation.

4.12 TELEMETRY AND SCADA SYSTEMS

Telemetry (synonymous with telematics) is a technology that allows remote measurement and reporting of information. It enables an automated communications process by which data are collected from instruments located at remote or inaccessible points and transmitted to receiving equipment for measurement, monitoring, display, and recording. A typical application of telemetry systems in urban water management is the monitoring of flows in pipe networks in real time. Telemetry receivers and telemetry transmitters are data acquisition components used to gather information from remote locations via wireless communication.

Wireless devices are small enough to easily carry or wear, have sufficient computing power to perform productive tasks and can communicate with the Internet and other devices. They include PDAs, cellular phones, smart phones. Wireless Application Protocol (WAP) is the standard that enables wireless devices with tiny display screens, low bandwidth connections and minimal memory to access Web-based information and services.

Table 4.5 Some of the common commercial GIS software packages

System Name	Developer	Web Extension	Geometric Storage
ARCMAP	ESRI	ArcIMS / ArcGIS Server	Vector and Raster
MAPINFO	Mapinfo Corporation	Mapinfo MapX	Vector and Raster
GEOMEDIA	Integraph	GeoMedia WebMap	Vector and Raster
IDRISI Taiga	IDRISI	IDRISI Taiga	Vector and Raster
MANIFOLD	Manifold Net Ltd	Manifold IMS (Internet Map Server)	Vector and Raster

Microbrowsers are Internet browsers with a small file size that can work within the low-memory constraints of wireless devices and the low bandwidths of wireless networks.

A Pager is a one-way, wireless messaging system; it alerts the user when it receives an incoming message.

SCADA stands for Supervisory Control and Data Acquisition. It refers to a system that collects data from various sensors (from either close or remote locations) and then sends this data to a central computer which then manages and controls the data. SCADA is a term that is used broadly to portray control and management solutions in a wide range of industries. Some of the industries where SCADA is used are water control systems, electric power, traffic signals, mass transit systems, environmental control systems, manufacturing systems, etc.

For urban water systems management, supervisory control is an invaluable tool as it provides:

- Monitoring status of control gates, control valves and pumps
- Monitoring key water levels and flow rates
- Enabling safe and reliable local and remote operation of control equipment
- Annunciating alarm conditions to operations personnel

Also, data acquisition provides a record for an analysis of operations by:

- Collecting key operational information such as water levels and flow rates
- Storing historical operational information for use by operations personnel
- Providing record of various operational decisions

A SCADA system includes various elements such as signal hardware (input and output), controllers, networks, user interface, communications equipment and software. One of the key processes of SCADA is the ability to monitor an entire system in real time. This is facilitated by data acquisition including meter reading, checking statuses of sensors, etc that are communicated at regular intervals depending on the system.

4.13 KEYWORDS

By the end of the chapter you should have encountered the following terms; ensure that you are familiar with them!

Data collection	Flow monitoring	Remote sensing
Digital terrain	GIS	SCADA

4.14 QUESTIONS

How can a digital terrain model be set up for an urban area?
What is the best way to measure flows in sewers?
What are the advantages and disadvantages for using satellites for collecting data on urban areas for water management?

4.15 REFERENCES

Abdullah A. F., Vojinović Z. and Price R. K. (2009). Lidar filtering algorithms and DTM generation for urban flood modelling applications: review of current algorithms and filters test. In: *8th International Conference on Urban Drainage Modelling*, Tokyo, Japan, 7–12 September.

AIAA (1995). *Assessment of Wind Tunnel Data Uncertainty*. American Institute of Aeronautics and Astronautics, Reston, Virginia, S-071.

ANSI/ASME (1985). *Measurement Uncertainty: Part 1, Instruments and Apparatus*. American National Standards Institute (ANSI)/American Society of Mechanical Engineers, Washington, DC, PTC 19.1-1998.

ASME (1998). *Test Uncertainty: Instruments and Apparatus*. American Society of Mechanical Engineers, New York, PTC 19.1-1998.

CEC (2000). *Directive Concerning Water Framework – Integrated River Basin Management for Europe*, 2000/60/EEC.

Choi J., Engel B. and Farnsworth R. L. (2005). Web-based GIS and spatial decision support system for watershed management. *J. Hydroinf.*, **7**(3), 165–174.

Einfalt T., Denoeux T. and Jacquet G. (1990). A radar rainfall forecasting method designed for hydrological purposes. *J. Hydrol.*, **114**(3–4), 229–244.

Fletcher T. D. and Deletic A. (eds.) (2007). *Data Requirements for Integrated Urban Water Management*. Taylor & Francis, Paris.

Gillespie T., Chu J., Frankenburg E. and Thomas D. (2007). Assessment and prediction of natural hazards from satellite imagery. *Progr. Phys. Geogr.*, **31**(5), 459.

ISO (1993). *Guide to the Expression of Uncertainty in Measurement*. International Organisation for Standardisation, Geneva.

Metternicht G., Hurni L. and Gogu, R. (2005). Remote sensing of landslides: an analysis of the potential contribution to geo-spatial systems for hazard assessment in mountainous environments. *Rem. Sens. Environ.*, **98**(2–3), 284–303.

Chapter 5

Modelling Paradigms

5.1 BACKGROUND TO MODELLING

Modelling is at the heart of hydroinformatics and therefore of this book. As we understand it, modelling is about quantifying the physical and social processes in the physical world through the use of underlying mathematics. We assume that all such processes can be represented individually or in an integrated manner by some form of symbolic equations or relationships, whether based on good science and deduction or working with measured data and induction. This means that we can create a virtual mathematical world that is conceived within our mental world and substantiated ultimately in the digital functionality of our current computing technology. With the aid of computers we are able to manipulate digital representations of the physical world in order to explore the possibilities for improving the performance of the physical systems for urban water, and for identifying and working with the social systems that initiate the demand for water, use it, and seek to sustain the environment.

In modelling the physical world we first identify why we want to model it. What is the purpose of the model? In particular, do we want to investigate an intervention in the physical world, such as extending a network, redeveloping an urban area, or rehabilitating a sewerage system? Or is our purpose non-structural: do we want to assess future demand for water supply in a region of a city, to forecast the operation of a combined sewer overflow, or to identify when fish kills will occur in receiving waters? In the first case, we need to work with a model that accurately represents the physics of the situation, whereas in the second case we may be more dependent on a model that learns directly from relevant data. So we appear to have at least two model paradigms: one that is deductive from physical (and possibly, social interaction) principles, and the other inductive from data. However, not all models turn out to be purely deductive or inductive: deductive models have inductive components, and vice versa. So we adopt an alternative way of categorizing models: one which broadly supports the distinction between the example applications above.

5.1.1 Model categories

We recognize that much of our modelling is normally concerned with water interacting with physical objects in physical space (though we also have reason to address below the modelling of social processes in social 'space'). In particular, we can talk about the 'model space'. This is the physical geometric domain potentially occupied by the water: the pipe or channel network, the groundwater strata containing an aquifer, the above-ground surface topography for overland flow, the sub-catchment draining to a pipe, the regions of a city with a demand for water, or the bathymetry of the coastal waters. Are we concerned with the motion of the water in the particular space, or are we more interested in the space containing the water? Such questions lead us to consider two categories of model (see Table 5.1): in one, we define mathematical equations formulating the physical laws for the flow of water *at a point* (in 1D, 2D or 3D space), we then adopt a one-to-one mapping of the physical space onto the model space and discretise the model space consistent with finding a numerical solution to the analytical equations (see Table 5.2); in the other, we compartmentalize the model space according to logically deduced physical boundaries and assume appropriate mathematical constructs to relate the output to the input for each compartment (see Table 5.3). In the former case the focus is on how the actual physical geometry affects flows at all points in the fluid, and how changes in the geometry affect these flows. In the latter case, there may be little if any interest in the detailed flows with respect to the actual geometry, but only in the relationship of the output to the input. The mathematical constructs may be derived from quasi-physical principles, resulting in analytical equations that apply to the compartment as a whole. Alternatively, the constructs may have little if anything to do with the physics, but are designed to provide a consistent connection between the output and the input; that is, the construct learns about the nature of the connection from data; see Table 5.1.

Table 5.1 Categories of model

Categories of model	Description
Physically (or process) based (or white box)	Numerical solution of equations describing the physical (and chemical or biological) processes for the fluid as a continuum in the model space, which is a one-to-one mapping of the actual physical space and is deconstructed into cells or elements appropriate for the numerical solution
'Conceptual' (lumped or grey box)	Numerical solution of equations describing the physical (and chemical or biological) processes for the fluid as a continuum in the model space which is a simplification of the actual physical space or a lumped quantity in one or more conceptual objects linking output to input in the model space or in compartments for the model space which are identified on physical principles
Data driven (or black box)	Numerical solution of equations derived for a particular mathematical construct that includes a number of parameters, whose values are trained to replicate the connection between the output and input observed for the model space or for compartments for the model space which are identified on physical principles
Rule based	An inference rule is a statement that has two parts, an if clause and a then clause
Deterministic	The model predicts the same output each time it receives the same input
Stochastic	The model predicts a different output each time it receives the same input

Table 5.2 Physically based modelling

Physically based modelling
Deconstructs the model space into compartments (cells/elements) consistent with the numerical solution of the algorithmic equations

Table 5.3 Compartmentalised modelling

Compartmentalised modelling
'Conceptual' Lumped Black box Grey box
Deconstructs the model space into compartments consistent with geophysical concepts and connects the output from each compartment with the input

What is common to both categories of model is that we have to identify the model space in terms of the physical objects that make it up. The model space may be defined by a network of water distribution pipes. For the first category of model we have to include in the model space each significant object that contains the flow. This means that we include each pipe as a separate object, with nodes at the junctions between the pipes, source reservoirs, pumping stations, air vessels, valves, and demand outlets. Each of these objects has its own attributes, including physical dimensions, material, roughness, upstream and downstream node references, and so on. All of the objects play some part in contributing to the whole of the distribution system. It is important to make this collection of objects complete, so that nothing is overlooked that might be important. At the same time there is a need for parsimony in selecting the objects, for example, in assuming that pipes less than a certain size can be ignored, that the pipes have a uniform cross section, and that a pump can be defined by a pumping rate and we do not have to model the flow inside the pump itself. Similarly, it is important to identify all of the physical processes that contribute to the determination of the flow. It turns out, for example, that head-losses at bends and valves along a pipe can contribute to the energy loss of the flow.

Meanwhile, for the second category of model, we are not so interested in the details of the model space, instead the model space is a simplification of the physical space, such as an equivalent uniform reach corresponding to an irregular natural reach, one or more reaches along an open channel in the Muskingum routing method (see below) or a sequence of storage tanks in a tank model for rainfall-runoff from a catchment, or a *zero dimensional* model consisting simply of a time series for the inflow of potable water to the distribution network and the time series consisting of all the flows at the demand points that are supplied with water. In the latter case, given a model that relates demand to inflow, it can be used to explore issues such as the leakage from the network and the reliability of the service with changing demand patterns, and so on.

As an alternative, consider an urban sub-catchment contributing to a given outlet or conduit in a drainage network. The sub-catchment may include its own sub-network of pipes. What we are interested in is the relationship between the rainfall falling on the sub-catchment and the runoff. Whereas we could use the first category of model in simulating the local rainfall losses and the runoff over the 2D surface, the high resolution required makes the model cumbersome and accuracy is prejudiced. In practice, only the second category of model is applicable to this situation, especially as we are focused on the outfall of the sub-catchment and not on the details of the flows in the sub-catchment. There are different types of model that may be used, depending on whether some physical information about the sub-catchment is needed or not. Those models dependent on physical information require the sub-catchment to be defined by the geographical boundary that forms the (approximate) watershed of the sub-catchment. The sub-catchment then has its own unique values for certain parameters that may be important for the model, such as the total area, the percentage paved area, and the characteristic slope. There are however, models that are purely driven by the data and simply need the time series for the rainfall and the runoff (if available). There is no need for any other physical information. Obviously this last type of model is heavily dependent on adequate data for the rainfall and runoff.

As another example of the first category of model consider a pipe network defining the model space. Analytical equations can be derived for the time dependent flow at a point in 3D. But it can be argued that the flow in the pipes can be approximated as being one dimensional in that the length scale along the pipe is considerably greater than the scale of the cross section, and therefore the three dimensional equations are integrated over the uniform cross section of a pipe to deduce a 1D version of the equations. We then solve these equations numerically within the geometric domain for the flow in each pipe. In particular, the model space in a pipe is discretized into equal lengths according to the size of the pipe, with the dependent variables for the flow defined at the sections between each length; see Figure 5.1.

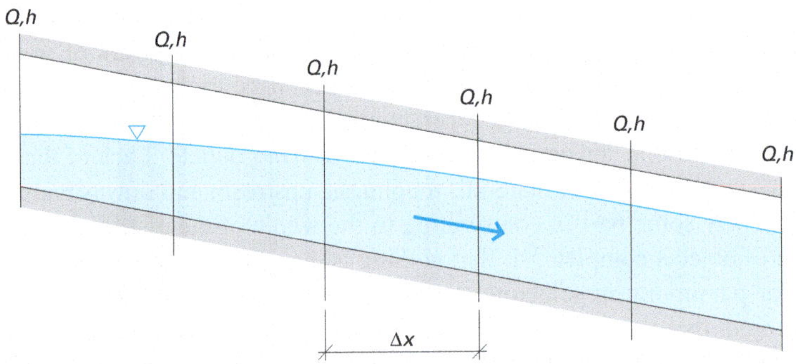

Figure 5.1 Discretization of 1D model space for a conduit

In line with normal practice we call the first category, 'physically-based' modelling. However, the second category does not have a universally accepted nomenclature. For convenience we call it 'compartmentalized' modelling. Other terms used under the category of compartmentalized modelling are: conceptual, lumped, black box, grey box, and data-driven. 'Conceptual' is not used here as a separate category because of the generic use of the word to identify processes and objects. 'Lumped' reflects the word 'compartmentalized', so despite its popularity it is not used here as an independent category in this book. The primary sub-categories of compartmentalized modelling recognized here are: 'data-driven' of 'black box' and 'grey box' modelling. Data driven modelling uses mathematical constructs that have minimal dependence on knowledge about the processes and objects in the model space. Included in this sub-category are machine learning methods such as artificial neural networks (ANNs) (Hassoun, 1995), model trees (MTs) (Quinlan, 1992), nearest neighbour methods (Toth *et al.*, 2000), fuzzy logic generators (Lobbrecht and Solomatine, 1999), evolutionary programming (Savic *et al.*, 1999) etc. Also included under data driven modelling are regression (Yevjevich, 1972), unit hydrograph (Sherman, 1932) and ARMA-type (Box *et al.*, 1976) methods. In contrast, grey box modelling includes all those approaches that include some knowledge about the processes and objects in the model space. For example, hydrological methods for catchment runoff that are based on the notion of a linear (or non-linear) reservoir would be included in this sub-category, as would the rational and time-area methods for determining the peak discharge and time dependent runoff respectively from a catchment; see Figure 5.2. Also, included in the compartmentalized category is the Muskingum method for flow routing referred to above; see McCarthy (1938). This method deconstructs a length of conduit or a reach of river into one or more routing compartments for which it is assumed that the storage in each compartment is dependent on a linear combination of the input and output discharge. Interestingly, provided the linear combination is correctly formulated, the Muskingum method can be shown to be consistent with an approximation of the 1D gradually varying free surface flow equations as the length of the compartment goes to zero (Price, 2009).

To summarise, with a physically based model we are interested in what happens in the physical geometric domain and how it affects the output, or the outputs at a range of locations within the model space. With a compartmentalised model, which is zero dimensional, we are only interested in the output and not in the changes internal to the model space.

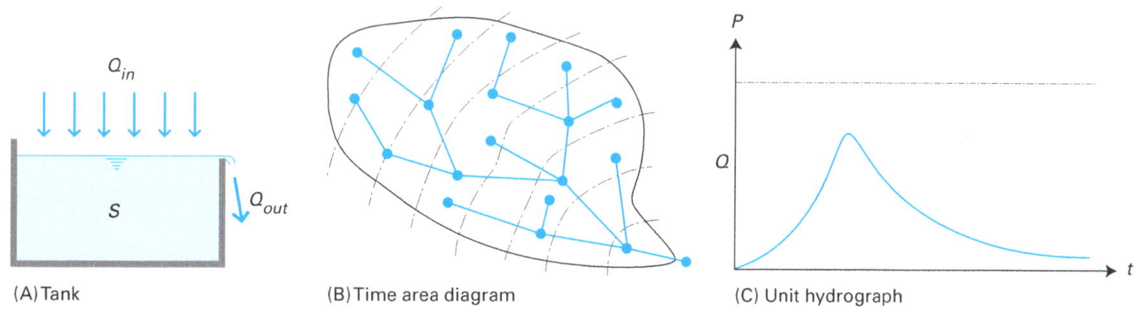

Figure 5.2 Alternative rainfall-runoff model concepts

5.1.2 What is a model?

Before we proceed further we need to be more precise about what we mean by a model; and to begin with we focus on physically based models: we discuss data driven models separately. As described at the beginning of this chapter, modelling is about quantifying the physical and social processes in the real world through the use of underlying mathematics. For us, a model is essentially a mathematical object that generates output data from input data. And yet it is more than that. The object represents an aspect of reality, and must necessarily be a simplification of that reality. Further, the model is designed such that when operating properly it provides insights into particular attributes of a system, whether it be physical, biological, economic or social (USEPA, 2008).

All of the models described in this book are deterministic in that the same input always produces the same output; in other words, there is no random component in the models. Necessarily, assumptions are made in the derivation of the mathematical algorithms for all physically based models. In making these assumptions we introduce the notion of parameters for the model, which are usually closely tied to particular physical processes to do with energy loss, such as the frictional effect of a boundary surface, or the highly turbulent and contorted flow through structures including gates, orifices, weirs and pumps. The parameters may acquire fixed numerical values through analysis of experimental laboratory or field data, or their values be left open in order to be defined separately for each model application. The values of these open parameters may be defined in terms of qualitative descriptions of physical structure, such as the roughness of a boundary surface. But there may be situations where guidelines are given as to a range of appropriate values but the precise value is unknown. In these circumstances the modeller has to go further in 'calibrating' the parameter, possibly by adjusting the output of the model for the prescribed input such that the output closely resembles observations of the output variables.

5.1.3 Calibration of a model

Calibration is a process that can easily be abused. The desire of the modeller is to have a physically based model that reproduces the values of an observed variable as closely as possible, and sometimes the parameters will be adjusted unreasonably in order to achieve this. Such an adjustment fails to recognise that a model is always going to be an idealised version of the real world, and therefore incorporates a number of approximations that generate errors in the model output even with the correct values of the parameters. Therefore, setting aside the fact that there are likely to be errors in the data for the observed input and output variables, it is important that the modeller learns to live with the limitations of the model, and to avoid falling into the trap of force-fitting the model by adjusting the parameters outside their normal ranges of values. The penalty of not observing this limitation is the probable prediction of incorrect results in applying the force-fitted model to design and prediction. Of course there are situations where there is little experience available in identifying values for certain parameters in particular contexts. This is often the case for diffusion coefficients in advection-diffusion processes for pollutants or rate coefficients in chemical and biological processes; even rainfall-runoff coefficients for a catchment may be difficult to quantify. Even then the modeler should endeavour to seek published values of the parameters for similar models. In such cases calibration becomes more meaningful as the modeller adjusts the parameter values to replicate the observed variable data.

Calibration should therefore be handled with care, and in general optimisation should not be used to refine parameter values with physically based models. Where calibration is involved, many modellers like to verify the parameter values that they have selected by running the model with another set of input data and demonstrating that the results continue to be reasonable. Unfortunately, modellers tend to use words like 'verification' and 'validation' somewhat loosely. As pointed out by Oreskes, Shrader-Frechette *et al.* (1994), such words have to do with truth, and it is very difficult to justify claims to truth in modelling. The authors suggest that a better word than either 'verification' or 'validation' is 'confirmation'. This book uses the word 'confirm' as the act of gaining confidence that a model is fit-for-purpose, often through peer review. Nevertheless, verification and validation will continue to be referred to because of their common use by the modelling community.

5.1.4 Confirming a model

As stated, the words 'validate' and 'verify' have been generally used as a sequel to calibration. 'Verification' has also been used in the UK to refer to the process of confirming that the integrity of the objects in the model space for urban drainage networks in particular, which replicate the structural objects that make up the physical space, is complete; that is to say, there are no additional or missing objects, they are correctly connected and the geometric data for each object is acceptably accurate. In most hydraulic modelling situations this form of verification is not an issue. However, it happens that in urban drainage especially, there are numerous occasions when there is missing information about the below-ground network, and even with expensive manhole and CCTV surveys it is difficult to be sure that all the details of the pipe network have been accurately recorded. It is necessary therefore for the integrity of the network structure replicated in the model space to be 'verified'. This is done in the normal way by running the model for observed events and comparing the results of the model with the observed flows. Through both a quantitative and/or a qualitative analysis it is possible to identify deficiencies in the structural data and then to confirm or disprove the existence of such deficiencies through field survey. A difficulty with this approach is that there may also be the need to calibrate the model parameters at the same time, and therefore confusion can arise between what is justifiably a problem with the structural data and what is the consequence of an adjustment to one or more model parameters. This topic is discussed more thoroughly in Chapter 6.

Another note of caution about physically based models should be raised at this point, and this concerns the fact that almost all such models depend on numerical techniques for the solution of the mathematical algorithms. This implies the development of modelling software to implement the solvers of the algorithms in a solution space that is based on a particular generic deconstruction of the model space. In other words, the modelling software is generic and can be applied to any situation that satisfies the class of problem for which the software is designed, whether for distribution or collection networks of pipes or open channels, groundwater, river networks, and so on. A physically based model is the instantiation of the appropriate modelling software with the geometry and input data for the given situation. The modelling software on its own is not 'the model'.

So far as data driven models are concerned there are strict rules for training and verifying them. A data driven model is entirely dependent on training a selected mathematical construct between observed output and input data. The particular underlying mathematical algorithm forming the connection has nothing to do with the physics of the situation, and therefore there is no dependence of the model on the physical structure of the model space. Where the physics becomes important is in the choice of the input variables and the use of the data values. Such a choice may be affected by factors such as the shape and characteristic geometry of the 'compartment' (the catchment, the wastewater treatment process, etc), the nominal 'lag time' between the input and output response, or the decision to deconstruct the physical process into components and to model them separately, such as separating the base flow in the runoff for a catchment from the response to rainfall. The instantiation of different types of data driven model is discussed at the end of this chapter.

Obviously, the development of a model can be highly subjective and is dependent on the individual choices made by the modeler on what to include and what not to include, the selection of time dependent data for calibrating and verifying the model, and the way in which the model should be applied. Therefore it is important that there should be a peer review of the model in order to confirm the approach adopted by the modeler and the outcome.

Finally, all users should be aware of the limitations of models. They are designed to replicate the physical world; and whether we are concerned with physically based or compartmentalized models, they are only as good as the underlying mathematics, the data for the model space, the temporal input (boundary) data, the calibration or training of the model parameters, and the user's expertise. They are therefore limited in their capability to explain and predict, and the wise user will always bear that in mind when interpreting the output from such models.

5.1.5 Modelling phases for urban water systems

Before we explore the physically based equations for the flow in particular geometric domains, we need to conceptualise the different phases of the flow occurring in water distribution and drainage (or collection) systems. For water distribution we can identify water flowing under pressure in a looped network of pipes, with inflow from a pump or storage tank with regulated water level to a number of demand points, and controlled by a set of valves and pumps interior to the network. The pressures at all junction nodes of the network are inter-related, and therefore the network forms one phase. There can be inter-dependent calculations for the water quality, including the age of the water. In addition, calculations may be required to prescribe the size of the pipes in the network (probably using optimization to minimize the construction cost), or to forecast the future demand at existing and conceptual nodes introduced to allow for the development of the urban area and therefore the extension of the water distribution network. On some occasions there may be the need to model the interaction of groundwater with infiltration to a pipe when the operating pressure has dropped considerably. In such cases the water in the pipe may become polluted and unusable for drinking.

The situation with drainage is very different. Here there are a number of distinct phases including monitored forecast or design rainfall, rainfall-runoff from urban areas, free surface and pressurized flows in dendritic or looped networks of underground conduits, storage at manholes, including above ground storage associated with surface flooding, infiltration from groundwater, wastewater inflows from housing and industry, overflow operation, storage tanks and ponds, diversion structures, outfalls to receiving streams, discharges to treatment works, rainfall-runoff from rural catchments to urban streams, free surface flow in urban streams and rivers, flood flows from urban streams to urban areas, and overland flood flows interaction of overland flood flows and underground pipe flows; see Figure 5.3. Similarly, there can be interdependent calculations for water quality in terms of sediment and dissolved pollutants for surface runoff, pipe flow, above ground flow, and urban streams and rivers. As for water distribution, calculations may be required to prescribe the size of the pipes in dendritic or looped drainage networks. Note that different types of models may be used. Whereas a physically based model may be adopted to simulate the flow in each pipe and manhole, compartmentalised models may be adopted for the rainfall-runoff.

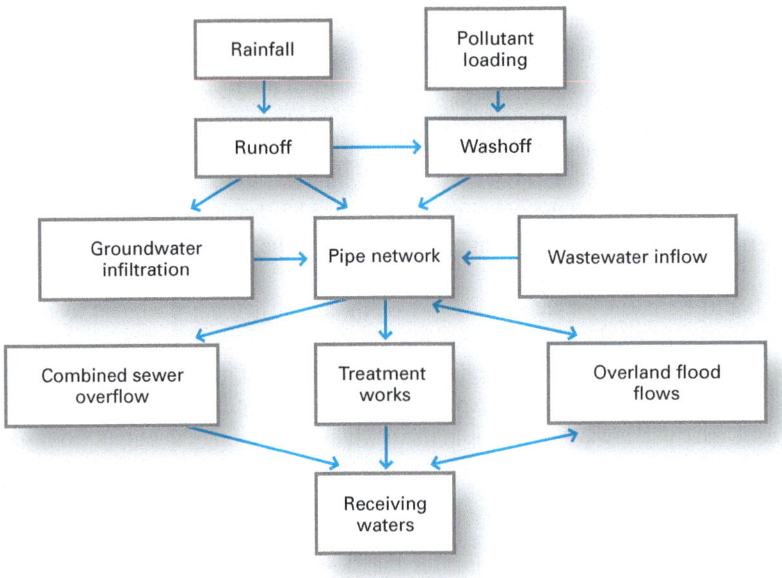

Figure 5.3 Urban drainage system phases

In order to appreciate the contribution of the physical processes to the form of the analytical equations we need to review some relevant physical concepts associated with water.

5.1.6 Some physical concepts associated with water

Water is a most remarkable fluid. Besides having a maximum density at about 4°C water becomes a solid, namely ice, at 0°C. This feature ensures that ice, which is less dense than the surrounding water, remains at the surface rather than sinking to the bottom of the water body. Water also requires considerable pressure to be compressed; so much so, that in the natural environment it can be regarded for almost all practical purposes as being incompressible. This greatly simplifies the formulation of the equations governing the flow of water. The only exception which we need be concerned with in this book is the phenomenon of water hammer, when pressure waves are created in surcharged pipes with water flowing under pressure.

Further, water is a real Newtonian fluid; that is, it resists deformation such that the shear stress is proportional to the velocity gradient normal to the stress. The constant of proportionality is the viscosity. The resulting time dependent three dimensional flow equations are the Navier-Stokes equations. Given that the water particles immediately next to a rigid boundary do not move (the no-slip condition) the shear stress in the flow is locally large near to the boundary. This generates a strong rotation, or vorticity, in the fluid. Physically, vorticity is generated along the boundary, and it can be shown that the vorticity diffuses away from the boundary into the flow where it is convected further away. The region near the boundary where the vorticity is large is termed the boundary layer. The thickness of this conceptual layer is proportional to the viscosity and inversely proportional to the square root of the Reynolds number, UL/ν, where U is the scale of the velocity, L is the length scale and ν is the kinematic viscosity.

The boundary layer is important for sediment transport and the introduction of random fluctuations, or turbulence, into the flow. Whereas it is possible to formulate the Navier-Stokes equations for laminar flow rigorously from first principles, turbulence introduces further dependent variables, namely the Reynolds stresses, with the need for additional equations to generate a unique solution for the mean flow variables. The closure problem is circumvented in open channel hydraulics by assuming that the energy loss in the flow is proportional to the square of the velocity averaged over depth (in two spatial dimensions, 2D) or over the cross section (in one dimensional flow, 1D).

The equations derived in this way for 2D or 1D flow are termed the Saint Venant equations. The additional assumption made in the derivation is that the flow has a gradually varying free surface such that the vertical accelerations of the water particles are negligible, and the pressure in the flow is then essentially hydrostatic, that is, the pressure is the same as if there were no flow. Although corresponding equations can be derived when the pressure deviates to first order away from being hydrostatic, these equations do not feature in the modelling of urban water systems considered in this book. Besides free surface flow, equations are needed for surcharged flow under pressure in pipes. The equations are essentially similar to the free surface flow equations with the exception that the discharge is uniform along a given pipe at any instant due to the assumed incompressibility of the fluid.

The fact that water can have a well defined surface with the atmosphere is due to the bonds between the hydrogen atoms. The surface tension that is the result of these bonds enables insects to walk on water, a meniscus to form in the contact of the free surface of water with a hard surface (which benefits osmosis in plants), and raindrops to form on dust particles.

Finally, water is an excellent solvent, which means that it dissolves and carries with it a range of substances, as well as eroding and transporting sediment. The concentration of a particular substance is advected with the mean velocity of the flow, diffused due to the turbulence (and the Brownian movement) and dispersed according to the three dimensional structure of the flow. The interaction between various organic and inorganic substances and the processes of sediment transport introduce the complex analysis of water quality in urban water systems.

In what follows we deconstruct the physics of the flow of water in conduits, over a 2D surface and in 3D groundwater aquifers or treatment processes, as appropriate for urban water systems. Our concern is to formulate the analytical equations based on the physical processes involved. Once these have been defined we can then address the issue about how to solve the equations numerically in different contexts. Subsequently we explore various forms of data driven modelling techniques that can be applied to urban water problems, and conclude with a discussion of intelligent agent methods.

5.2 MODELLING WATER QUANTITY

Fundamental to any analysis of urban water systems is the ability to simulate the time dependent flow of the quantity of water in a network of pipes or channels or over a 2D surface, whether originating from a resource such as an aquifer, reservoir or river, or from rainfall.

5.2.1 Navier-Stokes equations

The physically based mathematical equations for fluid flow at a point have been generally agreed since about 1825. For laminar, incompressible flow of a Newtonian fluid treated as a continuum they are the Navier-Stokes equations; see Batchelor, 2000 and Temam, 2001.

$$
\begin{aligned}
&\frac{\partial u}{\partial t}+u\frac{\partial u}{\partial x}+v\frac{\partial u}{\partial y}+w\frac{\partial u}{\partial z}=-\frac{1}{\rho}\frac{\partial p}{\partial x}+\nu\left(\frac{\partial^2 u}{\partial x^2}+\frac{\partial^2 u}{\partial y^2}+\frac{\partial^2 u}{\partial z^2}\right)\\
&\frac{\partial v}{\partial t}+u\frac{\partial v}{\partial x}+v\frac{\partial v}{\partial y}+w\frac{\partial v}{\partial z}=-\frac{1}{\rho}\frac{\partial p}{\partial x}+\nu\left(\frac{\partial^2 v}{\partial x^2}+\frac{\partial^2 v}{\partial y^2}+\frac{\partial^2 v}{\partial z^2}\right)\\
&\frac{\partial w}{\partial t}+u\frac{\partial w}{\partial x}+v\frac{\partial w}{\partial y}+w\frac{\partial w}{\partial z}=-\frac{1}{\rho}\frac{\partial p}{\partial x}+\nu\left(\frac{\partial^2 w}{\partial x^2}+\frac{\partial^2 w}{\partial y^2}+\frac{\partial^2 w}{\partial z^2}\right)+g\\
&\frac{\partial u}{\partial x}+\frac{\partial v}{\partial y}+\frac{\partial w}{\partial z}=0
\end{aligned}
\tag{5.1}
$$

where u, v, w are the velocity components (m^3/s) in the x, y, z directions (m)
t is the time (s)
p is the pressure (kg/s^2/m)
ρ is the density (kg/m^3)
ν is the kinematic viscosity (m^2/s)
g is the acceleration due to gravity (m/s^2)

These essentially hyperbolic equations are symbolic for Newton's laws of motion, namely that the acceleration of the water is proportional to the applied force made up of gravity and the internal friction losses, and that the mass of water is conserved. In deriving the Navier-Stokes equations it is assumed that the fluid is Newtonian, that is, the shear stress in the flow is proportional to the velocity gradient normal to the direction of flow. There are four equations in four dependent variables u, v, w and p, so it is possible to determine a unique solution in the solution space (x, y, z, t). Note that we distinguish between the model space which is formed by the physical boundary to the flow, namely the interior of the pipe, say, and the corresponding solution space.

A major difficulty with these equations is in extending them for turbulent flow. The classical way of doing this is to introduce turbulent fluctuations in the dependent variables and averaging over the random fluctuations. This generates a similar set of equations but with six additional unknown 'Reynolds stresses' arising from the non-linear inertial terms in the momentum equations. A number of different assumptions have been made in order to define these Reynolds stresses in terms of the mean dependent variables; see Speziale, 1991. The assumptions become important in large scale 2D and 3D flows, such as in coastal waters or reservoirs, and in computational fluid dynamics applied to flow in overflow chambers or chlorine contact and activated sludge tanks. Turbulence is characterized by eddies of different sizes and randomness in the flow. Energy is dissipated in the flow as the larger eddies break down into smaller eddies, and so on, eventually converting into heat. There are three primary length scales associated with turbulence.

The length scale of the larger eddies is the integral length scale, l_0, given by

$$l_0 \propto \frac{k^{3/2}}{\varepsilon} \tag{5.2}$$

where k is the turbulent kinetic energy (m^2/s^2)
ε is the energy dissipation rate (m^2/s^3)

The turbulence Reynolds number associated with these large eddies is defined as

$$\mathrm{Re}_L = \frac{k^{1/2} l_0}{\nu} = \frac{k^2}{\varepsilon \nu} \tag{5.3}$$

The Kolmogorov length scale, η, is for the size of the smallest eddies that are responsible for dissipating energy when the Reynolds number is 1, that is,

$$\eta = \left(\frac{\nu^3}{\varepsilon}\right)^{1/4} = l_0\,\mathrm{Re}_L^{-3/4} \tag{5.4}$$

The Taylor length scale, λ, is for eddies in the inertial sub-range, which are intermediate between the large and small scale eddies:

$$\lambda = \left(\frac{10\nu k}{\varepsilon}\right)^{1/2} = \sqrt{10}\, l_0\,\mathrm{Re}_L^{-1/2} \tag{5.5}$$

The different length scales are used to determine the resolution of the numerical grid for the solution of the equations in a given boundary domain.

Boussinesq followed the assumption of the shear stress in laminar flow being proportional to the velocity gradient normal to the shear stress and proposed that the turbulent Reynolds stresses can similarly be assumed proportional to the mean velocity gradient normal to the stresses, with the coefficient of proportionality being the eddy viscosity. Kolmogorov and Prandtl suggested that the turbulent viscosity is given by

$$\nu_{turb} = C_\mu \frac{k^2}{\varepsilon} \tag{5.6}$$

where C_μ is an empirical constant
and k and ε are given by

$$\frac{\partial}{\partial t}(\rho k) + \frac{\partial}{\partial x_j}(\rho k u_j) = \frac{\partial}{\partial x_j}\left[\left(\mu + \frac{\mu_t}{\sigma_k}\right)\frac{\partial k}{\partial x_j}\right] + P_k - P_b - \rho\varepsilon - Y_m + S_k$$

$$\frac{\partial}{\partial t}(\rho \varepsilon) + \frac{\partial}{\partial x_j}(\rho \varepsilon u_j) = \frac{\partial}{\partial x_j}\left[\left(\mu + \frac{\mu_t}{\sigma_\varepsilon}\right)\frac{\partial \varepsilon}{\partial x_j}\right] + \rho C_{1\varepsilon}\frac{\varepsilon}{k}(P_k + C_{3\varepsilon}P_b) - C_{2\varepsilon}\rho\frac{\varepsilon^2}{k} + S_c$$

$$\mu_t = \rho C_\mu \frac{k^2}{\varepsilon} \tag{5.7}$$

$$P_k = -\rho\overline{u_i u_j}\frac{\partial u_j}{\partial x_i} = \mu_t S^2 = 2\mu_t S_{ij} S_{ij}$$

$$P_b = \beta g_i \frac{\mu_t}{\mathrm{Pr}_t}\frac{\partial T}{x_i}$$

where β is the volume expansion coefficient
Pr_t is the turbulent Prandtl number (~0.85)
g_i is the gravitational acceleration in the ith direction (m/s^2)
T is the temperature
σ_k, σ_ε, $C_{1\varepsilon}$, $C_{2\varepsilon}$ and $C_{3\varepsilon}$ are empirical constants

In these equations, μ_t is the turbulent viscosity, P_k represents the generation of turbulence kinetic energy due to the mean velocity gradients, calculated in same manner as standard k-ε model. P_b is the generation of turbulence kinetic energy due to buoyancy, calculated in same way as standard k-ε model. Some of the coefficients take the following values:

$$C_{1\varepsilon} = 1.44,\ C_{2\varepsilon} = 1.92,\ C_\mu = 0.09,\ \sigma_k = 1.0,\ \sigma_\varepsilon = 1.3$$

The coefficient of thermal expansion, β, is defined as

$$\beta = -\frac{1}{\rho}\left(\frac{\partial \rho}{\partial T}\right)_p \tag{5.8}$$

The interested reader should consult Mathieu and Scott, 2000.

5.2.2 Saint Venant equations

Given a gradually varying free surface to the flow such that the vertical accelerations of the water particles are negligible compared to the acceleration due to gravity, and assuming that the viscous terms in the third equation above can be ignored, then the momentum equation in the vertical direction becomes

$$0 = -\frac{1}{\rho}\frac{\partial p}{\partial z} + g \tag{5.9}$$

Integrating with respect to z:

$$p = p_0 + \rho(h - z) \tag{5.10}$$

where p_0 is the atmospheric pressure (kg.m^2/s) at the free surface. The pressure is therefore (approximately) a linear function of depth, and is identical to the pressure when the water is at rest (the hydrostatic pressure). The viscous terms involving the second order derivatives of the horizontal velocities in x and y can be regarded as negligible compared with the remaining viscous term, as can the derivatives of the Reynolds stresses in x and y. Integrating the two remaining momentum equations over depth and assuming that the Boussinesq coefficients formed by the integration of the non-linear convection of momentum terms are approximately unity:

$$\begin{aligned}
&h\left(\int_0^h u^2 dz\right)/\left(\int_0^h udz\right)^2 \approx 1\\
&h\left(\int_0^h v^2 dz\right)/\left(\int_0^h vdz\right)^2 \approx 1\\
&h\left(\int_0^h uvdz\right)/\left(\int_0^h udz\right)\left(\int_0^h udz\right) \approx 1
\end{aligned} \tag{5.11}$$

gives

$$\begin{aligned}
&\frac{\partial u}{\partial t} + \frac{\partial (u^2)}{\partial x} + \frac{\partial (uv)}{\partial y} + g\frac{\partial h}{\partial x} - \frac{1}{\rho}(\tau_{x,surf} - \tau_{x,bed}) = 0\\
&\frac{\partial v}{\partial t} + \frac{\partial (uv)}{\partial x} + \frac{\partial (v^2)}{\partial y} + g\frac{\partial h}{\partial y} - \frac{1}{\rho}(\tau_{y,surf} - \tau_{y,bed}) = 0\\
&\frac{\partial h}{\partial t} + \frac{\partial (uh)}{\partial x} + \frac{\partial (vh)}{\partial y} = q
\end{aligned} \tag{5.12}$$

where $\tau_{x,surf}$ is the viscous and Reynolds stress in the x-direction at the free surface (kg/s^2/m)
$\tau_{x,bed}$ is the viscous and Reynolds stress in the x-direction at the bed (kg/s^2/m)
$\tau_{y,surf}$ is the viscous and Reynolds stress in the y-direction at the free surface (kg/s^2/m)
$\tau_{y,bed}$ is the viscous and Reynolds stress in the y-direction at the bed (kg/s^2/m)

Assuming that the stress at the surface is zero (eg there is no wind) and that the bed stresses can be replaced by

$$\begin{aligned}
\tau_{x,bed} &= f\rho\frac{u\sqrt{u^2 + v^2}}{4h}\\
\tau_{y,bed} &= f\rho\frac{v\sqrt{u^2 + v^2}}{4h}
\end{aligned} \tag{5.13}$$

where f is the Darcy friction factor (see below)
then

$$\frac{\partial u}{\partial t} + \frac{\partial (u^2)}{\partial x} + \frac{\partial (uv)}{\partial y} + g\left(\frac{\partial h}{\partial x} + f\frac{u\sqrt{u^2+v^2}}{4gh}\right) = 0$$

$$\frac{\partial v}{\partial t} + \frac{\partial (uv)}{\partial x} + \frac{\partial (v^2)}{\partial y} + g\left(\frac{\partial h}{\partial y} + f\frac{v\sqrt{u^2+v^2}}{4gh}\right) = 0 \qquad (5.14)$$

$$\frac{\partial h}{\partial t} + \frac{\partial (uh)}{\partial x} + \frac{\partial (vh)}{\partial y} = q$$

Here q is the inflow per unit area (m/s).

Note that the relationships in Eq (5.13) are only approximate, but they are good enough for practical purposes.

5.2.3 1D Saint Venant equations

The 1D Saint Venant equations for gradually varying free surface flow in uniform open channels can be deduced from Eq (5.7) with the adjustment that h in the friction term is replaced by R, the hydraulic radius:

Conservation of momentum:

$$\frac{\partial u}{\partial t} + u\frac{\partial u}{\partial x} + g\left(\frac{\partial h}{\partial x} + f\frac{u|u|}{4gR}\right) = 0 \qquad (5.15)$$

Conservation of mass:

$$\frac{\partial h}{\partial t} + u\frac{\partial h}{\partial x} + \frac{A}{B}\frac{\partial u}{\partial x} = 0 \qquad (5.16)$$

where A is the flow area (m^2)
B is the surface flow width (m)
R is defined as A/P (m), where P is the wetted perimeter (m)

See Figure 5.4. The assumptions underlying these equations include:

1. The fluid, namely water, is a continuum.
2. The water surface gradients are gradually varying in space and time such that the pressure is hydrostatic; that is, the vertical accelerations of water particles in the flow are negligible compared to the acceleration due to gravity.
3. The Boussinesque coefficients modifying the convection of momentum terms in the momentum equation, and arising from the integration of the 3D Navier-Stokes equations over a cross-section to obtain the 1D averaged equations, are approximately constant and equal to unity.
4. The fluid is incompressible.
5. The resistance to the flow due to the shear stress at the boundaries and the dissipation of energy due to turbulence is proportional to the square of the mean velocity at a section, and the coefficient of

proportionality can be derived in terms of the geometry of the cross-section, a measure of the boundary roughness and the Reynolds number.

6. The forms for the resistance (or friction) force, which are derived from laboratory experiments under steady flow conditions, are applicable to unsteady flow.

The Saint Venant equations can be written in different ways. They have different forms and different approximations. Consider first alternative forms.

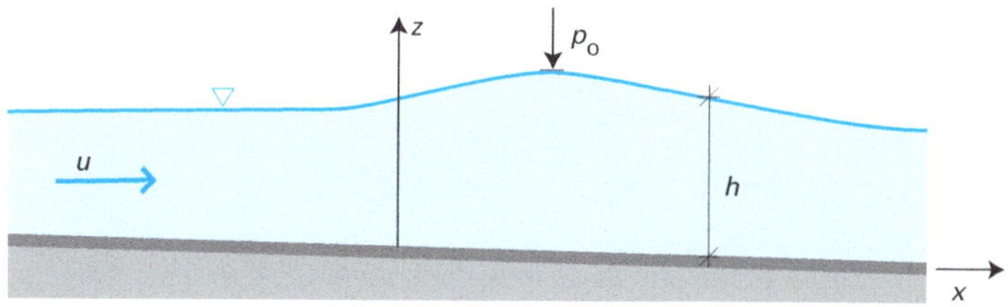

Figure 5.4 Flow in a channel

Algorithmic form

This is written as:

$$u\frac{\partial u}{\partial t} - g\frac{\partial h}{\partial t} + \left(u^2 - g\frac{A}{B}\right)\frac{\partial u}{\partial x} + gus_f = 0$$

$$-\frac{A}{B}\frac{\partial u}{\partial t} + u\frac{\partial h}{\partial t} + \left(u^2 - g\frac{A}{B}\right)\frac{\partial h}{\partial x} - g\frac{A}{B}s_f = 0 \tag{5.17}$$

provided the Froude number, $Fr \equiv u/\sqrt{gA/B} \neq 1$. This formulation is important when forming the 6-point implicit finite difference scheme below.

Characteristic form

Another important formulation depends on the hyperbolic nature of the equations, giving the characteristic form for a uniform channel:

$$\left[\frac{\partial}{\partial t} + \left(u \pm \sqrt{\frac{gA}{B}}\right)\frac{\partial}{\partial x}\right]\left(u \pm \int_0^y \sqrt{\frac{gB}{A}}dy'\right) - g(s_0 - s_f) = 0 \tag{5.18}$$

where y is the depth (m) defined by $h = y - s_0 x$

s_0 is the bed slope (m/m)

There are two characteristic curves passing through any point in the solution space (x, t) given by the lines:

$$\frac{dx}{dt} = \left(u \pm \sqrt{\frac{gA}{B}}\right) \tag{5.19}$$

The characteristic curves carry information into the solution space from its boundaries; see Figure 5.5. This stresses the importance of the boundaries in determining uniquely what happens in the space. Note that here a 'boundary' is denoted by what happens in time to the flow at the upstream and downstream ends of the pipe, and the initial state of the flow along the pipe: we are not referring to the physical boundary of the flow formed by the interior of the pipe.

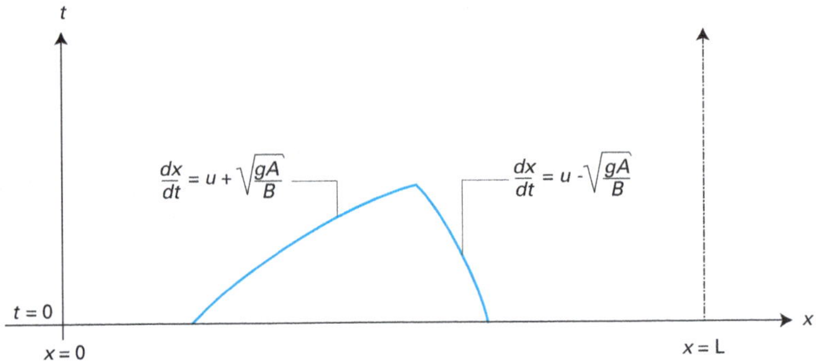

Figure 5.5 Characteristic curves in solution space.

The two equations in Eq (5.18) refer to the propagation of two sets of dynamic waves. Because $u + \sqrt{gA/B}$ is always positive for $u > 0$, every point in the domain receives information from the upstream boundary (or from the initial condition at time t = 0). However, $u - \sqrt{gA/B}$ can be either negative or positive for $u > 0$. When it is negative, that is $Fr < 1$, the corresponding characteristic curve conveys information from the downstream boundary (or from the initial condition at time t = 0). If $Fr = 1$, information does not come from upstream or downstream, in which case we have a 'critical' flow condition. For $Fr < 1$ the flow is said to be sub-critical, and for $Fr > 1$ the flow is super-critical. In the latter case, no information comes from downstream: all the information comes from upstream. At points where $Fr = 1$ there is a discontinuity in water levels. This is referred to as a hydraulic jump; see Figure 5.6 and Figure 5.7. Momentum and mass are conserved across the jump, and according to the second law of thermodynamics energy is converted into other forms such as heat and sound. Conditions on the upstream and downstream velocity and depth can be found such that Eq (5.18) apply upstream and downstream of the trajectory of the jump in the solution space. In other words, the trajectory of the jump marks the boundary where Eq (5.18) has a singularity but where conditions can be imposed at the trajectory to ensure continuity of the solution across it.

Discharge form

The momentum equation, Eq (5.15), with u replaced by Q/A, where Q is the discharge, and the continuity equation Eq (5.16), become

$$\begin{aligned} &\frac{\partial Q}{\partial t} + \frac{\partial}{\partial x}\left(\frac{Q^2}{A}\right) + gA\left[\frac{\partial h}{\partial x} + f\frac{Q|Q|}{4gRA^2}\right] = 0 \\ &\frac{\partial A}{\partial t} + \frac{\partial Q}{\partial x} = q \end{aligned} \tag{5.20}$$

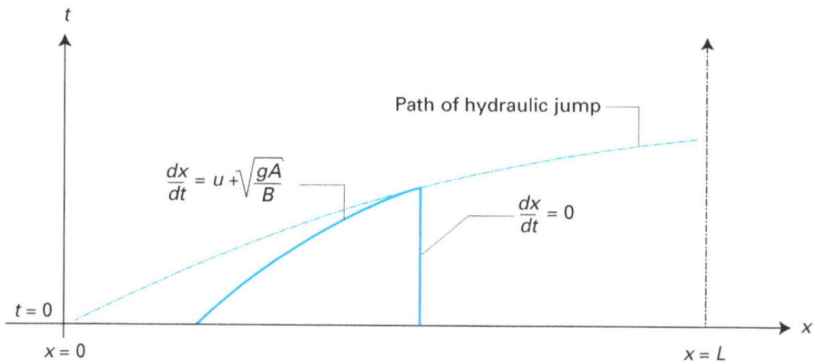

Figure 5.6 Path of hydraulic jump in solution space

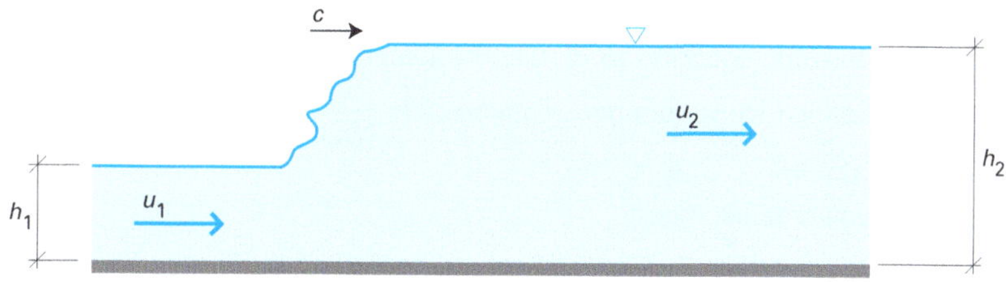

Figure 5.7 Hydraulic jump

Typically, in pipe applications for drainage, say, f is given by the Darcy friction factor, where

$$\frac{1}{\sqrt{f}} = -2\log_{10}\left(\frac{k_s}{14.8R} + \frac{2.51}{\text{Re}\sqrt{f}}\right) \tag{5.21}$$

where R is the hydraulic radius ($=A/P$) (m)
Re is the Reynolds number
k_s is the Nikuradse equivalent sand roughness height (mm)

k_s can be defined for the internal surface of a pipe, though it may be difficult to prescribe accurately when the pipe is in poor condition with cracks and badly aligned joints, or when the pipe is penetrated by tree roots. In such cases, k_s may need to become a calibration parameter. This is discussed further in Section 6.6.4.

When the Saint Venant equations are applied to rivers or coastal waters, modellers prefer to use simpler forms for f including Manning:

$$f = \frac{4n^2}{R^{2/3}} \tag{5.22}$$

or Chezy:

$$f = \frac{4}{C^2} \tag{5.23}$$

Here n is the Manning coefficient (taking a typical value of 0.03 for an urban river)
C is the Chezy coefficient (taking a corresponding value of about 35).

The conservation of volume equation normally has a source term q. Unlike lateral inflow along a river reach q is generally regarded as being zero along an urban conduit, and any lateral inflow is assumed to enter at the upstream boundary for flow in the conduit.

Approximate forms

The conservation of momentum equation has a number of terms which may take different orders of magnitude in alternative circumstances. For example, we have already discussed the case of gradually varying flow in open channels where the vertical acceleration of water particles is small compared to the acceleration due to gravity, resulting in the assumption of hydrostatic pressure in the flow. A careful scale analysis of the terms in Eq (5.20) shows that the momentum equation becomes

$$\varepsilon\gamma\frac{\partial Q}{\partial t} + \varepsilon\gamma\frac{\partial}{\partial x}\left(\frac{Q^2}{A}\right) + gA\left[\varepsilon\frac{\partial y}{\partial x} - s_0 + f\frac{Q|Q|}{4gRA^2}\right] = 0 \tag{5.24}$$

where ε and γ are inserted into the equation to denote the magnitudes of the associated terms and are such that ε is the characteristic surface slope number given by $\varepsilon \approx \left|\frac{\partial y}{\partial x}\right| / |s_0|$, and γ is the square of the Froude number given by $\gamma \approx |u^2B/gA| \equiv |Fr^2|$.

If $\varepsilon \approx 0$, the momentum equation becomes:

$$Q = \frac{1}{\sqrt{f}}A\sqrt{4gRs_0} \tag{5.25}$$

When combined with the conservation of mass equation this gives:

$$\frac{\partial Q}{\partial t} + \frac{1}{B}\frac{dQ}{dy}\frac{\partial Q}{\partial x} = \frac{1}{B}\frac{dQ}{dy}q \tag{5.26}$$

Here B is the surface width (m).

Eq (5.26) describes a kinematic wave, which translates with speed $c(Q) = (1/B)dQ/dy$ downstream but without any change to the peak (assuming there is no lateral inflow). If we assume that $0 < \varepsilon << 1$ then in place of the kinematic wave equation we obtain:

$$\frac{\partial Q}{\partial t} + c(Q)\frac{\partial Q}{\partial x} + \varepsilon c(Q)\frac{\partial}{\partial t}\left(\frac{a(Q)}{c^2(Q)}\frac{\partial Q}{\partial x}\right) = c(Q)q - \varepsilon c(Q)\frac{\partial}{\partial t}\left(\frac{Qq}{2gAs_0}\right) + O(\varepsilon^2) \tag{5.27}$$

where $a(Q)$ is termed the attenuation parameter.

$$a(Q) = \frac{Q}{2s_0B}\left\{1 - \frac{B}{gA}\left(c - \frac{Q}{A}\right)^2\right\} \tag{5.28}$$

Eq (5.27) is the analytical equation for the Muskingum-Cunge method of flood routing. The second derivative term in this equation modifies the kinematic wave, especially in giving a decrease in the peak discharge. This attenuation is induced by gravity, which tends to dampen the wave disturbance. If $\gamma \approx 0$ then we obtain the non-inertial equations for a uniform channel:

$$\begin{aligned}
&\frac{\partial Q}{\partial t} + c(Q,h)\frac{\partial Q}{\partial x} - a(Q,h)\frac{\partial^2 Q}{\partial x^2} = 0 \\
&B\frac{\partial h}{\partial t} + \frac{\partial Q}{\partial x} = q \\
&\frac{\partial h}{\partial x} + \frac{Q^2}{K^2} = 0 \\
&c(Q,h) = \frac{Q}{BK}\frac{dK}{dh} - \frac{a(Q,h)}{B}\frac{dB}{dh} \\
&a(Q,h) = \frac{K^2}{2BQ}
\end{aligned} \tag{5.29}$$

Note that the equation for the discharge is an advection–diffusion equation, namely a mixed hyperbolic-parabolic equation, which occurs in a variety of other circumstances, such as the movement of pollutants in water and the diffusion and advection of vorticity away from a fixed boundary.

5.2.4 Boundary conditions for pipe flow

Where the Saint Venant equations are applied to a pipe it is important to establish the appropriate boundary conditions for the solution space. Given that the pipe has a bed gradient $s_0 \neq 0$ then we can define upstream and downstream nodes for the pipe. Because of the hyperbolic nature of the two equations, and they have first order derivatives in x and t, they need four boundary conditions: one upstream and one downstream (for sub-critical flow) or two upstream boundary conditions (for super-critical flow), and two initial conditions along the pipe. Note that Eq (5.27) only needs one upstream condition and one initial condition, whereas Eq (5.28) needs an upstream condition and a downstream condition as well as an initial condition. At the upstream boundary the flow can be sub-critical (a) or critical (b); see Figure 5.8. The water level in the upstream manhole can be above the outgoing soffit, but where the pipe is not flowing full air may be entrained in the form of bubbles and brought in from the manhole. Additional energy is therefore lost as the flow is forced to contract into the pipe downstream.

Similar conditions apply at the downstream boundary; see Figure 5.9. In this case the flow in the pipe can be sub-critical or super-critical. If it is sub-critical there can be a critical flow point before the outflow reaches the water level in the downstream manhole (a), or alternatively, the water level is high enough to drown out the critical flow condition (b). If the flow is super-critical in the pipe then the flow at the downstream boundary will be at normal depth and there may or may not be a hydraulic jump in the manhole depending on the water level there (c). There is one other condition to note, and that is the case when the discharge is greater than the pipe full discharge. In this case the downstream boundary condition is such that the water level is at or above the outgoing soffit; see (d) and (e).

Note that head-losses can occur both at the upstream and downstream boundaries. There is generally a contraction of flow into the pipe at the upstream boundary and an expansion of flow at the downstream boundary. These head-losses need to be included in the analysis, and are defined by:

$$\Delta H = K\frac{u^2}{2g} \tag{5.30}$$

where K is an energy loss coefficient.

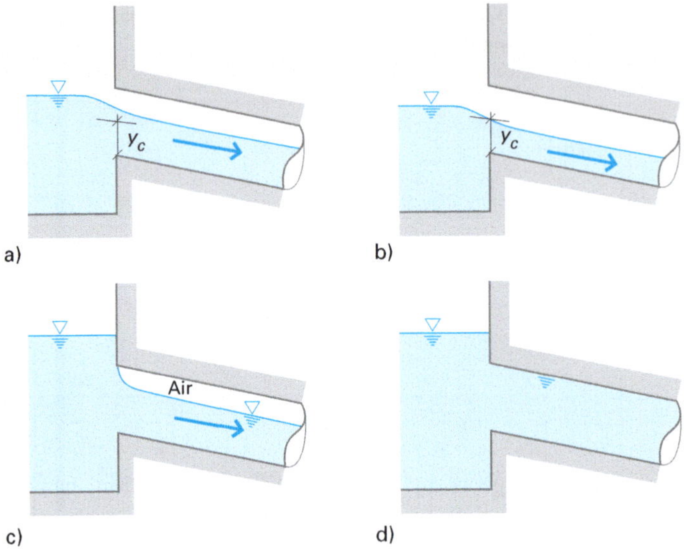

Figure 5.8 Upstream boundary conditions for flow in a pipe (y_c is the critical depth) (after Kellagher, 1993)

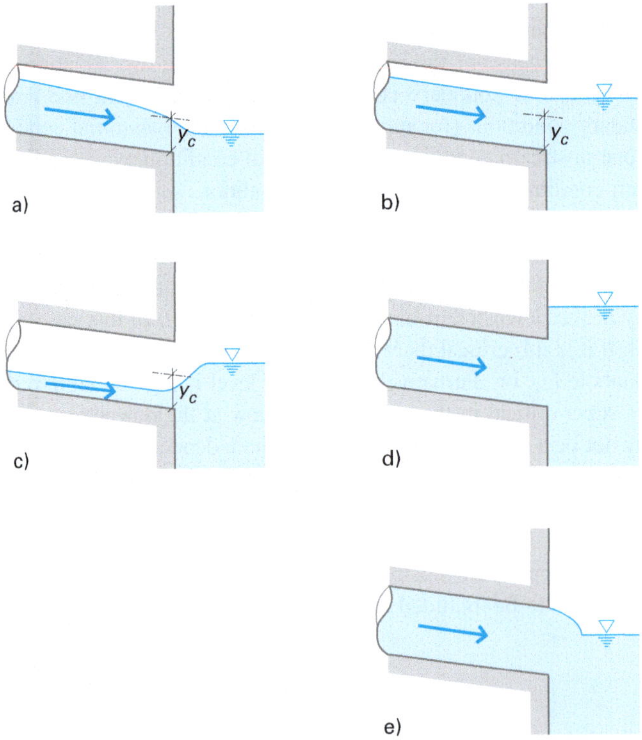

Figure 5.9 Downstream boundary conditions for flow in a pipe (y_c is the critical depth) (after Kellagher, 1993)

See Figure 5.10. The actual headlosses that occur across a manhole vary according to the shape of the manhole, any benching designed to reduce headlosses, the alignment of the incoming and outgoing pipes, shaping of the entrance to the pipe, the height of the water level above the outgoing soffit, possible air entrainment, and so on. In other words, there is a large combination of possible factors affecting the calculation of the headloss at the upstream and downstream ends of a pipe. A considerable number of laboratory experiments have been done to identify the headloss coefficients for ranges of different manholes, but the experiments are by no means complete; see Yen, 2004.

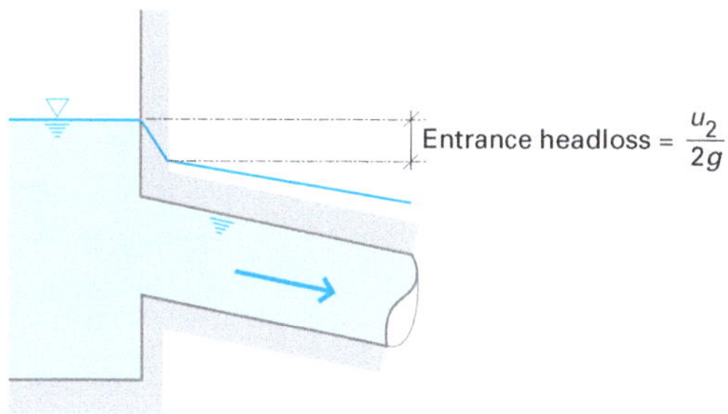

Figure 5.10 Headloss at a manhole

5.2.5 Pressurised flow

The pipe can flow partially and completely full. When the transition between free surface and pressurized flow takes place at some point along the pipe, there are a number of possibilities; see Figure 5.11. Of the 10 options, three of them: (e), (f) and (h), cannot exist under steady flow conditions.

The transition between free surface and pressurised flow needs to be considered separately. Assume that the water surface is parallel to the bed of the pipe, that is, normal depth flow applies. The relationship between the discharge and the water level for free surface flow in a circular pipe is as shown in Figure 5.12. Here, the proportional discharge and velocity are defined in terms of the pipe full discharge and velocity. An interesting fact is that the velocity takes the same value at the middle depth in a circular pipe as at the top (soffit/crown) of the pipe. As the discharge is increased (which is difficult to conceive in practice because we are talking about steady flow) then the water depth must 'jump' and fill the pipe suddenly. Sufficient head has then to be built up at the upstream end of the pipe in order to drive the flow through the pipe. Similarly, as the discharge is decreased from a condition that the pipe is fully surcharged the water level suddenly 'drops'. This is descriptive of a hysteresis effect; see Figure 5.13. The pipe completely surcharged is shown in Figure 5.14. In practice it is assumed that the depth-discharge relationship is unique and follows the rising limb of the hysteresis loop. Note that when the flow into a pipe is increasing such that the discharge exceeds the pipe full capacity there will be a shock travelling along the pipe downstream and the water level in the upstream manhole increases rapidly. Provided the discharge keeps on increasing (or at least not decreasing) the shock travels quickly down the pipe and the whole pipe becomes surcharged. Examination of the time series for the water level in the upstream manhole can reveal a sharp oscillation, which is made complicated by air entrainment and the trapping of pockets of air in the pipe. Surcharging can also occur when the water level in the downstream manhole increases above the outgoing soffit. There is backup into the pipe.

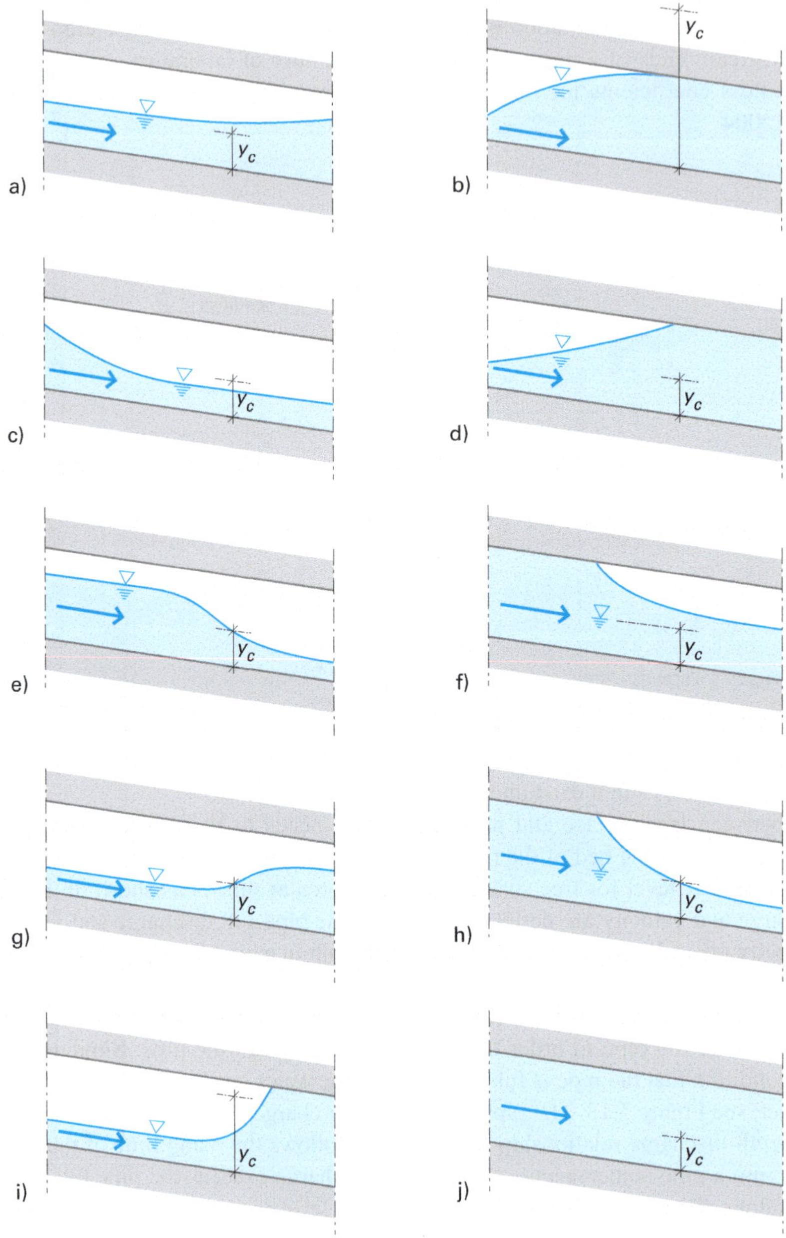

Figure 5.11 Free-surface and pressurised flow conditions in a pipe (y_c is the critical depth) (after Kellagher, 1993)

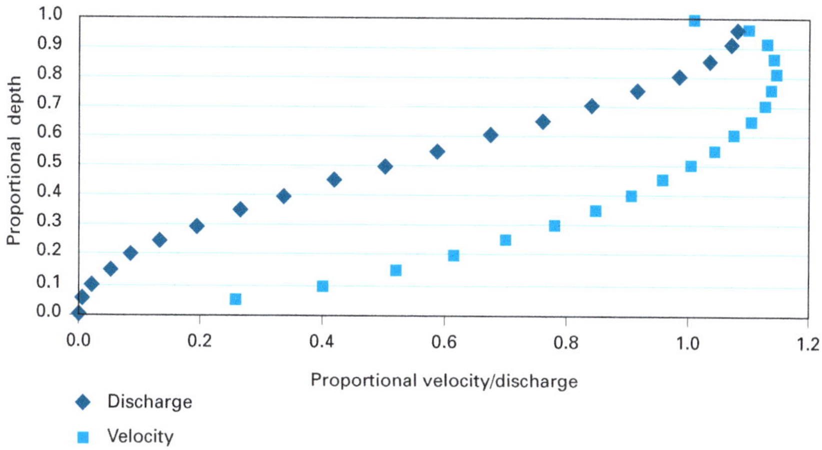

Figure 5.12 Proportional discharge and velocity curves for a circular pipe

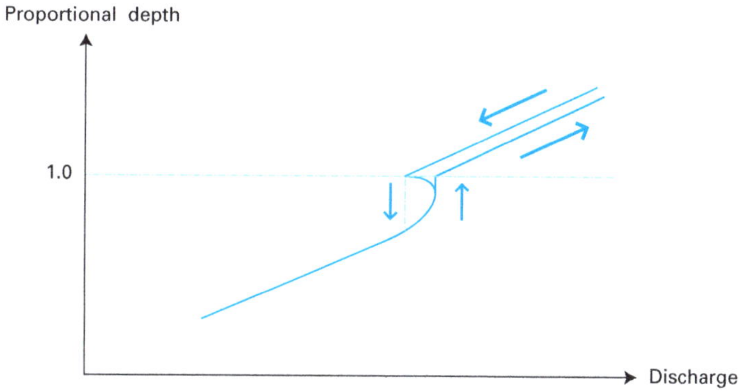

Figure 5.13 Hysteresis effect between free-surface and pressurised flow in a pipe

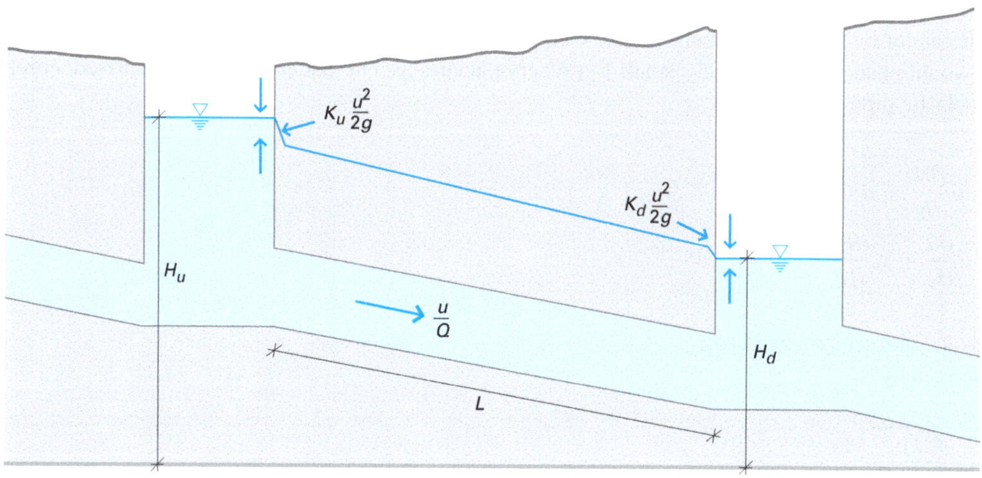

Figure 5.14 Flow in a surcharged pipe (after Kellagher, 1993)

The issue now arises about how to switch between the two modes: free surface and pressurised flow, in a pipe. Free surface flow can be represented by the 1D Saint Venant equations. Assuming similar equations apply to the pressurised flow along a pipe, they become:

$$\begin{aligned} &\frac{\partial u}{\partial x} = 0 \\ &\frac{\partial u}{\partial t} + \frac{g}{L}(H_k - H_i) + \left(\frac{f}{D} + \frac{K_u + K_d}{L}\right)\frac{u|u|}{2} = 0 \end{aligned} \tag{5.31}$$

where L is the length of the pipe (m)

H_k is the water level in the downstream node (m)

H_i is the water level in the upstream node (m)

K_u is the headloss coefficient upstream

K_d is the headloss coefficient downstream

D is the pipe diameter (m)

The dependent variable u is independent of x. This would not matter if the flow in the pipe always switches from having a free surface along the whole length of the pipe to being pipe full at the next instant. But this does not happen: at any one time part of the pipe can be surcharged and the flow in the remainder may have a free surface. Although surcharging due to the discharge exceeding the capacity of a closed conduit is very significant in identifying throttles in a network, conduits also surcharge because of backup from downstream. In this case a conduit can have the stable situation of a free surface flow upstream and pressurised flow downstream. (A stable situation with pressurised flow upstream and free surface flow in the pipe downstream cannot occur).

Preissmann slot

When it comes to solving the two categories of equation numerically, the solution algorithm has to be switched between free-surface and pressurised flow. A considerable amount of book keeping has to be done to keep track of the segments of the pipe that are surcharged or not. To get around this problem, Preissmann (Cunge and Wegner, 1964) came up with the ingenious idea of assuming that the flow always has a free surface by assuming a conceptual slot of small width in the top of the pipe; see Figure 5.15.

A major issue is the width of the slot. Obviously it has to be small to reproduce what happens in a surcharged pipe, but it cannot be too small, otherwise the Courant number would be very large and as a consequence the time step would need to be extremely small to preserve accuracy. The equations for pressurised, compressible flow in an elastic pipe are:

$$\begin{aligned} &\frac{\partial H}{\partial t} + u\frac{\partial H}{\partial x} + u\sin\theta + \frac{a^2}{g}\frac{\partial u}{\partial x} = 0 \\ &\frac{\partial u}{\partial t} + u\frac{\partial u}{\partial x} + g\frac{\partial H}{\partial x} + f\frac{u|u|}{D} = 0 \end{aligned} \tag{5.32}$$

where H is the piezometric head (m)

and a is defined by

$$a = \sqrt{\frac{K/\rho}{1 + \frac{D}{e}\frac{K}{E}}} \tag{5.33}$$

where K is the fluid bulk modulus ($kg/m/s^2$)
e is the thickness of the pipe wall (m)
E is the elastic modulus ($kg/m/s^2$)

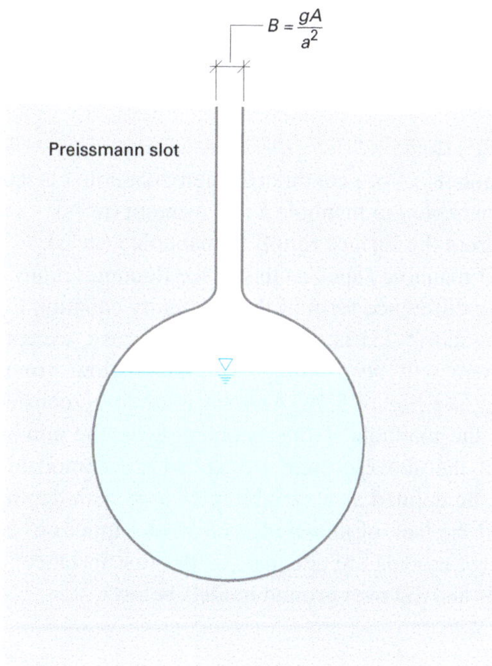

Figure 5.15 Preissmann slot

See Chaudhry (1987). Pressurised flow is well known to produce a rapid transmission of disturbances along a conduit related to the speed of sound in water. Therefore in marked contrast to the propagation speed of disturbances under free surface flow conditions, the propagation of disturbances under pressurised flow in a 1 m circular conduit 100 m long can be effectively instantaneous, whereas it may take 60 s for a disturbance under free surface flow to travel along the same pipe.

The unsteady flow equations relating transient flow of a compressible fluid within an elastic pipeline can be seen to be analogous to the Saint Venant equations with the piezometric head replacing the depth and the surface width, B, replaced by the expression:

$$B = \frac{gA_{pf}}{a^2} \tag{5.34}$$

where A_{pf} is the pipe full cross-sectional area (m^2)
a is the velocity of the pressure wave in water (m/s)

Eq (5.34) is derived from the expression for the dynamic wave speed $c = \sqrt{gA/B}$. Typically, values of a are greater than 100 m/s. Note that there is a smooth transition from the pipe to the slot. This is important in damping down the effect of the shock that is known to propagate along the pipe when the discharge is increasing beyond pipe full. Care has also to be taken so that the storage in the slot is compensated for by adjusting the storage in the manholes.

5.2.6 Manhole storage

Normally, in a distribution network, there is no storage at the nodes representing the junctions between the pipes. In the collection systems however, manholes are provided for inspection at all junctions. These provide a certain amount of storage in the system. The water levels in the manholes change in time according to the net inflow or outflow:

$$\frac{dV_j}{dt} - \sum Q_{ij} + Q_{jk} - Q_{j,in} + Q_{j,out} = 0 \tag{5.35}$$

where V_j is the volume in manhole j (m^2)

Q_{ij} is the discharge into manhole j from conduit ij, where manhole i is upstream of manhole j (m^3/s)

Q_{jk} is the discharge from manhole j to manhole k downstream (m^3/s)

$Q_{j,in}$ is the inlet discharge from the surface runoff to manhole j (m^3/s)

$Q_{j,out}$ is the discharge out of manhole j such as to surface flooding, if any (m^3/s)

Note that Eq (5.35) is the finite difference form of the continuity equation Eq (5.16)

The manhole storage equation Eq (5.35) is straightforward to use, except that the volume is normally a function of depth. Generally there will be a storage chamber below ground with a narrow access shaft leading up to the ground surface; see Figure 5.16. Another issue to be considered at this point is the storage of floodwater discharging from the manhole if the surcharging in the conduits is extreme. It is customary therefore, to include a model of the above-ground storage to accommodate excess water issuing from the manhole. This can, for example, be defined as a variable plan area with depth, though in practice the scarcity of data on the ground profile and the lack of knowledge on what happens to emerging flood water means that fairly simple above ground storage models are adequate in the first instance; see Figure 5.16. We discuss the integration of 1D network models and 2D over ground models below.

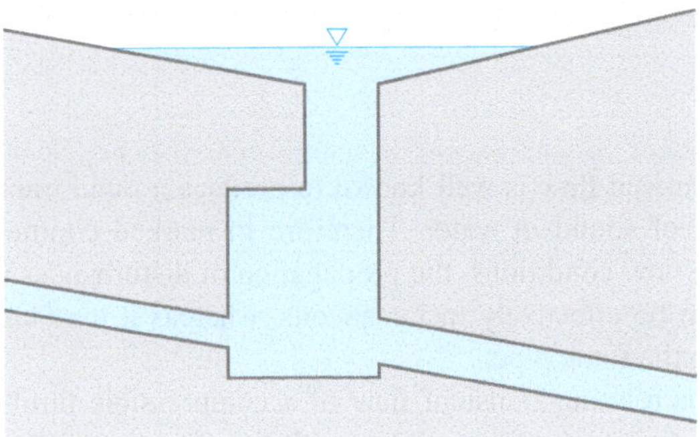

Figure 5.16 Manhole storage

5.2.7 Ancillary structures

Besides pipes and manholes there is a range of other structures, termed ancillary structures, that are associated with urban drainage systems. One of the most common structures is the weir, which comes in different forms such as a side weir, leaping weir, and the traditional weir across the direction of flow; see Figure 5.17. The basic equation for modular flow over a weir normal to the flow is given by:

$$Q = C_d L \sqrt{g} h^{3/2} \tag{5.36}$$

where C_d is the discharge coefficient
L is the length of the weir (m)
h is the height of the water level upstream above the weir crest (m)

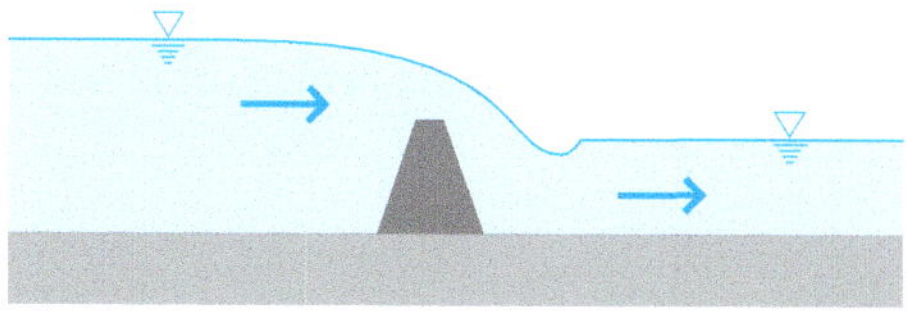

Figure 5.17 Weir

Eq (5.36) assumes that the flow is modular, that is, the flow over the weir has a critical flow at some point. This implies that the water level downstream has no influence on the water level upstream and therefore on the flow over the weir. If the water level downstream is sufficiently high then the flow may no longer be modular and the discharge over the weir is a function of the downstream water level as well. In this case the discharge given by Eq (5.36) is modified by a factor that is a function of the height of the downstream water level above the weir crest, as well as the height of the upstream water level above the weir crest. The equation for a side weir is more complicated than Eq (5.36). See Butler and Davies (2000) for further details.

Another important ancillary structure is the sluice or vertical gate; see Figure 5.18. Here the discharge through the sluice is given by

$$Q = C_d L h_d \sqrt{g} h^{1/2} \tag{5.37}$$

where C_d is the discharge coefficient
L is the width of the sluice (m)
h_d is the height of the opening of the sluice (m)
h is the height of the upstream water level above the top of the gate opening (m)

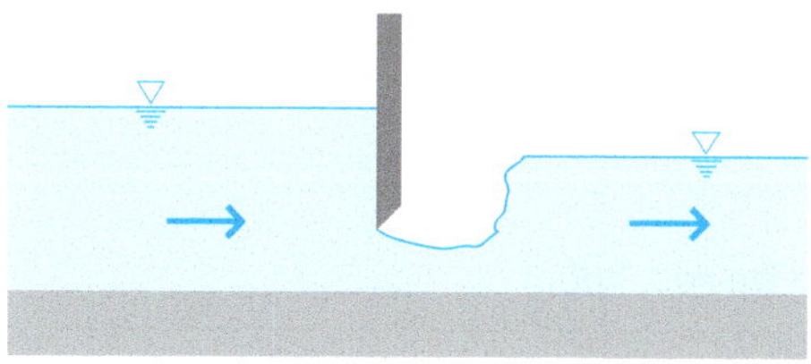

Figure 5.18 Sluice

Note that the water level downstream is such that the flow immediately downstream of the sluice is supercritical. If the downstream water level is sufficiently high then the downstream water level drowns out the sluice reducing the discharge through the structure. Again, the discharge given by Eq (5.37) is reduced by a factor which reflects the magnitude of the downstream water level above the top of the gate opening as well as the height of the upstream water level above the top of the gate opening.

Orifices (or valves) form yet another common structure. The basic equation for an orifice is

$$H_u - H_d = k_v Q^2 \tag{5.38}$$

where H_u is the head upstream (m)
H_d is the head downstream (m)
k_v is the headloss coefficient for the flow through the orifice (s^2/m^5)

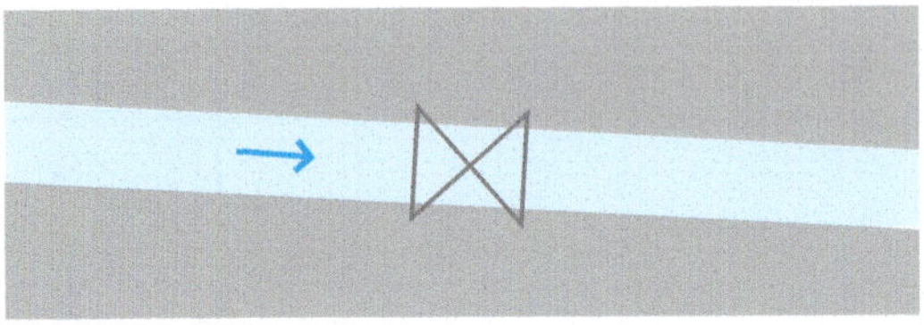

Figure 5.19 Valve

There is a large range of different pumps which are normally arranged as one or more pumps in a pumping station; see Figure 5.20

$$h_d - h_u = k_p Q^2 \tag{5.39}$$

where k_p is the discharge coefficient for the pump (s^2/m^5)

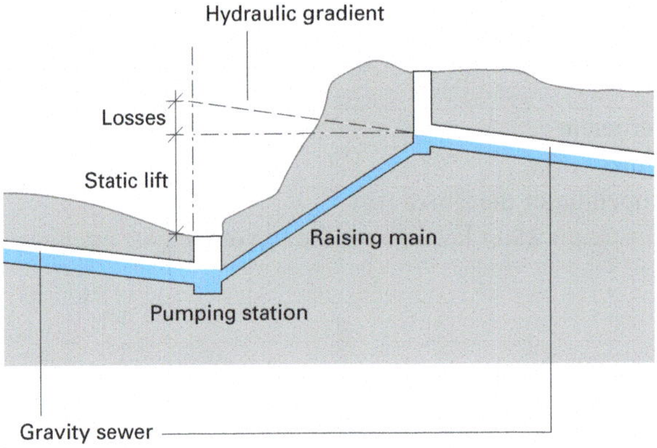

Figure 5.20 Pump (after Butler and Davies, 2000)

5.2.8 Incompressible pressurised flow in inelastic pipes

Returning to the equations for pressurised flow in pipes, if we accept that the fluid is incompressible and the pipes are inelastic then we obtain:

$$\begin{aligned} &\frac{\partial u}{\partial x} = 0 \\ &\frac{\partial u}{\partial t} + \frac{g}{L}(H_k - H_j) + \frac{f}{D}\frac{u|u|}{2} = 0 \end{aligned} \tag{5.40}$$

where H_j and H_k are now the piezometric heads (m) at the 'upstream' and 'downstream' ends of the pipe, namely nodes j and k.

Eq (5.40) applies to unsteady pressurised flow in pipes. However, if there is no storage at the 'upstream' and 'downstream' nodes as is the case in a network of distribution pipes then changes at the boundaries of a distribution network, namely the inflow and demand points, are transmitted instantaneously throughout the network. In this case, Eq (5.40) resolves into an equation for steady flow.

$$\begin{aligned} &H_k - H_j + \frac{1}{2gA_{kj}^2}\left(\frac{fL_{kj}}{D_{jk}} + K_{jk}\right)Q_{kj}|Q_{kj}|^n = 0 \\ &\sum_j Q_{kj} - d_k = 0 \end{aligned} \tag{5.41}$$

where K_{jk} is a headloss coefficient for minor losses along the pipe due to bends and fittings

d_k is the demand and the losses (m^3/s)

The second equation in Eq (5.41) relates the inflows to the node from all connecting pipes to the demand D_k at node k. Eq (5.41) greatly simplifies the computational process for water distribution networks. Note that a similar equation to Eq (5.41) is obtained for the headloss across a pump, but the headloss is proportional to Q^{n+1} where n is an exponent that may be different from 1.

5.3 MODELLING WATER QUALITY

The modelling of water quality in pipes and natural channels is particularly difficult to quantify and regulate because of the diverse nature of the pollutant inflows, the development of pollutants such as bio-films in water distribution pipes, and the intermittent and varied nature of overflow discharges. Whereas the chemical and biological processes are very complex, simplifications have to be made in order to model the pollutant behaviour with some level of confidence.

5.3.1 Sediment transport

It is assumed that we can distinguish between sediment and chemical and biological pollutants. Fine sediment occurs in two layers: an active layer in which the sediment is un-consolidated and available for pick-up when the bed shear stress exceeds a certain threshold, and a passive layer with more permanent deposits on the bed of the sewer. Given a sufficient velocity fine sediment deposited in the active layer during quiescent periods can be brought into suspension. This is related to what is termed the 'first foul flush'. The passive layer can contribute to a decrease in the pipe's capacity, though in most cases the passive layer is assumed to have a uniform depth along the pipe. The capacity of the flow in a conduit to transport fine-grained or low-density particles in suspension can be calculate from the Ackers (1993) equation, which is based on the Ackers-White sediment transport equation (Ackers and White, 1973) for alluvial channels adapted to apply to conduits. The transport capacity can be expressed as the volumetric sediment concentration as the ratio of the flow rate of sediment to the flow rate of the water:

$$C_v^{\max} = J\left(\frac{W_e R}{A}\right)^{\alpha}\left(\frac{d_{50}}{R}\right)^{\beta}\lambda_c^{\gamma}\left[\frac{|u|}{\sqrt{g(s-1)R}} - K\lambda_c^{\delta}\left(\frac{d_{50}}{R}\right)^{c}\right]^{m} \tag{5.42}$$

where W_e is the effective width of the deposited sediment bed (m)
λ_c is the composite Dacry-Weisbach friction factor of the flow
A is the cross-sectional area (m^2)
R is the hydraulic radius (m)
u is the velocity of the water (m/s)
s is the specific gravity
d_{50} is the mean particle size (mm)
α, β, γ, δ, c and m are constants
The coefficients depend on the sediment size:

$$D_{gr} = d_{50}\left[\frac{g(s-1)}{\nu^2}\right] \tag{5.43}$$

where ν is the kinematic viscosity of the water

See May (1993). For fully suspended sediment transport, D_{gr} should not be greater than unity.

The effective width W_e is related to the actual width, W_b of the sediment bed by

$$W_e = \left[0.2 + \frac{10}{3}\left(\frac{y_b}{D} - 0.01\right)\right]W_b \tag{5.44}$$

where D is the diameter of the pipe (m)
y_b is the depth of the deposited bed (m)
W_e has a maximum value of $0.5W_b$ and a minimum value of $0.02D$

It is assumed that the suspended sediment is well mixed with the water and that the concentration C is advected by the flow with negligible dispersion according to

$$\frac{\partial C}{\partial t} + u\frac{\partial C}{\partial x} = 0 \tag{5.45}$$

The model described here does not include the transport of coarser granular sediment as bed-load, but the hydraulic effects of the deposited materials can be taken into account. The roughness of the bed can be determined as a function of the shear stress using a method devised by May (1993) for which $d_{50} = 1.0$ mm and $s = 2.6$. The composite friction factor λ_c takes into account the different roughnesses of the pipe walls and the deposited sediment bed according to their relative wetted perimeters.

The sediment will be suspended and transported according to Eq (5.42). If the suspended load exceeds the transporting capacity of the flow, it is assumed that the particles settle at their fall velocity onto the invert of the pipe or any permanent layer of deposited sediment.

5.3.2 Chemical pollutants

Pollutants such as Total Suspended Solids (TSS), Biochemical Oxygen Demand (BOD), and Total Nitrogen (TN) are either dissolved in the water or attached to the sediment according to a potency factor defined separately for each pollutant. (Note that Ammonia only appears in the dissolved state). Consequently, a pollutant is advected with the water and diffused according to the turbulent structure of the flow. This diffusion is made up of molecular diffusion (Brownian movement), turbulent diffusion and dimensional effects. The net effect is a degree of mixing of the pollutant. In addition, there are processes of dilution, flocculation as electro-chemical forces bring sediment particles together, deposition of sediment when the shear stress is sufficiently low, erosion when the shear stress exceeds a threshold, thermal effects and re-aeration. The dynamic concentration can be described by the advection-diffusion equation:

$$\frac{\partial C}{\partial t} + u\frac{\partial C}{\partial x} - D\frac{\partial^2 C}{\partial x^2} = S \tag{5.46}$$

where C is the concentration of pollutant (kg/m^3)
u is the speed of the water (m/s)
D is the diffusion coefficient (s^2/m)
S is the source term which includes the range of processes above (kg/m^3/s)

The source term may take a variety of forms such as

$$\begin{aligned} &(i)\ S = KC \\ &(ii)\ S = KC^n \\ &(iii)\ S = \frac{KC}{a + C} \end{aligned} \tag{5.47}$$

(iii) is the Michaelis-Menton form.

The biochemical processes include aerobic and anaerobic oxidation, nitrification, adsorption and desorption of metals and other toxic compounds. In addition, we may have microbiological processes, including the growth and die-off of species and toxicant accumulation. Oxygen balance in sewers is given by:

$$\frac{\partial C_O}{\partial t} + u\frac{\partial C_O}{\partial x} - D\frac{\partial^2 C_O}{\partial x^2} = K_{LA}(C_{O,S} - C_O) - (r_w + r_b + r_s) \tag{5.48}$$

where $C_{o,s}$ is the saturation dissolved oxygen concentration (kg/m^3)
C_o is the dissolved oxygen concentration (kg/m^3)
K_{LA} is the volumetric re-aeration coefficient (1/s)
r_w is the oxygen consumption rate in the bulk water (kg/m^3/s)
r_b is the oxygen consumption rate in the biofilm (kg/m^3/s)
r_s is the oxygen consumption rate in the sediment (kg/m^3/s)

There are various important chemical processes involving oxygen, nitrogen, phosphorus and carbon. Oxygen is dissolved through the free surface from the atmosphere into the water and is diffused through the flow. There is a saturation level for oxygen in the water. In sewers Pomeroy and Parkhurst (1972) showed that

$$K_{LA} = 0.96\left(1 + 0.17\left(\frac{u^2}{gd_m}\right)\right)\gamma(s_f u)^{3/8}\frac{1}{d_m} \tag{5.49}$$

where u is the mean velocity (m/s)
d_m is the hydraulic mean depth (m)
γ is the temperature correction factor (1.00 at 20°C)
s_f is the hydraulic gradient

There are a variety of different chemical and ecological processes going on in a river or pipe. They have different spatial and temporal scales; see Figure 5.21.

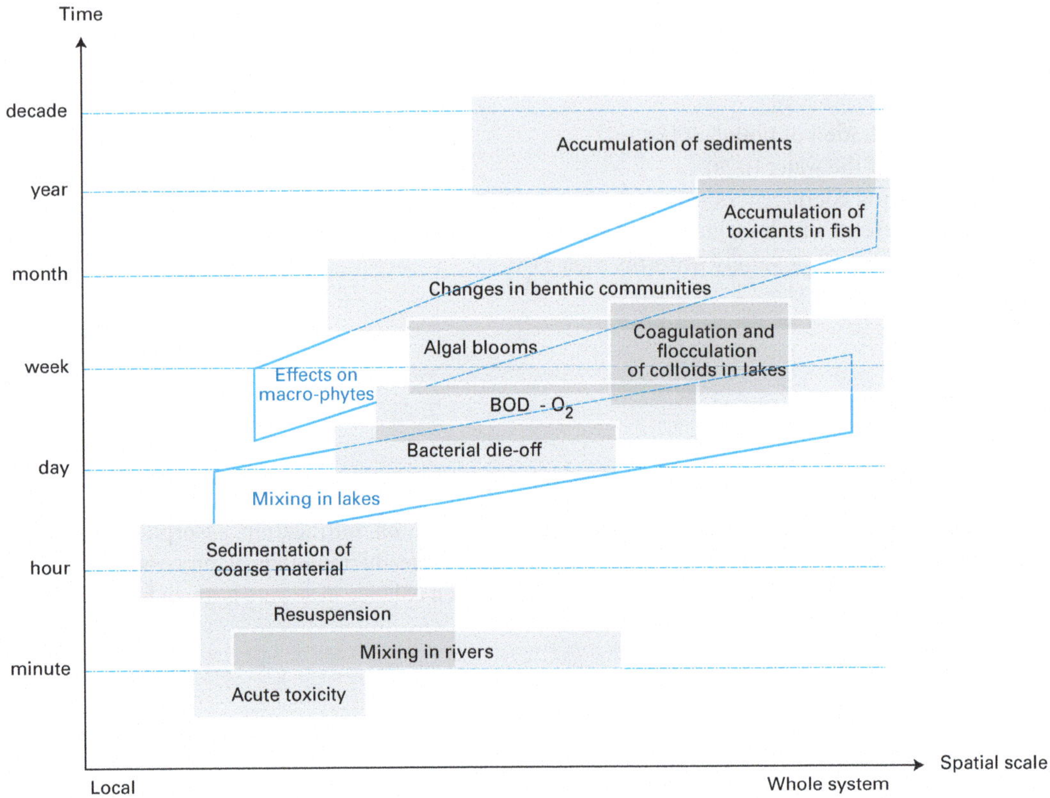

Figure 5.21 Temporal and spatial scales for water quality processes (adapted from Butler and Davies, 2000)

Similarly there is a wide range of biological processes; see Figure 5.22. Typical ordinary differential equations for these processes were developed by Lotka (1925) and Volterra (1926) such as:

$$\begin{aligned} \frac{dX}{dt} &= \alpha X - \beta XY \\ \frac{dY}{dt} &= \gamma\beta XY - \delta Y \end{aligned} \tag{5.50}$$

where X is the concentration of the prey species (1/l)

Y is the concentration of the predator species (1/l)

α is the prey species' growth rate (1/s)

β is the predation rate of the predator species on the prey species (l/s)

γ is the assimilation efficiency of the predator species
δ is the mortality rate of the predator species (1/s)

Obviously, the models for both the chemical and biological processes can be extremely complex. In practice the processes have to be limited to those that are perceived as being important.

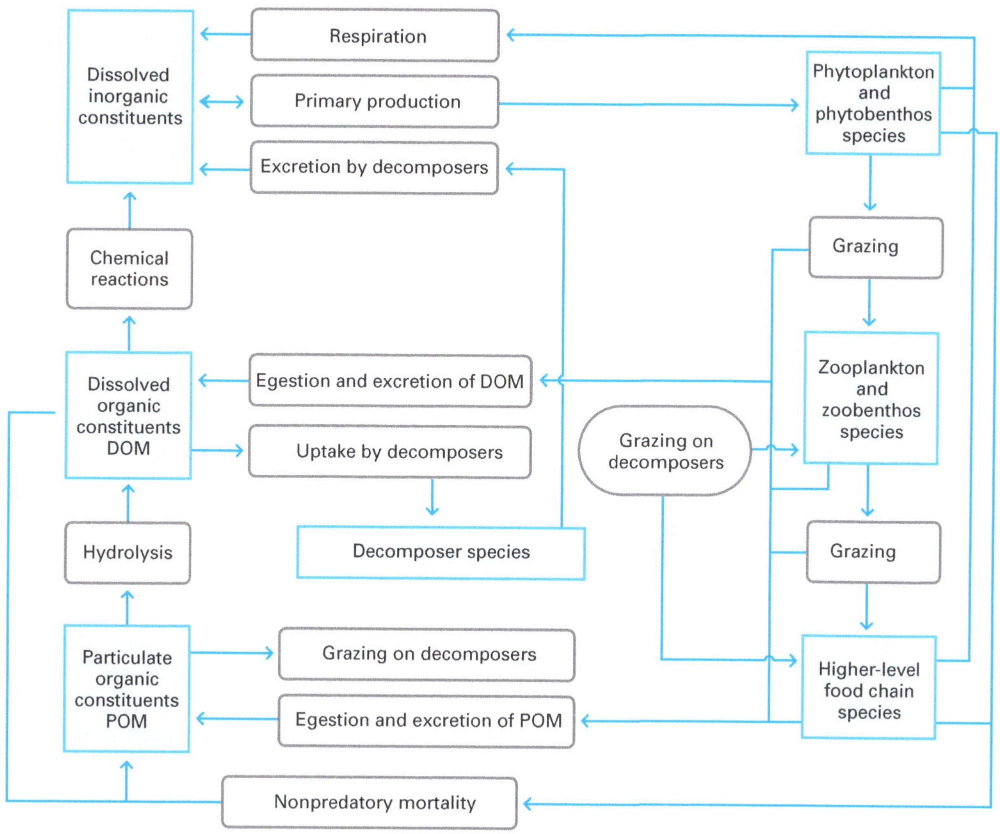

Figure 5.22 Ecological processes

5.4 GROUNDWATER

The Richards equation for flow in an aquifer is given by:

$$S_s \frac{\partial h}{\partial t} - \frac{\partial}{\partial x}\left(k_x \frac{\partial h}{\partial x}\right) - \frac{\partial}{\partial y}\left(k_y \frac{\partial h}{\partial y}\right) - \frac{\partial}{\partial z}\left(k_z \frac{\partial h}{\partial z}\right) = -G \tag{5.51}$$

where S_s is specific storage (volume that can be released from a unit volume of a saturated aquifer by a unit reduction in hydraulic head)
h is the groundwater potential (m)
k_x, k_y, k_z are the hydraulic conductivities (m^2/s)
G is the source term (m/s)

The assumptions made in the derivation of Eq (5.51) include:

- the aquifer material is incompressible,
- the water is of constant density (incompressible),
- any external loads on the aquifer (e.g., overburden, atmospheric pressure) are constant,
- the groundwater is flowing slowly (Reynolds number less than unity), and
- the hydraulic conductivity (k) is an isotropic scalar.

The two-dimensional form of the equation for an unconfined aquifer is

$$S_y \frac{\partial h}{\partial t} - \frac{\partial}{\partial x}\left(kh\frac{\partial h}{\partial x}\right) - \frac{\partial}{\partial y}\left(kh\frac{\partial h}{\partial y}\right) = -N \tag{5.52}$$

where N is the recharge (m/s)

S_y is the specific yield of the aquifer

H is now the head (m)

The interaction between an aquifer and a river can be derived as:

$$q = 5\alpha kP(h - h_r) \tag{5.53}$$

where q is the lateral inflow (m^2/s)

α is a constant which is approximately 1.0

P is the wetted perimeter formed by the aquifer in contact with the river channel (m)

h_r is the stage in the river (m)

See Miles, 1985.

5.5 PIPE NETWORKS

Distribution and collection systems consist of a number of pipes connected in a network. A distribution network could resemble the branches of a tree (in two spatial dimensions) with the demand points at the ends of the branches. In fact, engineers prefer to provide interconnections between branches so that demands can be balanced from one part of the network to another; see Figure 5.23. Similarly, with a looped rather than a dendritic (tree like) structure engineers can readily investigate leakage from a part of a distribution network by having the ability to disconnect the part under investigation from the rest of the network using valves placed strategically.

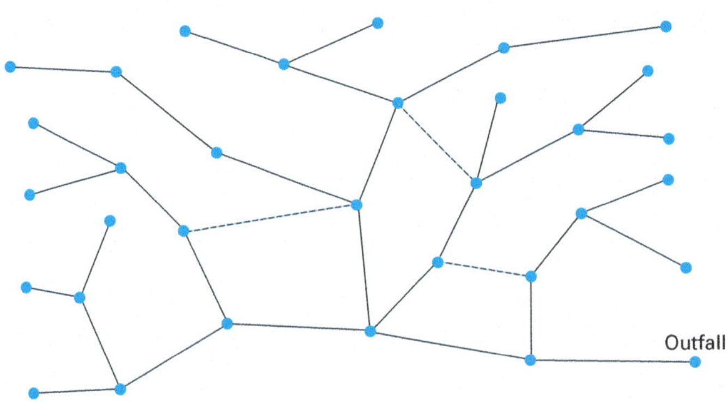

Figure 5.23 Looped network

A collection network is more suited to a dendritic structure because the aim of the drainage engineer is to design a system that drains under gravity, so far as that is possible. Connections between different branches may be introduced, especially in the larger networks, in order to balance the loading on the network generated by spatially varying rainfall-runoff, or to bypass pipes of limited capacity. Diversion structures are also common in sewerage networks, so that excess loads on particular parts of a network may be limited. For example, overflows are used to divert excess flows from entering a wastewater treatment works.

5.6 NUMERICAL SOLUTION OF THE SAINT VENANT EQUATIONS

The solution space of the 1D Saint Venant equations for a pipe occupies the domain in (x, t) space bounded by the lines $t = 0$, $x = 0$ and $x = L$. A numerical solution of the equations is based on a discretization of this domain. The discretization can be done using finite differences, elements or volumes in different coordinate systems (Cartesian, curvilinear, cylindrical, radial). A finite difference representation is the most popular method, though the finite volume technique is being used more widely.

There are a large number of different ways of forming a finite difference representation of the basic equations, but only a limited set provide a satisfactory solution. In particular, the solution should be stable (that is, it should not exhibit rapidly growing oscillations that have a period of twice times the time increment), be consistent with the original equations, and converge to the right solution as the space and time increments tend to zero (as can be investigated through what happens to the truncation error between the numerical and analytical equations). Also, the solution should be economical to obtain. In general, provided consistency and convergence are satisfied, attention is focused on the stability and economy of the scheme used. Note that almost all numerical solutions of simulation equations for water flow focus on finding values for the dependent variables (such as water level or discharge) in physical space for a time step rather than calculating the dependent variables for every time step at a succession of space steps. In other words, the solution is found throughout the physical space at successive time levels with increasing time.

There are two major categories of finite difference schemes: explicit and implicit. An explicit scheme calculates the dependent variables at the new time level solely in terms of the dependent variables at the old time level (or even at a previous time level or the half time level). The Lax-Wendroff scheme (see Morton and Mayers, 2005) is a well known explicit scheme; see Figure 5.25. To explore this scheme, consider the advection equation:

$$\frac{\partial u}{\partial t} + c\frac{\partial u}{\partial x} = 0 \tag{5.54}$$

where c is the advection speed (m/s)

The Lax-Wendroff scheme uses the following finite difference representations for the first order differentials:

$$\begin{aligned}\frac{\partial u}{\partial t} &\approx \frac{u_{j+1}^{n+1} - u_{j+1}^{n}}{\Delta t}\\ \frac{\partial u}{\partial x} &\approx \frac{u_{j+1}^{n} - u_{j-1}^{n}}{2\Delta x}\end{aligned} \tag{5.55}$$

See Figure 5.24. Suppose we take the Taylor expansion of $u(x,t+\Delta t)$ to second order:

$$\begin{aligned} u(x,t+\Delta t) &= u(x,t) + \Delta t\frac{\partial u}{\partial t} + \Delta t^2\frac{\partial^2 u}{\partial t^2} + O(\Delta t^3) \\ &= u(x,t) - \Delta t c\frac{\partial u}{\partial x} + \frac{1}{2}\Delta t^2\left[-\frac{\partial c}{\partial t}\frac{\partial u}{\partial x} + c\frac{\partial}{\partial x}\left(c\frac{\partial u}{\partial x}\right)\right] + O(\Delta t^3) \end{aligned} \tag{5.56}$$

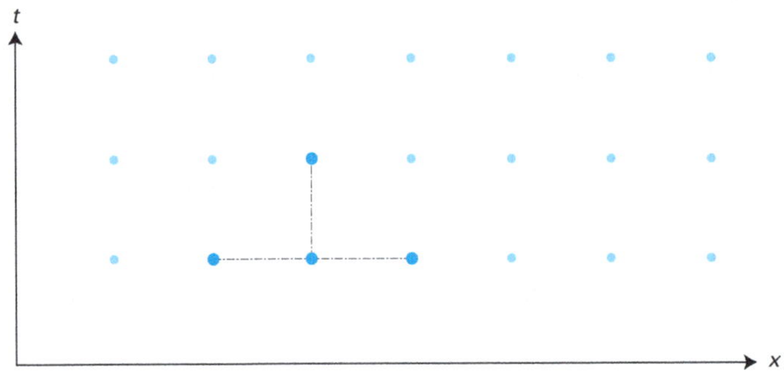

Figure 5.24 Lax-Wendroff explicit scheme

We can write the finite difference form as:

$$\begin{aligned} u_j^{n+1} &= u_j^n - \Delta t c_j^n\left(\frac{u_{j+1}^n - u_{j-1}^n}{2\Delta x}\right) - \frac{1}{2}\Delta t^2 (c_t)_j^n\left(\frac{u_{j+1}^n - u_{j-1}^n}{2\Delta x}\right) \\ &+ \frac{1}{2}\Delta t^2 c_j^n\left[(c_x)_j^n\left(\frac{u_{j+1}^n - u_{j-1}^n}{2\Delta x}\right) + c_j^n\left(\frac{u_{j+1}^n - 2u_j^n + u_{j-1}^n}{\Delta x}\right)\right] + O(\Delta t^3) \end{aligned} \tag{5.57}$$

If we now write $u_j^n = \alpha^n e^{i\beta j\Delta x}$ and assume that c is constant then

$$\begin{aligned} \alpha &= 1 - \frac{c\Delta t}{2\Delta x}(e^{i\beta\Delta x} - e^{-i\beta\Delta x}) + \frac{c^2\Delta t^2}{2\Delta x^2}(e^{i\beta\Delta x} - 2 + e^{-i\beta\Delta x}) \\ &= 1 + \frac{c\Delta t}{\Delta x}i\sin(\beta\Delta x) + \frac{c^2\Delta t^2}{\Delta x^2}(\cos(\beta\Delta x) - 1) \end{aligned} \tag{5.58}$$

Therefore

$$|\alpha| = \left[1 - \frac{c^2\Delta t^2}{\Delta x^2}(1 - \cos(\beta\Delta x))\right]^2 + \frac{c^2\Delta t^2}{\Delta x^2}\sin^2(\beta\Delta x) < 1 \tag{5.59}$$

The Lax-Wendroff scheme is unconditionally stable if $|\alpha| < 1$, that is, if $c\Delta t/\Delta x < 1$; otherwise, any perturbations in the solution will grow uncontrollably. The dimensionless expression $c\ \Delta t/\Delta x$ is termed the Courant number.

An implicit scheme, on the other hand, calculates the dependent variables at the new time level in terms of other adjacent, dependent variables at the new time level as well as dependent variables at the old time level. There are two well known implicit schemes.

5.6.1 6-point implicit scheme

One is the 6-point staggered grid scheme with h and u identified at alternate computational points along a pipe; see Figure 5.25. In this case the algorithmic equations are written as:

$$\begin{aligned} &u\frac{\partial u}{\partial t} - g\frac{\partial h}{\partial t} + \left(u^2 - g\frac{A}{B}\right)\frac{\partial u}{\partial x} + gus_f = 0 \\ &-\frac{A}{B}\frac{\partial u}{\partial t} + u\frac{\partial h}{\partial t} + \left(u^2 - g\frac{A}{B}\right)\frac{\partial h}{\partial x} - g\frac{A}{B}s_f = 0 \end{aligned} \tag{5.60}$$

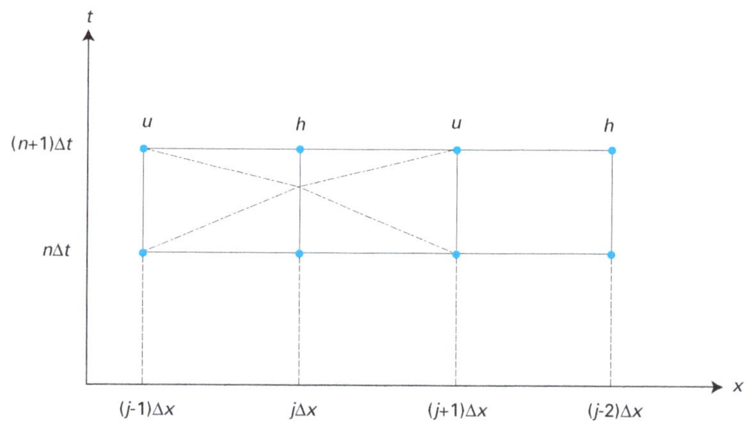

Figure 5.25 6-point scheme

The corresponding finite difference expressions for the derivatives are given by

$$\begin{aligned} \frac{\partial h}{\partial t} &\approx \frac{h_j^{n+1} - h_j^n}{\Delta t} \\ \frac{\partial h}{\partial x} &\approx \theta\frac{h_{j+2}^{n+1} - h_j^{n+1}}{\Delta x} + (1-\theta)\frac{h_{j+2}^n - h_j^n}{\Delta x} \\ \frac{\partial u}{\partial t} &\approx \frac{u_{j+1}^{n+1} - u_{j+1}^n}{\Delta t} \\ \frac{\partial u}{\partial x} &\approx \theta\frac{u_{j+1}^{n+1} - u_{j-1}^{n+1}}{\Delta x} + (1-\theta)\frac{u_{j+1}^n - u_{j-1}^n}{\Delta x} \end{aligned} \tag{5.61}$$

where h_j^n refers to the depth (m) at the point $(j\Delta x, n\Delta t)$
Δx is the space step (m)
Δt is the time step (s)
θ is the weighting factor for stability $(0.5 < \theta < 1.0)$

This scheme is consistent with a 1D subset of the standard 2D finite difference scheme on a rectangular grid; see Figure 5.26 and Abbott and Ionescu, 1967. The focus of the scheme is on evaluating the derivatives consistently near the centre point in terms of space and time. The space derivative is deliberately defined off centre and biased to the new time level according to the value of $\theta > 0.5$. This device introduces a small numerical diffusion that dampens down any oscillations that might occur, and therefore preserves the stability of the calculations. There are other devices that are introduced to deal with small depth flows and the tendency of the standard finite difference formulation for the friction term to introduce more than one solution. These details are discussed below.

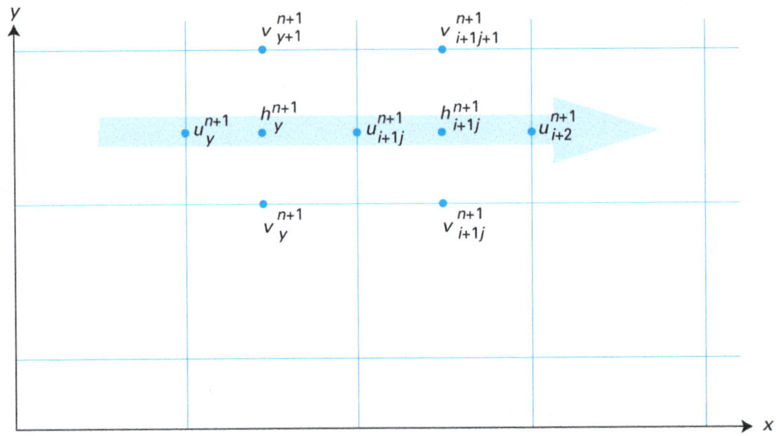

Figure 5.26 2D finite difference grid

5.6.2 4-point implicit scheme

The other major implicit scheme is the 4-point scheme with h and u defined at each of the computational points along a pipe; see Figure 5.27. In this case the finite difference expressions for the derivatives are given by:

$$\begin{aligned}
\frac{\partial h}{\partial x} &\approx \theta \frac{h_{j+1}^{n+1} - h_j^{n+1}}{\Delta x} + (1-\theta)\frac{h_{j+1}^{n} - h_j^{n}}{\Delta x} \\
\frac{\partial u}{\partial x} &\approx \theta \frac{u_{j+1}^{n+1} - u_j^{n+1}}{\Delta x} + (1-\theta)\frac{u_{j+1}^{n} - u_j^{n}}{\Delta x} \\
\frac{\partial h}{\partial t} &\approx \frac{h_j^{n+1} - h_j^{n}}{2\Delta t} + \frac{h_{j+1}^{n+1} - h_{j+1}^{n}}{2\Delta t} \\
\frac{\partial u}{\partial t} &\approx \frac{u_j^{n+1} - u_j^{n}}{2\Delta t} + \frac{u_{j+1}^{n+1} - u_{j+1}^{n}}{2\Delta t}
\end{aligned} \tag{5.62}$$

Again, $\theta > 0.5$ for the stability of the scheme. The same precautions in dealing with small depth flows and non-unique solutions as for the 6-point scheme are also needed here.

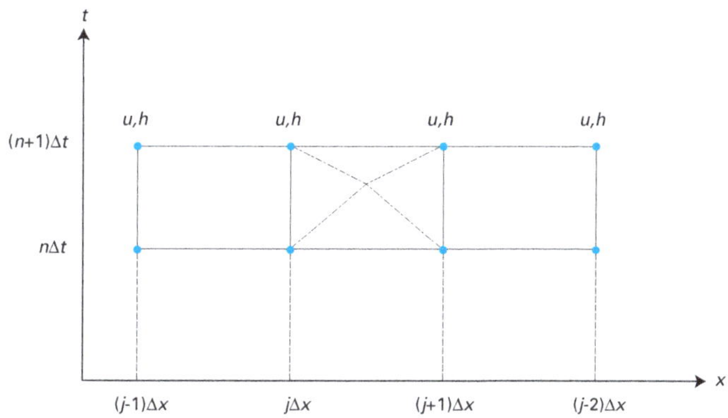

Figure 5.27 4-point scheme

5.6.3 Double sweep algorithm

Obviously, in both cases the finite difference equations are non-linear. Therefore, to solve them they need to be linearised in terms of the increments to the estimates for the dependent variables. Such an approach opens up the possibility of using an iterative Newton-Raphson approach (Tjalling, 1995) and Abbott and Minns (1998) whereby a solution is obtained for the increments to the estimates for the dependent variables; the estimates of the dependent variables are then updated, and new increments are calculated. The process is repeated until there is acceptable convergence for the solution of the dependent variables at the new time level $(n+1)\Delta t$. Given a single pipe the solution for the increments is obtained by inverting a tri-diagonal or four-diagonal matrix. This involves a double sweep (Abbott and Minns, 1998); the first sweep down the diagonal involves the elimination of certain elements in the lower triangle of the matrix (see the •) and then solving for the most downstream increment, before returning back up the diagonal and solving for the remaining increments; see Figure 5.28.

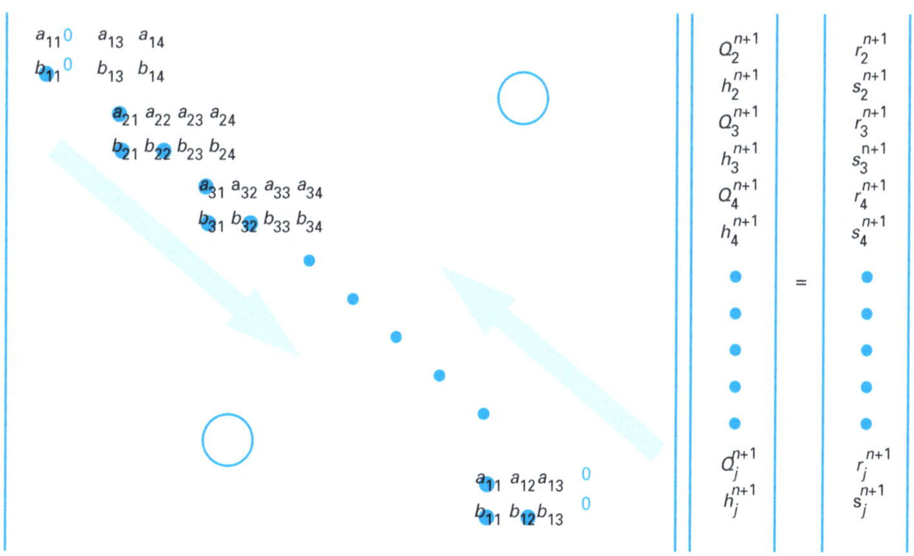

Figure 5.28 Double sweep algorithm (*r* and *s* are the residues of the two equations)

If there is failure to converge then the time increment can be reduced and the process repeated. Alternatively, if convergence has been achieved for an adequate number of successive time increments then the time increment can be increased.

5.6.4 Network of pipes or channels

Where there is a network of pipes then more work has to be done in order to get a solution. This is because the solution for one pipe can be dependent on the solution for adjacent pipes. Before we introduce the solution method however, we need to indicate how headlosses at manholes and at ancillary structures (such as pumps, weirs, sluices, valves, etc) are included.

The headloss at, say, the exit of a manhole into a pipe is defined explicitly as an energy loss between two sections zero distance apart. The same concept is used at the exit of the pipe into the (downstream) manhole. For the headloss for the flow into pipe *ik* (as with the 4-point scheme):

$$\begin{aligned} h_{ik,0}^{n+1} &= h_{ik,1}^{n+1} + K_{ik}\frac{(u_{ik,0}^{n+1})^2}{2g} \\ u_{ik,0}^{n+1} &= u_{ik,1}^{n+1} \end{aligned} \tag{5.63}$$

where the index 0 and 1 with h refers to two sections at the entrance to the pipe.

An ancillary structure is treated in a similar manner. It is assumed that there are computational nodes either side of the structure and the headloss generated by the structure occurs between the two nodes. In effect, this defines a dummy pipe (or link). The distance between the two nodes can be non zero. This can affect the mass conservation equation. So far as the relationship between the water levels is concerned, this is given by

$$\frac{1}{2}(Q_{ik,1}^{n+1} + Q_{ik,2}^{n+1}) = F(h_{ik,1}^{n+1}, h_{ik,2}^{n+1}) \tag{5.64}$$

(again for the 4-point scheme) with the mass conservation equation providing a second equation linking the values for the discharge and water level.

Given the difference equations for each pipe then it can be shown that partial elimination using the double sweep process along the pipes results in a matrix equation for the increments in the water levels at the nodes of the network; see Abbott and Minns, 1998. This sparse matrix can be inverted and the solution found for the water levels at the nodes. Once the water levels (or the increments in the water levels) are determined at the nodes, it is a simple matter to calculate the water level and discharge increments along each pipe, and therefore to update the water levels and discharges.

5.6.5 SWMM

The most significant of the explicit methods used in urban drainage modelling is the modified Euler method adopted by SWMM-EXTRAN; see Roesner *et al.*, 1983. The momentum equation is rewritten as

$$\frac{\partial Q}{\partial t} = -\frac{gn^2}{1.49^2 R^{4/3}}Q|u| + 2u\frac{\partial A}{\partial t} + u^2\frac{\partial A}{\partial x} - gA\frac{\partial h}{\partial x} \tag{5.65}$$

where the Manning form of the friction slope is used. The corresponding finite difference form of this equation for the whole pipe is written as:

$$Q_{jk}^{n+1} = Q_{jk}^{n} - \frac{k\Delta t}{R^{4/3}} Q_{jk}^{n} |u_{jk}^{n}| + 2u_{jk}^{n} \frac{\Delta A}{\Delta t} \Delta t + (u_{jk}^{n})^{2} \frac{A_2 - A_1}{L_{jk}} \Delta t - gA \left(\frac{h_2 - h_1}{L_{jk}} \right) \Delta t \tag{5.66}$$

Similarly, the finite difference form for the continuity equation is

$$h_j^{n+1} = h_j^n + \frac{1}{P_j} \sum Q_{jk}^n \Delta t \tag{5.67}$$

where P_j is the plan area of the jth manhole or node.

Eqs (5.66) and (5.67) are solved sequentially to determine the discharge in each conduit and the water level in the manholes over a time increment. Roesner actually calculates values for the water level and pipe discharges at the half time step:

1. Compute the discharge at $t + \Delta t/2$ in all conduits based on the water levels in the nodes at time t.
2. Compute the water levels at all nodes at $t + \Delta t/2$ based on the average of the discharges in the conduits at time t and time $t + \Delta t/2$.
3. Compute the discharge at $t + \Delta t$ in all conduits based on the water levels in the nodes at time t.
4. $t + \Delta t/2$.
5. Compute the water levels at all nodes at $t + \Delta t$ based on the average of the discharges in the conduits at time t and time $t + \Delta t$.

For stability this scheme requires

$$\Delta t \le \frac{L_{jk}}{\sqrt{gD_{jk}}}$$

$$\Delta t \le C^* \frac{P_j h_{\max,j}}{\sum_k Q_{jk}} \tag{5.68}$$

where C^* is a dimensionless constant = 0.10

$H_{max,j}$ is the maximum water surface rise (m) in time step Δt

Note that the finite difference scheme is applied to a complete pipe, and therefore the pipe length must be neither too long or too short. Long pipes can be divided into a number of shorter pipes.

Roesner, Shubinski *et al.* (1983) states that time increments the order of 10 secs are typical for applications of the method. Some adjustments are made to the way in which the flooded surface area is assigned to the nodes in special cases, such as when there is back-up from the manhole downstream and there is no flow in the pipe from the upstream manhole.

Other than in SWMM explicit schemes have not been used widely, though there is growing interest in schemes such as those proposed by McCormack (Garcia-Navarro and Saviron, 1992) and Roe (Priestley, 1989) because of their ability to model discontinuous flows accurately.

5.6.6 Small depth problem

When we come to form the finite difference expressions for the momentum equation, there are two difficulties. One is that without care in formulating the finite difference expression for the friction slope two solutions for the water depth can be obtained for a given discharge. The numerical solution can therefore tend to oscillate between two flow states; see Cunge, 1980. This can be avoided by modifying the finite difference representation of the conveyance term in the friction slope. The average of the upstream and downstream values is replaced by a simple upstream weighting. In addition, a small base flow can be introduced throughout the pipe network using a synthetic slot in the bottom of the pipes (for example, see MOUSE) or alternatively a small artificial flow can be injected at the upstream end of a pipe and removed at the downstream end (for example, see InfoWorks CS). Such a concept avoids the issue about what to do with the wetting of a previously dry bed.

5.6.7 Treatment of sub- and super-critical flows

The remaining numerical problems with the solution of the Saint Venant equations for gradually varying flow in conduits concern the transition from sub- to super-critical flow and the set up of the initial condition through a pipe system. As shown above in Section 5.2.3 the end boundary conditions for sub-critical flow include an upstream boundary condition (preferably the discharge) and a downstream boundary condition (preferably the water level or a functional relationship between the downstream water level and the corresponding discharge). The functional relationship at the downstream boundary could be the normal or critical depth conditions. For super-critical flow the situation is very different: there need to be two upstream boundary conditions. Inevitably, this means that if the numerical solution is faithful to the basic analytical equations then the numerical solution algorithm has to be changed and rigorous book keeping is needed to track the propagation of the critical point(s). This is complicated in the extreme, so most modelling software systems make the assumption that the algorithm for sub-critical flow does not change its structure. To achieve this, the non-linear acceleration of momentum terms in the conservation of momentum equation are reduced to zero as the Froude number approaches unity. The resulting non-inertial equations can be applied to super-critical flow conditions in a conduit, but the velocity head is essentially zero.

5.6.8 Generation of the initial condition

It is desirable to have a consistent initial condition for flows in a pipe network before simulating the response to driving forces such as rainfall or extraneous time dependent inflows. In theory arbitrary water levels and discharges can be defined in each conduit at every computational point and the simulation be allowed to run until a steady state is achieved. In practice this can take some time, and if the system contains pumps controlled by water levels in their wet-wells then there will probably be no steady state solution. Consequently, two particular actions may be taken. One is to increase the numerical diffusion by temporarily weighting the space derivative to the forward time level. This accelerates convergence to the (possible) steady state. The second action is to modify the pump discharges artificially, again temporarily, to preserve a steady state condition at the pumps.

5.6.9 Groundwater

The solution of the Richards equation (Eq (5.51)) is normally done using a fully forward difference scheme with the second spatial derivatives evaluated at the forward time level and the time derivative in terms of the difference between the groundwater potential h at the new and the old time level. This is the approach adopted in MODFLOW, which is one of the most widely used physically based modelling software package for groundwater problems.

5.6.10 Solving the pollutant transport equations

The solution of the advection-diffusion equation is complicated by the transport term. Generally, it is best to solve for the transport (advection) term using the Galerkin approach along the characteristic (see, for example, Hughes *et al.*, 1989) and to solve for the parabolic diffusion terms separately using the Crank-Nicolson method; see Crank, 1947, 1996.

5.7 1D MODELLING OF NATURAL RIVERS

The digital solution to the 1D Saint Venant equations is also relevant to the 1D flow in natural river channels. Unlike the pipe situation separate cross-sections are defined at each node where h is calculated. The cross-sections may include flood plains, though where there are embankments along the river channel, or the flow is strictly two-dimensional on the flood plain, then a 2D flow model is needed in interaction with the 1D model for the river channel. Alternatively, a full 2D model should be used for both the channel and the flood plain. For the channel A-h and B-h relationships are established for each model cross section, and a rating curve is more likely to be used as the downstream boundary condition.

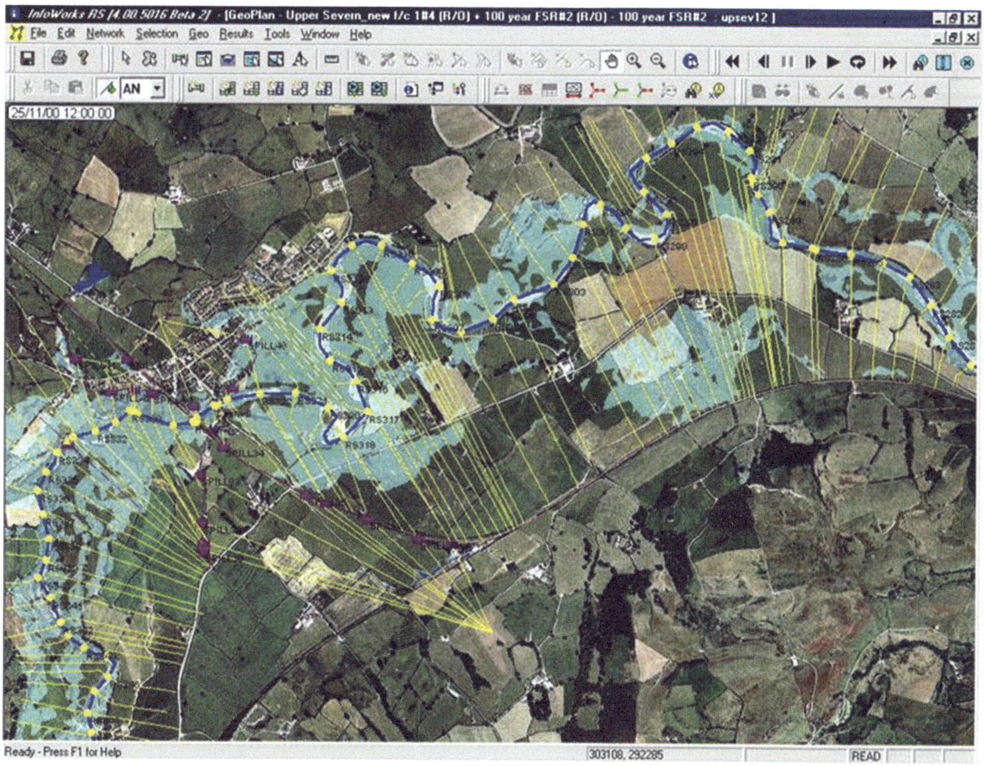

Figure 5.29 Illustration of a detailed 1D model of flooding (with permission of MWH Soft)

5.8 2D ABOVE GROUND FLOW MODELLING

Besides the 1D modelling of flow in networks of conduits, significant modelling is also needed for overland flood flows originating from excess rainfall-runoff prevented from entering the 1D network, discharges emerging from the surcharged 1D network, or over-bank spillage from urban streams or rivers. Such overland flows are typically shallow and can therefore be treated as being two dimensional. Necessarily the flows are controlled by a complex topography formed by the streets and buildings of the urban area. Simulating flood flows on inundated flood plains is difficult enough, let alone the artificial features of curbs and cambers on roads, the flows round or through buildings, and the natural flow paths generated by the general ground terrain as if there were no buildings or structures. In addition, there may be a large number of discrete points or lines at which there are discrete headlosses. It would take enormous modelling effort to include such headlosses explicitly in any 2D model, and the normal practice is not to include them directly, but to assume that they are incorporated implicitly in the boundary roughness (such as the Manning or Chezy forms) variation over the domain.

The equations describing the flow are

$$\begin{aligned}
&\frac{\partial u}{\partial t} + u\frac{\partial u}{\partial x} + v\frac{\partial u}{\partial y} + g\left(\frac{\partial h}{\partial x} + f\frac{u\sqrt{u^2+v^2}}{4gh}\right) = 0 \\
&\frac{\partial v}{\partial t} + u\frac{\partial v}{\partial x} + v\frac{\partial v}{\partial y} + g\left(\frac{\partial h}{\partial y} + f\frac{v\sqrt{u^2+v^2}}{4gh}\right) = 0 \\
&\frac{\partial h}{\partial t} + \frac{\partial (uh)}{\partial x} + \frac{\partial (vh)}{\partial y} = q
\end{aligned} \tag{5.69}$$

These equations are typically applied to flows in coastal waters or wide lakes. Because the urban overland flows are typically shallow and can be super-critical, the standard finite difference solution of Eq (5.74) can become extremely complex. Therefore the inertia terms (or the non-linear inertia terms) in the momentum equations are best neglected.

5.8.1 Numerical solution of the 2D equations

As with the 1D equations, finite differences, elements and volumes can be used to solve the 2D equations numerically. Such a solution is needed increasingly to model flood flows in urban areas as well as coastal waters receiving overflow and outfall discharges from drainage networks. A straight forward finite difference approach is to use a rectangular grid as shown in Figure 5.30. The ground levels can be defined at the centres of the resulting cells. An alternative is to define the ground levels at the mid-points of the boundaries for each cell. This has an advantage that a natural channel crossing the cell from one side to the opposite side can be accommodated by the geometry. Further, if the ground topography is defined at a finer resolution than the computational grid then the *A-h* and *B-h* relationships have to be defined at the boundaries of each cell and similarly the *V-h* relationship for each cell.

It is straightforward to set up an explicit finite difference form of Eq (5.69) for a rectangular grid. So far as urban situations are concerned there is little value in retaining the non-linear inertial terms in the momentum equations, so Eq (5.70) shows an implicit finite difference form of the reduced equations. The resulting finite difference equations are however, still non-linear, so the Newton-Raphson iterative method can be used to solve them at each time step. This results in a set of linear equations in the increments to the water levels in each cell. The resulting matrix can be inverted and the solution for the increments to the water levels can be found. Iteration can be carried out if the magnitude of the maximum increment is too large. The iterations are concluded if the magnitude of the maximum increment is acceptable. The calculations then proceed to the next time step.

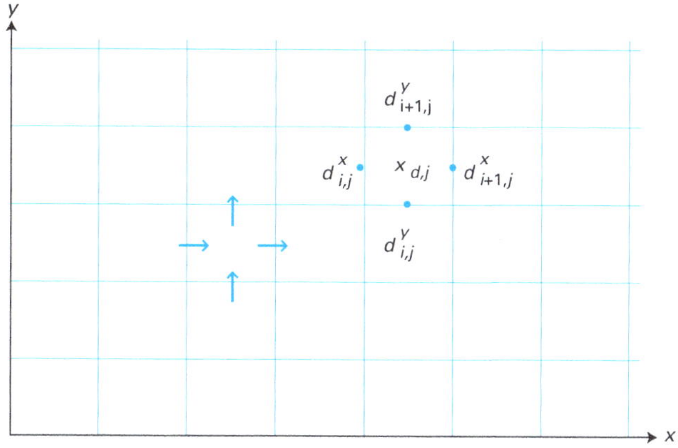

Figure 5.30 2D finite difference grid

A problem with this solution is that the number of arithmetical operations involved in inverting the matrix, and therefore the computational time, is at least proportional to N^3, where N is the smaller number of cells in the x- or y-directions. An alternative numerical solution can be derived using the alternating direction (ADI) method whereby the velocities in the x-direction are calculated at the half time step, and the velocities in the y-direction at the full time step; see Figure 5.31.

$$
\begin{aligned}
&\frac{1}{\Delta t}\left(\frac{Q_{i+1,j}^{n+\frac{1}{2}}}{A_{i+1,j}^{n+\frac{1}{2}}}-\frac{Q_{i+1,j}^{n-\frac{1}{2}}}{A_{i+1,j}^{n-\frac{1}{2}}}\right)+g\theta\frac{h_{i+1,j}^{n+\frac{1}{2}}-h_{i,j}^{n+\frac{1}{2}}}{\Delta x}+g(1-\theta)\frac{h_{i+1,j}^{n-\frac{1}{2}}-h_{i,j}^{n-\frac{1}{2}}}{\Delta x}\\
&+\theta\frac{1}{K_{i+1,j}^{n+\frac{1}{2}\,2}}\frac{Q_{i+1,j}^{n+\frac{1}{2}}}{A_{i+1,j}^{n+\frac{1}{2}}}\sqrt{\left(\frac{Q_{i+1,j}^{n+\frac{1}{2}}}{A_{i+1,j}^{n+\frac{1}{2}}}\right)^2+\bar{v}_{i+1,j}^{n\,2}}+(1-\theta)\frac{1}{K_{i+1,j}^{n-\frac{1}{2}\,2}}\frac{Q_{i+1,j}^{n-\frac{1}{2}}}{A_{i+1,j}^{n-\frac{1}{2}}}\sqrt{\left(\frac{Q_{i+1,j}^{n-\frac{1}{2}}}{A_{i+1,j}^{n-\frac{1}{2}}}\right)^2+\bar{v}_{i+1,j}^{n\,2}}=0\\
&\frac{1}{\Delta t}\left(\frac{R_{i,j+1}^{n+1}}{A_{i,j+1}^{n+1}}-\frac{R_{i,j+1}^{n}}{A_{i,j+1}^{n}}\right)+g\theta\frac{h_{i,j+1}^{n+1}-h_{i,j}^{n+1}}{\Delta x}+g(1-\theta)\frac{h_{i,j+1}^{n}-h_{i,j}^{n}}{\Delta x}\\
&+\theta\frac{1}{K_{i,j+1}^{n+1\,2}}\frac{R_{i,j+1}^{n+1}}{A_{i,j+1}^{n+1}}\sqrt{\left(\frac{R_{i,j+1}^{n+1}}{A_{i,j+1}^{n+1}}\right)^2+\bar{u}_{i+1,j}^{n+\frac{1}{2}\,2}}+(1-\theta)\frac{1}{K_{i,j+1}^{n\,2}}\frac{R_{i,j+1}^{n}}{A_{i,j+1}^{n}}\sqrt{\left(\frac{R_{i,j+1}^{n}}{A_{i,j+1}^{n}}\right)^2+\bar{u}_{i+1,j}^{n+\frac{1}{2}\,2}}=0\\
&\frac{1}{\Delta t}\left(V_{i,j}^{n+\frac{1}{2}}-V_{i,j}^{n-\frac{1}{2}}\right)+\theta\frac{Q_{i+1,j}^{n+\frac{1}{2}}-Q_{i,j}^{n+\frac{1}{2}}}{\Delta x}+(1-\theta)\frac{Q_{i+1,j}^{n-\frac{1}{2}}-Q_{i,j}^{n-\frac{1}{2}}}{\Delta x}+\frac{R_{i,j+1}^{n}-R_{i,j}^{n}}{\Delta y}=q\\
&\frac{1}{\Delta t}\left(V_{i,j}^{n+1}-V_{i,j}^{n}\right)+\frac{Q_{i+1,j}^{n+\frac{1}{2}}-Q_{i,j}^{n+\frac{1}{2}}}{\Delta x}+\theta\frac{R_{i,j+1}^{n+1}-R_{i,j}^{n+1}}{\Delta y}+(1-\theta)\frac{R_{i,j+1}^{n}-R_{i,j}^{n}}{\Delta y}=q
\end{aligned}
\tag{5.70}
$$

Here $Q_{i+1,j}^{n+\frac{1}{2}}$ is the discharge across the cell boundary in the x-direction

$R_{i,j+1}^{n}$ is the discharge across the cell boundary in the y-direction

The numerical solution then resolves into solving for the water levels in the cells for a series of 1D channels, first in the x-direction and then in the y-direction. The solution procedure for a single channel is outlined above. The non-linear difference equations can be solved using Newton-Raphson, but solving for the whole domain before testing the maximum increment in water level in order to decide whether or not to repeat the iteration in order to refine the estimates of the solution for the water levels. Just as with the 1D model the 2D model can adjust its time step, halving it if the increments in water level fail to converge within a required number of iterations, or doubling the time step if the convergence has been satisfied for a further number of time steps; see Seyoum *et al.*, 2010.

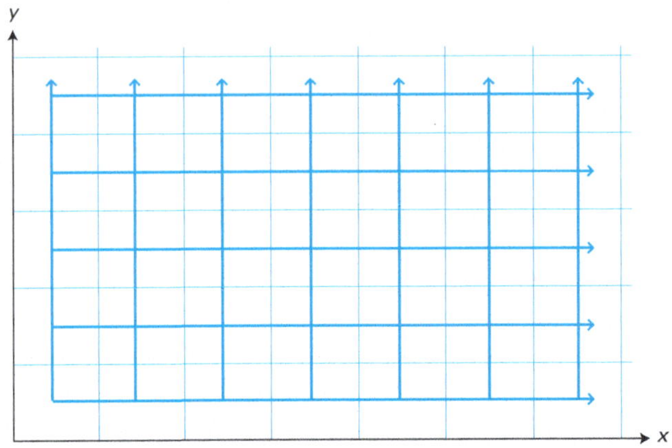

Figure 5.31 ADI approach

The ADI approach has arithmetic operations the order of N^2, and is therefore considerably faster than the first approach. For this reason, almost all finite difference methods for solving 2D flow use either a straightforward explicit or implicit ADI approach.

5.8.2 Integrating 1D and 2D models

The integration of the 1D pipe network model with the 2D overland flow model needs to be done with care. Ideally, they should be integrated implicitly so that there is a consistency in the water level at the nodes of the network that are flooded, that is, where and when water emerges from the network nodes above ground level and can generate overland flow. A similar link can be envisaged between flow spilling over the banks of an urban stream and flowing through the urban area.

One method for linking the two models is to assume that flow emerging from a manhole spills out radially into an above-ground 2D cell via a circular orifice, with the flow determined by the difference in water level between the manhole and the cell. The discharge emerging from the manhole during the time step is input as a source to the 2D model; see Figure 5.32. The 2D model then computes the resulting overland flow for a number of time steps for a duration equal to the 1D time step (normally the 2D model time step will be smaller than the 1D time step). The calculation then returns to the 1D model using the latest water levels at the manhole and in the cell containing the manhole. See Figure 5.33 for an illustration of the effectiveness of this integrated approach.

5.8.3 Wetting and drying

The 2D model has to be able to cater for the flow emerging from a manhole and flowing over a dry surface. The condition at the front is important. Here it is assumed that the front propagates across a cell with a velocity at any instant equal to the velocity u of the inflow to the cell; see Figure 5.34. The cell is assumed to be wet (and the front can move to the next dry cell) when

$$L = \sum_t u\Delta t < \Delta x \tag{5.71}$$

where L is the cumulative distance (m) that the front travels in the cell since first arriving

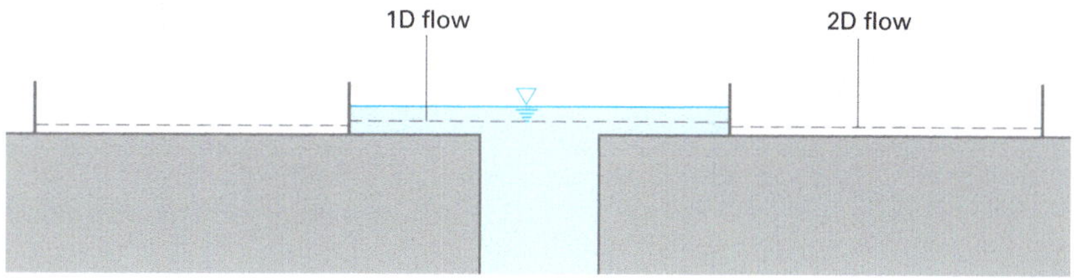

Figure 5.32 Interaction of 1D and 2D flow at a manhole

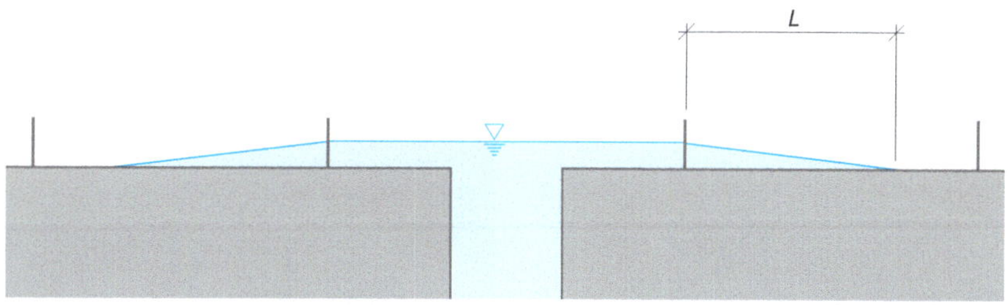

Figure 5.33 2D flow of water from a manhole on a flat surface

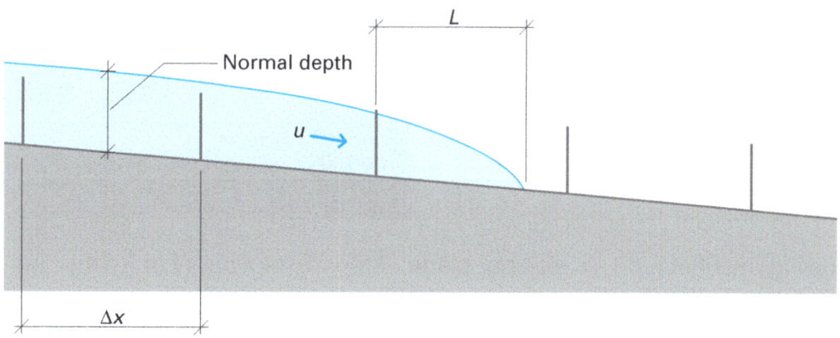

Figure 5.34 Propagation of a front over a dry bed on a sloping surface

Note that on a sloping surface the front propagates at the speed associated with the (approximate) normal depth upstream. Such a normal depth cannot exist on a flat surface, but the propagation is determined by the local velocity, which is a function of the time history of the forcing flow from the manhole(s), and possibly the rainfall runoff. See Figure 5.35 and Figure 5.36 for 2D calculations for flow from a manhole on flat and sloping beds with a 1 m*1 m and a coarse 5 m*5 m grid. For further details see Seyoum *et al.*, 2009.

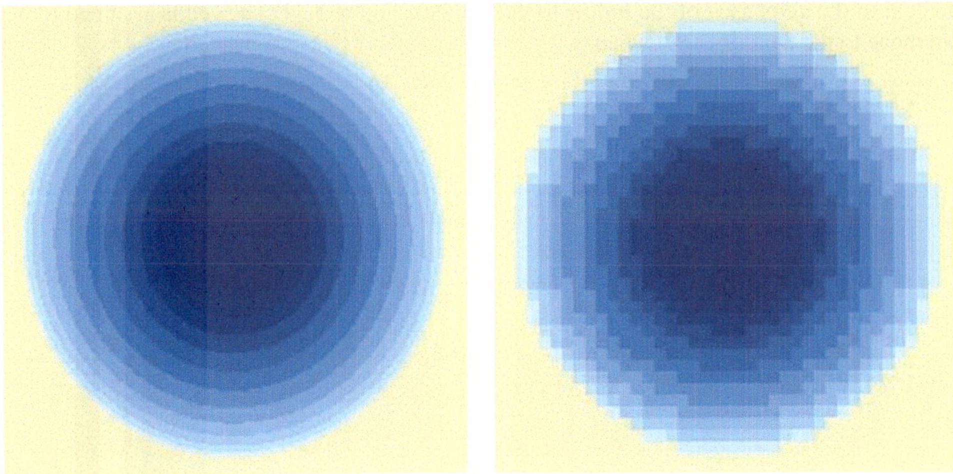

Figure 5.35 Overflowing manhole on a flat plain: Fine resolution (1 m × 1 m) and coarse resolution (5 m × 5 m)

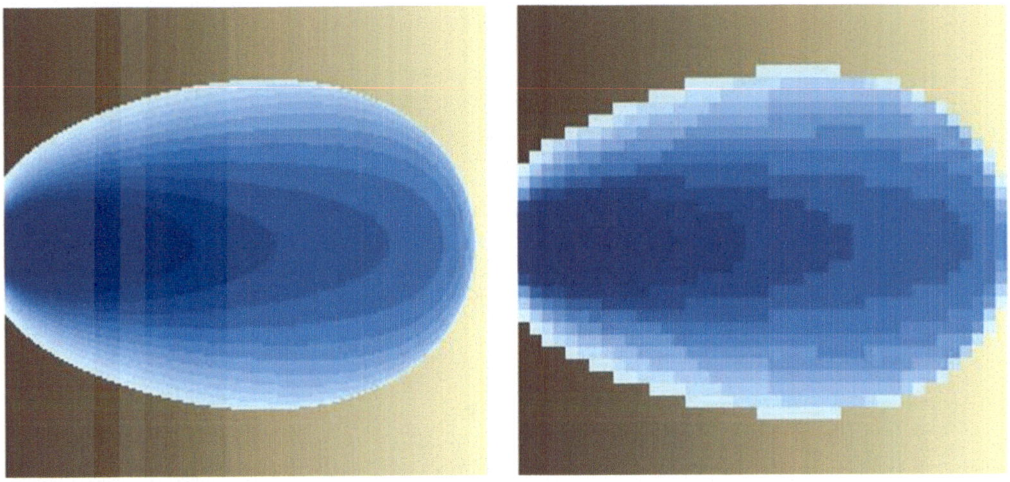

Figure 5.36 Overflowing manhole on a sloping plane: Fine resolution (1 m × 1 m) and coarse resolution (5 m × 5 m)

5.9 SOLVING THE WATER DISTRIBUTION EQUATIONS

The water distribution equation for a single pipe with steady state flow is

$$H_k - H_{lj} + \frac{8L_{kj}}{\pi^2 D_{kj}^4 g}\left(\frac{f_{kj}}{D_{kj}} + K\right)Q_{kj}|Q_{kj}|^n = 0 \tag{5.72}$$

where L_{kj} is the length of pipe (k, j) (m)
D_{kj} is the diameter of pipe (k, j) (m)
f_{kj} is the friction coefficient for pipe (k, j)

At any node j the sum of the discharges to or from all pipes is zero:

$$\sum_i Q_{ij} = d_j \tag{5.73}$$

where d_j is the demand at node j (m^3/s)

Swamee and Jain (1976) came up with an explicit approximation of the implicit Colebrook-White equation:

$$f_{kj} = \frac{0.25}{\left[\log\left(\frac{k_{kj}}{3.71D} + \frac{5.74}{\text{Re}^{0.9}}\right)\right]^2} \tag{5.74}$$

where k_{kj} is the roughness height for pipe (k, j) (mm)

Because of the relative simplicity and accuracy of this equation it has been introduced into a number of modelling tools for water distribution system analysis, such as the widely available hydraulic analysis software EPANET (Rossman, 2000).

The Hazen-Williams equation, originally introduced in 1902 (Liou, 1998) is still used by many water supply engineers to calculate the headloss along a pipe based on the flow in the pipe and the physical properties of the pipe. The equation is empirical and in SI units is:

$$H_j - H_k = 10.67 \frac{Q_{kj}^{1.85}}{C_{kj}^{1.85} D_{kj}^{4.87}} L_{kj} \tag{5.75}$$

where, C_{kj} is the Hazen-Williams friction coefficient for the pipe, which indicates the roughness of the interior surface. Note that the coefficient in Eq (5.75) is dimensional and therefore a numerical conversion coefficient needs to be introduced if different units for headloss, flow and diameter are used.

For any two nodes in the water distribution network the energy equation can be written as:

$$H_j - H_k = \sum_{path_{j-k}} h_f = \sum_{path_{j-k}} RQ|Q|^{n-1} \tag{5.76}$$

where $path_{j-k}$ includes all the set of pipes on the path from node j to node k
H_j, H_k are the heads at nodes j and k (m)
R is the resistance coefficient for *each pipe* along $path_{j-k}$
n is the exponent from the headloss equation (2 for Darcy-Weisbach and 1.85 for Hazen-Williams)

For a closed loop, that is, one beginning and ending at a same node, the energy loss is zero.

For specified pipe characteristics, demand patterns and reservoir heads the system of non-linear equations (Eqs (5.72), (5.73) and (5.74)) has a unique solution defined by the flows and heads in the whole network. There are several iterative techniques available for solving the above system which are incorporated into modern simulation tools. Over the last two decades, there has been a significant increase in the number of network analysis software packages both in the commercial (e.g., WaterGEMS®, H_2ONET®, InfoWorks™ WS, MIKE URBAN) and in the public domains (e.g., EPANET). These tools allow network analysts to concentrate solely on building realistic representation of the water distribution network thus enabling easier development of models. However, it should be pointed out here that most of the standard water distribution analysis methods are built on the assumption that demand is fixed at a demand node and fully satisfied regardless of the pressure at the node, i.e., the demand driven analysis. This assumption is reasonable for well designed systems and under normal operating conditions. However, when networks operate under abnormal conditions, e.g., due to pipe burst or excessive demands, the demand-driven approach gives unrealistic results in terms of delivered water to customers. In that case, a pressure-driven analysis method (Bhave, 1988); (Todini and Pilati, 1987) should be used.

5.9.1 Steady-state models

Conditions in real water distribution systems constantly change with time due to varying demands and other operating characteristics. However, steady-state models assume that demands and operations are constant. This simplification is acceptable when particularly important instantaneous conditions within a network need to be investigated, e.g., flows and pressure at a particular snapshot of time. In such a case, computer models are normally used to calculate how much water takes which of the numerous alternative routes through a looped water distribution system and the head losses along each route.

Hardy-Cross method

The first mathematical method for calculating flows in complex networks was proposed by Hardy Cross (Cross, 1936). This is a manual, iterative procedure that was later implemented in a number of software packages. As such, it was used throughout the water industry for almost four decades before it was replaced by improved solution methods capable of solving large problems with improved convergence characteristics.

The Hardy-Cross method starts with enumeration of all closed simple (independent) loops in the network system. Then an initial guess of the direction and magnitude of pipe flows satisfying the principle of continuity at each node is made. At the next step, the head losses around each independent loop are calculated, which normally results in the need to change assumed flows until the sum of the head losses around each loop is zero (actually it is close to zero within a pre-specified tolerance limit). A correction of the assumed flows, ΔQ, is computed for each independent loop in the network. Iterations are performed such as that:

$$\Delta Q_{ij}^{(k)} = \frac{-\sum \Delta H_{ij}^{(k)}}{n\sum \dfrac{\Delta H_{ij}^{(k)}}{Q_{ij}^{(k)}}} \tag{5.77}$$

where $Q_{ij}^{(i)}$ is the flow in a pipe ij at iteration k (m^3s^{-1}),

ΔQ_{ij} is the corrective flow passing through a pipe ij in a loop (m^3s^{-1}),

ΔH_{ij} is the (friction) headloss in pipe ij (m)

n is the exponent from the headloss equation and the summation operator is applied to all pipes in a loop.

This gives the new flow estimate in a pipe for the next iteration:

$$Q_{ij}^{(k+1)} = Q_{ij}^{(k)} + \Delta Q_{ij}^{(k)} \tag{5.78}$$

The corrective flow is equal for each pipe in a loop. For pipes belonging to multiple loops, the correction is taken to be the sum of individual loop corrective flows. When the corrective flow gets lower than a predefined error threshold the flows in each pipe are found. Considering that the headloss in each pipe is known, the nodal heads are simply calculated by starting at known heads, e.g., fixed-head supply sources.

The method works for networks with a small number of loops, it depends on the initial estimate of the flows and could suffer from slow convergence. Instead of adjusting flows around each individual loop in the network, modern techniques based on the Newton-Raphson method adjust the flows, or heads, or flows and heads along all loops simultaneously, which improves convergence over the Hardy-Cross method and the current modelling software relies almost exclusively on the Newton-Raphson approach.

Todini-Pilati method

Todini and Pilati (1987) introduced a matrix-formulation and an efficient gradient algorithm for solving the system of nonlinear equations found in Eqs (5.72), (5.73) and (5.74). For the gradient solution method the system of equations can be written as the following matrix equation:

$$\begin{bmatrix} A_{11} & A_{12} \\ A_{21} & 0 \end{bmatrix} \begin{bmatrix} Q \\ H \end{bmatrix} = \begin{bmatrix} -A_{10}H_0 \\ -q \end{bmatrix} \tag{5.79}$$

where Q are the unknown pipe flows/discharges (m^3s^{-1}),
H and H_0 are the unknown and known nodal heads, respectively (m), and
q are the known nodal demands (m^3s^{-1}).

Eq (5.79) is a non-linear matrix equation with the diagonal matrix $\mathbf{A}_{11}$ defined (for pipes) as follows:

$$A_{11}(i,i) = R_{ij}|Q_{ij}|^{n-1} \tag{5.80}$$

The matrix $\mathbf{A}_{12}$ is the *unknown* head topological incidence matrix with elements defined as follows:

$$A_{12}(i,j) = \begin{cases} -1 & \text{if the flow in pipe } j \text{ leaves node } i \\ 0 & \text{if pipe } j \text{ is not connected to node } i \\ +1 & \text{if the flow in pipe } j \text{ enters node } i \end{cases} \tag{5.81}$$

The matrix $\mathbf{A}_{10}$ is a *fixed* head node topological incidence matrix (hence defined similarly to $\mathbf{A}_{12}$). Note that $\mathbf{A}_{21} = \mathbf{A}_{12}^{T}$.

The Todini-Pilati solution method begins with an initial estimate of flows in each pipe that normally does not satisfy flow continuity. In the next step, the recursive Newton-Raphson methodology is applied to the system of equations Eqs (5.72), (5.73) and (5.74). The iterative process of calculating unknown heads and flows continues until a convergence criterion is reached. The method has excellent convergence characteristics and is incorporated into a number of software packages currently available as commercial or public-domain software (e.g., EPANET).

5.9.2 Unsteady flow models

Steady-state network analysis models have been developed for checking the performance of an existing or a proposed network by performing a network simulation based on fixed design conditions, such as the network topology, the pipe lengths and diameters, the peak demand, etc. For example, the analysis then verifies whether the pipe diameters are large enough to guarantee water delivery at each demand node with sufficient pressure under several operating conditions.

Extended Period Simulation (EPS) models the slow variation in flow conditions with time due to the diurnal changes in demand, changes in reservoir levels or pump and valve operations. The standard approach to EPS modelling is to alternate in time a succession of steady-state analyses at fixed time intervals with a succession of mass balance computations in each of the service reservoirs (i.e., variable head water storage locations sometimes referred to as 'tanks' in EPANET terms).

Rapid changes in the fluid velocity within a pipeline are commonly referred to as waterhammer, pressure transients or pressure surges. Due to their sudden and violent nature these pressure transients can cause a catastrophic failure of pipes or equipment, or contribute to fatigue failure of system elements. These pressure surges are often generated by the rapid opening/closure of valves or switching/failure of pumps within the network system. It is thus important to understand the occurrence of pressure surges and how pipe networks can be designed to avoid the damaging effects caused by them.

Extended period simulation

The EPS model is commonly solved as a quasi-steady system, i.e., the pipe system is modelled as if it is going through series of steady states (snapshots). For each snapshot, a system of equations Eqs (5.72), (5.73) and (5.74) is solved for unknown pipe flows and junction heads. At each time step, the head at tank i is updated using the following continuity equation:

$$A_{T,i}\frac{dH_{T,i}(t)}{dt} = \sum_{j\in T_i} Q_{ij}(t) \tag{5.82}$$

where $H_{T,i}$ is water level (head) in tank (service reservoir) i at time t, (m)
$A_{T,i}$ is the cross-sectional area of the i-th tank, (m^2)
Q_{ij} is the flow in link ij connected to tank i, (m^3/s)
T_i is the set of links connected to tank i, t is the extended period simulation time.

Eq (5.86) is typically solved using forward Euler's method. Assuming constant tank cross-sectional areas, the following equation can be written for each tank:

$$H_{T,i}(t+\Delta t) = H_{T,i}(t) + \frac{\Delta t}{A_{T,i}}\sum_{j\in T_i} Q_i(t) \tag{5.83}$$

It is assumed that the head, i.e., the water level at each tank in the system is known at the initial time step (i.e. at $t = t_0$). Source heads are assumed constant throughout the simulation (or a given function of time).

Transient models

The governing equations for solving the transient flow problems are based on the conservation of momentum and mass laws (Wylie and Streeter, 1978):

$$\frac{\partial H}{\partial x} + \frac{1}{g}\frac{\partial V}{\partial t} + J = 0 \tag{5.84}$$

$$\frac{\partial H}{\partial t} + \frac{c^2}{g}\frac{\partial V}{\partial x} = 0 \tag{5.85}$$

where $H(x,t)$ is the piezometric head (m)
$V(x,t)$ is the fluid velocity (m/s)
c is the wave speed (celerity) (m/s)
g is the gravitational acceleration (m/s^2)
J is the friction slope.

The above equations are valid for one-dimensional flow, constant pipe and fluid properties (diameter, wave speed, temperature, etc.), relatively small convective terms ($\partial H/\partial x$ and $\partial V/\partial x$) and relatively small pipe slopes.

Normally, the above transient flow equations cannot be solved analytically, but only numerically. The most widely used numerical solution method is the fixed-grid Method of Characteristics (MOC), in which two governing partial differential equations are first transformed into two equivalent total differential equations, each valid along the appropriate characteristic line (Chaudhry, 1987):

$$C^{+}: dH + \frac{a}{gA}dQ + (J_s + J_u)dx = 0 \quad \left(\text{valid along: } \frac{dx}{dt} = +a\right) \tag{5.86}$$

$$C^{-}: dH - \frac{a}{gA}dQ - (J_s + J_u)dx = 0 \quad \left(\text{valid along: } \frac{dx}{dt} = -a\right) \tag{5.87}$$

where Q is the pipe flow (m^3/s)
A is the pipe cross-sectional area (m^2)

The last two equations need to be integrated, each along its characteristic line. The problem is the last friction term which cannot be integrated analytically. Various numerical schemes exist for the integration of the quasi-steady friction term (Karney and Ghidaoui, 1997).

5.9.3 Water quality models

Drinking water utilities around the world face the challenge of providing water of good quality to their consumers on a daily basis. Disinfection of drinking water by chlorination is used to prevent water quality deterioration once it enters the distribution system. Chlorine models represent transport of chlorine (or any pollutant) within pipes, mixing at nodes and storage reservoirs and reaction kinetics in the bulk liquid phase and at the liquid-pipe wall boundary with a set of mathematical equations. Modelling of the movement of chlorine within the distribution system is one of the most often performed water quality analysis. This modelling is based on three main assumptions:

1) conservation of mass within differential lengths of pipe,
2) complete and instantaneous mixing of the water entering pipe nodes, and
3) appropriate kinetic expressions for the decay of chlorine through pipes and storage facilities.

A calibrated EPS model is used to ensure that the first of the above modelling assumptions is satisfied. In other words, it is assumed that the velocities in the pipes are known (i.e., from the solution to a hydraulic model of the network) before water quality modelling commences. This also means that the usefulness and reliability of a water quality model depends critically on the accuracy of the underlying hydraulic model. Although it has been shown that the assumption of complete and instantaneous mixing at nodes is not fully justified (Romero-Gomez *et al.*, 2008), it has been accepted as a reasonable simplification for one-dimensional water quality modelling. Thus, this assumption indicates that the concentration of a substance in water leaving a node is simply the flow-weighted sum of the concentrations in the inflowing pipes. The third assumption relates to the disappearance of residual chlorine due to reactions with materials in the aqueous phase at different residence times in the network. The first-order model for chlorine decay is most often used to represent this process:

$$\frac{dC}{dt} = -kC \tag{5.88}$$

where C is the volumetric constituent concentration at time t (m/s^3);
k is the reaction rate constant (1/s).

Therefore, taking all three assumptions into account the change in concentration can be expressed by the advection equation with a rate term:

$$\frac{\partial C}{\partial t} = -v\frac{\partial C}{\partial x} + kC \tag{5.89}$$

where v is the velocity of flow.

5.10 PHYSICALLY BASED MODELLING SOFTWARE

The proper development and robustness of software packages for physically based modelling is obviously a key to the success of such modelling. The software is the responsibility of its creator(s) and depends critically on his (the alternative of 'her' is assumed) scientific knowledge and mathematical skills, as well as his coding ability and the testing of the software in a variety of different circumstances. One way of testing the software is to trace all possible logical paths through the code. This would form the basis of a mathematical validation of the software. However, the number of paths through the software is effectively infinite, and therefore an alternative approach has to be found. IAHR came up with a paper-based approach for validating modelling software, which identified a number of well-defined steps (see Dee, 1994):

1. Represent the generic physical laws in terms of mathematical algorithms
2. Replicate the resulting equations in terms of a digital representation of the algorithms
3. Discretize the model space in a way that is consistent with the difference equations
4. Solve the resulting difference equations within particular boundary constraints and conditions
5. Design, code and test the numerical procedures
6. Design, produce and test the resulting software system that can be linked to other systems such as databases, GIS, CAD, 2D and 3D graphics, and so on

Recall that a clear distinction should be made between modelling software packages and instantiated models. The former are frequently termed 'models', but this claims more for the packages than is justified: data for the particular instance for which the modelling software is being applied needs to complement the software. Such data includes the data discretising the model space and providing the (temporal) boundary conditions for the solution of the numerical equations. Once the data for the model space and boundary conditions are input to the computer running the software, and parameter values have been selected while building and calibrating the model, the model of the particular instance can be said to be 'instantiated'.

5.11 MODEL INSTANTIATION

The instantiation of a physically based model is yet another important process:

1. Decide on the purpose of the model
2. Review existing data and collect other necessary data
3. Build the model infrastructure and select initial values of the parameters
4. Carry out consistency tests of the model
5. Calibrate parameters and confirm the model
6. Produce the base performance against which to test improvements to the physical system

A fuller description of the model instantiation process is given in Chapter 6.

The models are constructed by first identifying the physical geometric space in the physical world that contains the flowing water, and which the model is going to replicate. This physical space becomes the model space. So the next step is to deconstruct the physical space into (idealised) geometric objects that comprise the corresponding objects in the model space. Similarly, the flow of water is governed by particular generic processes. A hydroinformatics model in Abbott's sense is one that consists of the integration of a collection of indicative signs that point to the geometric objects and physical processes and expressive signs that are the graphical output objects. Note that all of these are designed by the developer of the software (the developer can of course be a team of people) who works from his point of view; see Figure 5.37. He deconstructs the physical world from his own perspective, which certainly is not necessarily that of the potential user. Consequently there are risks associated with the use of the software by somebody other than the developer. Nevertheless these risks can be minimised by the user understanding the limitations and constraints of the software and not using it outside the limits for which it was designed. Remember, much is left up to the user concerning how to structure his model, what data to select, how to calibrate the model, how to interpret the results, and so on, such that two modellers working independently are more than likely to arrive at different (yet similar) models for the same situation. Generally, the user has behind him yet a third person who is even more remote from the modelling process, but is ultimately concerned with what the model produces in terms of data that will assist in the decision making process. All of these different types of people (and there are others) play important roles in hydroinformatics.

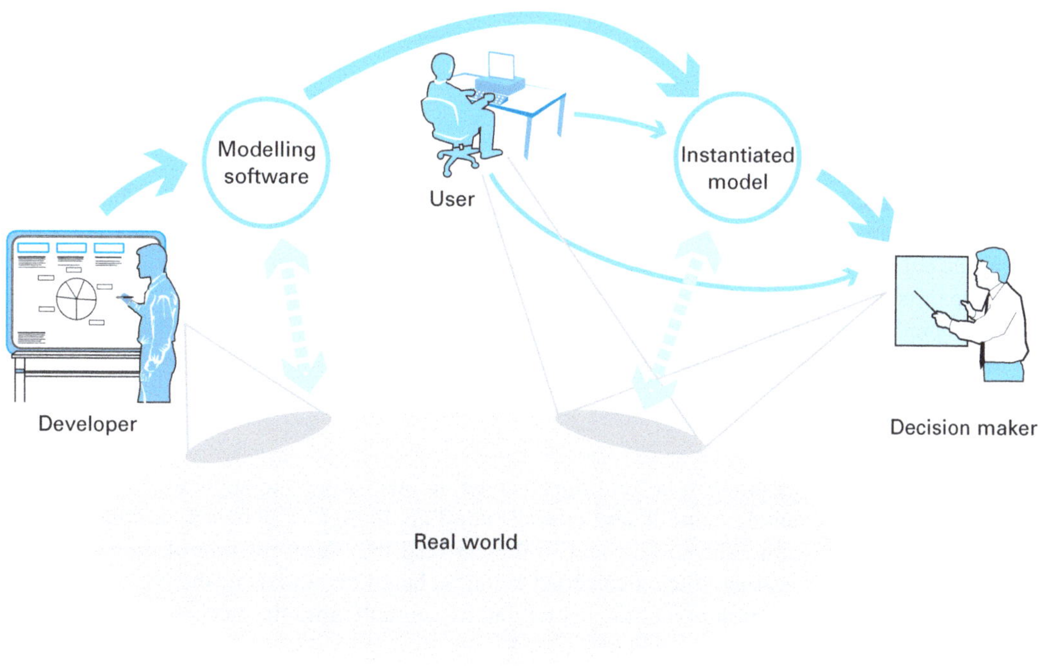

Figure 5.37 Modelling relationships

5.12 COMPARTMENTALIZED MODELLING

As indicated above there are two ways of approaching the modelling process:

- Derive generic equations at a point and discretize the model space in a way that is consistent with finding a numerical solution of the governing equations

- Deconstruct the model space into compartments based on physical boundaries and relate the output to the input for each compartment using any suitable mathematical construct; for example, see Figure 5.38.

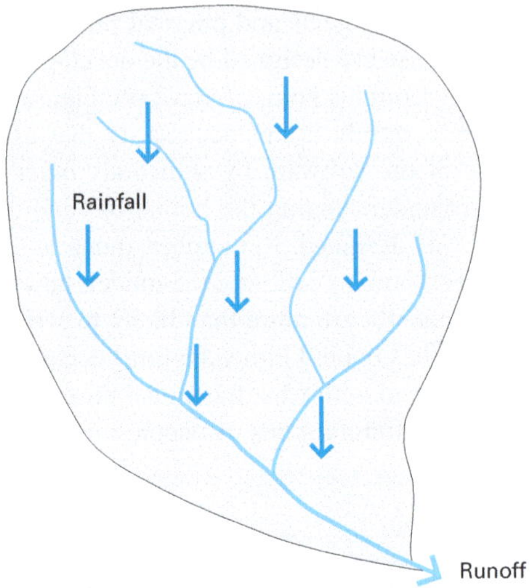

Figure 5.38 Rainfall-runoff from a (sub-)catchment

The first category of model is said to be physically based, while the second category applies to lumped, machine learning, intelligent agent, and black and grey box models. Strictly, intelligent agent models are a different paradigm again, but for simplicity they can be included in the second category here.

Some compartmentalized models in urban water modelling include

- Rainfall prediction and forecasting
- Rainfall-runoff
- Pollution surface loading
- Pollution wash-off
- Treatment works processes

In each case, the models are assumed to be zero dimensional, in other words, to apply over a domain defined by a compartment. So, for example, rainfall, and rainfall-runoff apply over a catchment contributing to a pipe in the network. Instead of treating the surface as two dimensional the transformation of rainfall to runoff is done using a mathematical construct. Such a construct could be based on notions of the physics of catchment runoff, including the perceived translation of the net rainfall hyetograph times the catchment area to the outfall hydrograph and the corresponding attenuation of the peak.

5.12.1 Reservoir concept

The translation and attenuation can be induced through the storage concept, which is introduced as a reservoir. In fact, rainfall-runoff models based on the tank or reservoir concept are very popular with hydrologists. In its simplest form, the reservoir can be linear, that is, the storage is proportional to the outflow:

$$\frac{dS}{dt} = Q_{in} - Q_{out} \tag{5.90}$$

$$S = kQ_{out}$$

See Figure 5.39. One important feature of the linear reservoir solution is that the peak discharge for the outflow coincides in time with a point on the recession of the inflow hydrograph. This does not in general occur in practice. Therefore an extension of the simple linear reservoir model is called for. Nash (1957) suggested that a sequence of identical linear reservoirs removes the need for the peak of the outflow to be on the recession of the inflow hydrograph. Such a sequence is termed the Nash cascade:

$$u(t,k,n) = \frac{1}{k\Gamma(n)}\left(\frac{t}{k}\right)^{n-1} e^{-\frac{t}{k}} \tag{5.91}$$

where $\Gamma(n)$ is the gamma function.

k is the storage constant (1/s)

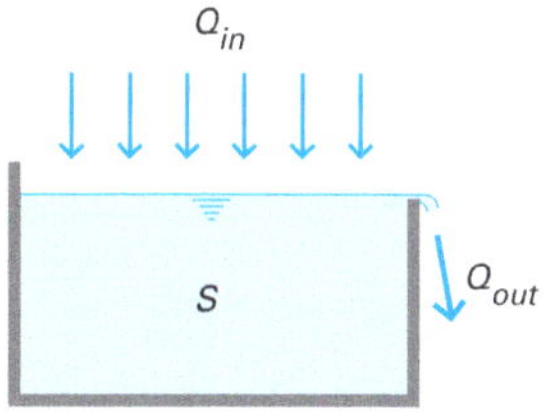

Figure 5.39 Linear reservoir

So the Nash cascade is described by two parameters: namely k and n the number of reservoirs. k and n can be chosen to match the peak output and the lag between the peak input and output.

5.12.2 Unit hydrograph

Another mathematical device is to assume that unit impulse of rainfall produces a standard response (unit) hydrograph at the outfall; see Sherman, 1932. As the rainfall hyetograph is made up of a sequence of such impulses, the assumption can be made that the response hydrographs to each impulse are independent of each other, in other words, the responses are linear, and that the ordinates of the unit hydrograph $\{U_i\}$ are simply multiplied by the net rainfall $\{P^n\}$; see Figure 5.40. The outflow hydrograph is the convolution sum of the ordinates of the rainfall hyetograph with the ordinates of the unit hydrograph:

$$Q^i = \sum_{n=1}^{i} P^n U_{i-n+1} \tag{5.92}$$

where P is the rainfall intensity (mm/hr)

U is the ordinate of the unit hydrograph (3.6^{-6}m^2)

Given sufficient data it is possible to derive the unit hydrograph for a catchment.

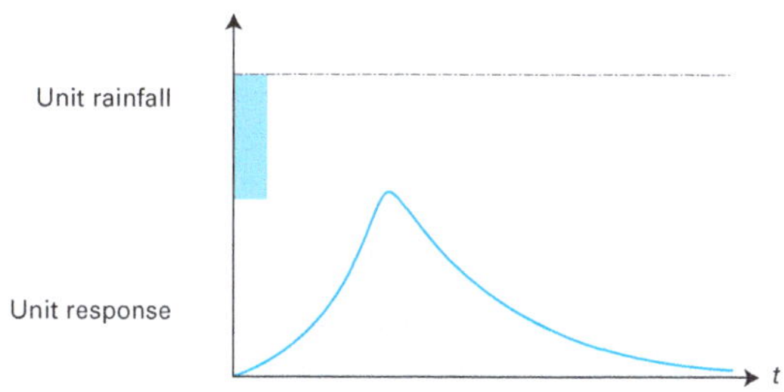

Figure 5.40 Unit hydrograph

5.12.3 Time-area diagram

Yet another device is to assume incrementally increasing travel times for rainfall on the catchment reaching the outfall according to the distance from the outfall. This produces the time-area diagram; see Figure 5.41. Again, a convolution integral is needed to implement this concept.

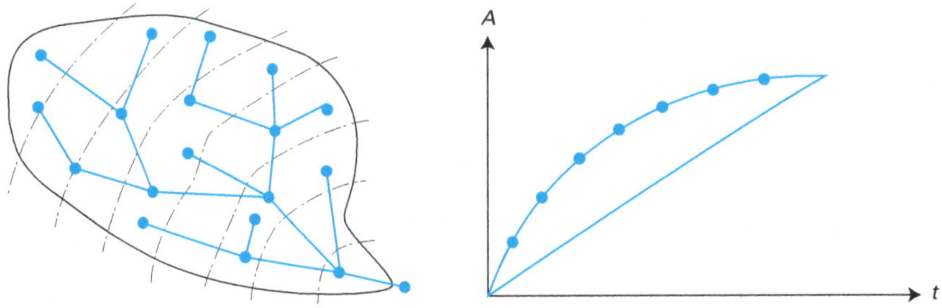

Figure 5.41 Time area diagram

5.13 DATA DRIVEN MODELS

Data driven models can be defined as models that connect the system state variables (input, internal and output) without the knowledge about the systems physics. Typical examples include various regression models, artificial neural networks (ANNs or simply NNs), support vector machines (SVMs), etc. These models induce underlying dependences between the system variables using training data sets. They are called data-driven models in that they depend on the availability of data (e.g., examples, patterns, observations, measurements, etc.). If there is no available data, a data-driven model cannot be derived. NNs and SVMs are used to solve regression and classification problems and some of their applications in urban water management are: classification (e.g., water quality parameters, etc.), time series prediction and forecasting (e.g., flows, water levels, overflow spills, rainfall, etc.), process control (nonlinear and multivariable control of water/wastewater plants, etc.).

Data-driven modelling starts with a body of data and searches for a set of generalisations to describe connections between the data parsimoniously and to explain the process under consideration. A data-driven model is used to map N inputs $\mathbf{u}_i$ ($i = 1, \ldots, N$) gathered in an n-dimensional input vector $\mathbf{u}_i = [u_1, u_2, \ldots, u_n]^T$ to M outputs $\mathbf{y}_i$ gathered in an m-dimensional output vector $\mathbf{y}_i = [y_1, y_2, \ldots, y_m]^T$. This form of model is called the *multiple-input multiple-output* (MIMO) model; see Figure 5.42.

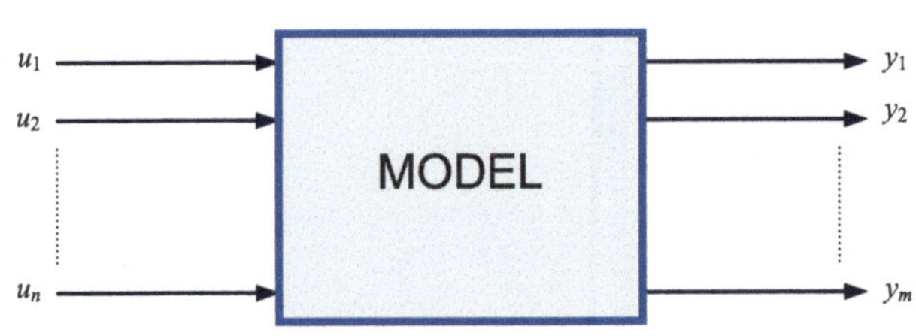

Figure 5.42 A general MIMO model

Also, a MIMO model can be decomposed into m different *multiple-input single-output* (MISO) models, which have the advantages that they can be easier to understand and apply, the accuracy of each MISO model can be adjusted separately, and different optimisation techniques can be applied to each MISO problem.

A traditional data-driven model is linear in a static (Eq (5.93)) or dynamic form (Eqs (5.95) to (5.97)). These include regression or an autoregressive moving average models. Linear models are easy to develop and to implement, and to understand and interpret. Furthermore, they are the best tools for modelling linear processes. The penalty for this convenience is that they may be entirely inappropriate for even moderately nonlinear processes. They are characterised by a small number of parameters and therefore they are also classified as parametric methods. Figure 5.43 demonstrates that a typical linear model with two inputs represents a plane. When the dimension of an input vector is greater than 2 (i.e., $n > 3$) a linear model represents a hyperplane.

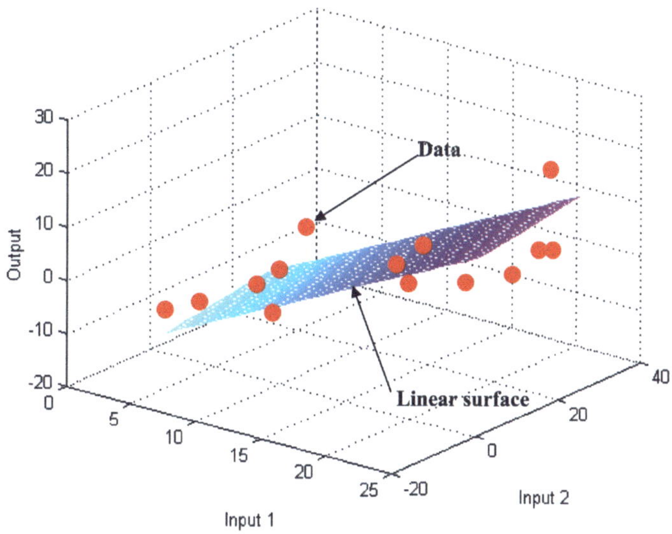

Figure 5.43 A linear model: $z = 5x + \frac{1}{2}y$ with two inputs (x,y) and one output (z) which simply finds the best linear fit to the data. Here, we seek the linear function of x and y that minimises the sum of squared residuals from z

Multiple linear regression (MLR) is one of the most popular and useful modelling techniques for quantitative analysis (Marquez *et al.*, 1991). This technique allows the analyst to build a model or to fit a curve to a set of data points through minimising the squared distance from the data points to the population mean, commonly called the least squares estimation. Such an approximation is presented by a function relating one or more input variables to an output variable, see Figure 5.43.

The two most important properties of linear models are that the interpolation and extrapolation behaviour is linear and that the parameter optimisation can be performed efficiently by the least squares algorithm. If the modelled dependencies are truly linear, then the overall accuracy is high. A typical multiple linear regression model is given by:

$$\mathbf{o} = \mathbf{W} \cdot \mathbf{x} \tag{5.93}$$

where $\mathbf{o}$ is an output vector, $\mathbf{W}$ contains the unknown model parameters w_{ij} ($i = 1, \ldots m, j = 1, \ldots n + 1$, m is the dimension of the output space, and n is the dimension of the input space) and $\mathbf{x}$ stands for the known inputs. More specifically, in the case where there are two input variables and one output variable the Eq (5.93) can be written as:

$$o = w_{11}x_1 + w_{12}x_2 + w_{13} \tag{5.94}$$

where the input vector **x** is augmented with the bias input $x_3 = 1$. MLR given in Eq (5.93), i.e. Eq (5.94) can be shown graphically in Figure 5.44.

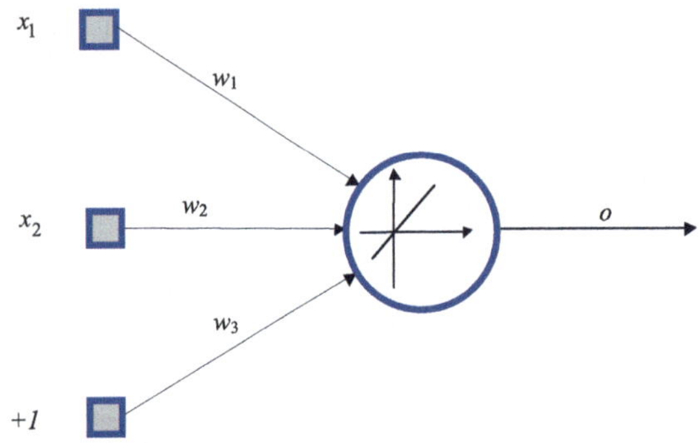

Figure 5.44 Linear Model Development – (2 inputs and 1 output)

In Eq (5.93) the multi-dimensional input vector **x** (multiple features data set) is used to approximate a one-dimensional output. The form of the functional relationship is linear and as such there is always a pre-specified number of parameters that drives the model.

In terms of dynamic linear regression models, a typical autoregressive (AR) class of linear models can be described as:

$$\hat{y}(k) = \sum_{n=1}^{N} a(n)y(k-n) + e(k) = y(k) + e(k) \tag{5.95}$$

where, $e(k)$ represents the external input series, $y(k)$ represents the observed series, N denotes past values of series and a stands for model coefficients.

The model given in Eq (5.95) is an Nth order autoregressive model. The component e can be either a controlled or uncontrolled input (noise) to the system. If there is an external input to the process u, then this model is called an ARX (AR with exogenous input) model:

$$y(k) = \sum_{n=1}^{N} a(n)y(k-n) + \sum_{n=1}^{N} b(n)u(k-n) + e(k) \tag{5.96}$$

By adding the moving average (MA) component to the original AR model, the result is an ARMA(N, P) model which can be presented as:

$$y(k) = \sum_{n=1}^{N} a(n)y(k-n) + \sum_{p=1}^{P} b(p)e(k-p) \tag{5.97}$$

The MA component represents a classical convolution filter: the new series y is generated by a linear filter with coefficients b. In statistics this component is referred to as generating P-th order moving average models, whereas in the engineering community, these models are referred to as finite impulse response (FIR) models because their outputs are guaranteed to go to zero. Weigend and Gershenfeld (1994) provide a summary which characterises MA(P) behaviours in terms of discrete Fourier transforms of the output.

If the ARMA model has an external (exogenous) input series u, then the same model is called an ARMAX model (with exogenous input component coefficients c) and can be presented as:

$$y(k) = \sum_{n=1}^{N} a(n)y(k-n) + \sum_{p=1}^{P} b(p)e(k-p) + \sum_{n=1}^{N} c(n)u(k-n) \tag{5.98}$$

As an alternative to the static as well as dynamic linear models, the nonparametric and nonlinear modelling techniques, such as NNs or SVMs, can be applied to the same tasks. The major benefits of applying such nonlinear models come when the underlying function possesses a dstinct nonlinear character. An example of a nonlinear model with two inputs and one output is given in Figure 5.45.

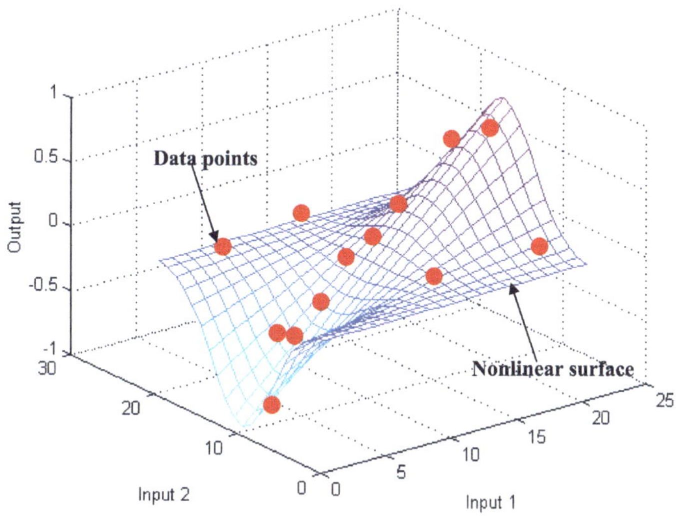

Figure 5.45 A nonlinear model: ($z = x_e^{-4y^2}$) *with two* inputs (x, y) and one output (z) which finds the best nonlinear fit to the data. In this case, we seek the nonlinear relationship of x and y that minimises the sum of squared residuals from z

5.13.1 Neural networks

NN models are based on a (simple) model of the brain. The brain consists of a very large number of neurons that are massively interconnected. Each neuron is a specialised cell that can transmit an electrochemical signal. The neuron has a branched input structure (dendrites), a cell body and a branched output structure (axons). The axons of one cell connect to the dendrites of other cells via a synapse. When a neuron is activated it fires an electrochemical signal along its axon, which crosses the synapse to other neurons, which may in turn fire themselves. A neuron only fires if the total signal received by the cell exceeds a certain level called the firing threshold. The strength of the signal received by a neuron depends critically on the nature of the synapse. Each synapse consists of a gap across which the neurotransmitter chemicals are poised to transmit the signal. Learning consists of altering the strength of the synaptic connections. In this way, by using a large number of very simple processing units, the brain appears to be able to carry out extremely complex tasks. An NN is a very much simplified model of this understanding of the human brain.

Since the 1980s there has been an explosion in research activities in NN related techniques. Nowadays, there are about 50 different types of NNs in use. In general terms, we can distinguish between *feed forward* and *recurrent* NNs. In feed forward NNs, the information moves only forward, from the input nodes, through the hidden nodes, to the output nodes. There are no cycles or loops in the network. Recurrent networks propagate data from later processing stages to earlier stages.

In general terms, an NN model can be described as:

$$o = \sum_{i=1}^{N} w_i \varphi_i(\mathbf{x}\mathbf{v}_i) \tag{5.99}$$

where $\varphi_i(\mathbf{x}\mathbf{v}_i)$ is a set of given activation functions. These functions represent the hidden layer neurons and they can be sigmoidal functions such as a logistic function or tangent hyperbolic function, or radial basis functions such as Gaussian, splines or multiquadratic functions. The existence of the hidden layer neurons introduces the difference between the linear (e.g., MLR) and the nonlinear model (e.g., NNs or SVMs) characteristics. The hidden layer neurons enable the neural network to capture nonlinear patterns in the data. The neural network technique is a nonparametric technique in the sense that there is no *a priori* knowledge of the form of function being estimated. The final number of parameters (weights) is driven by the data set itself.

The most popular type of the neural network models is the multi-layer perceptron (MLP) neural network model which became prevalent with the development of backpropagation. This is by far the most explored and applied neural network technique to date. The training of a MLP involves the estimation of the best set of weights. This requires the solution of a nonlinear optimisation problem which is usually solved by the application of a gradient descent in the weight-space. The weights are incrementally adjusted to decrease the error. This process is repeated until the error is minimised. The major problem in nonlinear optimisation is slow convergence and the difficulty in locating the global minimum.

Figure 5.46 shows the typical architecture of a MLP neural network model. J corresponds to the number of hidden layer neurons, $\mathbf{x}$ stands for input vector, $\mathbf{V}$ denotes the hidden layer weights matrix and $\mathbf{w}$ stands for output layer weights vector. The bias $b = +1$ can simply be integrated into a hidden layer as its last neuron has a constant activation $y_{j+1} = +1$. The bias term may be used in any neural network structure but it is always needed in the MLP structure.

Besides MLP NN models which are applied in majority of NN applications to date, there are other NN structures such as *radial basis function* (RBF) NNs, which have gained in popularity over MLP NNs. See Broomhead and Lowe (1988) and Girossi and Poggio (1989). Kecman and Pfeiffer (1994) have shown that RBF NNs are a special case of an *fuzzy logic* inference system.

5.13.2 Choosing parameters of a NN model

In designing the neural network model for a particular task, the user needs to carefully choose the key design parameters. These parameters are:

1. number of input variables,
2. number of hidden layers and hidden layer weights,

3. size of the training data set,
4. sampling frequency.

There are also certain other parameters that may have an impact on the model's accuracy (e.g. activation function, optimisation method, etc.).

A general review of NNs , their training and application is given by Zijderfeld (2003).

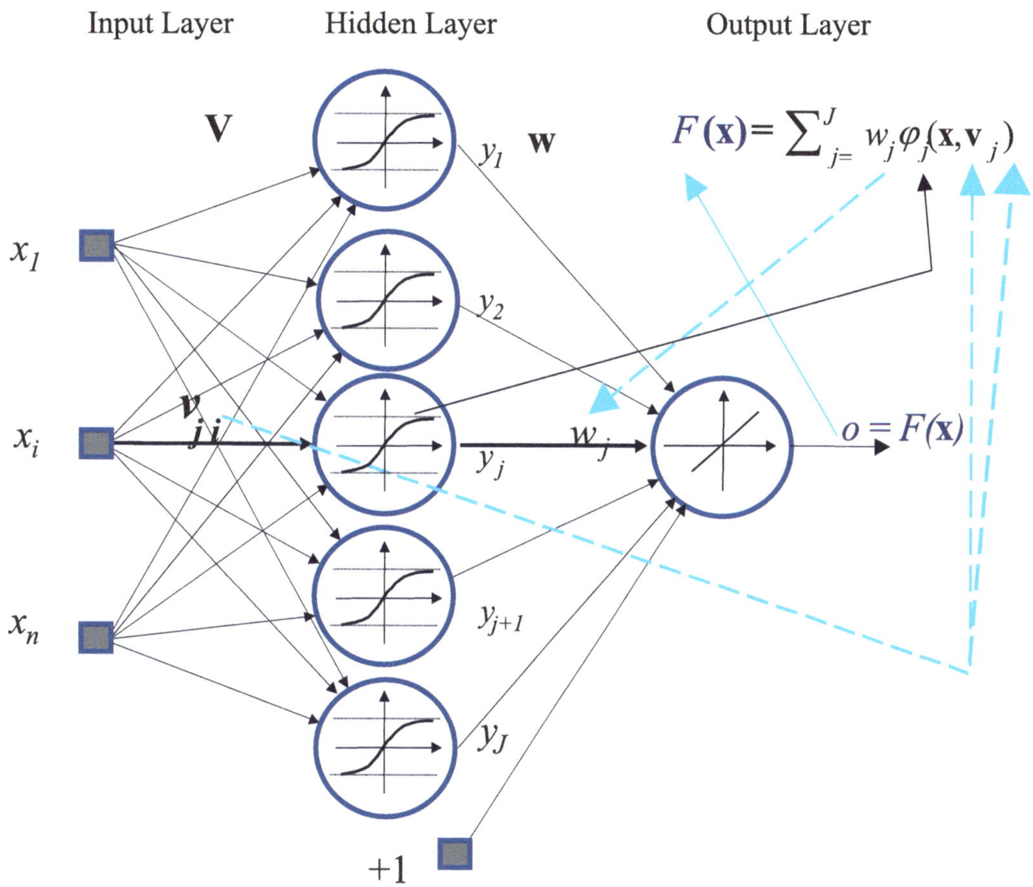

Figure 5.46 A multi layer perceptron neural network

5.13.3 Support vector machines

SVM is a learning machine that uses data to find the approximating function (in regression problems) or the separation boundary (in classification, pattern recognition problems). Generally speaking, there is no difference between NN models and SVM models in their representational capacity, i.e., in a structure; however, there is an important difference in learning. The learning process in SVM is typically performed by solving a Quadratic Programming (QP) problem with linear inequality and equality constraints rather than by solving a non-convex, unconstrained minimization problem, as in standard NN techniques (Osuna *et al.*, 1997), see also Kecman, 2001.

5.13.4 Chaos theory and nonlinear dynamics

Chaos theory and non-linear dynamics can be used to support data-driven modelling for time series prediction when the time series data are sufficient length and posses adequate information about the behaviour of the system. An important characteristic of chaos is the high sensitivity of results to the initial conditions. Such behaviour of systems was first described by (1963). Chaos consists of a class of signal intermediate between regular sinusoidal or quasiperiodic motions and unpredictable, truly stochastic behaviour. The main reason for applying chaos theory is the existence of methods that enable the prediction of future positions of the system in the state space.

The foundation behind the methods of non-linear signal processing and chaos theory is an embedding theorem by a Dutch mathematician F. Takens (Takens, 1981). It shows that from a single measured variable x_k with k_0 being some starting time and n the sampling time it is possible to generate a multi-dimensional space that is a proxy for the full multivariate state space of the observed system.

The use of nonlinear autoregression (e.g., NAR, NARX, NARMA, NARMAX, NARIMA, NARIMAX) can be described by the Takens' theorem (Takens, 1981 and Packard *et al.*, 1980). This theorem implies that there is an existence of a diffeomorphism (one-to-one differential mapping) between a finite window of the time series:

$$[x_k, x_{k-1}, x_{k-2}, \dots x_{k-n}] \tag{5.100}$$

and the underlying state of the dynamic system which gives rise to the time series. This means that there exists a nonlinear function of the form:

$$x_{k+1} = \varphi[x_k, x_{k-1}, x_{k-2}, \dots x_{k-n}] \tag{5.101}$$

which can model the series precisely (assuming no noise). Thus, the data-driven model (e.g., NN or SVM) forms an approximation to the ideal function $\varphi(\cdot)$.

The practical application is modelling a time series for finding the optimal embedding dimension $-m$ (Figure 5.47), the time delay $-\tau$ (Figure 5.48), and the number of nearest neighbours $-k$.

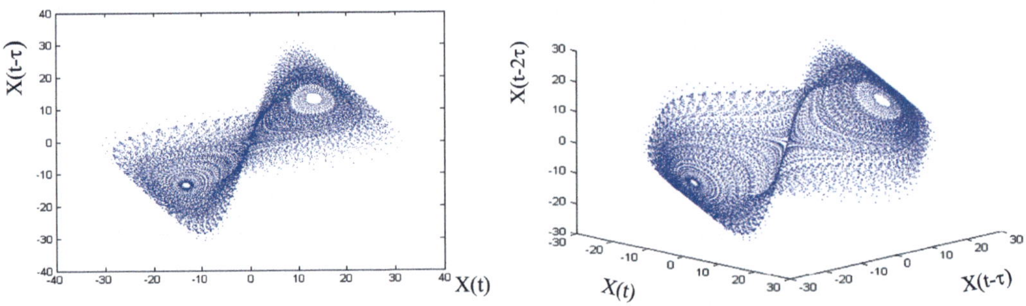

Figure 5.47 Embedding dimension example: plot X(t) vs. X(t − τ) – left; plot X(t) vs. X(t − τ) vs. X(t − 2τ) – right

The commonly used methods for finding the time delay (τ) are the autocorrelation function (Mees *et al.*, 1987) and average mutual information (AMI) (Fraser and Swinney 1986), and the method of false nearest neighbours (FNN), (Kennel *et al.*, 1992) is used to find the optimal embedding dimension. The method of false nearest neighbours measures the percentage of false neighbours as m increases. Farmer and Sidorowich (1987) have addressed the issue of finding the number of nearest neighbours k. The time delay τ must be sufficiently large so that independent information about the system is in each component of the vector (Abarbanel *et al.*, 1993). If the time delay is too short, the vector components will not be independent enough and will not contain any new information.

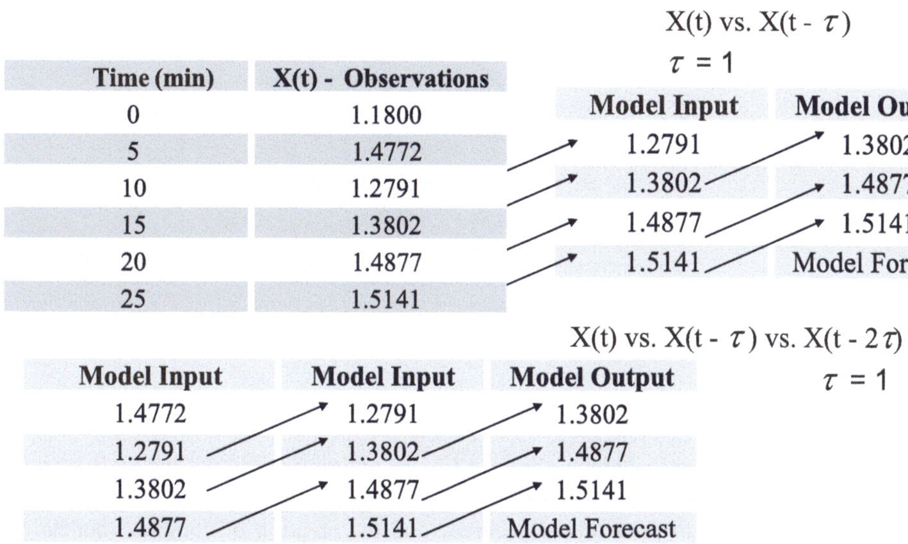

Figure 5.48 Time series modelling

5.13.5 Genetic programming

Genetic programming (GP) represents an extension of the Darwinian computational model of learning and was initially introduced by Koza and Rice (1992). While Genetic Algorithms (GA) (Goldberg, 1989) typically operate by combining binary strings which encode independent variables, in the case of GP the symbolic expressions are subject to genetic operators. In other words, it differs from the traditional GA in that it typically operates on parse trees instead of bit strings. A parse tree is built up from a terminal set (the variables used in the problem) and a function set (the basic operators used to form the function). An example of such a parse tree is given in Figure 5.49. The tree size of this expression is 4, where the tree size represents the maximum node depth of a tree and node depth represents the minimum number of nodes that must be traversed to get from the root node of the tree to the selected node; see Figure 5.49.

A key difference of GP when compared to other data-driven modelling approaches is that it does not assume any *a priori* functional form of the solution. For instance, in a classic regression method, the model structure is pre-specified in advance and the model coefficients are determined. For other techniques such as NNs, the network structure has to be identified and then the coefficients (weights) are found by the learning algorithm. For GP, the building blocks (the input and target variables and the function set) are defined initially, and then the learning method finds both the optimal structure of the model and its coefficients. Furthermore, since GP evolves an equation or formula relating inputs to outputs, it has the ability to select input variables that contribute beneficially to the model and to disregard those that do not. Therefore, it can reduce substantially the dimensionality of the input variables.

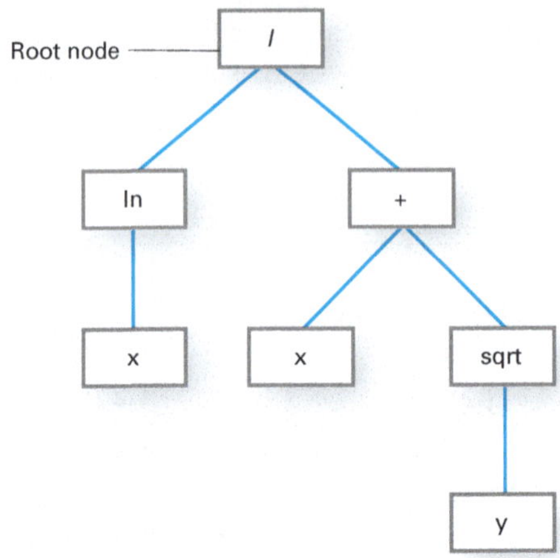

Figure 5.49 GP parse tree representing ln(x)/(x+sqrt(y))

As with any data-driven modelling technique, the selection of appropriate model inputs is very important. This is especially important for modelling dynamic systems where lagged input variables are used. The inclusion of irrelevant inputs leads to poor model accuracy and the creation of complex models, which are more difficult to interpret compared to simpler models.

5.13.6 Bayesian learning

Another learning paradigm is Bayesian learning which was named after Thomas Bayes, a British mathematician, who died in 1691. This is also termed the 'Bayesian approach' or 'Bayesian statistics' or even 'Bayesian statistical modelling'. It is based on the premise that all forms of uncertainty can be expressed and measured by probabilities (Bernardo and Smith, 2001). A simple application of the conditional probability definition allows us to derive Bayes' rule:

$$P(A|B) = \frac{P(B|A)P(A)}{P(B)} \tag{5.102}$$

where A and B denote two events (or proposition).

If an additional event/proposition C is introduced then:

$$P(A|BC) = P(A|C)\frac{P(B|AC)}{P(B|C)} \tag{5.103}$$

With this representation we have a mathematical formulation of the process of learning. $P(A|C)$ is the *prior probability* for event A when we know event C. $P(A|BC)$ is the *posterior probability*, when acquiring new information B. Typically, A refers to some hypothesis or theory (or model), B refers to observations, and C

refers to the background information (or knowledge) about A before getting the observations B. The same equation can be written as:

$$Posterior = \frac{Prior \times Likelihood}{\sum Prior \times Likelihood} \tag{5.104}$$

5.13.7 Fuzzy logic models

One of the most popular rule-based models in recent times is the fuzzy logic system. Fuzzy logic is a technique that can enable the embedding of human structured knowledge (or reasoning which is approximate rather than exact) into workable algorithms (Kecman and Pfeiffer, 1994). Human structured knowledge is structured in the form of IF-THEN rules. Fuzzy logic technique has been used to design intelligent systems whish have been shown to be more efficient than those based on conventional approaches. They are also used in combination with other techniques such as NNs to design robust adaptive intelligent control systems.

Fuzzy logic is treated as a fuzzy set theory of classes with vague or fuzzy boundaries; see Figure 5.50. The theory of fuzzy sets was introduced by Zadeh (1965) as a theory of graded concepts.

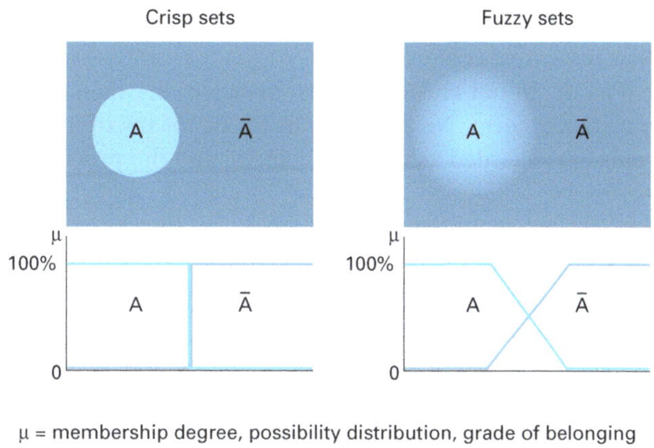

Figure 5.50 Crisp and fuzzy sets

Figure 5.50 implies that with crisp sets an element can be a member of only one of the two sets, whereas with fuzzy sets an element can be a member of two (or more) sets at the same time.

A general approach in all fuzzy logic modelling can be summarised as follows:

1. Define the variables of relevance, interest or importance,
2. Define the subsets' intervals,
3. Choose the shapes and the positions of fuzzy subsets,
4. Set the rule basis i.e., IF-THEN rules,
5. Perform calculations/model simulations

Such a modelling system consists of input-output membership functions, fuzzy rules and an inference engine. Crisp inputs are fuzzified, the fuzzy rules are applied and the inference engine is used to recover a crisp output; see Bardossy and Duckstein, 1995.

Fuzzy relations in different product spaces can be combined with each other by *composition* which also results in a fuzzy set. The three most common compositions are: MAX-MIN, MAX-PROD and MAX-AVE; see Kecman, 2001.

The methods for obtaining a crisp set from the fuzzy set (referred to as defuzzification) are center-of-gravity, first-of-maxima and middle-of-maxima.

5.13.8 Nearest neighbour

This is a simple method to explore the relationship of data points to each other. Vectors are identified in a multidimensional feature space $X = \{\boldsymbol{x}_1, \ldots., \boldsymbol{x}_n\}$where each data point $\boldsymbol{x}_j$ is a P-dimensional vector. This space is partitioned into c clusters $\{C_i\}$ by minimizing the objective function

$$J = \sum_{i=1}^{c} \sum_{j \in C_i} \left\| \boldsymbol{x}_j - \boldsymbol{v}_i \right\|^2 \tag{5.105}$$

where $\|\cdot\|$ is the Euclidean norm

$\boldsymbol{v}_i$ is the coordinate of the centre of the ith cluster

The best partitions of N samples into the $\{C_i\}$ clusters are found by minimising the total distance between the clustered samples and their respective centres.

First, c points are selected as the initial cluster centres. Instances are then assigned to their closest centre according to the Euclidean distance norm. Next, the centroids or means of all the instances in each cluster are calculated. These centroids are regarded as the new centres, and the process is repeated until J becomes stable.

The distances from the test instance to all the stored instances are then computed and the k closest instances are selected. A major drawback to using this technique is that the classes with the more frequent examples tend to dominate the prediction of the new vector. This is because they tend to come up in the k nearest neighbors when the neighbors are computed, due to their large number. One way to overcome this problem is to take into account the distances of each of the k nearest neighbors from the new vector that is to be classified and to determine the class of the new vector based on these distances.

The best choice of k depends on the data; generally, larger values of k reduce the effect of noise on the classification, but make the boundaries between classes less distinct. A good value of k can be selected by various heuristic techniques, for example, cross-validation. The special case where the class is predicted to be the class of the closest training sample (i.e. when $k = 1$) is called the nearest neighbour algorithm; see Figure 5.51.

The accuracy of the k-NN algorithm can be severely degraded by the presence of noisy or irrelevant features, or if the feature scales are not consistent with their importance. The naive version of the algorithm is easy to implement by computing the distances from the test sample to all stored vectors, but it is computationally intensive, especially when the size of the training set grows; see Shakhnarovish, 2005.

5.13.9 Model trees

The idea behind this approach is to divide a complex modelling problem into a number of simple tasks, and then to build a simple model for each task. To do this the data set is divided into a number of subsets, and separate models are built for each subset. The subsets are formed through splitting the input data into regions that are ultimately connected like the leaves on a tree, hence the name ‘model tree’; see Figure 5.52.

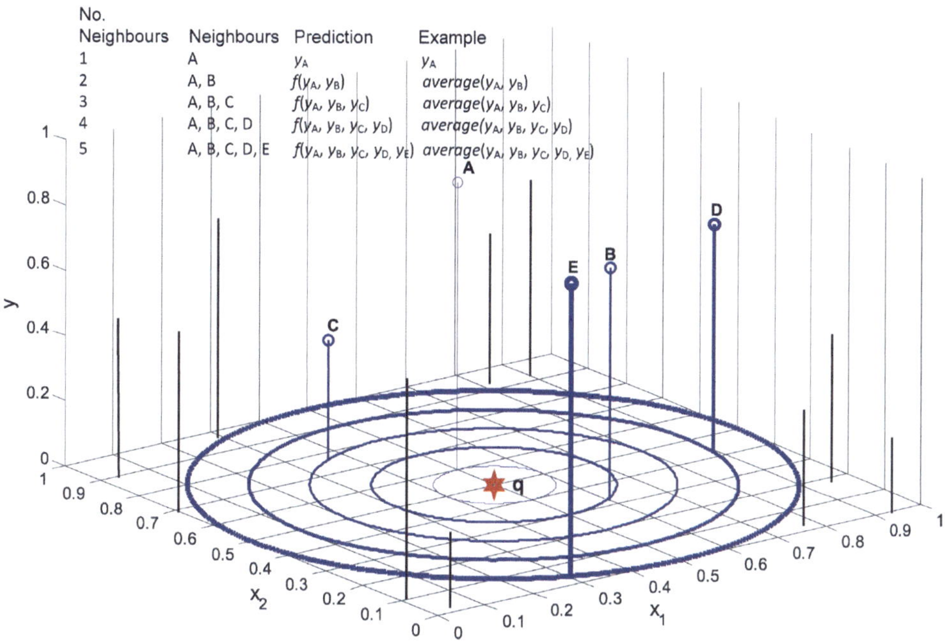

Figure 5.51 Example of prediction with KNN. q is the query point. 1 nearest neighbour is A and it predicts the query point q. 2 nearest neighbours are A and B; the prediction of q is a function of the outputs of A and B. If 3 neighbours are used, points A, B and C are used to predict q

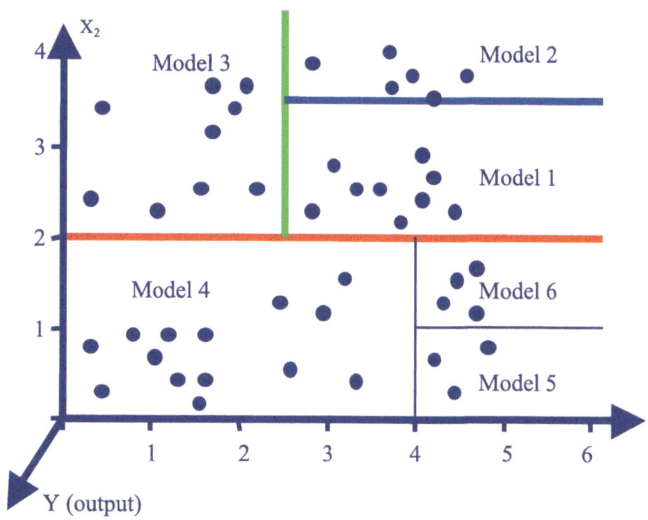

Figure 5.52 Splitting the input space $X_1 \times X_2$ by the M5 model tree algorithm; each model is a linear regression model: $y = a_0 + a_1x_1 + a_2x_2$ (Bhattacharya and Solomatine, 2005)

The M5 algorithm to construct a model tree was developed by Quinlan (1992). This algorithm starts with a set of N data pairs $\{\boldsymbol{x}_i, y_i\}$, $i = 1, \ldots, n$ where $\boldsymbol{x}$ is a p dimensional vector and y is the target. The model tree maps the input vector to the corresponding target by generating simple linear regression equations at the leaves of the tree. First, the attribute of $\boldsymbol{x}$ is selected with an appropriate splitting value of the attribute. The splitting criterion is based on the standard deviation of the set T of target values that reach a node. The standard deviation reduction (SDR) is calculated for each input attribute of the T data pairs reaching the node from:

$$SDR = sd(T) - \sum_i \frac{|T_i|}{|T|} sd(T_i) \tag{5.106}$$

where T_i is the subset of T that results from splitting the node for attribute i

sd is the standard deviation

$|T_i|/|T$ is the weight that represents the fraction of the instances belonging to subset T_i

M5 selects the split that maximises SDR. The splitting process of the training pairs is done recursively on the subsets, and terminates when the target values of all of the data pairs that reach a node vary only slightly (or only a few data pairs remain). This process tends to produce over-elaborate structures that must be pruned, for example, by replacing a sub-tree with a leaf. Finally, smoothing is introduced to compensate for the sharp discontinuities that occur between adjacent linear models at the leaves of the pruned tree. Witten and Frank (2000) provide details of the pruning and smoothing processes.

The model tree learns efficiently and can have high dimensionality. In addition the results are transparent and interpretable. New, improved versions of the M5 algorithm have been proposed by Solomatine and Xue (2004). Model tree is generally very effective for rainfall-runoff modelling (see Solomatine and Dulal, 2003; Solomatine and Siek, 2006 and Stravs and Brilly, 2007).

5.14 COMPARISON BETWEEN PHYSICALLY BASED AND DATA DRIVEN MODELLING

Obviously, there are advantages and disadvantages in using either a physically based or data driven model. Table 5.4 lists some of these.

5.15 AGENT BASED MODELLING

There remains a third modelling paradigm that is gradually assuming greater significance in urban water management. This is agent based modelling whereby individual members of a population, of people, fish, plants, or even raster areas of a city, are assumed to have independent intelligence that obeys certain rules related to the overall environment and to their neighbours.

A particular form of agent based modelling used in urbanisation studies is cellular automata. A cellular automaton (CA) consists of a regular lattice or irregular grid of cells with one or more discrete variables at each cell with a finite number of states (Wolfram, 1986). Normally, the CA evolves in discrete time steps such that the variables are updated simultaneously, according to the values of the variables in a given neighbourhood of each cell at the old time level and prescribed local rules on the variables. Ilachinski (2001) gives the following characteristics of a CA model:

1. Discrete lattice of cells
2. Homogeneity of the cells
3. Discrete states for each cell

4. Local interactions by each cell
5. Discrete dynamics in time for all cells

The primary difficulty about implementing CA is in determining the (simple) rules for the CA. What is so attractive about CA is that simple local interactions can create complex global patterns. For an example application of CA to aquatic plant dynamics modelling, see Li, 2009.

Table 5.4 Physically based and data driven models

Modelling paradigm	Physically based	Data driven
advantages	Detailed 1D, 2D and 3D modelling by taking account of the physical processes Identifies explicit consequences of particular model space features and associated flow phenomena within the flow domain and dissociated with the boundary conditions Good basis for flow forecasting, especially outside the calibration range Can be applied with a reduced amount of boundary condition detail Excellent in determining in detail flows within the flow domain	Pattern identification for knowledge discovery (on monitored data or model results) Excellent at linking monitored 'output' data to 'input' data Recovery of missing data Forecasting water levels, discharges operation of overflows Replication of complex (physically based) models Input-output relationships (CSOs, rainfall-runoff, water quality, etc) Optimisation Sensitivity/uncertainty Learning system Complex unknown phenomena Error management in physically based models
disadvantages	Need adequate model space data Less confidence in extrapolating beyond calibration range, but can be done Cannot deal adequately where the phenomena are unknown or cannot be adequately formulated algorithmically Difficult to identify as a learning system unless data assimilation is incorporated Problems in handling errors (mismatches between predictions and observations)	Cannot extrapolate easily outside the training range Cannot accommodate changes to the model space, unless explicitly included as an input parameter Difficulty in relating to physical phenomena Need adequate training and verification data Requires good understanding of the physics in order to correctly identify input and output

5.16 KEYWORDS

By the end of the chapter you should have encountered the following terms; ensure that you are familiar with them!

Accuracy
Advection-diffusion*
Alternating direction algorithm
Ancillary structures*
Boundary conditions
Calibration*
Characteristic equations
Colebrook White equation
Collection network*

Compartmentalised model*
Conceptual
Confirmation*
Consistency
Continuity equation
Courant number
Darcy friction factor
Data driven model*
Deterministic
Double sweep algorithm
Drainage network*
Ecological balances*
Finite differences
Finite elements
Finite volumes
Free surface flow*
Froude number
Groundwater
Headlosses
Hydraulic jump
Networks*
Intelligent agent model*
Lumped model
Model instantiation*
Model space
Modelling
Model trees
Momentum equation
Navier Stokes equations
Nearest neighbour
Physically based model
Pollutant transport*
Pressurised flow*
Preissmann slot
Reynolds number
Reynolds stress
Saint Venant equations*
Sediment transport*
Solution space
Stochastic Boundary layer
Stability
Water distribution network*
Water quality modelling*
1D equations*
2D equations*
3D equations *

The items with an asterisk are dealt with in more detail in later chapters.

5.17 QUESTIONS

What are the three primary modelling paradigms? What are their relative advantages and disadvantages?
Describe what can be done with the 1D Saint Venant equations: in terms of rewriting them in different convenient forms, and in simplifying the momentum equation.
What opportunities are there for modelling 2D above ground flow; how is the flooding caused, and what adjustments have to be made to the roughness of a 2D model as the computational grid is made coarser?
Discuss the boundary conditions for flow in a pipe?
What equations are derived for water quality?
Describe the relative merits of ANNs and Model Trees.

5.18 REFERENCES

Abarbanel H., Brown R., Sidorowich J. and Tsimring L. (1993). The analysis of observed chaotic data in physical systems. *Rev. Mod. Phys.*, **65**(4), 1331–1392.
Abbott M. and Minns A. (1998). *Computational Hydraulics*, 1st edn. Pitman, London/Longman, London/Wiley, New York; 2nd edn. Ashgate, Aldershot, UK/Brookfield, USA.
Abbott M. B. and Ionescu F. (1967). On the numerical computation of nearly horizontal flows. *J. Hydraul. Res.*, **5**(2), 97–117.
Bardossy A. and Duckstein L. (1995). *Fuzzy Rule-Based Modelling with Applications to Geophysical, Biological, and Engineering Systems*. CRC, BocaRaton Il.
Batchelor G. (2000). *An Introduction to Fluid Dynamics*. Cambridge University Press, Cambridge.
Bernardo J. and Smith A. (2001). Bayesian theory. *Meas. Sci. Tech.*, **12**, 221–222.
Bhattacharya B. and Solomatine D. (2005). Neural networks and M5 model trees in modelling water level-discharge relationship. *Neurocomputing*, **63**, 381–396.
Bhave P. (1988). Calibrating water distribution network models. *J. Environ. Eng.*, **114**, 120.
Box G., Jenkins G. and Reinsel G. (1976). *Time Series Analysis: Forecasting and Control*. Holden-Day, San Francisco, California.
Broomhead D. and Lowe D. (1988). *Radial Basis Functions, Multi-Variable Functional Interpolation and Adaptive Networks*. Royal Signals and Radar Establishment Malvern, England.

Butler D. and Davies J. (2000). *Urban Drainage*. Spon Press, London.

Chaudhry M. (1987). *Applied Hydraulic Transients*, 2nd edn. Van Nostrand Reinhold Co., New York, p. 521.

Crank J. and Nicolson P. (1947). A practical method for numerical evaluation of solutions of partial differential equations of the heat conduction type. *Proc. Camb. Phil. Soc.*, **43**, 50–67.

Crank J. and Nicolson P. (1996). A practical method for numerical evaluation of solutions of partial differential equations of the heat-conduction type. *Adv. Comput. Math.*, **6**(1), 207–226.

Cross H. (1936). Analysis of flow in networks of conduits or conductors. Engineering Experiment Station, University of Illinois, Bulletin No. 286.

Cunge J. A., Holly F. M. and Verwey A. (1980). *Practical Aspects of Computational River Hydraulics*. Pitman Publishing Inc., Marshfiled, MA.

Cunge J. and Wegner M. (1964). Intégration numérique des équations d'écoulement de Barré de Saint Venant par un schéma implicite de différences finies. *La Houille Blanche*, **1**, 33–39. (Numerical integration of the flow equations due to Barre de Saint Venant by an implicit finite difference scheme.).

Dee D. P. (ed.) (1994). *Guidelines for Documenting the Validity of Computational Modelling Software*. Special publication, International Association for Hydraulic Research.

Farmer J. and Sidorowich J. (1987). Predicting chaotic time series. *Phys. Rev. Lett.*, **59**(8), 845–848.

Fraser A. and Swinney H. (1986). Independent coordinates for strange attractors from mutual information. *Phys. Rev. A*, **33**(2), 1134–1140.

Garcia-Navarro P. and Saviron J. (1992). McCormack's method for the numerical simulation of one-dimensional discontinuous unsteady open channel flow. *J. Hydraul. Res.*, **30**(1), 95–105.

Girossi F. and Poggio T. (1989). Representation qualities of neural networks: Kolmogorov's theorem is irrelevant. *Neural Comput.*, **1**(4), 465–469.

Goldberg D. (1989). *Genetic Algorithms in Search and Optimization*. Addison-Wesley MA.

Hughes T., Franca L. and Hulbert G. (1989). A new finite element formulation for computational fluid dynamics: VIII. The galerkin/least-squares method for advective-diffusive equations* 1. *Comput. Meth. Appl. Mech. Eng.*, **73**(2), 173–189.

Ilachinski A. (2001). *Cellular Automata: A Discrete Universe*. World Scientific Publishing Company, Singapore.

Karney B. and Ghidaoui M. (1997). Flexible discretization algorithm for fixed-grid MOC in pipelines. *J. Hydraul. Eng.*, **123**(11), 1004–1011.

Kecman V. (2001). Learning and soft computing: support vector machines. In: Neural Networks, and Fuzzy Logic Models. MIT Press, Cambridge, Massachusetts, pp. 121–191.

Kecman V. and Pfeiffer B. (1994). *Learning fuzzy rules equals to radial basis function neural network training*, In: *Proc. 5th IEEE Nat. Conf. on Fuzzy Systems*. 8–11 September. New Orleans, Vol. 1, pp. 213–219.

Kellagher R. (1993). Sewer modelling for rehabilitation. In: *EU SPRINT Project Reports*. HR Wallingford, Wallingford, UK.

Kennel M., Brown R. and Abarbanel H. (1992). Determining embedding dimension for phase-space reconstruction using a geometrical construction. *Phys. Rev. A*, **45**(6), 3403–3411.

Koza J. and Rice J. (1992). *Genetic Programming*. MIT Press, Cambridge, MA.

Li H. (2009). Spatial pattern dynamics in aquatic ecosystem modelling. PhD thesis, UNESCO-IHE, Delft, p. 181.

Liou C. (1998). Limitations and proper use of the Hazen-Williams equation. *J. Hydraul. Eng.*, **124**(9), 951–954.

Lobbrecht A. and Solomatine D. (1999). Control of water levels in polder areas using neural networks and fuzzy adaptive systems. In: *Water Industry Systems: Modelling and Optimization Applications*, Vol. 1, Research Studies Press, Baldock UK, pp. 509–518.

Lorenz E. (1963). Deterministic nonperiodic flow. *J. Atmos. Sci.*, **20**, 130–141.

Lotka A. (1925). *Elements of Physical Biology*. Williams & Wilkins Company, Baltimore.

Marquez L., Hill T., Worthley R. and Remus W. (1991). Neural network models as an alternative to regression, *International Journal of Forecasting*, **10**(1), 5–15.

Mathieu J. and Scott J. (2000). *An Introduction to Turbulent Flow.* Cambridge University Press, Cambridge.

McCarthy G. (1938). *The Unit Hydrograph and Flood Routing*, Conference of the North Atlantic Division, US Army Corps of Engineers, 24 June.

Mees A. I., Rapp P. E. and Jennings L. S. (1987). Singular-value decomposition and embedding dimension. *Phys. Rev. A*, **36**(1), 340.

Miles J. (1985). The representation of flows to partially penetrating rivers using groundwater flow models. *J. Hydrol.*, **82**(3), 341–355.

Morton K. and Mayers D. (2005). *Numerical Solution of Partial Differential Equations: An Introduction*. Cambridge University Press, Cambridge.

Nash J. (1957). The form of the instantaneous unit hydrograph. *Int. Assoc. Sci. Hydrol., Publ.* 45, **3**, 114–121.

Oreskes N., Shrader-Frechette K. and Belitz K. (1994). Verification, validation, and confirmation of numerical models in the earth sciences. *Science*, **263**(5147), 641.

Osuna E., Freund R. and Girosi F. (1997). An improved training algorithm for support vector machines. In: *Proceedings of IEEE NNSP'97*, Citeseer, Amelia Island, Florida.

Packard N., Crutchfield J., Farmer J. and Shaw R. (1980). Geometry from a time series. *Phys. Rev. Lett.*, **45**(9), 712–716.

Pomeroy R. and Parkhurst J. (1972). Self purification in sewers.

Price R. (2009). Volume-conservative nonlinear flood routing. *J. Hydraul. Eng.*, **135**, 838.

Priestley A. (1989). Roe-type schemes for super-critical flows in rivers. *Numer. Anal. Rep.*, **13**, 89.

Quinlan J. (1992). *Learning with continuous classes*. In: *Proc. 5th Australian Joint Conf. on Artificial Intelligence*, 16–18 November, World Scientific, Singapore.

Roesner L., Shubinski R. and Aldrich J. (1983). *Storm Water Management Model User's Manual Version III: Addendum I, EXTRAN*. EPA Report, Cincinnati, Ohio (in press).

Rossman L. (2000). *EPANET 2 Users Manual*. US Environmental Protection Agency, Cincinnati, Ohio.

Savic D., Walters G. and Davidson J. (1999). A genetic programming approach to rainfall-runoff modelling. *Water Resour. Manage.*, **13**(3), 219–231.

Seyoum S. D., Vojinović Z. and Price R. K. (2009). Integrated modelling of urban wastewater systems. In: *8th International Conference on Urban Drainage Modelling*, Tokyo, Japan 7–11 September.

Seyoum S. D., Vojinović Z. and Price R. K. (2010). Urban pluvial flood modelling: development and application. In: *9th International Conference on Hydroinformatics, HIC 2010*, Tianjin, China 7–10 September.

Shakhnarovish G., Darrell T. and Indyk P. (2005). *Nearest-neighbor methods in learning and vision. Theory and Practice*. The MIT Press, Cambridge, MA.

Sherman L. (1932). Streamflow from rainfall by the unit-graph method. *Eng. News Rec.*, **108**, 501–505.

Solomatine D. and Dulal K. N. (2003). Model trees as an alternative to neural networks in rainfall–runoff modelling. *Hydrolog. Sci. J.*, **48**(3), 399–411.

Solomatine D. and Siek M. B. (2006). Modular learning models in forecasting natural phenomena. *Neural Networks*, **19**(2), 215–224.

Solomatine D. P. and Xue Y. (2004). M5 Model trees and neural networks: application flood forecasting in the upper reach the Huai River in China. *J. Hydrol. Eng.*, **9**(6), 491–501.

Speziale C. (1991). Analytical methods for the development of Reynolds-stress closures in turbulence. *Annu. Rev. Fluid Mech.*, **23**(1), 107–157.

Stravs L. and Brilly M. (2007). Development of a low-flow forecasting model using the M5 machine learning method. *Hydrolog. Sci. J.*, **52**(3), 466–477.

Swamee P. and Jain A. (1976). Explicit equations for pipe-flow problems. *J. Hydr. Div.*, **102**(5), 657–664.

Takens F. (1981). Detecting strange attractors in turbulence. *Lect. Notes Math.*, **898**(1), 366–381.

Temam R. (2001). *Navier-Stokes Equations: Theory and Numerical Analysis*. American Mathematical Society, New York.

Tjalling J. (1995). Historical development of the Newton–Raphson method. *Soc. Ind. Appl. Math.*, **37**(4), 531–551.

Todini E. and Pilati S. (1987). A gradient method for the solution of looped pipe networks. *Computer Applications in Water Supply*, **1**, John Wiley & Sons, 1–20.

Toth E., Brath A. and Montanari A. (2000). Comparison of short-term rainfall prediction models for real-time flood forecasting. *J. Hydrol.*, **239**(1–4), 132–147.

USEPA (2008). *Guidance on the Development, Evaluation and Application of Environmental Models*. Council for Regulatory Environmental Modelling, USEPA, Washington, DC.

Volterra V. (1926). Fluctuations in the abundance of a species considered mathematically. *Nature*, **118**(2972), 558–560.

Weigend A. and Gershenfeld N. (1994). *Time Series Prediction: Forecasting the Future and Understanding the Past*, Proc. NATO Advanced Research Workshop on Comparative Time Series Analysis, 14–17 May, Santa Fe.

Witten I. H. and Frank E. (2000). *Data Mining: Practical Machine Learning Tools with Java Implementations*. Morgan Kaufmaan, San Francisco, California.

Wolfram S. (1986). Theory and applications of cellular automata. Advanced Series in Complex Systems, World Scientific Publishers, Singapore.

Wylie E. and Streeter V. (1978). Fluid transients, McGraw Hill, New York.

Yen B. (2004). Hydraulics of sewer systems, Chapter 4. In: *Urban Stormwater Management Tools,* L. W. Mays (ed.), McGraw-Hill Professional, New York.

Yevjevich V. (1972). *Probability and Statistics in Hydrology.* Water Resources Publications Littleton, Colorado.

Zadeh L. A. (1965). Fuzzy sets. *Inform. Contr.*, **8**(3), 338–353.

Zijderveld A. (2003). *Neural Network Design Strategies and Modelling in Hydroinformatics*. Technical University of Delft, Delft.

Chapter 6
Decision Support Systems

6.1 INTRODUCTION

An intervention in the physical world usually involves making a choice between possible alternative forms of the intervention. Consequently a decision has to be made. Often the information needed to make a good decision is extensive, and yet decisions have to be made regularly on the basis of limited and uncertain data. Whereas decision makers endeavour to gather as complete a set of relevant data as possible they will often be frustrated by the cost and time taken to collect the data they want, the difficulties in obtaining accurate and reliable data, the impossibility of acquiring historic data that has not been recorded, and when they have appropriate data it is affected by the frequent structural changes that are made to the system and alter its performance. This is especially true of urban water systems.

Although monitored data is preferred, where there is a shortage of data it is possible to use models, especially physically based models, as a viable complementary tool to augment the data. They can do this by replicating the historical performance of the physical system or predicting the performance of a physically modified system. Models, together with recorded data, are therefore designed to provide better (more complete, less uncertain) information, provided the models are developed in such a way that they provide safe and reliable results. The information produced by a model can be considerable, and the data should be analysed in order to distill the essential information that is required. A variety of tools has been developed to facilitate such an analysis. In effect, the model generates a virtual digital world, and we need to view that world in different ways. A traditional subjective approach is to present the results graphically so that trends and patterns can be recognised visually by the user, for there is a real danger of being confused by the raw data as sets of numbers. The human brain is still one of the best facilities for recognizing patterns, though its conclusions often need to be quantified by the computer.

The combination of databases and models to interpolate in time and space between monitored data or to extend the range or scope of the data, together with a user-friendly interface, form the basis of what has come to be called a decision support system (DSS).

Decision support systems are a category of information systems that support decision making by an individual or group of people within an organization. The idea behind a DSS is to provide decision makers with facilities to collect relevant information from a range of sources, including databases, monitors, models, documents, case studies and personal knowledge, to analyse such data in order to address and solve particular problems, and to make safe and reliable decisions. Given a model it is possible to generate alternatives for physical intervention. In this way the decision support system may enable the selection of the 'best' alternative, which in turn may involve optimization.

6.2 COMPONENTS OF DECISION SUPPORT SYSTEMS

There are three essential components of a DSS: database, model and user interface. The interface assumes the user(s) as another important component of the DSS. Other possible components include on-line access to monitoring and real time control systems, together with a reasoning process incorporating the data and model(s) and aimed at a specific set of goals. In urban water management, such a reasoning process is often based on procedures that are distilled from human experience. Rules of thumb can be captured by such procedures, and codified knowledge can be encapsulated in specific tools such as simulation models. The procedures are a way of structuring human mental and practical activity, and of ensuring that things are done in the appropriate order. Such an order can be sequential or parallel or even looped, such as during an iterative process. The benefit of such procedures is that proven steps and sequences of steps can be taken which are known to lead to safe and reliable decisions, provided the steps are properly taken and made.

A systems approach to making decisions centers first on clearly defining the objectives associated with an issue or problem. In order to compare the objectives we need to agree measures, whether quantitative or qualitative. Alternative solutions or scenarios should be generated to meet the objectives, and these should be assessed in order to qualify one or more alternatives that are included in making the decision.

6.3 DECISION MAKING UNDER UNCERTAINTY

Two of the greatest problems faced by decision makers are the incompleteness in the relevant data set and uncertainty in the data that is available. Incompleteness of the data can be overcome by acquiring more data, but where this applies to historic data the only possible way left open is to use a simulation model to replicate what happened historically, provided the relevant input data is given. The treatment of uncertainty in the data is another major issue affecting a decision. In fact, there is uncertainty not only in the data but in the tools used to analyse and make use of the data. We can distinguish between *inherent* or *aleatory uncertainty*, which consists of random (stochastic) fluctuations that are intrinsic to the problem being studied, and *epistemic uncertainty*, which results from a limited knowledge of fundamental phenomena; see Korving *et al.*, 2002. An example of aleatory uncertainty is the temporal and spatial variability of water demand or of rainfall over an urban area. Epistemic uncertainty can be classified as *model uncertainty* (due to a lack of understanding of the physics) and as *statistical uncertainty* (due to a lack of sufficient data), where the latter can be reduced when the relevant knowledge becomes available. So, decision making itself is fraught with uncertainty.

Uncertainty can be represented in terms of the two common concepts of probability and possibility; see Maskey and Price, 2004. Traditionally, uncertainty has been represented by probability, though since the 1970s possibility theory has been used increasingly as an alternative. The theory of possibility is similar to, yet conceptually different from, the theory of probability. Probability depends on the frequency of occurrence of an event, while possibility focuses on the meaning of an event. It does not imply that a high degree of possibility leads to a high degree of probability. If, however, an event is not possible, it is also improbable. Therefore, possibility can be viewed as an upper bound for probability. This weak connection between probability and possibility is defined by Zadeh (1999) as the *possibility/probability consistency principle*.

In probability theory all uncertainties are assumed to be random and therefore they can be represented by probability distributions of events. Uncertainty in possibility theory however, is quantified in terms of imprecision and vagueness, and is represented by a possibility distribution. Possibility is a *fuzzy measure*, which is a function taking a value between 0 and 1. This indicates the *degree of evidence* or *belief* that a certain element x belongs to a set Zadeh (1999) and Dubois and Prade (1988).

Uncertainty can be characterised as either quantitative or qualitative. Quantitative uncertainty is expressed using numerical values. Typically, these numerical values have a meaning from a probabilistic point of view. For example, '*the probability for the water level at the road junction to exceed a value of 24.3 m during a 10 year rainfall event, is 4%*'. In the quantitative approach, the laws that describe the physical phenomena are assumed to be known. However, the value of some of the parameters, for example, surface runoff coefficients or pipe roughness, may not be known precisely. The uncertain parameters and variables are treated as random variables, that is, variables which have known probabilities to exceed (or not) certain values. The random variables are characterised by a probability density function (pdf) and its moments (mean, variance, etc.).

Qualitative uncertainty can be expressed using qualitative terms. It is used when the sources of uncertainty cannot be estimated precisely using numerical values. For example, '*the water level is very likely to be high*'. In this statement, the probability for the water level to be high is not represented in numerical terms, but using what is called a linguistic variable. The most widely used method to deal with qualitative uncertainty is fuzzy set theory (a case of possibility theory), in which the linguistic variables are represented by a membership function. The membership function is the representation of the degree of belief in the statement.

The two most commonly used methods for estimating uncertainty in the time series generated as the output from a simulation model are Monte Carlo Simulation (MCS) and the First Order Second Moment (FOSM) method, both of which are based on probability theory.

6.3.1 Monte Carlo simulation method

The concept behind this method is that there are probabilistic uncertainties about the input variables and parameters, and that these uncertainties can be quantified in terms of probability distributions with standard properties, including mean, variance, etc. By defining the distributions in terms of their parameters and randomly sampling from the distributions as input to the model, running the model a large number of times for different sampled inputs, and statistically analysing the output of the model, it is possible to relate the uncertainty in the input to the uncertainty in the output. The following approach can be adopted:

1. Use data and expert judgement to select input distributions for model parameters. Particular attention should be given to the quantity of information at the tails of the distributions.
2. Provide detailed information on the input distributions selected. This information should identify whether the input represents largely variability, largely uncertainty or some combination of both.
3. Randomly sample uncertain parameter variables X_i from the relevant distributions
4. Run the model $y = g(x_i)$ with the set of random values and store the model output y
5. Repeat steps 3 and 4 a number of s times
6. Derive the cdf and other statistical properties of Y with the model realisations $y_1, \ldots, y_s$
7. Discuss the presence or absence of dependencies and correlations

See Shrestha, 2009. There are a number of derivatives from the Monte Carlo approach, such as the Markov chain Monte Carlo method (Metropolis *et al.*, 1953 and Hastings, 1970) and Latin hypercube sampling (McKay, 1992).

6.3.2 First order second moment method

The FOSM is an analytical technique and is one of the most widely used techniques for uncertainty analysis in civil engineering applications. It uses the first-order terms of the Taylor series expansion about the mean value of each input variable. Considering a function y that relates the input random variables $X_1, \ldots X_n$ to the output random variable Y, the expressions for the moments of Y will be:

$$E(Y) = y(\mu_{x1}, \ldots, \mu_{xn}) \tag{6.1}$$

$$Var(Y) = \sum_{i=1}^{n}\left(\frac{\partial y}{\partial X_i}\right)^2 Var(X_i) + 2\sum_{i=1}^{n}\sum_{j=1}^{n-1}\left(\frac{\partial y}{\partial X_i}\right)\left(\frac{\partial y}{\partial X_j}\right) Cov\,(X_i, X_j) \tag{6.2}$$

where $Cov\,(X_i, X_j)$ is the covariance between X_i and X_j. If the input variables are statistically independent (uncorrelated), the second term of Eq. (2) vanishes and the variance of Y becomes

$$Var(Y) = \sum_{i=1}^{n}\left(\frac{\partial y}{\partial X_i}\right)^2 Var(X_i) \tag{6.3}$$

In this way the parameters of the probability distribution can be estimated.

6.3.3 Qualitative method: fuzzy set theory with expert judgment

This qualitative method was first proposed by Sundararajan (1994, 1998) and is based on fuzzy set theory. In particular, it uses assessments made by experts about uncertainty. The assessments are carried out on the Quality (Q) and Importance (I) of the uncertain parameters for their uncertainty contributions to the output. *Quality* refers to how good the parameter is, and *Importance* refers to how important a parameter is, in its contribution to the uncertainty of the output. Both Q and I are evaluated qualitatively using a set of linguistic variables. For example, *Excellent*, *Good*, *Acceptable*, *Bad* and *Poor* (for quality) and *Enormous*, *Large*, *Moderate*, *Low* and *Minute* (for importance). Each of these variables is represented by a membership function.

The higher the quality of a parameter, the lower is the resulting uncertainty. The contribution of a parameter to the total uncertainty is therefore proportional to the so-called 'mirror image', $\tilde{Q}$, of Q. If the membership function of Q is defined by couples $[(b_1, s_1), \ldots, (b_n, s_n)]$, the membership of its mirror image is given by $[(1 - b_1, s_1), \ldots, (1 - b_n, s_n)]$, where $b_1, \ldots, b_n$ are base values defined in the scale of 0 to 1 and $s_1, \ldots, s_n$ are belief levels (Maskey and Guinot, 2002). The total uncertainty in the output is then given by:

$$U_Y = \sum_{All\ i}^{Fuzzy\ sum} I_{Xi} \otimes \tilde{Q}_{Xi} \tag{6.4}$$

where U_Y is the uncertainty in the output (forecast),

I_{Xi} is the importance and

$\tilde{Q}_{Xi}$ is the mirror image of the quality of the uncertainty parameter X_i represented by fuzzy numbers (membership functions).

The symbol $\otimes$ represents multiplication between fuzzy numbers. Readers are referred to Maskey (2004) for the details of this method. Good illustrations of various fuzzy calculi can be found in Kaufmann and Gupta (1991).

6.3.4 Qualitative uncertainty scale

A qualitative scale is prepared for the interpretation of the uncertainty result represented by a membership function, using the concept of the *best case scenario* and the *worst case scenario*. The best case scenario corresponds to the case where qualities of all the sub-parameters are assigned as *Very Good* (the best quality) and the worst case scenario corresponds to the case where all the qualities are taken as *Very Bad* (the worst quality). In both cases the Importance of the parameters is taken from the assessment of the experts. If a scale of five divisions is used, the qualitative indexes can be *very small uncertainty* (VSU), *small uncertainty* (SU), *moderate uncertainty* (MU), *large uncertainty* (LU) and *very large uncertainty* (VLU). The two extremes in the scale VSU and VLU correspond to the best and worst case scenario respectively. The fuzzy scale can be further transferred into a crisp scale. This is achieved by defuzzifying the fuzzy membership function of each uncertainty index by a suitable method; see Ross (1995) for different defuzzification methods.

Despite the fact that the two methods presented above are based on different theories (probability and possibility), they share some similarities. The basic forms of Eqs. (6.3) and (6.4) are same. Both equations are the summations of quantities resulting from the product of two quantities. The two quantities are the square of sensitivity and the variance in the quantitative method and the Importance and the Quality (its mirror image) in the qualitative method. Therefore, in these two equations, the sensitivity of a parameter is equivalent to the importance of a parameter and the variance of a parameter is equivalent to the mirror image of the quality of the parameter. The difference is only in the way the two multiplying quantities are derived and represented.

6.3.5 Improved uncertainty methods

Maskey and Guinot (2002, 2003) proposed an improvement to the traditional FOSM method. Krzysztofowicz (1999) introduced a Bayesian framework for uncertainty quantification in flood forecasting. Romanowicz and Beven (1998) introduced a method called the Generalised Likelihood Uncertainty Estimation (GLUE), which has been applied by several other researchers.

GLUE is based on the Monte Carlo method with constraints on the admissible parameter vectors. The assumption is made that there are many different model structures and each model structure has many different sets of values for the parameters that are acceptable in reproducing the observed behaviour of the real world system. This is called equifinality; see Beven and Freer, 2001. For each model run with a set of parameter values the output is assigned a likelihood value related to the performance of the model. All runs with likelihood value above a threshold are retained for further analysis. These have their likelihoods rescaled such that the sum of the likelihoods is equal to unity and the uncertainty of the model output is estimated by deriving the likelihood weighted cumulative distribution function of the output from the simulations; see Figure 6.1.

The UNEEC method was developed by Shrestha (2009) to estimate the uncertainty associated with the model structure and the parameter set of values by analysing the model residuals. In this case the method determines the uncertainty for all sources of error. The main steps in the UNEEC method are:

1. Cluster the input data to identify subsets of more-or-less similar points
2. Estimate the probability distribution of the model residuals for each subset
3. Construct the machine learning model (using, say, an ANN or Model Tree) of the probability distribution of the model error

In general, the UNEEC method gives consistently narrower uncertainty bounds than the analytical approach; see Figure 6.1.

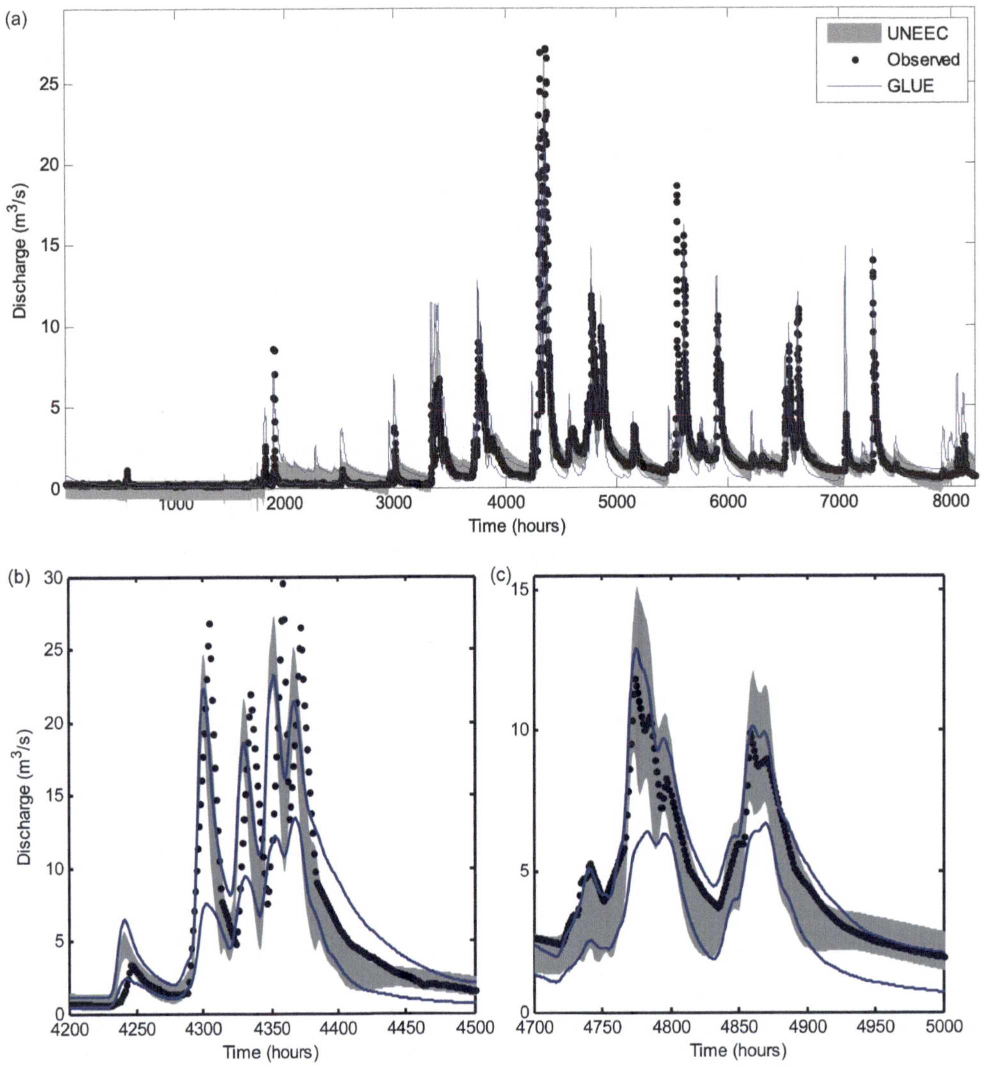

Figure 6.1 A comparison of 90% prediction bounds estimated with the UNEEC, GLUE method in the validation period (a) 1995/06/24 05:00 to 1996/05/31/13:00 (b) enlarge view of the period of about 12 days including the highest peak flow at 1995/12/22 and (c) the relatively small event after about one week later (Shrestha, 2009)

6.4 DECISION MAKING WITH OPTIMISATION

Many urban water management decisions seek the best from a number of options, such as from alternative scenarios for rehabilitating a drainage system, or an optimum of a particular quantity, such as the minimum cost for a particular rehabilitation scenario. Optimization can therefore be an important aspect of decision making. The application of optimization techniques ranges from identifying optimal values of model calibration parameters to training a data driven model to designing the optimal sizes of pipes in a network.

In order to apply one of the optimisation methods there must be a well-defined statement of the problem, a clear definition of the objectives and constraints, specified interrelationships between feasible regions for the solution, alternative courses of action, etc. Subsequently a mathematical model must be formulated. In decision-making, preferences are defined using a payoff function called the objective function. The decision-maker's problem can be presented as one of choosing an action among feasible decision variables that maximize or minimize the value of this function. The selection of decision variables can be subject to restrictions or constraints, and therefore, a set of models must also be defined representing such constraints.

A decision maker can transform the problem into a mathematical statement such as:

$$\underset{x}{Min}\, f(x) \quad \text{for}\, x \in S \tag{6.5}$$

where $f(x)$ is a payoff or objective function,

x is a decision variable and a vector $(x_1, x_2, x_3, \ldots, x_n)$

S is the feasible region in which the selection of solutions is made.

This region is defined depending on the constraints in the problem.

An optimization method can be applied to solve the decision-making problem. There are several such methods, depending on the nature of the objective function, the decision variables and the constraints. Usually a good practice is to perform a sensitivity analysis to determine the critical variables that can affect the solution due to their uncertainty.

6.4.1 Traditional optimisation methods

Each optimization method has a different algorithm to solve it. If the objective function is known analytically, five main approaches are commonly used: Linear Programming (Hiller *et al.*, 1995), Integer Programming (Nemhauser and Wolsey, 1999), Dynamic Programming (Howard, 1960), Non Linear Programming (Zangwill, 1967) and Stochastic Programming (Birge and Louveaux, 1997).

Optimization using linear programming is represented by:

$$Min\left(\sum_{i=1}^{n} c_i x_i\right) \quad \text{for} \sum_{\mathrm{i}=1}^{n}\sum_{j=1}^{nc} a_{ij} x_i < b_{Jn} \tag{6.6}$$

where x_i are the variables

a_{ij}, b_{jn} and c_i are coefficients

6.4.2 General optimization methods

As most of the urban water related optimization problems are non-linear with multiple extreme values (local minima), and which are not known analytically (so gradients cannot be calculated), some other techniques need to be used. In this respect, some of the potential techniques are: random search (McKay *et al.*, 1979), evolutionary computation (Holland, 1975 and Goldberg, 1989), simulated annealing (Kirkpatrick Jr, 1983 and Cerný, 1985), particle swarm optimization (Kennedy and Eberhart, 1995), ant optimization system (Dorigo and Di Caro, 1999), immune system (Coello, 2002, 2007), and clustering optimization algorithms (Solomatine, 1999).

6.4.3 Multi-objective optimization

In many cases decision makers have to deal with several objectives. Frequently, these objectives can be found conflicting between each other, and if one objective is achieved another becomes worse and vice-verse. Such problem can be tackled by using a multi-objective optimization technique. Multi-objective optimization involves the optimization of a vector of objective functions, which would give a set of acceptable solutions for decision makers (Osyczka, 1985). General Multi-objective Optimization Problems (MOP) can be posed mathematically as:

Find the vector $\mathbf{x}^*=[x_1, x_2, x_3, \ldots, x_n]$ that optimizes the vector function:

$$f(x) = |f(x_1), f(x_2), f(x_3), \ldots, f(x_n)| \quad \begin{array}{l} \text{for } g_i(x) > 0 \quad \text{for } i = 1,2,3,\ldots,m \\ h_i(x) > 0 \quad \text{for } i = 1,2,3,\ldots,p \end{array} \tag{6.7}$$

The notion of an optimum in MOP is not like the case of single optimization where just one solution exists. In MOP the concept of optimal solution is a set of solutions that represents the trade-off between objectives. This concept was introduced by Edgeworth (1881) and generalized by Vilfredo Pareto (1896). A solution $\mathbf{x}^*$ is Pareto optimal if there is no other $\mathbf{x}$ that improves the solution of one objective without damaging another. Figure 6.2 shows a graphical interpretation of the one MOP with two objective functions and two decision variables.

Dominance is a very important concept in MOP: a vector $\mathbf{u} = [u_1, u_2, u_3, \ldots, u_k]$ is said to dominate the vector $\mathbf{v} = [v_1, v_2, v_3, \ldots, v_k]$ if and only if $\mathbf{u}$ is partially less than $\mathbf{v}$.

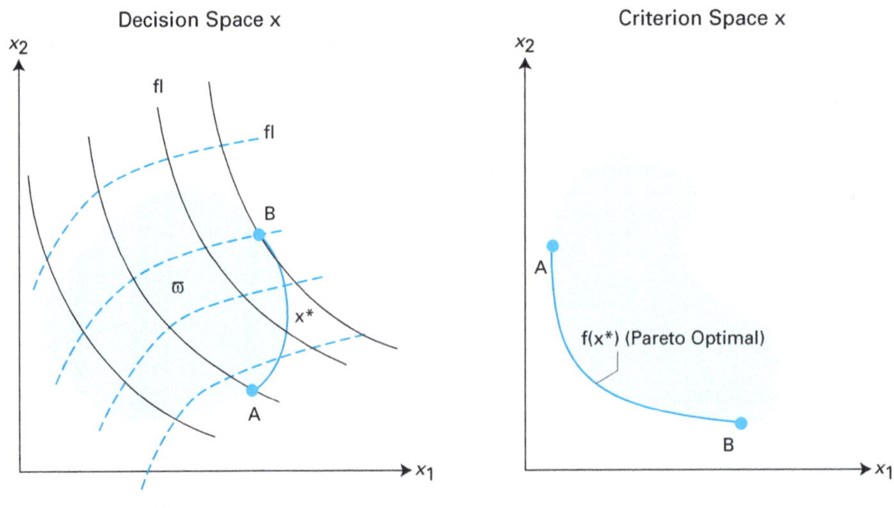

Figure 6.2 Interpretation of a MOP

6.4.4 Traditional methods for MOP solution

Methods to solve MOP are based on the same approach as for single optimization problems. However, the algorithm has to be adapted to find the optimal Pareto solution.

In order to solve Eq (6.7) several algorithms have been proposed. Early methods were based in aggregations such as weighted sums, distance to reference objective functions, the Keeney-Raiffa method, isovalues curves and the goal attainment method (Collette and Siarry, 2003).

6.4.5 Evolutionary algorithms

In the last two decades metaheuristic methods have became more prominent. There are several multi-objective genetic algorithms, and those most used are: Non-Dominated Sorting Genetic Algorithm (NSGA-II) and Epsilon Multi Evolutionary Algorithm (epsilon-MOEA).

NSGA-II was developed by Deb *et al.* (2002) as an improvement of its predecessor, NSGA, also developed by the same authors. In NGSA-II the offspring population Q_t is initially created using the parent population P_t, based on mating and mutation processes. These two populations are combined forming a new population R_t of $2N$ dimension. Then a non-dominated sorting algorithm is used to classify the entire population R_t. This classification is made using a ranking (fitness) function which can identify several fronts ($F_1, F_2, ..., F_n$).

If the new population is greater than N only the best ranked individuals remain. In this way the elitism is maintained. Usually, more than N individuals have the same rank (as they belong to same front); individuals are not deleted arbitrarily, they are selected in such way that they reside in the least crowded region using the concept of crowding distance which ensures diversity. Figure 6.3 illustrates the selection process.

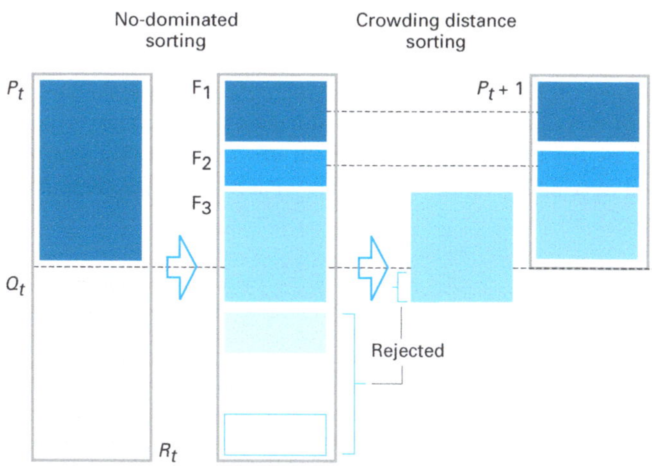

Figure 6.3 Selection process

Epsilon-MOEA, introduced by Laumanns *et al.* (2002) and Deb (2003), is based in the concept of ε-dominance, which does not allow two solutions with a difference of less than ε in each objective function. The search space is divided into cells and diversity is maintained by ensuring only one solution per cell. The convergence (elitism) is preserved by using an archive that contains all ε-non-dominated solutions. A 'normal' individual from the population is selected by a tournament (based on the dominance concept) and is combined with a randomly selected parent from the archive. The child is compared with each individual from the archive and if it is ε-non-dominated then it becomes part of the archive. Figure 6.4 shows the schematic diagram of Deb's Epsilon-MOEA.

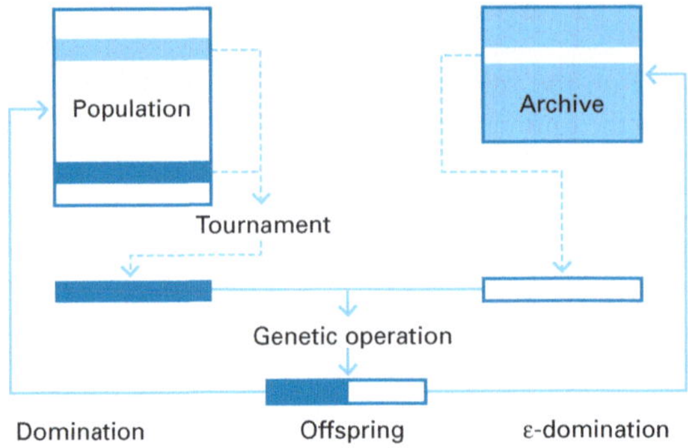

Figure 6.4 Schematic diagram for Deb's Epsilon-MOEA

6.4.6 Performance and Pareto comparison

One of the main issues in multi-objective optimization is how to establish the stopping condition. Another issue is the comparison between Pareto sets. These issues imply the need to select a suitable metric indicator.

There are several possible metric indicators that can be used to make a comparison between Pareto sets. However, Zitzler and Künzli (2004) have indicated that there is no unique quantitative indicator available that can suggest which Pareto set is better, and it is not always possible to decide about the advantages or disadvantages of such indicators. Some traditionally indicators are: cardinality, number of function calls, computational time, hyper-volume and ε-indicator.

Cardinality is the number of solution vectors in the Pareto set, which is an indirect measurement of the diversity. The number of function calls and the corresponding computational time is of great importance in urban water modelling due to the long running time of the underlying models. Hyper-volume is a measure of the hyper-space dominated by the Pareto front, and ε-indicator I_ε is a scale factor that indicates how much a Pareto set A must be moved or scaled in order to dominate a Pareto set B (the Pareto reference set):

$$I_\varepsilon(A, B) = \underset{z \in B}{Max}\ \underset{z \in A}{Min}\ \underset{1 \leq i \leq n}{Max}\ \frac{z_i^1}{z_i^2} \tag{6.8}$$

where I_ε is the epsilon indicator
A and B are the Pareto sets to compare,
z^1 and z^2 are the objective function vectors of A and B respectively and
n is the number of objective functions.

6.5 PROCEDURES FOR DECISION SUPPORT

One of the key aspects therefore, affecting the successful application of modelling software is the expertise of the user. In particular, the user has to conceptualise the physical world situation to generate the model space, to identify the physical processes for the flow of the fluid as well as the structural or asset objects defining the space, to select appropriate performance and structural or asset data, to calibrate the model and/or confirm the structural or asset data, to apply the model to develop scenarios, and to interpret the results sensibly. The aim is to produce safe and reliable information from the model for decision making. A procedure provides an accepted framework within which the user can build, develop and apply his model with confidence. Such a framework has other advantages. For example, some tasks can be made prescriptive, others can involve a strict sequence of steps, and still others can be open in that the user can follow his insight. The trend is to treat these tasks as objects with attributes, such as an associated knowledge base, documents, tools, sub-tasks, input, output and an audit. This approach formed the basis of SAFEDIS (see Ahmad and Price, 1998), which was conceived as a computer tool for managers and engineers working on sewerage rehabilitation; see Figure 6.5. Note that the audit capability is very important in that it provides the opportunity to carry out a formal quality assurance on the instantiated model (rather than just the software). A similar approach was adopted by the EC project HarmoniQuA, which resulted in a computer-based set of tools to assist the model developer take advantage of the wide experience of experts and is independent of the particular modelling software. Like SAFEDIS the tool takes into account the class of user (such as managers, modellers, auditors, stakeholders, laypeople). In particular the knowledge base system uses proven ontological techniques, making the knowledge base consistent, maintainable, accessible and flexible.

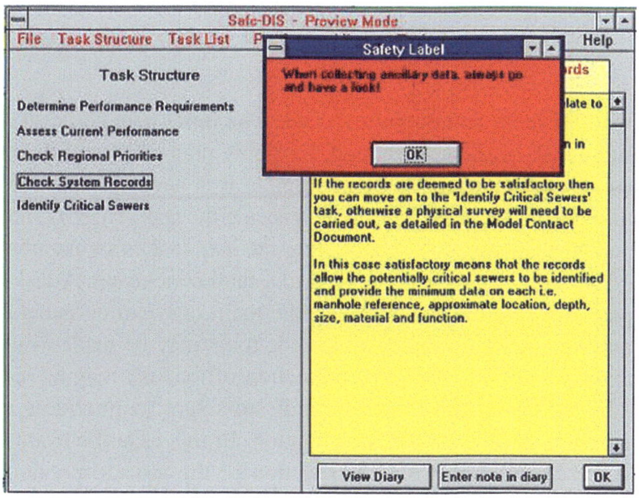

Figure 6.5 Safety labels for SAFEDIS

In this book we advocate decision support systems based on procedures such that they can be regarded as a series of tasks in a particular structure. This concept is like a rule based expert system, but different from it. Rule based expert systems were anticipated in the 1980s as providing the basis of a robotic society in which artificial intelligence would circumvent human inadequacy; this turned out not to be realizable. Such expert systems rely on a strict application of rational logic, which does not reproduce the flexibility with which human beings think and reason. Human thought and reasoning is often irrational as humans draw on their unconscious minds and embedded experience. Therefore a different, more flexible approach is needed, and instead of rules we adopt the notion of a procedure as a sequence of tasks.

6.5.1 Tasks and attributes

Tasks within procedures can be considered as objects with attributes. A task can be very simple with the requirement, for example, of getting the value of a variable. Other tasks can be extremely complex and involve a whole range of activities. Ideally, a task should consist of only a few activities. A list of attributes of a task could include

- Input
- Forgoing tasks
- Output
- Subsequent tasks
- Documented description including purpose and objective
- Knowledge based advisor
- Warnings and errors
- Activities
- Tools, including databases, models and GIS
- Criteria, standards and parameters
- References
- Case studies and example applications
- User comments and advice
- Tutorial

See Figure 6.6.

Necessarily there is a need for input and output to a task. The input may come from one or more previously executed tasks or from a future task in an iterative loop. Certainly the result(s) of the task needs to be output for consideration by subsequent tasks. For a task to be intelligible it is necessary for the user to have an accurate description of it, and to be provided with advice on how to execute the task. There is room for a rule-based system to provide such advice, coupled with warnings concerning the steps taken or parameter values outside certain ranges, and even the identification of errors that can lead to a fatal conclusion. A task may well have a range of activities that need to be performed before the output of the main task can be generated. Similarly, in order to assist in generating the resulting output a number of specific tools may be made available including databases, models, optimization techniques and GIS. The implementation of the task may be subject to agreed criteria or standards, or be facilitated by the evaluation of certain indicators such as provide a measure of the 'health' of a function or aspect of the particular asset under investigation. In this case the evaluation of the indicator may well be a sub-task. Then there are examples of the execution of the task on previous implementations of the procedures. These case histories can be analysed using case based reasoning (see Aamodt and Plaza, 1994) in order to evaluate the confidence that can be placed in the result(s) from the existing application. There should always be room for the user to add his comments on the nature of the task, its limitations and benefits. This confirms that any resulting system should have a learning capability. Finally, the user may need to be tutored in performing the more complex tasks.

6.5.2 Closed and open task structures

Some procedural task structures are closed, and can appear like a 'cook-book'. The coding of most computer programs is like this, namely a sequence of tasks that form a logical path that is pre-determined (deterministic) with no opportunity for random behaviour.

In contrast, there are open ended task structures in which there are many degrees of freedom open to the user. In both closed and open ended structures the equivalent of the digital computer is the human being, who follows the tasks identified in a procedure in order to reach a conclusion, that is, to make a decision. In other words, it is the human being who makes the decision. This means that the human individual brings his values and beliefs to bear unconsciously on his responses to the procedural approach.

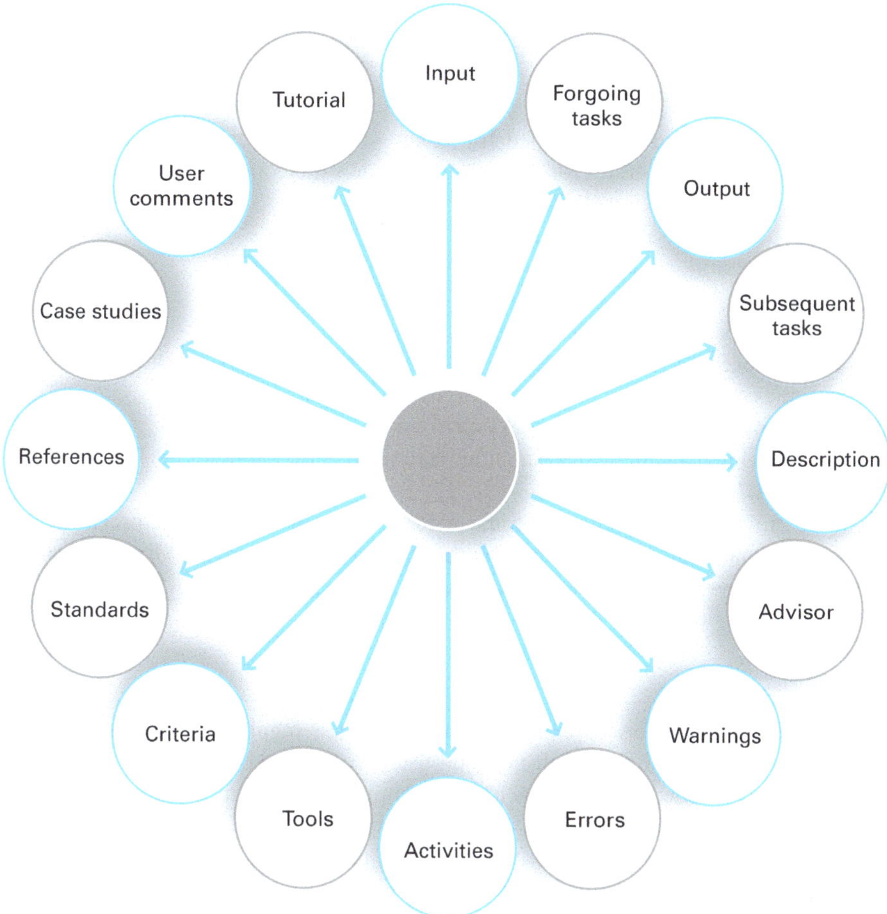

Figure 6.6 Task and its attributes

People making decisions depend on their own experiences to a large extent. Often they are working under stress due to time pressure. They know that their data is uncertain, and that the decision they make is associated with the notion of risk.

6.5.3 Making decisions

Making decisions is a complex human process that involves a continuous interaction of the human being with his or her environment. A decision is also the result of the attitudes, values and beliefs of the individual. It depends on the degree of use of logic and rational thought, and the explicit or tacit assumptions made. Experts apply their knowledge in a specific area when making decisions, though depending on the situation, data availability, and ambiguous circumstances they think intuitively more than following a rational structured approach. Decisions also involve making the 'best' choice between alternatives. This may well involve optimization.

6.5.4 Examples of procedures

There are a large range of procedures developed for urban water management, especially in the developed countries. Besides the important procedure for instantiating models there are well established procedures for

- Water distribution network design (WRc, 1990)
- Sewerage network design (Wallingford, 1981)
- Water distribution network rehabilitation (USEPA, 2009)
- Sewerage network rehabilitation (WRc, 1994) (USEPA, 2009)
- Urban water quality management (FWR, 1998)

Besides these widely accepted procedures there are a large number of best management practice (BMP) documents covering topics such as

- International Stormwater Best Management Practices (USEPA/ASCE, 1996)
- Urban water demand management and planning (Baumann *et al.*, 1998)
- SWITCH project (UNESCO-IHE, 2005)
- Similarly, there are quality assurance procedures for a range of tasks. These procedures are developed nationally and internationally, for example, by the International Standards Organization and the British Standards Institute

6.5.5 Updating procedures

There is a need to keep the decision support systems up-to-date regularly. Procedures should strictly be learning systems that benefit from the experience of their users. Every use of the DSS should potentially attract an improvement that can be encoded into the system. Such improvements can be made generally available after confirmation by a software administrator, over Internet if necessary.

Groups can fashion their own variants or procedures, and tailor them to meet the specific needs of their particular organisation. Indeed, this flexibility can be very important in meeting the needs of stakeholders.

6.5.6 Joint decision making

Whereas much decision making is done by individuals working alone, there are many situations where groups of people need to make decision jointly. There are various tools to assist this process. Besides the obvious means of voting and expressing the majority viewpoint, there are ways of collaborative working which draw people together and help them to appreciate different aspects and factors about a particular issue before the decision is made. Communities of practice can be run over the Internet, using tools for brainstorming and information sharing. Additionally, the Internet can be accessed over mobile phones implying that location or even the need of a laptop is not necessary to carry on discussions. Practitioners in the field can operate instantiated models remotely on a server, download results, and analyse the resulting data.

6.5.7 Benefits of DSS

There are a number of benefits that accrue to the individual and the organisation through the use of a well structured and user friendly DSS. These include:

- Improving personal efficiency in processing information for decision making
- Promoting learning or training through using the DSS
- Speeding up problem solving
- Encouraging curiosity on the part of the decision maker
- Generating new information affecting a decision and securing its application
- Revealing new approaches to thinking about the problem space
- Facilitating interpersonal communication, especially between managers and public
- Removing the routine from the managerial processes and therefore stimulating managers
- Increasing organizational control without stultifying initiative by others
- Creating competitive advantage over organisations and countries.

6.6 INSTANTIATING MODELS

The modelling software produced by its developer(s) becomes an object in its own right and with its own responsibilities. It is the invention of one or more developers, produced in accordance with their particular view of the world, including the way in which the physical objects for the model space are selected and related and how knowledge is encapsulated within them. The instantiation of the software for a particular situation is then the responsibility of the user. Indeed, the user may have no personal relationship with the developers and may therefore be entirely ignorant of the intentions of the authors behind the software other than what they make accessible through manuals, the interface and any experimental use made of the code. What is more, the user has his own view of the world, and particularly of the physical situation that he wishes to model. Again, the user will use the software to build and apply a model for the situation as he sees it. Indeed, no two users working independently will build exactly the same model, except perhaps in very simple circumstances.

6.6.1 Modelling as part of the knowledge management process

In Chapter 5 we discussed modelling from the point of view of information associated with an interaction with the physical world. It is apparent that modelling requires extensive knowledge, both implicitly in terms of the knowledge 'in hiding' within the software, and explicitly, such as is needed by the user. But the user is not in isolation. He is surrounded by others, including the developer(s), those responsible for data collection, the decision makers, and so on. There is an important communication process that needs to go on. This inevitably brings in the need for effective knowledge communication or transfer. We need to understand how professionals in different fields think. What is 'knowledge' for the engineer? How does the engineer communicate his knowledge effectively with another person? What makes effective communication? How can knowledge be shared accurately and concisely? Increasingly, various functions on Internet such as blogging, twittering, communicating through Facebook™ etc are opening up new ways of accessing and sharing knowledge. As mobile phone technology changes so it is to be expected that more immediate mobile communication will continue to enhance the practical application of decision support systems in the field as well as in the office.

6.6.2 Modelling within a project

The whole process of modelling is normally embedded in a project as defined in Figure 6.7. Modelling is always part of a larger objective. For simplicity we call the process of achieving the larger objective the *project*. What is more, computational hydraulic modelling is a serious business with commercial and public interests. It should be viewed therefore, as an activity that falls within best management practices (BMPs). There needs to be a clear structure to the practice of designing, building, calibrating or training, verifying, confirming and applying a model. This will enable a better development and application of models whose results can be trusted for decision making.

We can conceive of at least four primary phases (or 'steps' in HarmoniQuA terminology) in carrying out a modelling instantiation project; see Figure 6.7:

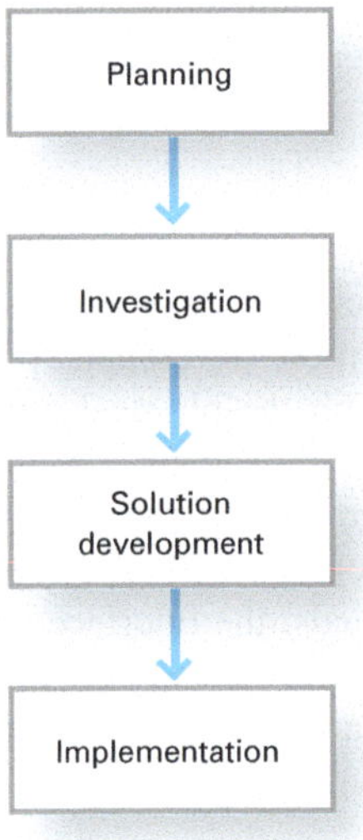

Figure 6.7 Modelling project tasks

Within the planning phase there are a number of tasks that need to be carried out, largely in sequence:

1.1 Initiate and maintain project audit documentation
1.2 Determine project objectives
1.3 Determine project scope
1.4 Assess (historical) data availability
1.5 Consult stakeholders about objectives and information
1.6 Identify the need for modelling
1.7 Select the modelling approach
1.8 Agree performance criteria for modelling
1.9 Identify additional tools
1.10 Procure modelling software/tools
1.11 Determine additional data requirements

We can regard the investigation phase as being centred around the development of one or more suitable models, though there may be other tasks that have to be carried out simultaneously. The key modelling tasks at the investigation stage include:

2.1 Plan and execute additional data collection
2.2 Build model by assembling data
2.3 Simplify model structure
2.4 Confirm the structural or asset data used in the model
2.5 Calibrate selected model parameters
2.6 Confirm instantiated model
2.7 Assess model sensitivity
2.8 Assess model uncertainty
2.9 Assess base performance

Note that tasks 2.2 to 2.6 are programmed here for a physically based model. Alternatives for a data driven model are given in Section 6.7 below.

The solution development phase builds on the initial results and conclusions drawn from the model:

3.1 Generate solution options
3.2 Select preferred option
3.3 Assess preferred option
3.4 Produce detailed design

See Figure 6.8.

6.6.3 Planning phase

Task 1.1 Initiate and maintain the project audit documentation

Description	Documentation that records the execution and completion of each task, including all relevant details
Input	Information on initiation, execution and completion of project tasks
Output	Complete audit trail and summary
Activities	Descriptions of tasks performed
Tools	Dedicated DSS that prompts for input and highlights missing data

There are a number of things to do before launching into a full-scale model investigation. The first is to initiate and maintain documentation that will provide an audit trail on all tasks, results and decisions made during the project. This is important for ensuring quality control, providing necessary information for stakeholders, decision makers and other users of the model, archiving the model, and facilitating its possible subsequent rehabilitation.

Resource questions:

- Is the documentation accessible to everybody?
- Are quality assurance procedures in place?

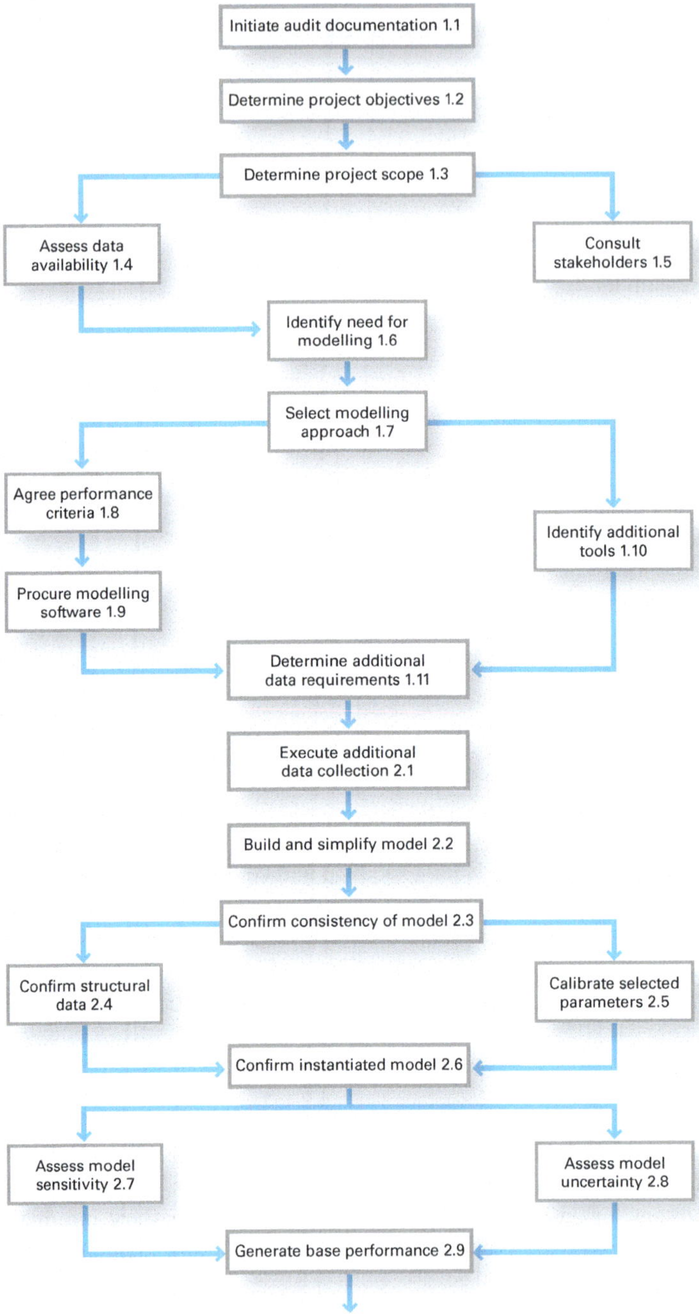

Figure 6.8 Procedure for building and using physically based models

Task 1.2 Determine the project objectives

Description	Identify and catalogue the objectives of the project and make explicit the arguments in support of these objectives
Input	Information from decision makers on purpose and objectives of the project
Output	Audit documentation and project scope
Sub-tasks	Inform team members
Tools	Brainstorming, Collaborative decision making, Case-based reasoning

Without clear project and modelling objectives, time can be wasted in doing unnecessary tasks. There should be confidence that a model is really needed for the purpose-in-hand. Generally a model is required for one of four purposes: to understand the relationships between different aspects of a complex system for operational and maintenance purposes, to assist in the assessment of the system performance for the design of a physical intervention, to forecast the system performance in real time, and to assess the conformity of the system with regulatory standards. Usually, the objective of the model will be precise and measurable. On occasions, however, such as in a research project, the objective may be open-ended. In either case, there is a real need to ensure that the end-model gives a good representation of the physical world system and that there will be a significant gain in understanding of how the system performs.

From a management point of view, the immediate application of the instantiated model must be clearly specified, and will normally result in the confirmation of the design of particular engineering works. A new distribution network may be functioning poorly. An old sewerage system may need rehabilitating. The level of pollution in a river, lake or coastal waters from CSOs may be unacceptably high. These are all reasons at least to consider using a numerical model to improve predictions of what will happen after civil engineering works have been completed. Such predictions may make the difference on whether such works may go ahead. More likely than not they will enable engineers to make improvements to the design, reducing the overall cost of the structural works and the impact on the environment or demand.

For drainage there are at least four major reasons for constructing a hydraulic simulation model:

i) Coarse Drainage Area Planning (Master planning)
ii) Detailed Drainage Area Planning (DAP)
iii) Detailed Design
iv) Sewer Quality

as well as other applications.

A coarse drainage area model would normally be used for master planning over a large catchment or watershed, or to screen plans for a number of distinct sewerage systems in order to determine priorities for investment. Such a model could also be used to determine flows at a coastal outfall, to calculate the output from a large sub-catchment contributing to a detailed model of a major catchment downstream, to assess the performance of trunk sewers, or to determine water level constraints in trunk sewers affecting the outfalls from one or more detailed models. This type of model is not suitable for the detailed representation of flooding problems as it requires a better resolution of the model space.

A detailed drainage area (DAP) model may be used for asset management planning, including the identification of the causes of hydraulic problems such as flooding and CSO spills, the evaluation of investment in specific construction projects, and the confirmation of the value of a given network. It is then used to develop integrated rehabilitation solutions to all the identified problems and needs within an overall planning strategy for improving the system performance.

A detailed design model is used for investigations where considerable detail is needed, such as for appraisal of schemes and their detailed design. Such models are restricted to specific areas within a DAP model where hydraulic upgrading schemes and other similar design projects are being considered.

Finally, a sewer water quality model becomes important when there are concerns about the environmental impact on receiving waters, particularly shell fisheries, bathing waters and rivers. It is necessary to estimate the pollution loading in combined sewer overflows, with a possible concern about trade effluent discharges. The deposition and accumulation of sediment in the sewerage network may also be of interest due the consequences for ongoing maintenance.

The objectives therefore should be clearly defined as these will determine what type of model is developed, the need for data collection, the calibration accuracy, the types of applications, and the scenarios to be considered. The objectives should be clearly recognised and understood by all members of the team working on the project. The team may well consist of engineers, planners, operations and maintenance staff, financial, legal and managerial staff, as well as particular end customers. Once the overall objective has been agreed then work can begin to identify what data is available.

Resource questions:

- Is it clear what the major reason is for constructing a hydraulic model?
- Are the objectives clearly stated?
- Is the audit documentation in place?
- Are all agreed on the objectives of the project?

Task 1.3 Determine the project scope

Description	Identify the scope of the project in terms of the application, the tools and the data
Input	Project objectives and information from key stakeholders
Output	Audit documentation, project scope
Sub-tasks	Consult decision makers and key stakeholders
Tools	Brainstorming, Collaborative decision making, Case-based reasoning

The next decision to be made is on the scope of the project. The objective will define what is going to be addressed. The scope looks at the implications and consequences of the objective. So, for example, the objective could be to assess whether pressures in a water distribution network satisfy minimum requirements under different demand scenarios. Consequently, the scope would be defined by the different scenarios and the extent of the network. Alternatively, consider the objective to rehabilitate a sewerage system. In this case, the scope could be to limit the frequency of surface flooding and to reduce the impact of CSO discharges on a receiving stream.

A scoping study should be carried out whatever the nature of the project. It would often be necessary to consult urban developers on future plans for housing and industrial development in the catchment. For water distribution it will be important to have information on land use and its consequences for demand. For wastewater drainage a major concern will be with consequences of urban development for the existing treatment works and to investigate the implications for water quality in the receiving waters. Stakeholders should be identified, with plans to involve them where possible in the project. The environmental requirements, including those of the regulator, should be acknowledged, together with a clear statement of the nature of the problem, the extent of the model and the data required. A consequence of the scoping study is the development of a conceptual model for the proposed components of the model space consistent with the physical system.

In the case of combined wastewater and storm drainage it may be necessary to assess the performance of the integrated system in determining how best to rehabilitate the sewerage system. This will mean bringing together sewerage, treatment works and river water quality experts to work as a team. Similarly it will be important to consult widely with relevant stakeholders so that their views can be taken into account, issues that may have been missed can be included, and that adverse legal consequences can be anticipated and avoided wherever possible.

Resource questions:

- Is the overall scope of the project clear?
- Are all agreed on the scope of the project?
- Have all stakeholders been included?
- Have their views been taken into account?

Task 1.4 Assess the historical data availability

Description	Appraise the existing historical data on the system, including both temporal and structural or asset data
Input	Project objectives and scope, information on data availability, and access to the data itself
Output	Audit documentation, assessment of historical data
Sub-tasks	Consult archives and personnel
Tools	GIS, databases

The data assessment process should cover all aspects of data associated with the project. Critical categories of data for a physically based model are:

- Surface topography and mapping
- Geology (for groundwater analysis)
- Channel, stream and pipe network data (as required)
- Ancillary structures, reservoirs, valve or overflow settings, pumps, gates, orifices, weirs
- Historical performance data, including rainfall and antecedent conditions, discharges and water levels, infiltration, pressures and leakage rates, flooding, pollutant concentrations and sediment
- Maintenance records
- Previous study documentation

Nowadays, water utilities possess large quantities of data for their networks stored in digital format, which are embedded within many organisational functions and are being used for a range of modelling and management tasks, including network analysis. For example, GIS maps could be used to generate models automatically directly from asset databases. Water utilities often want to integrate geospatial data with CAD drawings, accounting systems, maintenance programs, asset management, and other activities to improve efficiency and better meet their modelling needs. Figure 6.9 shows a GIS map view of a water distribution network alongside its hydraulic model.

The extent of the network or model space will already have been indicated by the project objectives and scope. Often however, the extent is also a function of the available data. Care should be taken to see that such data is not a limiting criterion.

For many 1D modelling applications for drainage the rainfall-runoff generation can be an important part of the whole modelling process. Consequently it may be necessary to have good records for the above-ground catchment(s) so that the modelling can be done reliably, achieve required levels of accuracy and minimise the degree of uncertainty. Similarly, a robust water distribution model is dependent on data for the demand at nodes throughout the network together with the appropriate performance data to deduce leakage.

Figure 6.9 GIS map (left) and corresponding hydraulic model (right)

In 2D coastal flow or 3D groundwater problems the boundary conditions may become a critical factor, particularly as it is usually impossible to have accurate information on velocities, water levels and pollutant concentrations at each and every boundary. Even if that were possible then errors in the data could lead to an unacceptable build-up (or draw-down) of water or pollutant within the flow domain. Consequently, a judicious choice of boundary conditions usually allows some of the boundaries to be 'free', that is, to permit the model to 'sort itself out' such that no significant information enters the flow domain. Another issue with the boundary conditions is the need to limit the possibility of any inappropriate reflected disturbance feeding itself back into the flow domain. If any significant information does enter the domain then there may be a need for inverse modelling to determine the (unknown) input (Cañizares and Torre, 1999).

Very often the trigger for a particular range of studies is the inadequate performance of the system in the past. Associated with these historical events are records that indicate the extent of flooding or damage, the changes in demand for potable water, or the magnitude of fish kills in a polluted river.

Resource questions:

- Is the structure of the model space complete for a physically based model?
- Is the performance data sufficient to confirm the reliability of the model?

Task 1.5 Consult stakeholders

Description	Involve stakeholders in advising on the implementation of the project
Input	Project objectives and scope, data availability
Output	Audit documentation, stakeholder feedback
Sub-tasks	Questionnaire to stakeholders, communication of project objectives and scope
Tools	Web based questionnaire, Internet site

Stakeholders, such as major water users and the local community represented by political leaders, should be involved in discussions with the managers of the urban water system and their technical experts to take advantage of indigenous knowledge, ambitions and preferences when making decisions. Such knowledge and information can be obtained through questionnaires, workshops, discussion groups, and other forms of consultation meetings. Internet tools can be used wherever possible, especially as the Internet can be the vehicle for virtual reality modelling to convey information on proposed solutions to specific problems to do with urban water systems. The model becomes an important means of communication between modellers, analysts and decision makers. The resulting models can even be operated by the intelligent layperson such that they can investigate their own concerns about the project. It is important that such users are given the right information to assure them of the accuracy and trustworthiness of the models used.

Resource questions include:

- Have all stakeholders been approached?
- What is the most important advice provided by the stakeholders in the light of their involvement with the model development?

Task 1.6 Identify the need for modelling

Description	Determine whether a model is needed and what sort of model should be used
Input	Project objectives and scope, data availability
Output	Audit documentation, need for modelling
Sub-tasks	Questionnaire to team members
Tools	DSS

Strictly, the need for modelling within the project depends on the appreciation of the complexity of the water-based system, and the existing ability to understand and predict the complex interactions that occur.

Many of the phenomena that can be identified in these systems are so complex in their generation and interaction that we have to resort to models to help us understand what is going on, and to make predictions or forecasts of what will happen. Here we regard forecasting as a possible sub-process of prediction. In other words, a prediction can be generated from any structural rearrangement or initial or other boundary condition, whereas a forecast is the prediction of what will happen from time 'now' assuming a known history of the structural or asset state(s) and boundary conditions.

Resource questions include:

- What are the boundaries of the physical domain that determine the model space?
- Have all the significant physical, chemical and biological processes been recognised?
- Is the scope of the model consistent with the objectives of the project?
- What are the accuracy criteria for the different types of data for the modelling project?
- Where are the critical points in the domain where information must be determined?

Task 1.7 Select the modelling approach

Description	Identify the scope of the model and the way of instantiating and applying the model
Input	Project objectives and scope, data availability, need for modelling
Output	Audit documentation, modelling approach
Sub-tasks	Assess staff skills and need for training
Tools	Benchmark the potential modelling software

In that most studies involve the modelling of the hydraulics, a key to obtaining reliable and robust results will be the proper instantiation of a hydraulic simulation model. Normally this will be based on a solution of the relevant non-linear equations, including the full Saint Venant equations for the flow in a drainage network of pipes or open channels.

Having established the scope of the model, that is, its physical extent, the range of processes that it should embrace and the frequency of events that it should consider, a decision should be made on whether a physically based or data driven model should be used. A comparison of the advantages and disadvantages of using data-driven (compartmentalised) or physically based modelling is given in Table 6.1:

Table 6.1 Comparison of data driven and physically based models

Modelling paradigm	Data-driven	Physically based
Advantages	Pattern identification for knowledge discovery (on monitored data or model results)	Detailed 1D, 2D and 3D modelling by taking account of the physical processes
	Excellent at linking monitored 'output' data to 'input' data 'Recovery' of missing data Forecasting levels, flows, currents Replication of complex (physically based) models input-output relationships (CSOs, rainfall-runoff, water quality, etc) optimisation sensitivity/uncertainty Learning systems Complex unknown phenomena Error management in physically based models	Identifies explicit consequences of particular structural or asset domain features and associated flow phenomena within the flow domain and disassociated with the boundary conditions Good basis for flow forecasting Can be applied with reduced amount of boundary condition data Excellent at determining in detail flows within the flow domain
Disadvantages	Cannot extrapolate easily outside range of training Cannot accommodate changes to the structural or asset domain, unless explicitly included as an input parameter Difficulty in relating to physical phenomena Need adequate training and verification data Requires good understanding of the physics in order to correctly identify input and output	Need adequate structural or asset domain data Less confidence in extrapolating beyond calibration range Cannot deal with situations where the phenomena are unknown or cannot be adequately formulated algorithmically Difficult to identify as a learning system Problems in handling errors (mismatches between predictions and observations)

Note that a data driven model may only be used without a physically based model if there is sufficient time series data for the necessary input and output available. Alternatively, there may be reason to use a data driven model to replicate the physically based model, for example, when there is a need to carry out an optimisation with a physically based model that is very time consuming, or there is a need to generate a quick and reliable forecast of flows in a network.

Given that a physically based model is going to be used then it is important to determine the level of detail that the model should reproduce. It is also important to identify where specific results are required, whether in terms of discharge, velocity, water level, pollution concentration, and so on. Once these factors are agreed then the next step is to identify all of the physical objects to be included in the model space. Necessarily this will affect the selection of data for the model.

A number of choices have to be made. These will include

- Physically based or grey box?
- Prediction or forecasting?
- 1D, 2D or 3D?
- Steady state or dynamic?
- Design (prescriptive) or simulation?
- Stationary boundary for the model space or dynamic (morphological)?
- Integrated or discrete?
- Finite difference, element or volume?
- Conservative or non-conservative (pollutants)?

See Table 6.2.

Table 6.2 Choice of model dimension

Model category	Application areas
Compartmentalized (0D)	rainfall-runoff, reservoirs, flood routing
1D hydrostatic, sub- or super-critical flow, free surface or pressurised 1D Boussinesq (first order non-hydrostatic) 1D compressible	rivers, flood plains, estuaries, irrigation, water distribution, wastewater and stormwater collection, treatment works, water hammer
2D hydrostatic, vertical (depth – layers, horizontally averaged) 2D hydrostatic, horizontal (spatial, depth averaged, turbulence, wind (for shallow depths) pressure, Coriolis density variations, wave stress) 2D Boussinesq (first order non-hydrostatic) 2D wave refraction, reflection, diffraction	lakes, wide river channels, flood plains, estuaries, coastal waters, groundwater
3D hydrostatic (turbulence, wind, pressure, Coriolis, thermoclines, density variations) 3D non-hydrostatic, (Navier Stokes plus turbulence)	lakes, river channels, channel-flood plains, coastal waters, estuaries, treatment processes almost anything, river channel-flood plain interaction, structures

If a water quality model is needed for the drainage network, the UPM manual (FWR, 1998) recommends three levels of modelling according to the amount of detail involved:

1. A full water quality model consisting of a detailed hydraulic model with a corresponding detailed water quality model, such as using InfoWorks CS or MOUSE TRAP
2. A full hydraulic model with a simplified water quality model, such as provided by SIMPOL (Dempsey *et al.*, 1997)
3. A simplified water quality model, such as SIMPOL

The decision to construct a detailed water quality model depends on the availability of adequate water quality data. Inevitably, the uncertainties in the water quality data are much larger than the flow data, and there is considerably more uncertainty in the mathematical formulation of the water quality processes, which is reflected in the resulting water quality model. Both calibrating and verifying the water quality model can be extremely difficult. Often the modeller has to resort to default values for the various parameters associated with the various water quality process equations.

Resource questions:

- Can the choice of model be justified?
- What are the principle processes and structural or asset objects making up the model space?
- Why have these process and structural or asset objects been selected?
- What possibly significant process and structural or asset objects have been ignored, and why were they not selected?
- What are the essential links between the structural or asset objects?
- Have the process objects been correctly linked?
- Is the mathematical system 'closed'?
- What assumptions have been made about the model structure?
- Benchmarking models for the water framework directive: Model characterisation scheme, model inventory, benchmark questions/criteria, benchmark process

Task 1.8 Agree the performance criteria

Description	Agree on the performance criteria that the modelling software and instantiated model should meet
Input	Range of performance criteria
Output	Audit documentation, performance criteria
Sub-tasks	Review previous applications of modelling software
Tools	Modelling software, benchmarking data

Another important decision needs to be made before proceeding with the study and that is to agree on suitable criteria for the acceptance of the instantiated model. It may be that the decision to construct a particular engineering work is sensitive to cost, in which case the safety factor that has to be imposed because of uncertainties in the model may be too high. Consequently, it is valuable to aim for an agreed uncertainty in input data and reliability of the model to anticipate any safety factor that might be imposed. These performance criteria become particularly important during the model calibration and verification phases.

There are a number of measures that have been devised for the performance of a model. These include the following:

Root mean square error

$$RMSE = \sqrt{\frac{\sum_{i=1}^{n}(P_i - O_i)^2}{n}} \tag{6.9}$$

r-correlation coefficient

$$r = \frac{\sum_{i=1}^{n}(O_i - \overline{O})(P_i - \overline{P})}{\sqrt{\sum_{i=1}^{n}(O_i - \overline{O})^2 \sum_{i=1}^{n}(P_i - \overline{P})^2}} \tag{6.10}$$

Reliability index

$$RI = \exp\sqrt{\frac{1}{n}\sum_{i=1}^{n}\left(\log\frac{Q_i}{P_i}\right)^2} \tag{6.11}$$

Absolute average error

$$AAE = \frac{\sum_{i=1}^{n}|P_i - O_i|}{n} \tag{6.12}$$

Modelling efficiency

$$ME = \frac{\left(\sum_{i=1}^{n}(O_i - \overline{O})^2 - \sum_{i=1}^{n}(P_i - \overline{P})^2\right)}{\sum_{i=1}^{n}(O_i - \overline{O})^2} \tag{6.13}$$

where P_i and O_i are the ith predicted and observed values respectively

$\overline{P}$ and $\overline{Q}$ are the averaged predicted and observed values respectively

The *RMSE* and *AAE* measure the size of the discrepancies between predicted and observed values. Values of these measures near zero indicate a close match, and therefore they provide an indication of model accuracy. The r-correlation coefficient measures the tendency of the predicted and observed values to vary together linearly. Values of r vary between -1 and $+1$. Negative values indicate that the predicted and observed values tend to vary inversely. The *RI* quantifies the average factor by which model predictions differ from observations. Ideally *RI* should be close to one. Finally *MF* measures how well a model predicts relative to the average of the observations. A value near one indicates a close match between predictions and observations. A value of zero indicates that the model predicts individual observations no better than the average of the observations. Values less than zero indicate that the observation average would be a better predictor than the model predictions.

Resource questions:

- Are the performance criteria realistic for the objectives?
- Is there an assessment of the acceptable uncertainty in key parameters?

Task 1.9 Procure the modelling software

Description	Select and procure modelling software and train staff
Input	Project objectives and scope, data availability, modelling approach
Output	Audit documentation, modelling software
Sub-tasks	Benchmark software
Tools	Modelling software product

The process of selecting a suitable modelling software product addresses a number of factors that have to be implicitly assumed through extensive experience, or preferably explicitly assessed.

The actual selection of the appropriate physically based modelling software raises a number of other important issues. Careful assessment must be made of the facilities identified in Table 6.3.

Table 6.3 Modelling software facilities

Facilities	Algorithmic equations solved (processes included)
	Finite scheme used Solution technique adopted Data requirements and handling Model build Grid generation (2D and 3D) Results display Results and data comparison Networking Links to databases, GIS, MIS
Ease of use	Display of information Economy of user actions Intuitiveness of screen layout and tasks/actions On-line help and documentation Tutorials
Reliability and quality assurance	Supporting documentation including full description of modelling theory including treatment of recognised problems: Small depths Pressurised flow Sub- and super-critical flow Free surface-pressurised flow transition Numerical dispersion Numerical stability Hydraulic instabilities (eg roll waves) Benchmarks Case histories Accreditation Independent assessment Supplier reputation User appreciation Access to demo software and ability to test run
Cost	Purchase price Familiarisation/experience required Training courses Support Upgrades Hardware availability and requirements

A general piece of advice is to make a check list of what you want from the modelling software in terms of

- facilities from the software
- services from the supplier

If the software contains most of the facilities then questions should be asked such as:

- Why does the modelling software not consider the facilities that are not included?
- Can the modelling software contain them?
- When they will be available?
- At what price? etc.

Generally it will be found that while a modelling software product may include all of the general features and facilities, there will be some particular facilities or features that it does not include. A decision then has to be made whether the absence of these facilities or features is significant or not. For example, they could be provided complementary by other products (e.g. data or results handling, no explicit links to GIS or standard databases), or even written in-house (though this route is usually more expensive in the long run). The best approach is to establish a good relationship with a reliable supplier who has a good range of products, and in whom you can have the confidence of a win-win relationship. Remember that in general a commercial supplier will provide the best and most reliable software. The aim of such a supplier is to provide top quality products and service at a reasonable price that ensures he can continue his business and make an acceptable return on investment. To continue to build the business the products must demonstrate their quality in terms of what the customer wants, and they should be backed up by service that enables the customer to use the products successfully. Remember too that the more sophisticated products require considerable experience on the part of the user if they are to be used successfully. Therefore the support provided by the supplier or some other provider is an important part of the assessment.

The (USEPA, 2008) recommends asking the following questions when selecting modelling software:

- Does sound science support the underlying hypothesis of the modelling software?
- Is the complexity of the modelling software appropriate for the problem at hand?
- Do the quality and quantity of data support the choice of modelling software?
- Does the modelling software structure reflect all the relevant inputs based on the underlying conceptual model?
- Has the modelling software code been developed? If so, has it been verified?

Resource questions:

- Has somebody reviewed the software you have selected?
- Have you benchmarked the software?
- Can you identify savings in time and space that may affect the end result?

Task 1.10 Identify additional tools

Description	Identify additional tools for handling input data, calibrating models, communicating results
Input	Project objectives and scope, data availability
Output	Audit documentation, need for modelling
Sub-tasks	Questionnaire to team members
Tools	GIS, DTM, MIS, CAD

Nowadays there is a range of useful tools in addition to the main software engine that facilitate the handling of data and results and may even provide decision support on how to develop a given model. Such tools include:

- Computer Aided Design (CAD)
- Geographic Information System (GIS)

- Digital terrain Model (DTM) (or Digital Elevation Model, DEM)
- Database
- Data preparation
- Decision support
- Spread-sheet
- Video graphics in 1D, 2D or 3D
- Graphing – presentation

Here it is important that data exchange between the different products can be done with little effort. Generally the selection of the appropriate additional tools should be done with the decision maker(s) in mind as well as the facility with which the project can be completed. Generally, decision makers require clear, unambiguous presentation of results, usually in graphical form.

Resource questions:

- Have you explored tools that are consistent with your modelling systems?
- Has adequate training been given to users of the tools?

Task 1.11 Determine the additional data requirements

Description	Identify additional data needed to complete structural or asset description and temporal data
Input	Project objectives and scope, data availability
Output	Audit documentation, additional data report
Sub-tasks	Cost collection of different data items
Tools	Data QA

Usually the historic data will be lacking in some areas. This could be because records were not taken originally, or records have been lost or destroyed. The existing data should be analysed carefully, and needed data be noted and included in the requirements. Particular attention should be given to data for calibration and verification, and for drainage network models data for verification of the integrity of the model space.

Resource questions:

- Have the missing data been identified and their acquisition costed?
- Do you have sufficient data for a formal calibration and validation?

6.6.4 Investigation phase

Task 2.1 Plan and execute the additional data collection

Description	Formulate a plan to collect additional data over a period of time commensurate with the project
Input	Project objectives and scope, data availability
Output	Audit documentation, additional structural or asset, temporal and social data
Sub-tasks	Carry out quality assurance on data
Tools	Statistical tool for data analysis

Once the assessment of data has been done then decisions can be made on what further data is required. This should be done in conjunction with the establishment of the conceptual model when defining the scope of the study.

There are usually several types of survey required, depending on the nature of the problem and what is regarded as input to the model.

- Physical asset survey
- Structure survey
- Catchment survey
- Flow survey
- Social survey

Flow and pressure measurements are also necessary for comparing a water distribution model with the corresponding physical system to verify that it behaves similarly. This is achieved through a field measurement exercise normally consisting of:

1) Carrying out simultaneous measurements of flows on selected locations in the system (e.g., source works, pump stations, reservoirs) and flow transfers to and from the network (at all network import and export points).
2) Carrying out simultaneous pressure measurements at nodes to achieve sufficient head loss between points with known heads for performing calibration, i.e., locate pressure measuring devices on the perimeter of the network, away from water sources and near points of high demand. Pressure measurement should be performed at as many key sites as possible, e.g. at pump stations, in known problem areas, at large diameter pipes, etc.
3) Monitoring reservoir levels.
4) Carrying out measurements of significant demands, e.g., consumption of major users.
5) Carrying out pump tests to confirm head/flow behaviour curves.

See WRc, 1989. The field measurement exercise should normally be carried out over a 48-hour period, at least, but preferably over a 7-day period to cover diurnal changes and within week differences in flows and pressures due to variation in demand.

It could be said that water demand is the key modelling input and driving force behind the dynamic behaviour of a water distribution system. It is, therefore, critical to represent accurately demands in network hydraulic systems. Since the demand for water is modelled to take place at nodes, consumer records and maps are inspected in order to enable spatial allocation of the total demand to network demand zones and finally to nodes.

The total demand is composed of a number of demand categories based on the type of users, e.g., domestic, commercial, industrial, and their location. The demand normally allocated based on the average daily usage is called the base demand. *Peaking factors* and *diurnal demand patterns* are then applied to this demand to account for variation in demand. Peaking factors are generally used to multiply the average-daily demand to obtain a specific critical demand, e.g., peak-hour demand (during the maximum hour of usage), while diurnal demand patterns are used to account for demand variation throughout the day.

The simplest way to distribute demands is to divide the total demand by the number of demand nodes in the network and apply the same demand to each node. This is also the most inaccurate methods. The method can be improved by applying a correction for known large metered users and redistributing the remainder of the total demand uniformly. Demand can also be allocated based on actual customer connections using GIS tools; see Figure 6.10.

It should be pointed out that automated GIS routines can be used to allocate service connections to the nearest distribution pipe and then to the nearest node on it. Care should be taken to avoid allocating service connections to dedicated transmission mains that may be closer to the water user location than the distribution mains, but do not have service connections on them.

In a 1D physically based model of a drainage network the situation is more complex. Here there is often doubt about the completeness or accuracy of the asset data, both below and above ground. This means that besides calibration of the rainfall-runoff parameters there is a need to ensure that the right asset data is used in the model. Because the focus can be more on the asset data than the calibration parameters, use is made of short term flow surveys. The emphasis is on collecting flow data at a large number of points in the network in a short period of time (say 2 months) to identify where there are deficiencies in the asset data. This procedure has been adopted for some years in the UK. Now however, there is the recognition that longer term surveys are important in order to include more rare events in the survey data.

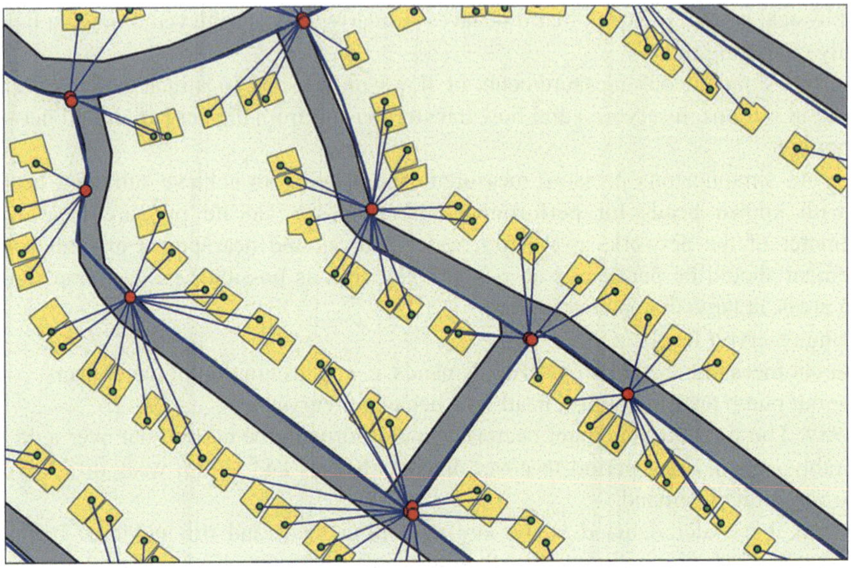

Figure 6.10 Distribution of service connections to nearest nodes using GIS

Resource questions:

- Has the data been collected to the required accuracy?
- Is there sufficient data to carry out both calibration and verification, and for drainage networks, to confirm the integrity of the data for the model space?

The following tasks apply to a physically based model. Corresponding tasks for a data driven model are described below.

Task 2.2 Build and simplify the physically based model

Description	Collect, assemble and validate the structural or asset data for the model and simplify if necessary
Input	Project objectives and scope, data
Output	Audit documentation, built model
Sub-tasks	Adjust compensation parameters for the simplified model, validate consistency of structural or asset data
Tools	Modelling software, Simplification tool

One of the first tasks of the model builder is to design the conceptual model by identifying the physical objects that are going to make up the model space: pipes, reservoirs or tanks, valves and demand points for a water distribution network, and pipes or channels, sub-catchments contributing to each conduit, manholes at the junctions between the pipes, ancillary structures such as overflows, outfalls, pumps, and storage tanks, and the urban surface topography for a drainage network, and bathymetry for a 2D coastal waters model. The objective is to include all objects that are significant, such that the scope of the model is established.

However, for drainage network models in particular there is the issue of the integrity of the network data in representing what is under the ground. Again, underground pipe networks are notorious for having features that have either been forgotten or been constructed illegally, such as overflows or underground streams connected to solve a previous construction problem.

Once the modeller is convinced that all significant objects are included, the next problem is to collect all the necessary data for each object. It is rare that the model space data is complete, and additional data may have to be collected or estimated. For example, a manhole survey should reveal many of the unknown features of a drainage network, but not all.

Besides the model space data, the boundary condition data for the solution space has to be established. In a water distribution network this will include pumped inflows, reservoir levels, valve settings and demands at demand points, in addition to discharges and heads at particular points in the network in response to the inputs and demands. For a drainage network the boundary data includes rainfall, pollutant loading rates, operating rules for structures, etc or discharges and water levels at points in a drainage network in response to recorded rainfall, probably collected through a short term flow survey.

The process of building a model can therefore be subjective and a function of the person(s) who does it. This is because in building a model a person has to make choices about the actual data he will use. In particular, he is likely to be faced with a decision about the simplification of the physical extent and detail of the model. Although experiential knowledge of experts can be documented and provided to the user, the choices to be made are so considerable that it is almost inevitable that no two models produced by different people will be the same.

In addition, the user should understand what the developer of the modelling software had in mind when producing the code. There should be a good transfer of knowledge between developer and user through manuals. ‘Rules of thumb’ embodied in best management practices (BMPs) should be clarified and used wherever possible. Such documents go a long way to improving consistency in modelling practice.

It is important to remember that data is always collected with a specific purpose in mind, and may not conform to the purpose of the current project. Therefore each data item should be considered from the point of view of the project. Note that most data driven model applications focus on relating output (runoff or overflow discharges to rainfall, or node demand or leakage to system inflow) and are not concerned with data for the model space.

In assembling the asset data all the major features identified in forming the conceptual model need to be included. Tests will need to be made to ensure that the data is consistent, that is, the various physical objects are correctly connected in the model and that the data is consistent between the objects. It should be born in mind that there can be large differences in the datum for separate site surveys and these datum may have not been correctly taken into account when referring the data to the model datum.

When assembling the event data it is important to ensure that the correct data is catalogued for each event. Like the domain data it is important that timing errors in the event data are minimised. Whereas it is obviously important to have good consistent flow data of sufficient density to cover the flow domain, in practice the modeller probably has more circumstantial evidence (qualitative data). This data can still be very important in enabling the modeller to appreciate the scale and nature of the problem. Indeed, apparent inconsistencies in such data may be resolved when the calibration process is underway.

Simplification of the asset data may be necessary to reduce the model complexity or to speed up the model runs. But there are limits to how far the model may be simplified in terms of the scope of the study, such that there are dangers in over-simplification. Pipes which are below certain size are normally either ignored or grouped together and replaced by equivalent pipes. Only those pipes that have a significant impact on the behaviour of the system need be included. An example of a water distribution network model generated by a GIS and its skeletonised (or simplified) version is shown in Figure 6.11.

Simplification of drainage network models is more difficult however. Simplification must take into account the storage that is removed with the pipes that are neglected. Compensation needs to be made for this storage by artificially enlarging the manholes adjacent to the neglected pipes. The WaPUG procedure (WaPUG, 2002) includes advice on how to compensate for the missing storage; see Chapter 10.

There are guidelines for the degree of simplification of drainage models according to their application. A coarse drainage area planning model includes a simplified representation of the core sewers, the sewered sub-areas, and the principle ancillary structures. Simplification levels are typically

1.0 to 3.0 ha/pipe (100–200 m) for core areas,

3.0 to 15.0 ha/pipe (200–500 m) for peripheral areas with 2 to 10 pipes/1000 population.

Figure 6.11 Network model (left) and corresponding skeletonised (simplified) model (right)

A detailed drainage area planning model includes major core sewers, simplified peripheral areas, all ancillary structures, and a detailed modelling of areas with known flooding and surcharging; Simplification levels are typically

0.5 to 2.0 ha/pipe (10–100 m) for core areas,

2.0 to 5.0 ha/pipe (100–300 m) for peripheral areas, with 10 to 25 pipes/1000 population for generally, and most pipes for problem areas, especially for flooding at low points in the system

The level of detail required for a detailed design model depends on the layout and density of pipes/manholes local to the scheme under consideration. However, these models normally include all relevant pipes and manholes, together with all ancillary structures.

In general, a water quality model must have a high degree of detail, generally in excess of drainage area planning (DAP) model standards, and be fully verified hydraulically for dry weather flows as well as for storm flows. This is discussed further below.

Resource questions:

- Are all important ancillary structures included in the model?
- Has the correct degree of simplification been established for the type of modelling exercise envisaged?
- Has the storage volume removed through simplifying the (drainage) network been compensated for?

Task 2.3 Confirm the consistency of the model

Description	Compare model results with flow survey data to confirm the structural or asset data
Input	Project objectives and scope, built model, flow survey data
Output	Audit documentation, confirmed structural or asset data for model
Sub-tasks	Survey physical system to confirm structural or asset deficiencies
Tools	DSS

Once the model is constructed, a number of checks are needed to identify possible data input errors. Tests are also made to ensure the model is stable in simulating the inputs (and demands for a water distribution model). A drainage model should be tested for rainfall events up to the maximum return period for which the model will ultimately be used. Routine checks following amendments to the model during the verification and design phases of the study are required to ensure the model remains stable throughout its life. What we are trying to do is to confirm the reliability of the model for the purpose 'in hand'.

Good quality assurance procedures throughout the construction of the model should enable a final quality control check on the completed model. The purpose of this check is to highlight any errors, omissions, or inconsistencies within the model data, as well as possible errors in the model's construction. Inspecting a plan view of the model network should confirm the replication of the connectivity of the physical network. Viewing long section profiles of sewer runs in a drainage model may also highlight errors. Any instances of steps in invert levels, flat or reverse gradients, reductions in pipe diameter downstream, or pipe soffits protruding above ground level, should be investigated to confirm their authenticity. Independent checks should be made on all hand calculations, such as the derivation of ancillary coefficients.

Not all simulation models may reproduce every aspect of a water distribution or sewerage network exactly. For instance, not many commercial software products for modelling drainage networks accommodate accurate models of low side weirs, and therefore additional checks must be carried out on all aspects where the software is limited.

It is important to check the stability of the model for all states covering the range of events and flows for which the model will be used. These should be performed to establish the stable performance of the model for a sequence of demands or a range of storm events for the critical duration for the catchment, including once in 1 year, 2 year, 5 year, 10 year, 20 year, 50 year, 100 year, and so on. The simulation period should be sufficient to accommodate the whole of the demand sequence for a water distribution model, or to allow storm flows within the system to drain down and for storage tanks to discharge in a drainage network model.

A check should be made to see that there are no numerical instabilities, denoted by oscillations of increasing amplitude that make the model 'blow up', or with a degree of persistence, particularly in drainage models. A true numerical instability will normally cause the computer program to break down, but occasionally the numerical instability can be constrained or damped subsequent to its first appearance. The difficulty is that such events may generate volume and therefore volume balances should be checked continuously at all points in the system. Typically, after the system has drained down the volume balance for the whole model and individual ancillaries should be within +/− 5%.

Known limitations, which may cause a drainage simulation to fail, are mainly related to features causing rapid changes in water level such as:

- pump wet wells with very small volume
- pump switch on/off levels set close to each other.
- rapid surcharge of small diameter pipes caused by flows backing up from the larger diameter sewers.
- small steep pipes
- steep to flat transitions
- small initialisation depths in large channels

The traditional measures of fit of a drainage simulation model with observed data are given in Table 6.4.

Table 6.4 Performance accuracy for modelling

Indicator	Range
Peak flow rate	+25%−15%
Flow volume	+20%−10%
Surcharge	+0.5 m–0.1 m
Hydrograph shape	timing of peaks, slope of recession curve, indications of throttling etc.
Flooding	location and extent of predicted flooding corroborated

Resource questions:

- Have the key parameters that are known to generate a sensitive response on the calculated flows been considered?
- Has the data for established sensitive parameters been checked for accuracy?

Task 2.4 Confirm the structural or asset data

Description	Compare model results with flow survey data to confirm the structural or asset data
Input	Project objectives and scope, built model, flow survey data
Output	Audit documentation, confirmed structural or asset data for model
Sub-tasks	Survey physical system to confirm structural or asset deficiencies
Tools	DSS

It has already been commented that the asset data can be ill-defined. This is less likely in river, estuary and coastal waters models where there are well-identified structures on view and where surveys of the geometry of the flow domain are well established. The problems normally arise with the underground networks associated primarily with wastewater drainage but also for water distribution. This is particularly the case with storm and wastewater drains for which knowledge of their geometric structural properties and even of the existence of some pipes may not be available. In the case of distribution networks, besides the possibility of having the wrong connections between pipes there is the problem of having the right valve settings controlling the distribution of the flow in the network. For drainage systems there are also well-documented instances of missing drawings, ancillary structures such as overflows and pumps not being recorded, and unknown stream inputs. Additionally the condition of many drainage pipes is often unknown, to the extent that pipes may be partially collapsed or have been penetrated by tree roots. This is one of the reasons why there is more concern in these cases with the shortage of knowledge about the completeness of data for the asset rather than the calibration of surface runoff models, pipe roughness or headlosses at structures.

A solution developed in the UK to this problem for drainage systems has been to develop generic models for surface runoff from urban areas. These models have been calibrated on data collected from a large number of different urban drainage and above ground catchments throughout the country. The models are sufficiently generic that they can be used with some degree of confidence anywhere in the UK (see below). The issue of roughness of pipes is less of a problem in that values of the equivalent sand roughness height, ks, in the Colebrook-White expression for the conveyance are comparatively easy to define and can also be used with some confidence. This leaves the problem of how to 'verify' (or rather confirm) that the asset data is correct.

The solution is to carry out what are called short-term flow surveys; see Chapter 4. These last for four to eight weeks, during which time it is hoped to collect rainfall and flow data for four or five significant events at a large number (say, 20 to 40) of monitoring points in the pipe network. A model for the drainage network is then established for the asset data initially available, and predicted results for the observed rainfall events are compared with the monitored flows. Where there is significant divergence between the observed and predicted results then a comparison of the results is used to identify where possible deficiencies in the asset data may be located. Any miss-match between predicted and observed hydrographs may then be used to deduce deficiencies in the asset data; see Figure 6.12 and Figure 6.13.

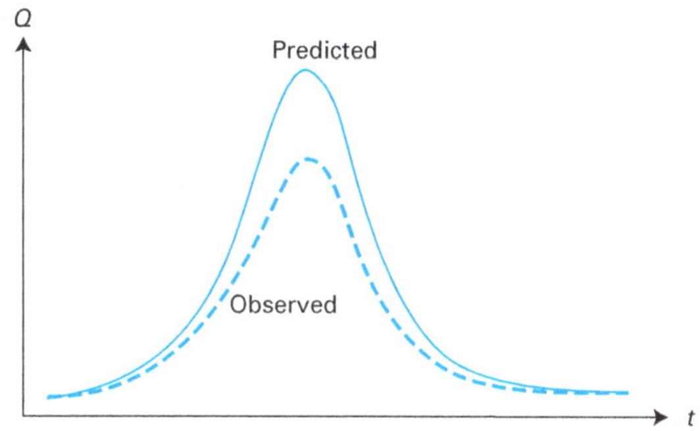

Figure 6.12 Excess connected area

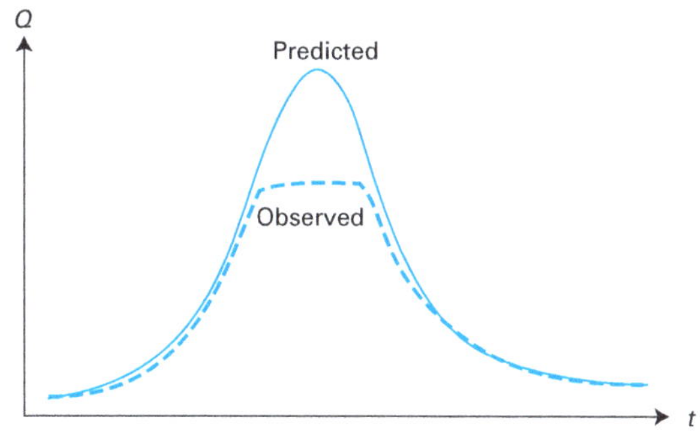

Figure 6.13 Missing overflow

These miss-matches are then confirmed by a manual survey of the particular parts of the asset including the above ground sub-catchments. If no alterations to the asset data can be discovered then resort has to be made to the original assumptions made about the surface runoff or even the roughness. In general however, it has been found on countless occasions that there are problems either with the asset or the above ground data.

The procedures developed in the UK for this task are very powerful but they depend significantly on having a reasonable surface runoff model that has already been calibrated for UK conditions. Unfortunately, the variance explained by the UK surface runoff model calibrated against available UK data is only about 50%. Also, the calibrated model has only been developed for a limited set of time series data and therefore does not include the more extreme events. Similarly, the short-term flow surveys will normally miss even the one year rainfall event. Consequently, any model with its asset data verified against such data cannot be expected to be verified for the more rare events. All of these considerations have promoted concern about the use of short-term flow surveys and the reliability of the calibrated surface runoff model. Increasingly engineers are looking to more limited long-term flow measurements in sewers to provide additional data both to calibrate the surface runoff model locally and to perform the verification of the asset data used in the network model. As yet this dual process of surface runoff model calibration and asset data verification has been done in an ad-hoc manner: a generally recognised procedure has yet to be introduced.

Table 6.5 Structural or asset verification issues

Structural or asset confirmation	Cause
Physical constraints	Missing or additional overflows/missing or incorrect pumps/loss of flood water at manholes
Infiltration/Base flow error	Missing or additional natural stream or groundwater inflows
Hydrograph scaling	Incorrect estimates of percentage runoff
Attenuation	Inadequate volume correction for network simplification
Time-shift error	Inadequate simplification/incorrect pipe roughness

The builder and developer of an instantiated model may not be the only person to take responsibility for the interpretation of the results. Indeed, it is desirable that the results should be made available to decision-makers who can then quickly assimilate them and make the appropriate decisions. Assimilation normally depends on effective use of images that embrace a (large) set of numbers. These images induce patterns that an experienced manager who is already familiar with the physical world system can appreciate. These images bring 'to life' understanding about the results. It should be borne in mind however that the procedures for interpretation are largely set by the developer of the software, even if the detailed appreciation of results for a particular instantiated model can only be done by the developer of the model. The possibility of having a third party review the instantiated model can be an important part of the model confirmation process. In this case the model should be subjected to a thorough, objective peer review, which pays attention not only to the form of the model output but also the quality of data used, and the degree to which the model replicates the system being modelled.

Once the structural integrity of the model is verified (as well as it can be), it is important to assess whether it can be used for the purpose it was intended.

Resource questions include:

- Have the structural or asset deficiencies been identified for all anomalies in comparing the predicted and observed flows?
- Have these structural or asset deficiencies been found within the physical system, and the model adjusted accordingly?
- Is the model detailed enough at the points of interest?
- Does the model perform well in surcharge if it is going to be used to assess the performance of the system under extreme events?
- Do the overflow structures operate in the model for some rainfall events (or time series)
- Do all of the pumps at a pumping station come into operation for at least one event or for part of the time series)?
- Will the model need to be altered to allow for planned changes in development in the area?
- Are there structures or situations that might occur which the model cannot replicate?
- How much uncertainty is there in the runoff model being used?

Task 2.5 Calibrate and validate selected parameters

Description	Calibrate roughness for pipes and/or hydrological model parameters
Input	Project objectives and scope, flow survey or historical event data
Output	Audit documentation, calibrated and validated model
Sub-tasks	Verify calibrated model
Tools	DSS

Before a model is used in anger, it must be adjusted to ensure that it will predict, with reasonable accuracy, the behaviour of the system it models, i.e., it must be calibrated, which is 'a test of a model with known input and output information that is used to adjust or estimate factors for which data are not (directly) available' (see Beck *et al.*, 1997). A number of studies on the calibration of water distribution models have been published; see Rahal *et al.*, 1980; Walski, 1983; Lansey and Basnet, 1991; Datta and Sridharan, 1994; Wu and Clark, 2009 and Savic *et al.*, 2009.

Manual calibration for water distribution

Calibration performed using modern simulation packages commences when input data including an initial estimate of the roughness values of all pipes is entered into the model. The model is then analysed and the results compared with the field test measurements; see Figure 6.14.

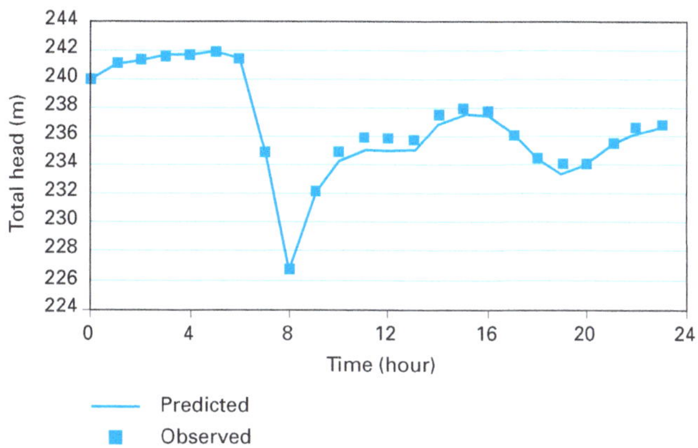

Figure 6.14 Visual Model Fit

Calibrations of this type are based on a tedious and time-consuming trial-and-error procedure where the parameter values are adjusted based on the hydraulic results and the analysis is repeated. This iterative process continues until operating specifications are satisfied or no viable change in input parameters can be found which improves agreement between observed and predicted values. In the latter case, the possibility of modelling anomalies, such as a reduced pipe diameter due to internal corrosion, or an incorrectly modelled open/closed valve, etc., should be investigated. Often this procedure will result in a less than optimum solution and may not be effective when a large number of variables and operating conditions are investigated.

Automated calibration for water distribution

The problem of calibrating a water distribution model, even if only for water quantity, (pressures and flows) is highly complex due to the large number of parameters examined and the non-linearity of the flow equations.

Several researchers have addressed this problem developing computational methods to minimise the difference between the values of the observed data and those computed by the network simulation model. These methods are based on:

(a) explicit (direct) approaches (Ormsbee, 1989 and Walski, 1983)
(b) implicit (indirect deterministic) approaches (Lansey and Basnet, 1991)
(c) Bayesian based (indirect stochastic) approaches (Kapelan *et al.*, 2007).

In the explicit approach, the hydraulic parameters to be estimated are solved directly as part of a system of n non-linear equations which describe the network hydraulics for a set of n unknown parameters. The major limitation of this approach is that the number of parameters to be estimated must be equal to the number of equations or measures. An implicit approach estimates the parameters by minimizing an error function which usually represents a measure of the difference between recorded and modelled data. The most promising calibration procedures are based on the implicit approach, namely, optimisation. The calibration process can be largely automated using an optimisation approach that is well suited to the problem. However, the success of traditional optimisation methods usually depends on linearising assumptions or the unrealistic calculation of partial derivatives. In addition, they are generally local optimisation procedures which tend to become entrapped in local minima or suffer from numerical instabilities associated with matrix inversion. Evolutionary computation methods (Holland, 1975 and Goldberg, 1989), of which the genetic algorithm (GA) is the best-known, are particularly suited to large combinatorial problems that do not suffer from the same drawbacks associated with classical mathematical programming techniques. A GA evolves solutions by sampling from all the possible solutions; the best of these solutions are then combined, using the genetic operators of *crossover* and *mutation* to form new solutions which are evaluated in turn. This process continues until some termination condition is fulfilled. The benefits of GAs stem from their conceptual simplicity (i.e., based on evolutionary principles with no need for simplifying assumptions) and their ability to converge rapidly on an optimal or near-optimal solution, having analyzed only a tiny fraction of the number of possible solutions available.

During the last 20 years, many researchers have used evolutionary computation methods to calibrate various hydraulic/hydrologic models. Wang (1991) investigated the use of GAs combined with fine-tuning by a local search method for calibration of a conceptual rainfall-runoff model. Models were calibrated by minimizing the residual variance defined as the sum of square of differences between computed and observed discharges. Duan *et al.* (1994, 1992) introduced the shuffled complex evolution method for a similar problem by hybridising a GA with the Simplex search method. The objective function used was the mean daily square root of the difference between the observed flows and simulated flows. Babovic *et al.* (1994) used GA and the hydrodynamic MOUSE package to fit Manning numbers to pipes while Mohan (1995) reported on use of GAs for estimating parameter values of some linear and non-linear flow routing and water quality prediction models.

The Bayesian-based calibration approach, e.g. Kapelan *et al.* (2007) represents the latest development in the area of the calibration of water distribution system models. It provides a natural, iterative type framework which can be used to (re)calibrate models when new information becomes available. Most importantly, this approach enables the simultaneous estimation of calibration parameter values and the associated uncertainties (i.e. errors) in a single model run as opposed to the aforementioned implicit calibration approach which aims to determine the calibration parameter values only. Even more, the Bayesian based methodology can determine the joint probability density functions of both calibration parameters and the associated model predictions in a single model run thus providing additional insights into quality of the calibration results obtained. The methodology is based on the Markov Chain Monte Carlo (MCMC) class of methods, e.g. the Shuffled Complex Evolution Metropolis (SCEM-UA) method (Vrugt *et al.*, 2003) used in Kapelan *et al.* (2007).

The SCEM-UA methodology does not require any derivatives to be calculated nor model linearity, parameter independence or similar assumption to be made. It also provides a natural way to specify prior information on calibration parameters (Kapelan *et al.*, 2004).

Water quality model calibration for water distribution

The calibration of a water quality network model requires an EPS analysis and a well calibrated hydraulic model. A demand-driven hydraulic network model requires the assignment of at least one temporal demand pattern that may be used to vary water flows during an EPS. The results from a water quality model calibration may be suspect if the demand patterns used for the hydraulic simulation are incorrect or contain significant errors. However, it has been shown by Jonkergouw *et al.* (2008) that the water quality model parameters may be correctly estimated by using an effective optimisation routine. They used the Shuffled Complex Evolution (SCE) algorithm developed by Duan *et al.* (1992) to calibrate model parameters.

Drainage model calibration

Unlike the calibration of water distribution models the calibration of drainage models can be caught up with the integrity of the structural data. For example, given estimates for the percentage paved areas contributing to different pipes, the percentages may need to be adjusted independently of any runoff coefficients, whereas altering the runoff-coefficients could be a viable possibility. The calibration of the various model parameters for the rainfall-runoff and for the pipe roughness values (especially for those pipes with tree roots penetrating them from outside) can be done in the normal manner by comparing the simulation model results with observed data. Note the caution expressed by Cunge (2003) who advocates that standard values of parameters should be used, at least to begin with, and these should only be modified in cases where the conditions are unusual (such as penetration of the pipes by roots). Certainly, optimisation of the parameters using a standard optimisation procedure should be avoided (optimisation may be used for other purposes such as prescriptive sizing of pipes or structures).

Parameters that may be eligible for calibration in a drainage model include:

- Rainfall-runoff loss coefficients
- Rate of runoff parameters
- Pipe roughness
- Ancillary structure headloss coefficients
- Infiltration
- Pump characteristics
- Rural catchment runoff
- Area of surface flooding at nodes

Water quality drainage model calibration

The basis of the calibration procedure for the water quality model component is the comparison of measured quality data with predicted quality. The procedure is structured so that standard values used in the construction of the model are replaced by measured values in stages, but only if necessary. This ensures that only essential data is collected and a catchment that is accurately represented using standard values can be modelled with the minimum of data collection. It also ensures that once a parameter has been measured, its value is fixed, and the situation in which several parameters are being adjusted simultaneously is thus avoided. There are two main steps in the calibration procedure:

1. dry weather flow confirmation
2. rainfall event confirmation

The dry weather flow depends on the following information:

- population per capita consumption
- commercial measured water consumption for major users
- actual/consented flow rates for trade effluent#
- infiltration rates
- potency factor
- base flow factor
- check load balance bounds, accumulation of sediment

During the dry weather flow calibration it may be considered necessary to adjust the friction factors for the bottom section of some pipes to represent the effects of debris, siltation and roots. In doing this, the following information should be analysed:

- historical blockage data,
- Closed Circuit Television (CCTV) reports if available,
- local knowledge and
- if the dry weather flow velocities exhibited by the model and flow monitors, show very low values.

During the process of model calibration it is important to take into consideration the degree of accuracy of the monitoring equipment and difference between the system geometry and model. As a general industry practice rule is that the dry weather depths recorded in pipes less than 300 mm should be viewed with engineering judgement.

The water quality model should be used to predict the dry weather flow quantity and quality over a 24 hour period, and to compare it with what is measured in the system. The 24 hour period should start just after the morning peak of dry weather flow so that any initial layer of fine mobile sediment is at its lowest level. The measured dry weather flows at the outfall of the system should be compared with those predicted by the model. This provides a check on the input data and on the deposition of pollutants in the system during dry weather flows. Disagreements should be resolved by checking the quality of consented discharges.

Once there is good agreement on the dry weather flows, model predictions can be compared with measurements during a rainfall event, preferably with a 5 day preceding dry period to allow for an adequate build-up of surface sediment.

For systems that are (theoretically) 100% separate, the calibration procedure can be simplified. For wastewater systems, there should be no need to collect or use rainfall data. A short antecedent dry period can be used when predicting the pollutant outflow from a wastewater system, because the build up and washout of sediments normally has a 24 hour cycle. If however, the model is used for operation and maintenance studies to study the build up of sediment over a long period of time, the antecedent dry period should be the time since the sewers were last cleaned. Note however, that the ingress of rainfall-runoff, either directly or by infiltration, is to be expected in most separate wastewater systems.

Where the water quality model is reduced to the use of SIMPOL, then this model has to be calibrated against the physically based water quality model.

Model validation

The validation of simulation modelling software is discussed in Chapter 5 and it is known to be difficult. It is even more urgent to validate an instantiated model for a particular physical system. Besides the verification of the software we are concerned with the validation of

- Asset data
- Performance data
- Model calibration
- Model application

There is no way in which we can prove the 'truth' of an instantiated model, anymore than we can the associated modelling software; we can only go as far as being satisfied about the viability or reliability of the results. 'Viability' may go so far as determining how acceptable the results are to the community on which the results of a particular model are going to be inflicted.

Validation is substantiating the claim that an instantiated model produces simulation results within the range of accuracy specified in the performance criteria for the study; see Dee, 1994. It is normal to calibrate a model with one set of data and to validate the calibrated values with the second set. Putting aside the issue of whether it is ever possible to validate a model, this particular procedure does raise some other objections. The chief one is to decide what to do if the model fails to reproduce the validation set of results.

The success in selecting suitable values for the model parameters to satisfy the calibration results means that some confidence can be placed in the model structure. However, a failure to reproduce the validation set raises questions as to what is missing, either from the distribution of the calibration values or in terms of some more fundamental factor affecting the structure of the model. What is apparent from the process above is that the instantiation of a particular model is very dependent on what the user does; that is, on the tasks that he/she performs. The 80–20 rule applies where 20% of the effort is put into getting 80% of the way towards a reliable model. Thereafter the law of diminishing returns comes into play. Together with reliability is the need by practising engineers for *robustness* such that there are no hidden surprises that may invalidate the results. Robustness is defined as the capacity of a model to perform equally well across the full range of environmental conditions for which it was designed; see Beck *et al.*, 1997.

If a model contains a large number of parameters then it could be that the solution obtained by one person involves a particular set of values for the parameters. Another person working with the same software and data may come with just a good a fit of their model with the data as the first person and yet have very different values for the parameters. This is the situation of equifinality discussed by (Beven and Freer, 2001). It is for this reason that the it would be good to reserve some survey data for a peer review test of whether the calibrated model performs well for what we can now call the validation set. Refsgaard and Storm (1996) discuss four basic categories of catchment modelling validation tests:

1. Split-sample test, where calibration is done against one set of data and the validation against another set with a period of similar length
2. Differential split-sample test, where calibration is done on data before changes are made to the catchment, with the adjustment of the model parameters to characterise the change and validation is based on a subsequent period
3. Proxy-basin test, where information is used from other gauged catchments
4. Proxy-base differential split-sample test, where calibration is done on another catchment but then the remainder of the validation is pursued on the study catchment.

Usually a split-sample test is sufficient on a model with few explicit processes, whereas the more complex tests are required for multi-process models.

Remember that a model can never be fully evaluated; we can only think in terms of degree of validation. For this reason the peer review test, which will include an examination of the audit trail for the model, is termed here the confirmation of the model.

Given that a model has been calibrated then it is important to bear in mind that the model has been applied during the calibration process to a range of performance data. Therefore confidence in the model is limited to its application with other data that are within the range of the original calibration data. In turn the specification of the range implies that there are no changes to the structure of the model space. This raises the difficulty of whether the model is appropriate to situations where there are such changes. Generally, we make the assumption that because of the generic nature of the processes and the idealised nature of any newly introduced structural changes to the model space we can use the model safely for situations that are outside the original range of the calibration. This is acceptable provided there are not too many structural changes and the basic character of the original system is retained.

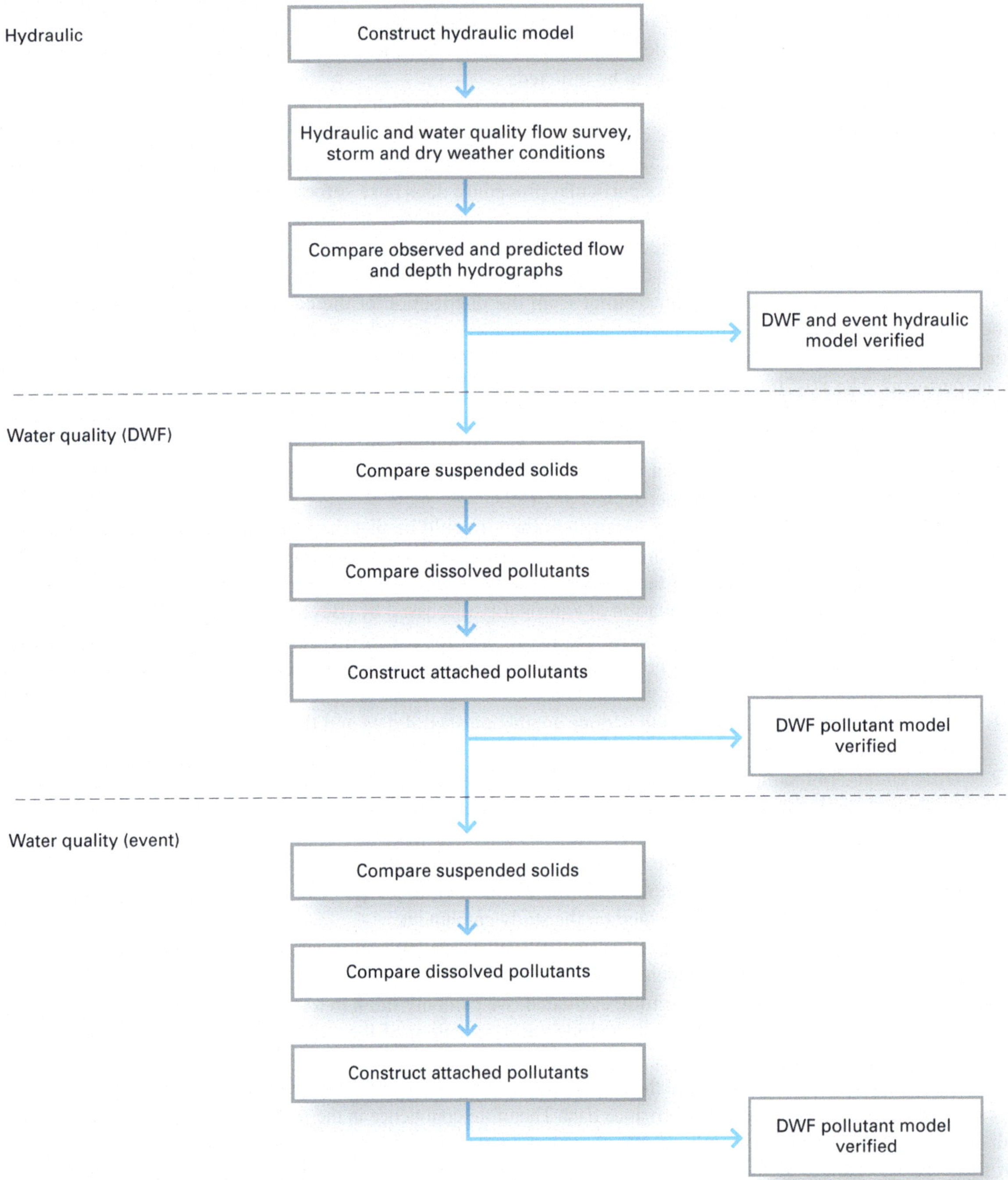

Figure 6.15 Procedure for instantiating a sewerage water quality model

Resource questions:

- Have the parameters been calibrated by 'force fitting' the model predictions to the observed data?
- Has the water quality model been calibrated separately for dry weather and wet weather flows?

Task 2.6 Confirm the instantiated model

Description	Carry out independent confirmation of the structurally validated and calibrated model
Input	Project objectives and scope, structurally validated and calibrated model
Output	Audit documentation, confirmed instantiated model
Sub-tasks	Seek independent review
Tools	DSS

Obviously, it is important that the model reflects the behaviour of flows in the physical network with acceptable accuracy for engineering decision making; for example, see Figure 6.16. The model then becomes a virtual laboratory for analysing the hydraulic performance of the physical network. In other words, various options can be considered with the model. The engineer could even go so far as to 'play' with the model in order to increase his understanding and appreciation of the intricate complexities of the network. In particular, a water distribution network model may be used to determine the pressures at the demand points in the network, or the leakage in certain sections of the network. Similarly, a drainage network model may be used to assess the peak surcharge in each link of the network during a rainfall event, or the peak flood level at each manhole. Given a frequency of occurrence for each rainfall event it is possible to assume a corresponding frequency of occurrence for surcharge or flooding levels at any point in the network. It is this type of analysis that can then generate information that can be linked to performance criteria. Strictly this analysis should be done with a time series of rainfall events and the frequency determined from an analysis of the results. In this way it is possible to develop a thorough description and understanding of the existing performance of the physical system. This can entail deducing the levels of service achieved throughout the network in terms of frequencies of surface flooding, operation of combined sewer overflows, filling of storage tanks, and so on. Such a measure of performance provides a base against which the performance of any modified system can be compared. This is invaluable when needing to justify, say, the cost-benefit information for the 'improved' system. The UK Sewerage Rehabilitation Manual (WRc, 1994) gives guideline criteria for the performance of a sewerage network.

The typical performance criteria for sewerage systems in developed countries is a frequency of damage to property by flooding from sewerage systems not greater than twice in ten years. Where this frequency is exceeded the owner of the sewerage system is expected to upgrade the system to a higher standard of performance.

Long term standards also exist for protecting the aquatic life in rivers. Although these standards are directly relevant to continuous effluent discharges from treatment work, the focus of this manual is primarily on overflow discharges and therefore the intermittent standards are more relevant. Water quality performance criteria applicable to the UK are described in The Urban Pollution Management Manual (FWR, 1998) and documents issued by authorities which stipulate particular requirements.

Resource questions:

- Is the existing performance of the system well defined?
- Are the deficiencies in the performance of the system clearly identified?

Resource questions:

- Has the geometry for the model space been verified?
- Has the calibration of the parameters been verified?
- Has the peer review been conducted by someone with experience of modelling?

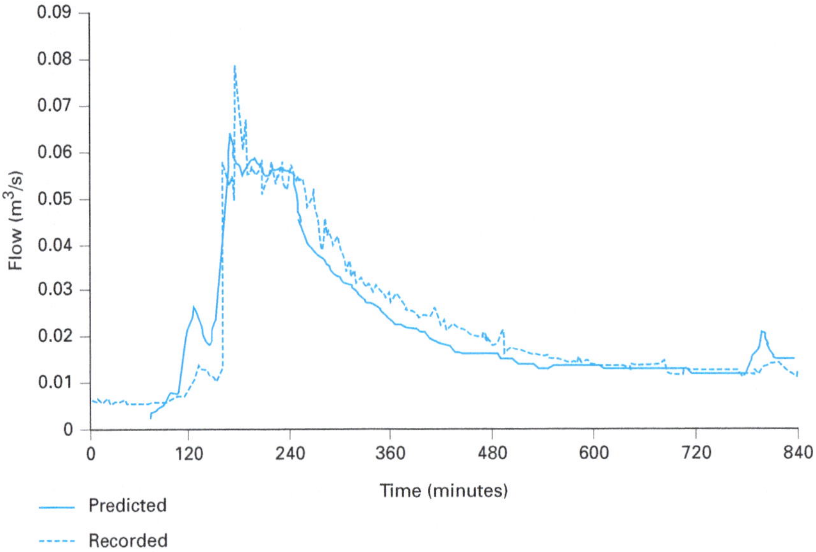

Figure 6.16 Comparison of Observed and Predicted flow hydrographs (after Kellagher, 1993)

Task 2.7 Assess the model sensitivity

Description	Carry out a sensitivity test of the model to parameter values, and refine values for any sensitive parameter
Input	Project objectives and scope, instantiated model
Output	Audit documentation, sensitivity report
Sub-tasks	Refine values for sensitive parameters
Tools	Sensitivity analysis

A sensitivity analysis is important for its identification of sensitive input data. Sensitivity analysis is the study of how the response of a model can be apportioned to changes in a model's inputs (Ratto *et al.*, 2006). A model's sensitivity describes the degree to which the model results are affected by changes in a selected input parameter (Beck *et al.*, 1997). Sensitivity analysis is recommended as the principal evaluation tool for characterizing the most and least important sources of uncertainty in environmental models. This may be carried further in the form of an uncertainty analysis. The idea is to determine the probability distribution for the input data item and to generate the corresponding probability distribution for the results. One way of doing this is to do a Monte-Carlo analysis (see below), where input data is sampled many times from the probability distribution for the item and the corresponding results derived. Inevitably this involves running the model many (thousands of) times. Where there are a number of significant input data items to be considered then the number of runs becomes impossible to manage, simply because of the sizeable running time of most models. Some limited Monte-Carlo analysis may be done if the model is sufficiently small, but faster models are required if this approach is to become feasible.

We define

- Model error: the deviation between the modelled and the observed state of the variable
- Model uncertainty: a statistical or fuzzy logic description of the model errors

We recall that the errors in a model may result from

- Identification of the physics, chemistry and biology of the problem
- Nature of the symbolic representation of the physical processes
- Numerical translation of the symbolic equations
- Numerical solution of the numerical equations
- Software written to implement the numerical solution and associated modelling concepts
- Definition and collection of asset data
- Definition and collection of temporal data
- Model building procedure (simplification etc)
- Calibration of model parameters
- Confirmation of asset data
- User application to base performance
- User application to identified problems
- Interpretation of results

This list is long and indicates the considerable opportunities for errors to creep into the analysis. Therefore it is very important that good quality assurance procedures are used in instantiating a model, in particular.

Sensitivity analysis applied to urban drainage modelling software has revealed that results (say, discharges at the outfall) are most sensitive to errors as shown in Table 6.6.

Table 6.6 Sensitivity to data in a sewerage model

Variable	Base value	Revised or change in value	Percentage change in peak discharge
Storm return period	2 year	1 year 5 year 10 year	−20 20 50
Percentage runoff	45%	40–50	±01
Percentage paved	0 to 100%	Nearest 10% Nearest 20%	10
Surface slope	Mainly mild	All mild All medium All steep	1
Number of gulleys per inlet	1 to 10	1 gulley per pipe	1
Pipe length	3 m to 120 m	±01%	10
Roughness height	0.3 mm	0.6 mm 0.15 mm	0.1
Headloss coefficient	0.2	±10%	2

The implication from at least a design point of view is that the rainfall and the surface-runoff processes are the most important for modelling in this case. Obviously pipe length is important, but given the layout the opportunities for errors are somewhat limited. Most effort should be put into collecting good (spatial) rainfall and surface runoff information.

Where models are directed at determining hazardous events then human and ecological health may be put at risk. In such cases a risk assessment should be carried out. However it should be born in mind that a quantitative analysis of uncertainty and variability may not necessarily improve the risk assessment.

The first consideration should be the major sources of variability and uncertainty. It is important that these are kept separate in the analysis. Questions should be asked as to whether there are time and resources to complete a complex analysis, and whether the project warrants the level of effort. Indeed, will a quantitative estimate of uncertainty improve the decision? How will the regulatory decision be affected by this variability and uncertainty analysis? What types of skills and experience are needed to perform the analysis? There should be an evaluation of the weaknesses and strengths of the method. Some view should be formed as to how the variability and uncertainty analysis will be communicated to the public and decision makers.

The analysis is more likely to be successful if there is an appreciation of the overall degree of variability and uncertainty and the confidence that can be placed in the analysis and its findings. It is important to recognise the key resources of variability and key sources of uncertainty and their impacts on the analysis. There should be an understanding of which assumptions are critical and which are unimportant and why. The decision maker also needs to know the extent to which plausible alternative assumptions or models could affect any conclusion. It is also helpful to be aware of key scientific controversies related to the assessment and a sense of what difference they might make regarding the conclusions.

Resource questions:

- Has the model sensitivity test highlighted particular parameters?
- What has been done to refine these parameters?

Task 2.8 Assess the model uncertainty

Description	Assess model uncertainty to input data
Input	Project objectives and scope, instantiated model
Output	Audit documentation, uncertainty report
Sub-tasks	Generate uncertainty model
Tools	Uncertainty analysis

'Effective uncertainty communication requires a high level of interaction with the relevant decision makers to ensure that they have the necessary information about the nature and sources of uncertainty and their consequences. Thus, performing uncertainty for environmental regulatory activities requires extensive discussion between analysts and decision makers.' (NRC, 2007) Uncertainty analysis investigates the lack of knowledge about a certain population or the real value of model parameters. Uncertainty is sometimes reducible through further study and with the collection of additional data. Existing Agency guidance (e.g., (USEPA, 2008)) distinguishes uncertainty analysis from methods that are used to account for variability in input data and model parameters. The variability in model parameters and input data can be better characterized through further study but is usually not reducible (USEPA, 2008).

Although uncertainty and sensitivity analysis are closely related, uncertainty is parameter specific, and sensitivity is algorithm-specific with respect to model 'variables.' By investigating the 'relative sensitivity' of model parameters, a user can become knowledgeable of the relative importance of parameters in the model. By knowing the 'uncertainty' associated with parameter values and the 'sensitivity' of the model to specific parameters, a user will be more informed regarding the confidence that can be placed in model results. There are a number of methods to assess model uncertainty as described in Section 6.3. It is better to assess total model uncertainty, probably using UNEEC (Shrestha, 2009). This gives bounds on the predictions at the demand points for a water distribution network or the outfall and overflows for a collection system.

Resource questions include:

- Have you decided how to use the uncertainty estimates?
- Does the uncertainty affect decision making?

Task 2.9 Generate the base performance

Description	Apply instantiated model to existing system to derive base performance for events of different frequency
Input	Project objectives and scope, instantiated model, design input data
Output	Audit documentation, base performance of model
Sub-tasks	Archive base performance
Tools	DSS

Once a model has been instantiated the model is then run with the existing structural or asset domain to generate the response for a wide range of events. The purpose is to produce a bench-mark of performance against which changes and improvements to the system can be assessed. Sometimes this assessment is done for one or two historical events of considerable significance. This is not sufficient for a true cost-benefit analysis; instead, there is a need for a bench-mark in terms of frequency of occurrence.

Resource questions:

- Is there anything unexpected about the base performance?
- What is missing from the base-performance (if anything)?

6.6.5 Solution development

Task 3.1 Generate solution options

Description	Identify possible solutions to the presenting problem
Input	Project objectives and scope, instantiated model, possible solutions
Output	Audit documentation, model results for possible options
Sub-tasks	Brainstorm possible solutions
Tools	DSS

There should be a number of possible solutions to solve the presenting problem. These solutions can be identified in similar projects that have been successfully completed. Attention should be given to resolving specifically difficult issues before time has gone by. The whole project team should be involved in brainstorming such solutions, and where possible stakeholders should be included.

Resources questions include:

- Have all ideas for possible solutions been sought and included?
- Have all necessary team members and stakeholders been included in the brainstorming?

Task 3.2 Select the preferred option

Description	Identify preferred option from the set of possible options
Input	Project objectives and scope, instantiated model, possible options
Output	Audit documentation, preferred option
Sub-tasks	Cost each option, find economic optimum
Tools	Multi-criterion optimisation

The selection of a preferred option from a set of solutions could involve minimisation (optimization) of, say, construction and damage costs. Data is obviously needed on the costs of constructing the solution, including installation and ongoing maintenance costs as well as the capital cost. Further, the damage costs are likely to include some estimates for the intangible as well as the tangible costs (such as human distress as well as depth of flooding or consequences of water shortage).

Resource questions:

- Has the optimisation (if used) been properly applied?

Task 3.3 Assess the preferred option

Description	Thoroughly test the performance of the preferred option
Input	Project objectives and scope, instantiated model, preferred option
Output	Audit documentation, performance of preferred option
Sub-tasks	Brainstorm tests, refine option
Tools	Cost analysis

A traditional way of doing an analysis with a given instantiated model is to run the model for single events. These events may be synthetic events in that the boundary inputs are generated from a statistical analysis of monitored data. For example, synthetic rainfall events may be derived with rainfall depths of different return periods, particular shape of profile, skewness, and duration. Because the word 'event' presupposes initial conditions that may be dependent on previous rainfall data (antecedent wetness) then this set of synthetic rainfall parameters may need to be complemented by the antecedent condition. The aim would be to simulate the response of the system to such a range of rainfall events and to be able to make some judgement, usually that the return period of the flow is the same as that of the rainfall depth, provided the other parameters are correctly defined. This is open to criticism even if it has been assumed satisfactory in the past. Increasingly these days, where the data is available, engineers are resorting to time series analysis. Instead of focusing on events the idea is that what is important is to input a continuous record of sufficiently long duration that the statistics of the input time series can be analysed sensibly. Correspondingly the output results can also be analysed statistically, and there should be a sufficiently long record that the statistics of the output can be compared sensibly with the statistics of the input; see Table 6.7.

Table 6.7 Cause and effect

Task	Cause	Model	System	Effect
Planning	√	√	?	√
RTC	?	√	√	√
Forecasting	√	√	√	?

The application of an instantiated model depends on its original objective. Whatever the primary objective, one a subsidiary objective is the knowledge enrichment of the user. Invariably a well posed model will provide the user with an additional insight into the behaviour of the system. For example, there may be a need to understand the interaction of all the different processes in fate of a contaminant. If we take the usual definition of the possible tasks then we can talk about

- Performance optimisation
- Planning
- Design and construction options

- Asset maintenance
- Real time control: design and on-line
- Forecasting/simulation
- Infiltration/inflow, leakage, demand management, trade waste, CSO, sedimentation,
- Prescriptive design

Resource questions:

- Have additional sensitivity and uncertainty tests been carried out?
- What has the user learned from this task?

Task 3.4 Produce detailed design

Description	Implement the preferred option in order to produce a detailed design
Input	Project objectives and scope, instantiated model, preferred option
Output	Audit documentation, detailed design
Sub-tasks	Prescriptive design of structures
Tools	Design tool

Resource questions:

- Has the model adequately represented the objects that are being designed in detail?

6.7 INSTANTIATION OF DATA DRIVEN MODELS

Data driven models work with dynamic input and output data that are observed in the physical system. There is no explicit acknowledgement of the physical and other processes that might be involved in linking the input and the output variables. Nevertheless they are included implicitly through the choice of the input parameters that are necessary to generate the target output parameters. The instantiation of a data driven model is intimately bound up in determining the structure of the model, whether it is in the hidden neurons in the ANN, the layout of the model tree, the clusters used in k-nearest neighbour, and so on. In what follows, the procedure for instantiating a data driven model follows that proposed by Zijderveld (2003) for ANNs. Although this author focuses on ANNs, the procedure proposed can be adapted for any data driven model. Figure 6.17 shows the replacement of the tasks in the second phase of the instantiation procedure for physically based models above.

Task 2.2ddm Analyse the system

Description	Select preferred data driven modelling tool
Input	Project objectives and scope, input and output time series data sets, accuracy criteria
Output	Audit documentation, recommended data driven modelling structure
Sub-tasks	Sub-divide the time series data
Tools	Data driven modelling tool, optimiser

As noted above, the instantiation of a data driven model is intimately bound up in determining the structure of such a model in line with the input and output parameters. Usually, the target output parameter is given, and the emphasis is on selecting the necessary input parameters to ensure that all of the significant physical and other processes determining the target output parameter are taken into account, while reducing the number of input parameters such that only those parameters are included which are essential for the model to achieve its required performance in terms of accuracy.

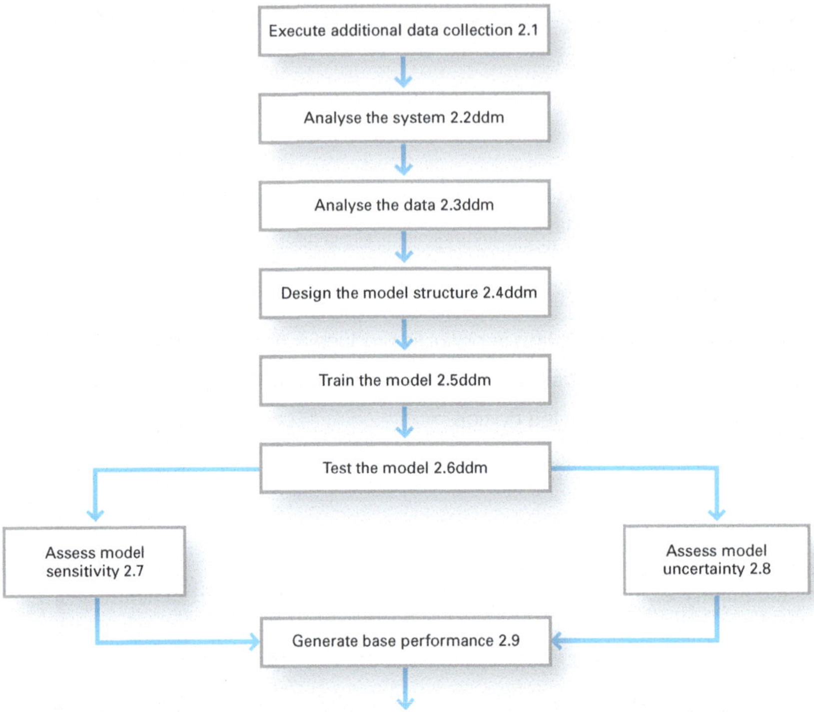

Figure 6.17 Procedure for the investigation phase for a data driven model

In general it cannot be proved that there is sufficient information of the underlying physical or other processes available in the input and output parameter data alone. The only way to test the sufficiency of the input and output data is to reserve a test set of observed data and to see if the model predictions for the target parameters are within an acceptable accuracy criterion. If the accuracy criterion cannot be met by the instantiated model then it is assumed that significant input parameters are missing. Identifying these parameters is part of the subsequent improvement of the model structure, provided the parameters can be recognized and there is data available.

In general, the structure of a data driven model is closely connected with the number of input parameters. Therefore, the optimization of the model structure and the number of input parameters is usually done at the same time.

Resource questions:

- Have the input parameters proved to provide an adequate model?
- Has the model structure been optimized?

Task 2.3ddm Analyse the data

Description	Select input variables
Input	Input and output time series
Output	Audit documentation, Choice of input variables
Sub-tasks	Correlation analysis, AMI analysis, PCA analysis, Weight analysis, Sensitivity analysis, Partial retraining analysis
Tools	Data driven modelling tool, Correlation analysis, AMI analysis, PCA analysis, Weight analysis, Sensitivity analysis, Partial retraining analysis

There are however some statistical methods available to support the selection of input variables. These include:

- Global estimation of available input information
- Correlation analysis
- Principle Component Analysis (PCA)
- Weight analysis (for ANNs)
- Sensitivity analysis
- Partial retraining

It should be noted that there is no unique model structure. This is the counterpart of the principle of equifinality for the values of hydrological model parameters identified by (Beven and Freer, 2001).

Experience with data driven models shows that there is a minimum number of inputs necessary to train a given data driven model. The number of input parameters must be greater than this minimum for the model to be viable.

Correlation analysis checks the relationship between the different parameters and the target parameter (cross-correlation) or the time dependencies between the parameters themselves (autocorrelation). A useful alternative to correlation is the search for the average mutual information (AMI) between the input parameters and the target parameter; see for example, Abebe, 2004. AMI is based on the information theory of Shannon (1948). Both correlation and AMI are able to estimate the lag between the input and output parameters, such as rainfall and outfall discharge. In turn, the choice of input parameters will focus on relating input parameters with the requisite lag time before the output parameter.

PCA analysis seeks the main patterns expressed in terms of the eigenvalues and eigenvectors derived from a correlation matrix describing the co-variance of the data set. The principle components are orthogonal, and the first few components are normally able to explain 80% of the variance of the original data set. This compresses the original data set and provides an insight into the drivers behind the original data.

Weight analysis for ANNs examines the weights between the input parameters and the hidden neurons. If when training the ANN using a robust training algorithm certain weights do not vary significantly from their initial small values then the corresponding input parameters may be removed without prejudicing the overall model.

Sensitivity analysis simply assesses whether a change in an input parameter affects the target variable. Where there is no demonstrable change then the input parameter can be removed.

Partial retraining assumes that a model has been trained on all available N input parameters, and that a new model based on N-1 input parameters demonstrates that the activity of the hidden layer does or does not change significantly. In the latter case the second model is to be preferred to the first, and yet another new model can be tested with N-2 input parameters.

Resource question:

- Has the input data been correctly identified?

Task 2.4ddm Design the model structure

Description	Determine structure of the data driven model
Input	Input and output data time series
Output	Audit documentation, Model structure
Sub-tasks	Refine the model structure
Tools	Data driven modelling tool

There are specific approaches to designing the model structure, that is, reducing the number of hidden neurons for an ANN, or pruning the branches of a model tree. Methods for pruning an ANN include Optimal Brain Surgeon, the optimization method of Snijder (1998), and cascade-correlation networks (Hsu *et al.*, 1995).

Resource question:

- Is the model structure determined satisfactorily?

Task 2.5ddm Train the model

Description	Determine the model parameters
Input	Input and output time series for training
Output	Audit documentation, trained model
Sub-tasks	Check with agreed model performance criteria
Tools	Data driven modelling tool, optimiser

The data set available for the model has to be partitioned into three distinct sets:

- training set
- validation set
- test set

The choice of these sets should follow certain heuristic rules:

- Each data set should contain events covering the operating range of the system. Where possible the test set should include extreme events to test the ability of the model to go beyond the training range of events
- The events in the training set should be equally distributed over the parameter space
- If the validation set is used to limit overfitting of the model the validation set should be at least 20% of the size of the training set
- A larger data set is needed if it is known that there is a significant amount of noise in the data
- If it is difficult to acquire sufficient data, the time series can be extended using statistical techniques (see Khu, van der Veer *et al.*, 2001)

A strategy for training an ANN is advocated by Zijderveld (2003). There are three important aspects:

- Initialization of the weights
- Selection of pattern or batch training
- Choice of stopping criterion

LeCun, Simard *et al.* (1993) recommends the 'rule of thumb' that a weight should be selected randomly from a uniform distribution with zero mean and a variance equal to the reciprocal of the number of connections of a neuron. In batch training the whole of the training data is presented to the model and the weights are updated according to the average error. In pattern training the weights are updated after each training pattern. Batch training is normally preferred because it is faster, though Zijderveld recommends that both approaches are tried. Finally, the stopping criterion is needed to halt the training process and to avoid overfitting; see Section 5.13.2.

Resource question:

- Has overfitting been avoided?

Task 2.6ddm Test the model

Description	Test the trained model against reserved test data
Input	Trained model, Test input and output time series data
Output	Audit documentation, model test report
Sub-tasks	Test uncertainty in model parameters
Tools	Data driven modelling tool, DSS

Once the model is trained it should be run with the test data. A healthy ANN model will have the majority of its weights around zero. The uncertainty associated with an instantiated model can be tested as in Task 2.8. However, the uncertainty in the ANN model weights can be assessed using the bootstrap procedure of (Heskes, 1997). The idea is that sequences of n data values are selected randomly (with replacement) from the training set and used to train the model each time. The ensemble of trained models can then generate the 95% confidence intervals around the mean prediction for the test data set. Obviously, the concern is to have a resulting ensemble of models for which the confidence intervals are small.

Resource question:

- Have the confidence intervals been derived, and are they sufficiently small?

6.8 MODELLING AS A DYNAMIC PROCESS

The building, development and application of a model are parts of a dynamic process involving the user(s). There is an analogy here with a young child learning to speak. For him language is a representation of the real world. Consequently the development of language in the child is related to his understanding of the world. In the same way the model develops in time with experience. We can thus talk in terms of the maturity of a model that has benefited from the experience of being applied to a range of inputs with adequate confirmation of the results.

6.9 DECISION SUPPORT SYSTEMS IN URBAN WATER MANAGEMENT

There exist a number of decision support systems for different aspects of water resources management at the river basin level and including urban development. The following systems are representative of what is currently available.

SRM	Sewer Risk Management, which was preceded by the Sewerage Rehabilitation Manual	WRc (2010)
Wallingford Procedure	For the design and analysis of urban storm drainage	HRW (2002)
HydroPlan	For risk-based asset management of both water distribution and sewerage systems	Vojinović (2005)
UPM	Urban Pollution Management	FWR (1998)
WaterWare	River basin planning, which integrates GIS, database management, modelling, optimisation and expert systems, and can be applied to explore the limits of sustainable development, the impact of new environmental legislation, the identification of new resources, etc	Jamieson (1996)
NERC-ESRC Land use Programme (NELUP)	Economic and environmental impacts of land use changes in river basins, which manages data for the river basin, predicts impacts using models and presents results for land use changes	O Callaghan (1995)
WaterStrategyMan	Assess the state of a water system in term of its sources, usage, water cycle and environmental quality, which is done using four modules: water availability, demand, allocation and evaluation	Peruffo (2004)

(Continued)

(*Continued*)

WHY	A web based DSS for watershed management, using the physical model L-THIA to estimate direct run-off and non-point source pollution and a web-based hydrologic GIS (WHY) including a real time watershed delineation capability, hydrologic data extraction function and a web GIS data publishing tool	Choi (2005)
MULti-sectoral, INtegrated and Operational DSS (MULINO)	Integrates social, economic and environmental modelling with GIS capabilities and a multi-criteria approach for water resources management,	Giupponi (2004)
Aquatool	Water resource planning and operational management at the watershed scale, using simulation and optimisation modules	Andreu (1996)

6.10 KEYWORDS

By the end of the chapter you should have encountered the following terms; ensure that you are familiar with them!

Calibration*
Confirmation*
Decision making*
Decision support system*
First order second moment method
Fuzzy set theory
GLUE
Instantiating models*
Monte Carlo simulation
Possibility
Probability*
Procedure
Sensitivity *
Uncertainty*
UNEEC
Verification*

The items with an asterisk are dealt with in more detail in later chapters.

6.11 QUESTIONS

What are the main steps in instantiating a physically based model?
How are the steps for instantiating a data driven model different?
What methods are available for assessing the uncertainty in a model and its results?
What are the benefits of a procedure?

6.12 REFERENCES

Aamodt A. and Plaza E. (1994). Case based reasoning: Foundational issues, methodological variations and system approaches, *AI Communications* (Vol 7(1), pp. 39–59), IOS Press, Amsterdam.

Abebe A. J. (2004). *Information Theory and Artificial Intelligence to Manage Uncertainty in Hydrodynamic and Hydrological Models.* UNESCO-IHE Institute for Water Education, Delft.

Ahmad K. and Price R. K. (1998). Safe hydroinformatics and the urban environment. In: *Hydroinformatics Tools for Planning, Design, Operation and Rehabilitation of Sewer Systems, Nato ASI, Harrachov, Czech Republic*, J. Marsalek, C. Maksimovic, E. Zeeman and R. K. Price (eds.). Kluwer, Amsterdam.

Andreu J., Capilla J. and Sanchis E. (1996). AQUATOOL, a generalized decision-support system for water-resources planning and operational management. *J. Hydrol.*, **177**(3–4), 269–291.

Babovic V., Larsen L. C. and Wu Z. (1994). Calibrating hydrodynamic models by means of simulated evolution. *Proceedings on the First International Conference on Hydroinformatics*, 9–13 September, Delft.

Baumann D., Boland J. and Hanemann W. (1998). *Urban Water Demand Management and Planning*. McGraw-Hill Professional Publishing, New York.

Beck M., Ravetz J., Mulkey L. and Barnwell T. (1997). On the problem of model validation for predictive exposure assessments. *Stoch. Hydrol.Hydraul.*, **11**(3), 229–254.

Beven K. and Freer J. (2001). Equifinality, data assimilation, and uncertainty estimation in mechanistic modelling of complex environmental systems using the GLUE methodology. *J. Hydrol.*, **249**, 11–29.

Birge J. and Louveaux F. (1997). *Introduction to Stochastic Programming*. Springer Verlag, New York.

Cañizares R. and Torre R. (1999). *On the Application of Data Assimilation in Regional Coastal Models*. Taylor & Francis, Balkema, Rotterdam.

Cerný V. (1985). Thermodynamical approach to the traveling salesman problem: an efficient simulation algorithm. *J Optimiz. Theory App.*, **45**(1), 41–51.

Choi J., Engel B. and Farnsworth R. (2005). Web-based GIS and spatial decision support system for watershed management. *J. Hydroinf.*, **7**(3), 165–174.

Coello C. and Carlos A. (2002). Theoretical and numerical constraint-handling techniques used with evolutionary algorithms: a survey of the state of the art. *Comput. Method. Appl. Mech. Eng.*, **191**(11–12), 1245–1287.

Coello C., Lamont G. and Van Veldhuizen D. (2007). *Evolutionary Algorithms for Solving Multi-Objective Problems*. Springer Verlag, New York.

Collette Y. and Siarry P. (2003). *Multiobjective Optimization: Principles and Case Studies*. Springer Verlag, New York.

Cunge J. A. (2003). Of data and models. *J. Hydroinf.*, **5**(2), 75–98.

Datta R. S. N. and Sridharan K. (1994). Parameter estimation in water-distribution systems by least squares. *J. Water Resour. Plann. Manage. Div. ASCE*, **120**(4), 405–421.

Deb K., Mohan M. and Mishra A. (2003). *A Fast Multi-objective Evolutionary Algorithm for Finding Well-Spread Pareto-Optimal Solutions*. Report 2003002, KanGAL. Indian Institute of Technology, Kanpur.

Deb K., Pratap A. Agarwal S. and Meyarivan T. (2002). A fast and elitist multiobjective genetic algorithm: NSGA-II. *IEEE Trans. Evolut. Comput.*, **6**(2), 182–197.

Dee D., Cunge J., Labadie G., Ruiz Mateo A., Mathiesen M., Price R., Santos M. and Warren R. (1994). *Guidelines for Documenting the Validity of Computational Modelling Software*. IAHR/AIHR, Delft.

Dempsey P., Eadon A. and Morris G. (1997). Simpol: a simplified urban pollution modelling tool. *Urban Storm Drainage*, **36**(8–9), 83–88.

Dorigo M. and Di Caro G. (1999). *The Ant Colony Optimization Meta-Heuristic, New Ideas in Optimization*. McGraw-Hill Ltd., Maidenhead, UK.

Duan Q., Sorooshian S. and Gupta V. (1992). Effective and efficient global optimization for conceptual rainfall-runoff models. *Water Resour. Res.*, **28**(4), 1015–1031.

Duan Q., Sorooshian S. and Gupta V. (1994). Optimal use of the SCE-UA global optimization method for calibrating watershed models. *J. Hydrol.*, **158**(3–4), 265–284.

Dubois D. and Prade H. (1988). *Possibility Theory: An Approach to Computerized Processing of Uncertainty*. Plenum Press, New York.

Edgeworth F. (1881). *Mathematical Psychics: An Essay on the Application of Mathematics to the Moral Sciences*. Augustus M. Kelley (Original work published in 1881. Kegan Paul, London). AMC.

FWR (1998). *Urban Pollution Management Manual*. Foundation for Water Research, Marlow.

Giupponi C., Mysiak J., Fassio A. and Cogan V. (2004). MULINO-DSS: a computer tool for sustainable use of water resources at the catchment scale. *Math. Comput. Simulat.*, **64**(1), 13–24.

Goldberg D. (1989). *Genetic Algorithms in Search and Optimization*. Addison-Wesley, Reading, MA.

Hastings W. K. (1970). Monte Carlo sampling methods using Markov chains and their applications. *Biometrika*, **57**(1), 97–109.

Heskes T. (ed.) (1997). Practical confidence and prediction intervals. In: *Advances in Neural Information Processing System*. MIT Press, Cambridge.

Hiller F., Lieberman G. and Lieberman G. (1995). *Introduction to Operations Research*. McGraw-Hill Publishers, New York.

Holland J. (1975). *Adaptation in Natural and Artificial Systems*. University of Michigan Press, Ann Arbor, MI.

Howard R. (1960). Dynamic programming and Markov process. MIT Press, Cambridge, MA.

Hsu K., Gupta H. V. and Sorooshian S. (1995). Artificial neural network modelling of the rainfall-runoff process. *Water Resour. Res.*, **31**(10), 2517–2530.

Jamieson D. and Fedra K. (1996). The 'WaterWare' decision-support system for river-basin planning. 1. Conceptual design. *J. Hydrol.*, **177**(3–4), 163–175.

Jonkergouw P., Khu S., Kapelan Z. and Savi D. (2008). Water quality model calibration under unknown demands. *J. Water Resour. Plann. Manage.*, **134**, 326.

Kapelan Z., Savic D. and Walters G. (2004). Incorporation of prior information on parameters in inverse transient analysis for leak detection and roughness calibration. *Urban Water J.*, **1**(2), 129–143.

Kapelan Z., Savic D. and Walters G. (2007). Calibration of water distribution hydraulic models using a bayesian-type procedure. *J. Hydraul. Eng.*, **133**, 927.

Kaufmann A. and Gupta M. (1991). *Introduction to Fuzzy Arithmetic: Theory and Applications*. Van Nostrand Reinhold Company, New York.

Kellagher R. (1993). *Sewer Modelling for Rehabilitation*. EU SPRINT Project Reports. HR Wallingford, Wallingford, UK.

Kellagher R. (2002). *Development of Guidelines on the Structural, Hydraulic and Environmental Rehabilitation of Sewers*. HR Wallingford, Wallingford, UK.

Kennedy J. and Eberhart R. (1995). Particle swarm optimization. *Proc IEEE Int. Conf. on Neural Networks*, Perth.

Khu S. T., van der Veer P. and Liong S. Y. (2001). Rainfall-runoff simulation with neural network and enlarged data set. In: *EuropIA'8, Advances in Design Sciences and Technology Conference*, Delft, The Netherlands, 25–27 April.

Kirkpatrick Jr. S., Gelatt C. D. and Vecchi M. P. (1983). Optimization by simulated annealing. *Science*, **220**(4598), 671–680.

Korving H., Van Gelder P., Van Noortwijk J. and Clemens F. (2002). *Influence of model parameter uncertainties on decision-making for sewer system management*. In I. D. Cluckie, D. Han, J. P. Davis and S. Heslop (eds.), *Proceedings of the 5th International Conference on Hydroinformatics* (vol. 2, pp. 1361–1366). 1–5 July, Cardiff.

Krzysztofowicz R. (1999). Bayesian theory of probabilistic forecasting via deterministic hydrologic model. *Water Resour. Res.*, **35**(9), 2739–2750.

Lansey K. and Basnet C. (1991). Parameter estimation for water distribution networks. *J. Water Resour. Plann. Manage.*, **117**(1), 126–144.

Laumanns M., Thiele L., Deb K. and Zitzler E. (2002). Combining convergence and diversity in evolutionary multiobjective optimization. *Evol. Comput.*, **10**(3), 263–282.

LeCun Y., Simard P. and Pearlmutter B. (1993). Automatic learning rate maximization by on-line estimation of the Hessian's eigenvectors. *Adv. Neural Inform. Process. Syst.*, **5**, 156–163.

Maskey S. (2004). *Modelling Uncertainty in Flood Forecasting Systems*. Taylor & Francis Group plc, London, UK.

Maskey S. and Guinot V. (2002). Improved estimation of uncertainty in flood forecasting. In: *Proceedings of the 5th International Conference on Hydroinformatics*, 1–5 July, Cardiff, UK.

Maskey S. and Guinot V. (2003). Improved first order second moment method for uncertainty estimation in flood forecasting. *Hydrolog. Sci. J.*, **48**(2), 183–196.

Maskey S. and Price R. K. (2004). Assessment of uncertainty in flood forecasting using probabilistic and fuzzy approaches. In: *6th International Conference on Hydroinformatics*. World Scientific Publishing Company, 21–24 June, Singapore.

McKay M. D. (1992). Latin hypercube sampling as a tool in uncertainty analysis of computer models. In: *The Winter Simulation Conference*, 13–16 December, pp. 557–564, Arlington, Virginia.

McKay M. D., Beckman R. J. and Conover W. J. (1979). A comparison of three methods for selecting values of input variables in the analysis of output from a computer code. *Technometrics*, **21**(2), 239–245.

Metropolis N., Rosenbluth A. W., Rosenbluth M. N. and Teller A. H. (1953). Equations of state calculations by fast computing machines. *J. Chem. Phys.*, **26**(6), 1087–1092.

Mohan S. and Loucks D. P. (1995). Genetic algorithms for estimating model parameters. In: *Integrated Water Resources Planning for the 21st Century, Proceeding of the 22nd Annual Conference*, ASCE, 7–11 May, Cambridge, Massachusetts.

Nemhauser G. and Wolsey L. (1999). *Integer and Combinatorial Optimization*. Wiley, New York.

O'Callaghan J. (1995). NELUP: an introduction. *J. Environ. Plann. Manage.*, **38**(1), 5–20.

Ormsbee L. E. (1989). Implicit network calibration. *J. Water Resour. Plann. Manage.*, **115**(2), 243–257.

Osyczka A. (1985). Multicriteria optimization for engineering design. In: *Design Optimization* (pp. 193–227), Academic Press, New York.

Pareto V. (1896). *Cours d'économie politique professé à l'université de Lausanne*, 3 volumes, Lausanne (Course on political economy taught at the University of Lausanne).

Peruffo A., Todini E. and Bologna I. (2004). The WaterStrategyMan decision support system. In: *Modelling and Control for Participatory Planning and Managing Water Systems*, 28 September–1 October, Venice, Italy.

Rahal C. M., Sterling M. J. H. and Coulbeck B. (1980). Parameter tuning for simulation models of water distribution networks. *Proc. Inst. Civ. Eng.*, **69**, 751–762.

Ratto M., Young P. C., Romanowicz R., Pappenberge F., Saltelli A. and Pagano A. (2006). Uncertainty, sensitivity analysis and the role of data based mechanistic modeling in hydrology. *Hydrol. Earth Syst. Sci. Discuss.*, **3**, 3099–3146.

Refsgaard J. C. and Storm B. (1996). Construction, calibration and validation of hydrological models. In M. B. Abbott and J. C. Refsgaard (eds.), Distributed Hydrological Modelling (pp. 41–54). Kluwer, Dordrecht.

Romanowicz R. and Beven K. (1998). Dynamic real-time prediction of flood inundaton probabilities. *Hydrolog. Sci. J.*, **43**(2), 181–196.

Ross T. J. (1995). *Fuzzy Logic with Engineering Applications*. McGraw-Hill, New York.

Savic D., Kapelan Z. and Jonkergouw P. (2009). Quo vadis water distribution model calibration? *Urban Water J.*, **6**(1), 3–22.

Shannon C. E. (1948). A mathematical theory of communication. *Bell Syst. Tech. J.*, **27**, 379–423, 623–656.

Shrestha D. (2009). *Uncertainty Analysis in Rainfall-Runoff Modelling: Application of Machine Learning Techniques*. CRC Press/Balkema, Leiden.

Shrestha D. L., Kayastha N. and Solomatine D. (2009). A novel approach to parameter uncertainty analysis of hydrological models using neural networks. *Hydrol. Earth Syst. Sci.*, **13**, 1235–1248.

Solomatine D. (1999). Two strategies of adaptive cluster covering with descent and their comparison to other algorithms. *J. Global Optimiz.*, **14**(1), 55–78.

Sundararajan C. (1994). Uncertainties in piping frequency analysis. *Fuzzy Set. Syst.*, **66**, 283–292.

Sundararajan C. (1998). Fuzzy models for uncertainty assessment in engineering analysis. In: *Uncertainty Modelling and Anlysis in Civil Engineering*, B. Ayyub (ed.). CRS Press, pp. 43–53.

UNESCO-IHE (2005). *The SWITCH Project*. UNESCO-IHE, Delft.

USEPA (2008). *Guidance on the Development, Evaluation and Application of Environmental Models*. Council for Regulatory Environmental Modeling, USEPA, Washington, DC.

USEPA (2009). *Rehabilitation of Wastewater Collection and Water Distribution Systems*. Office of Research and Development, National Risk Management Research Laboratory – Water Supply and Water Resources Division, Edison, New Jersey.

USEPA/ASCE (1996). *International Stormwater BMP Database*. http://www.bmpdatabase.org/index.htm (accessed 1 October 2010).

Vojinović Z., Price R. K. and van den Broek W. (2005). Hydroplan-EU knowledge management framework for urban water asset management. In: *International Conference on Urban Drainage*, 21–26 August, Copenhagen, Denmark.

Vrugt J. A., Gupta H. V., Bouten W. and Sorooshian S. (2003). A shuffled complex evolution metropolis algorithm for optimization and uncertainty assessment of hydrologic model parameters. *Water Resour. Res.*, **39**(8), 1201–1217.

Wallingford (1981). *Design and Analysis of Urban Storm Drainage – The Wallingford Procedure*. Department of Envrionment, Natioinal Water Council, Standing Technical Committee Reports, 28. Hydraulics Research Limited, Wallingford.

Walski T. (1983). Technique for calibrating network models. *J. Water Resour. Plann. Manage.*, **109**(4), 360–372.

Wang Q. (1991). The genetic algorithm and its application to calibrating conceptual rainfall-runoff models. *Water Resour. Res.*, **27**(9), 2467–2471.

WaPUG (2002). *Code of Practice for the Hydraulic Modelling of Sewers*. CIWEM, London.

WRc (1989). *Network Analysis – A Code of Practice*. WRc Engineering, Swindon.

WRc (1990). *Design of Water Distribution Networks*. WRc Engineering, Swindon.

WRc (1994). *The Sewerage Rehabilitation Manual*. WRc Engineering, Swindon.

WRc (2010). *Sewerage Risk Management: Managing the Risks of Service Failure*. http://srm.wrcplc.co.uk/ (accessed 1 October 2010).

Wu Z. and Clark C. (2009). Evolving effective hydraulic model for municipal water systems. *Water Resour. Manage.*, **23**(1), 117–136.

Zadeh L. (1999). Fuzzy sets as a basis for a theory of possibility* 1. *Fuzzy Set. Syst.*, **100**, 9–34.

Zangwill W. (1967). Non-linear programming via penalty functions. *Manage. Sci.*, **13**(5), 344–358.

Zijderveld A. (2003). *Neural Network Design Strategies and Modelling in Hydroinformatics*. Technical University of Delft, Delft.

Zitzler E. and Künzli S. (2004). *Indicator-Based Selection in Multiobjective Search*. Springer, Berlin.

Chapter 7

Involving Society in Urban Water Management

7.1 INTRODUCTION

Urban water systems serve specific urban communities. These communities are the beneficiaries of the services provided by the systems, and therefore the people concerned are those who should have a strong interest in their operation. This is reinforced by the fact that either explicitly through direct charges or implicitly through taxes the local population fund the management of their urban water systems. If potable water of good quality is not delivered through stand pipes in slum areas or directly to houses, public buildings and industrial units in developed areas the people concerned are quick to complain. Similarly if raw sewage does not flow away in streams in poor areas, or appears on the streets due to inadequate drainage in more affluent areas, the local population will express their concern to the city authorities. It follows that water projects are more successful if the people affected are involved in their planning and implementation, and have ownership of the operation of the resulting systems.

Different cases of urban water management practices in developed and developing countries are studied in this chapter. The issues cities are facing have to be analysed in order to devise suitable urban decision making structures to deal with them. Therefore it is important to establish first the individual and community needs concerning water, flooding and sanitation. These needs are met in the context of particular institutional and urban governance structures, which give rise to an integrated urban water management (IUWM) approach to deal with the issues raised using certain tools. In turn there are some ethical issues related to urban water, flooding and sanitation, which need to be acknowledged.

7.2 INDIVIDUAL AND COMMUNITY NEEDS

Maslov (1943) analysed basic human needs and defined a hierarchy of such needs. He differentiated between the more basic physiological needs, such as breathing and excretion, and higher needs, such as self-esteem and self-actualisation. When considering the water sector, it is apparent that drinking is a physiological need, while fear of flooding is at a somewhat higher level of experiencing safety. To assure their safety, communities living along rivers or in deltas, for example, take steps to organize themselves against flooding. Their leaders usually play a significant role in the organizational process, and they may be driven by their need for self-esteem or self-actualisation or their acknowledgement of the community's need for water and a safe and acceptable environment (including wastewater collection and stormwater drainage). Also the desire to strengthen the community and the sense of belonging to it, and the effects and consequences of population growth and migration need to be taken into account.

People live in urban areas for a variety of reasons, but especially to benefit from access to different resources, whether it is security, employment, services, traded goods including food and artefacts, or culture. Among the common water services provided in cities are potable water, and wastewater and stormwater drainage. Priority is normally given to the provision of potable water, for without this it is difficult to sustain life at all; drainage of wastewater and stormwater usually take second place. Nevertheless, when account is taken of the long term degradation of human health and the environment where drainage is poor, the significance of better drainage is appreciated and the desire for a safe and acceptable environment leads to the introduction of improved drainage systems.

7.3 URBAN GOVERNANCE AND INTEGRATED URBAN WATER MANAGEMENT

Wherever communities of people emerge there is a need for the management of the community space and resources. Various approaches to governance have been adopted in different countries and cultures. Generally however, cities have evolved governance structures that depend on politically elected representatives of the community and career professionals who carry out the administration and management of various urban resources and services. The community can be said to delegate responsibility for managing such services to professionals; this is particularly the case for water supply and drainage. Ultimately it is the community that has responsibility for the civil engineering interventions that are made in the physical world in connection with the provision of urban services.

Besides the need for ongoing operation and maintenance of the urban services there are important issues concerning urban development and the consequences for urban existing services. There are two approaches to urban development. On the one hand, physical planners are concerned with the use or the categorisation of the land area and develop a view on how a city should develop; this results in a physical development plan. On the other hand, some urban managers follow a more hands-on approach and organize stakeholders to find out what can be achieved or how to solve a given problem (van Dijk, 2006).

Key issues that urban managers have to deal with are usually mobilizing municipal finance, the competitiveness between cities or their capacity to attract private sector investment and to promote employment, and the capacity to deliver public services in an efficient manner, including care for the environment. Certain issues are related to each other; the proposed solutions should also deal with the problems in an integrated way. It is up to the urban manager to coordinate the teams doing the integrated problem analysis and those responsible for the implementation of the proposed integrated solution.

It is important to distinguish between good governance and good management. UNCHS (1999) states that: 'Good urban governance involves participatory decision making'. It refers to 'the complex set of values, norms, processes and institutions by which cities are managed'. UNCHS and the World Bank, the leading international organizations in the field of urban development, stress the importance of good governance, as does Van Dijk (2006). Good management refers to the officials formally executing the policies which deliver the services required.

The structure of good governance is depicted in Figure 7.1. The emphasis here is on the norms and values of the urban manager, on the participation of the urban population and on the effective monitoring and control of what is going on. In a decentralized, transparent system, such urban managers are also accountable for what they do: are they achieving the stated objectives?

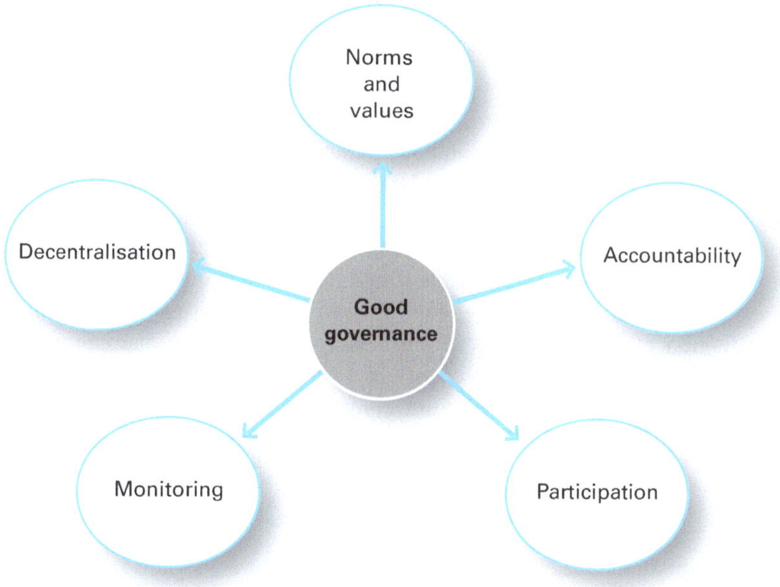

Figure 7.1 Elements of good governance

Good governance depends on certain norms. In particular, for cities this means care for the poor, support for people who are excluded or at risk, and concern for the environment. The question is often how to improve urban governance, and at the same time to deal with water related issues in an urban context. In discussions introduced by donors and donor organizations about good governance, there is often a cultural undertone as well. What is considered corruption in one country according to certain norms and values may be considered as providing incentives in another. However, good governance and more transparency is needed to increase the chances of the success of urban water and environmental projects. Urban governance is embedded in a larger system of regional, provincial, municipal and district level governance structures. Urban management requires more than the traditional master plan. It is necessary to develop a broader vision on what will be the city of the future, such as the opportunities for a city, given its competitive advantages. This requires a strategic planning exercise, which could be based on a preliminary action planning effort.

The private sector can be involved in different ways. One way would be to point to corporate responsibility of private firms. Companies have to make a contribution to society and provide services to the community. Typically urban management can help to achieve this, if it is clear how specialised tasks and functions are distributed between different levels of government, the private sector and the non-governmental sector. Appropriate governance structures need to be designed, but often the delegation of the responsibility for water supply and drainage activities is to professionals, either in the public or private sector or to partnerships involving both.

An integrated approach has to be adopted to urban water, flooding and sanitation issues. The issues are complex and interrelated, and problems can be acute and difficult to deal with implying a risk to society, if not handled appropriately. This integrated approach to urban water management is multi-disciplinary in nature, and therefore requires consultation and cooperation between different disciplines. It also involves the linking of physical events to socio-economic structures, and the development of a holistic approach to dealing with these issues in an integrated manner.

In that problems of water resources, water supply, drainage and flooding, sanitation and wastewater treatment are interrelated, the governance structures dealing with these issues are unfortunately often fragmented (located at different levels of government) and often autonomous authorities exist to deal with them. Each governance structure may refer to a different geographical area and have a different way of involving the people concerned. Hence integration requires involvement of stakeholders, a conception of the relevant issue as resulting from different developments, and collaboration between different disciplines and organizations. The actors may or may not share the objectives of achieving a sustainable solution to these many interlocking problems.

Integrated urban water management has developed its own tools, such as strategic planning, stakeholder consultation and the adoption of different types of partnerships. Other tools that can be used are mass flow analysis or a water balance, the calculation of the energy foot print and the health impact, doing an institutional analysis, preparing financial options, and carrying out equity analysis which help it to achieve its vision by constantly monitoring the progress.

7.4 INSTRUMENTS FOR URBAN WATER MANAGEMENT

There is a considerable benefit to society of good urban water services. Different instruments within the institutional framework can be used. Services can be undertaken by individuals, such as rainfall harvesting or extraction of water from an adjacent surface or groundwater resource for personal or family use, but these services are generally limited, as they can lead to conflicts between individuals extracting water from the same resource. Therefore, integrated urban water management is better conceived of as a collective responsibility, and this needs to be established by national, regional or local government. The appropriate authority should legitimise professional suppliers, whether public or private, to provide potable water of adequate quality on demand and to drain the urban area of wastewater and stormwater. The authorization is a political activity, and should be subject to checks and balances as is fitting in a transparent arrangement. In other words, society should delegate responsibility for urban water management to professionals even if such responsibility is muted. There should however, be a growing involvement of all stakeholders in urban water management.

Several techniques can be used for urban planning. One planning approach is to put the emphasis on successful examples elsewhere and to try and learn from these best practices. Also, action and strategic planning techniques can be used, and these may result in a list of priorities. Action planning basically refers to ways of solving an urban problem in a limited time frame and in an innovative or non-routine way with the involvement of the stakeholders. Action planning helps managers to focus on a relevant issue and suggests actions to deal with it. As a result, efforts in the 1990s helped to move urban planners towards a more participatory, performance-oriented approach. The 'action planning' approach has successfully been introduced as a tool for urban management. Strategic planning helps to focus on priorities and translates a strategic vision about the future into a strategic plan for an environment in which the actions of one player depend on and affect the actions of other players (Mintzberg, 1994). Finally, the master plan is no longer the most important planning instrument. Urban water management practices differ from the traditional approach of making a master plan, which is never really implemented.

The process of implementing an urban development project is often influenced by national political and economic circumstances in a particular country. Political statements may enhance the pressure put on the actors involved and contribute to the failure of an urban development project. Stren (1993) stresses that urban management is an integrating concept. This requires an understanding of the transition from 'physical planning' to urban management. Urban management implies an integrated approach of the major issues of cities in emerging economies. In many countries, intervention by local government is limited to 'zoning policies', and the attempt of authorities to impose some minimal building or housing standards. This limited role serves slum inhabitants, who want to build their own houses and have no further interferences. However, at a later stage an urban manager is required to intervene, because some problems (sewerage, pollution and health hazards, for example) cannot be solved individually.

As an example of the integrated approach, consider the building of houses, which not only solves a housing problem, but can also create employment if small contractors are involved. In turn, the construction influences the direction in which the city develops and the traffic flows that are generated. The urban management approach should take the most important issues in the city as the starting point. These could be poverty, crime or congestion. The literature rightly stresses the importance of the participation of stakeholders and the role of different financial actors, the commercial and non-commercial sectors. This is certainly a challenge for urban managers, who should forge public-private partnerships (PPP), or what are known in Indonesia as public-private-community partnerships (PPCP). These institutions together with civil society need to be involved. There also exist new opportunities to attract private finance.

Table 7.1 Instruments for urban water management

1. Legislation to authorise service delivery
2. National, regional and local policies and individual efforts in a regulatory context
3. Agreement on services to be provided
4. Formulation and imposition of operating, investment, and scientific standards
5. Stakeholder consultation
6. Strategic and action planning
7. Public, private, public-private partnerships
8. A good regulatory system
9. Society's monitoring and control
10. Monitoring performance of different aspects of urban water management, such as the use of environmental indicators

7.5 ETHICS OF URBAN WATER MANAGEMENT

It is an obvious question as to which urban (water) governance structures have worked and for what reasons. The good governance discussion is a step forward in urban water management, because urban politicians and managers are now more than ever accountable for what they do. Accountability is one of the key concepts of the New Public Management (NPM) theory, a theory that is revolutionizing public management by putting the emphasis on contracts and autonomy, while stressing the importance of market orientation and customer orientation. The challenge for urban managers is to make this theory work (van Dijk, 2006).

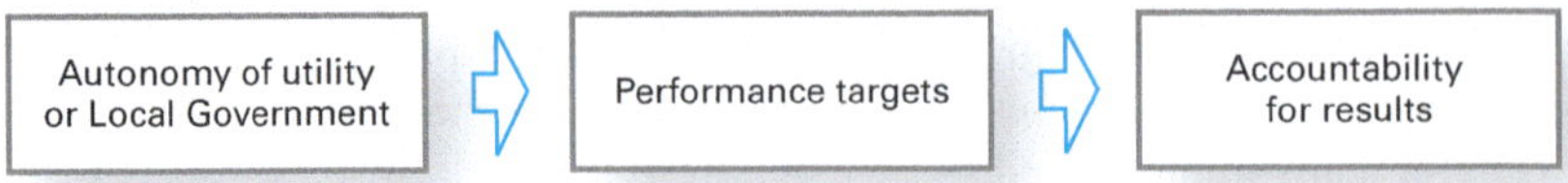

Figure 7.2 Framework for analyzing the relationship between an organization and its environment

Good governance requires a good basic information system and financial reporting. Local governments need budgets and to prepare them they need to take decisions what they want in the future. There is the complication that most municipalities in India, for example, are still using traditional (cash accounting) systems. The disadvantage of these systems is that debts are not reported and that rating agencies require more time to provide a rating if the municipality is accounting on a cash basis. This is an argument to move to more modern (double or accrual) accounting. The financial sector may not really care whether the customer has accrual or fund-based accounting. In the Indian situation the banks are at present only interested in providing finance to the major Municipal Corporations in India.

The position taken here is that there is a compelling need for an integrative ethic of urban water management: from resources, transportation, distribution, to consumption. We need to move from technical fixes to community management, that is, where people are informed and involved in making decisions. In some cities in developing countries there is the need for social marketing of hygiene, that is, bringing about behavioural change to achieve public health improvements, first for the individual, then the household, then the community, and finally city-wide. There should be a good working relationship between the necessary civil engineered infrastructure and the viability of the local regional and global ecological regimes. There should be a focus on small household and community projects as well as the import of advanced technology solutions. In other words, there is a social dimension to water resources development and management. It is important that local cultural traditions to facilitate, legitimise and support water supply, demand, distribution, consumption and waste management are recognised, acknowledged and incorporated in any solution. Entrepreneurial activity whether in consulting, vending water, plumbing, managing waste, etc should be encouraged and supported. In order to promote sustainability the emphasis should not be on short term localised profit, but on long term global consequences. Emphasis should be placed on the ethics of equity, community participation and social management, while integrating local regional and global elements to sustain society and the human habitat. This should be combined with ethical criteria for establishing priorities. We need to move away from vital resources being under the control of privileged few, who control the source of raw material and then process it into another usable form.

The situation is urgent: 2.3 billion people live today in water stressed areas, that is, where water supply is less than 1700 m^3/year. 48% of the population in Sub-Saharan Africa have no access to safe water; 35% have no access in South Asia. Every day 34,000 people die from water related diseases. It is generally recognised that a family needs a minimum of 150 l/day, whereas families in Africa receive on average 20 l/day. Whereas the industrial revolution led to a small family size in Europe, it is leading to an unprecedented population increase in poor countries. Unfortunately a powerful elite including water professionals believe in the emancipating role of technology and the market economy. This mitigates the ethics of sharing, compassion and empathy, and fails to acknowledge obligations rather than property rights or free market inequities. Social justice is a function of both utilities of individuals and common good, and is therefore a function of relationships, interactions, interdependency among people. It is also interdependent with nature, such as the fertility of the land, and the well being of the planet such as affecting climate change.

Values like cooperation across family lines, a public role for women, open and critical deliberation, a degree of solidarity between rich and poor, free association of individuals, instrumental dealing with time (planning) and a positive attitude towards change are involved (Easterly, 2006 and Sen, 1999). But cultural transitions need to be introduced consciously and responsibly (Wiarda, 2003). Can people find their own equilibrium between modern and traditional values and follow their own pathway into modernity? This combination of ethical questions, intercultural dialogue and constructive technology assessment is an innovative way of approaching infrastructure provision.

As an example of taking ethical decisions, consider building an ecological city. This intention would lead to the discussion of what should be or should not be in the ideal city. The positive points of environmentally friendly cities are such that they are attractive to live in and energy saving; they can promote integrated water and sanitation, regard wastewater as a resource rather than a liability, introduce better urban waste collection and processing, more gardens and trees, bio-diversity and better public transportation, and deal in a better way with climate change. Such cities would do away with air, water and soil pollution, congestion, flooding and the lack of green areas. Three normative questions can be raised:

1. What would the ecological city of the future look like?
2. To what extent do the existing examples satisfy the criteria for sustainable urban development formulated in the literature (for example, Kenworthy, 2006)?
3. What can we learn from these experiences for the ecological city of the future?

Table 7.2 Examples of terms referring to more ecological cities

Titles for eco city initiatives	Source
Eco-heaven, a model city near Shanghai	International Herald Tribune (24-6-2008)
Eco systems & biodiversity, the role of cities	UNEP & United Nations Habitat brochure
Garden city	Brochure Suqian city Jiangsu province China
Sustainable urbanization or cities	Van Dijk (2007)
Sustainable urban development network	SUD-network UN Habitat Nairobi
Cities of the future	Switch project Delft, the Netherlands[i]
Liveable and vibrant cities	National library Singapore
Sustainable living, best practices	NL Min. of Housing, Spatial Planning & Env.
Green urbanites	Strait Times (21-6-2008)
Climate resilient cities	World Bank primer, Washington Oct. 2008
Rotterdam, climate proof	Rotterdam municipality the Netherlands
Keeping cities alive	Strait Times (21-6-2008)
Green city philosophy, cooperation between Thailand and the Netherlands	Dutch Ministry of Agriculture, nature and food quality

[i]The Switch project (Sustainable Water Improves Tomorrow's Cities' Health) with support from the European Union (EU) is seeking a paradigm shift in urban water management. Its purpose is to make water treatment more sustainable and protect the quality of drinking water sources. In addition, it wants to reduce risks such as water related diseases, droughts and flooding.

To explore how the ecological cities of the future would look like, it is necessary to assess what the relevant dimensions for an ecological city are. What makes a city an ecological city and what does sustainable urban development really mean? Ten dimensions for sustainable city development in the Third World were developed by Kenworthy (2006, pp. 67–876). They provide a possible analytical framework to decide whether certain initiatives qualify for the ecological city label.

Ecological initiatives can be taken at three levels. The first level could be the city, and new towns or neighbourhoods where all kinds of ecological districts or neighbourhoods appear. The second level is that of buildings, such as ecological villas or block of houses/apartment towers with common heating/cooling systems or a common grey water re-use facility. Finally the third is the household level, which is spontaneously or triggered by incentives or price increases. There are a number of Chinese eco-city initiatives that are interesting and have been studied.

Table 7.3 Characteristics of a sustainable city

1. A compact, mixed urban form that protects the natural environment, biodiversity and food-producing areas
2. The natural environment permeates the city's spaces and embraces the city, while the city and its hinterland provide a major proportion of its food needs
3. Freeway and road infrastructure is deemphasized in favour of transit, walking and cycling infrastructure, with a special emphasis on rail. Car and motorcycle use are minimized
4. There is extensive use of environmental technologies for water, energy and waste management – the city's life support systems become closed loop systems
5. The central city and sub-centres within the city are human centres that emphasize access and circulation by modes of transport other than the automobile, and absorb a high proportion of employment and residential growth
6. The city has a high quality public culture, community, equity and good governance. The public realm includes the entire transit system and all the environments associated with it
7. The physical structure and urban design of the city, especially its public environments are highly legible, permeable, robust, varied, rich, visually appropriate and personalized for human needs
8. The economic performance of the city and employment creation is maximized through innovation, creativity and uniqueness of the local environment, culture and history, as well as the high environmental and social quality of the city's public environments.
9. Planning for the future of the city is a visionary debate and decision process, not a predict and provide computer-driven process
10. All decision making is sustainability-based, integrating social, economic, environmental and cultural considerations as well as compact, transit-oriented urban form principles. Such decision making processes are democratic, inclusive, empowering and engendering of hope

Some Chinese eco-cities and eco-provinces show how China tries to deal with urban environmental issues. These cities are introducing a very different, more integrated approach to a number of related environmental issues. Urban environmental policies in Asia could also be illustrated by the way in which Singapore is giving special attention to the issue of building in a sustainable way. Rotterdam in the Netherlands is dealing with climate issues in its Rotterdam Climate Proof plan (Stadshavens Rotterdam, 2008). Also Third World cities, particularly those located in deltas, are preparing themselves for climate change. These cases are based on research carried out in the framework of the SWITCH project, where work on modelling floods has been prominent, but where the question has also been asked how to link the results of such modelling efforts to the existing decision making structures.

7.6 ROLE OF HYDROINFORMATICS IN SOCIETY

In view of the discussion above it is worth returning to the contribution that hydroinformatics can make to the development of the urban water sector, particularly in facilitating stakeholders. Abbott has written extensively about the subject, and his recent writings emphasise the aspect of social justice; see Abbott, 2007; Abbott and Vojinović, 2010 and Abbott and Vojinović, 2010. He argues that although social justice for individuals, communities and groups has been sought throughout human history, the revolution in communication technologies and their deployment that has occurred over the last two decades has made it much more realisable. Abbott's argument focuses on the burgeoning development and application of the Internet and of net-connected mobile telephony, and how these developments promote stakeholder involvement in infrastructural projects, especially in the water sector. The involvement of stakeholders is a means of realising social justice through inducing changes in the built environment that are commensurate with positive changes in the social environment. Abbott refers to these environments as the 'outer world of infrastructure' and the 'inner world of infrastructure'. Thus on the one hand a water infrastructure project consists of the visible and tangible physical infrastructure of steel, concrete, stone and whatever other materials, formed by such tools as surveying equipment, global positioning systems, measuring devices and their radio links, supervisory control and data acquisition systems, numerical modelling systems with their seamless interfaces to geographical information systems, aerial photographic records and their interpreting equipment, satellite imaging devices and cartographic coordinate transformation systems. On the other hand there is an invisible and intangible *social* infrastructure sustained by the intentions, beliefs and codes of behaviour of people, most of whom are interested in, and some of whom are actively engaged in, this same project. This 'inner world' consisting of the minds of socialised people requires the means to transform the data, information and knowledge provided by the physical equipment into forms that they can easily assimilate, each person having his or her own individual interest and ability.

Abbott's argument is built on Heidegger's view of technology (Heidegger, 2009); (see http://www.english.hawaii.edu/criticalink/heidegger/index.html). Heidegger's thesis is that we are in real danger of being chained or trapped by technology, which we view as an instrument to get things done. By making technology better we aim to master it and solve the problems that accompany it. We need therefore to understand the essence of technology as having to do with the way in which we are oriented to the 'coming-into-existence' or the 'revealing' of the world (and therefore of 'truth'). This orientation is a form of an 'enframing' or a structured categorisation of everything in the world. As a consequence, we are all too ready to regard the world as a 'standing reserve', as a source of raw materials. Enframing therefore has within it the danger that by so ordering things our ability to view this 'coming-into-existence' or the 'revealing' is blocked, and this sets us on a destructive and self-destructive course. However, Heidegger argues that at the same time it makes enframing makes it apparent that we are part of the world and that the world 'needs' us to care for it. The way we can guard against the dangers of enframing and enter a free relationship with technology is by adopting an attitude that is more creative, formative, productive and active, that is, as something that is 'poietic' (from which we obtain our word 'poetry').

The urban area is the result of man's adventure *par excellence* with technology. The artificial environment formed by the water supply, treatment and distribution and wastewater and stormwater collection, treatment and discharge to receiving waters, let alone everything else pertaining to urbanisation, all too readily persuades us that the urban area is a 'standing reserve' that we have created and as such is for us to manage and exploit. Our monitoring and surveying, and modelling and decision support tools would reinforce the view that we can order and compute everything, even independently of stakeholders who on the whole in developed countries would admit that the underground water service networks are 'out of sight and out of mind'. These networks only become an issue when the services they provide fail. This is tantamount to being trapped by the technology. The existing knowledge/power structures referred to by Foucault (Foucault and Gordon, 1980) and reinforced by some water professionals can only be challenged by stakeholders if they have access to relevant information for the water services. We need therefore to involve stakeholders through the increasing facilities and power of yet another set of technologies, namely the World Wide Web and the Internet, using them not only as number crunchers but in providing new forms of communication between stakeholders and water professionals that effect a change in the social environment, and open up new ways of involving stakeholders in infrastructural projects, especially in the water sector. The involvement of stakeholders can therefore become a means of realising social justice through inducing changes in the built environment that are commensurate with positive changes in the social environment.

Abbott claims that what is at issue is the construction of web-based hydroinformatics environments that transport the inner worlds of the stakeholders to 'higher levels of consciousness', but then in such a way as to engender 'higher levels of conscientiousness' through drawing upon the inherent capacities of the stakeholders to exercise their 'active imaginations'. To support this further, the data and technical information that result from any technical work must be translated into a form of knowledge and understanding that can be assimilated by the non-technical stakeholders if they are to be able to contribute meaningfully and responsibly in the decision-making process. This can be realised through different metonymic structures that the hydroinformatician places upon the computer screens of the various classes of stakeholders to lead them to transmute the resulting allegories into their own most relevant symbols. It is argued, following experience in other fields, that such environments can make possible what would otherwise be impossible. In the most obvious case, it causes stakeholders to be move from a reactive participation to an interactive involvement. As with any group, this requires leadership, but such leadership cannot be imposed on the group by the project management, however well meaning, but must be allowed, and indeed encouraged, to emerge. The true leader needs to have the necessary theoretical and practical apparatus that can enable him or her to understand critical events and situations and promote the transformation of this understanding into well-informed judgements and decisions. This then becomes a sociotechnical endeavour since the urban water problems are neither purely social nor technical, but are for much the greater part sociotechnical. The study and practice of sociotechnology is essential if we want to understand how technical interventions have social consequences and how the resulting social changes react back on the technical developments, and so on, again and again, in what becomes a sequence of continuous feedback cycles.

As stated above, hydroinformatics is no longer simply about numbers, but it is also about the deep emotional and creative involvement of stakeholders as they employ the knowledge supply chains that the hydroinformatician provides. This demands new ways of working with data and models, with hydroinformaticians playing their part in facilitating the presentation of evidence in a clear unadulterated way, customised to the needs of the different stakeholders; this is especially the case when the evidence brings about conflicting points of view. Rather than the business strategy of application service providers in creating a dependence on the part of their customers, Abbott advocates a 'software as a service' paradigm which seeks to empower its users. Further, new ways have still to be found for enabling stakeholders to operate with models through simple, intuitive interfaces, implying the need for coherence and intelligibility of communications. What is also evident is that the social arrangements necessary to employ hydroinformatics tools must be rethought, with new knowledges and tools developed accordingly. Examples are given by Abbott and Vojinović of practical applications, especially to the tunnel and bridge connection between Denmark and Sweden, completed in 2000, and the St Maarten flood warning and prevention study established in 2007 and still ongoing. Abbott emphasises that this programme of development necessitates the introduction of a new kind of mathematics which is necessarily a 'mathematics of the qualities', together with the application of some aspects of category theory, as introduced in a discussion of the gender issue in hydroinformatics (Abbott, 2000) and in the way specific terms are defined in hydroinformatics (Abbott, 2002). He especially emphasises the deeper significance of what are called 'natural transformations' in category theory, recalling Pascal's aphorism that 'The natural is that which we cannot lose'.

The ambition to pursue social justice has of course its opponents including those who want to maintain stakeholders in a passive state, where they may only be consulted, informed or directed. Observing in particular that '[Scientism] dismisses human values as mere products of the emotions and rejects the notion of being in order to clear the way for facticity', Abbott decries the loss of spiritual values in the West which may compromise the use of such instruments as these, and quotes Rowan Williams the present Archbishop of Canterbury by way of a warning about their potential misuse:

> *'The city of man, the* civitas terrena, *is built in self love, out of the human desire to make order and intelligibility and rest on their achievements. And so two kinds of love have been constructed the two cities: the earthly city built be love of self leading to the contempt of God, the heavenly city by love of God leading to contempt of self. The earthly city is built by those who are confident of their control not only over their own biographies but over the whole human world. And this attitude Augustine rejects as fiercely in the social as in the individual sphere. Social good is good, peace and justice* are *to be sought with passion, but with the recognition that they are not to be considered as signs of human glory, human authority, human capacity to tame the world. If sought and 'enjoyed', loved for their own sake rather than .... directed to a greater end...., their precariousness follows for those who put their trust in them. Those who look for a home of their own making secure only their own deeper deepest alienation and homelessness'*

(Williams, 1978/1990/2002). This quotation is apt for the subject of this book which addresses water especially in the city.

Abbott concludes by drawing attention to a warning he has given before, and which was mentioned in his original seminal book on hydroinformatics (Abbott, 1991), that although hydroinformatics has emerged as a European phenomenon, the decay in the west of spiritual values, and its secular approach to life generally, means that an Asian hydroinformatics must now increasingly take the lead.

7.7 CONCLUSIONS

Dealing with complex issues in an urban context requires an integrative approach to water related projects. Some aspects of such integrated approach are reflected in the framework of urban management (van Dijk, 2006). In this chapter, issues discussed are the integration of the different sectoral interests, the role of planning and management, and the importance of economic, financial, social and environmental criteria. In addition, questions are addressed such as who are the decision makers and how do we bring about sustainable urbanisation by involving stakeholders in the decision making process. Finally, the discussion goes into the development and installation of sociotechnical arrangements to enable interactive stakeholder participation, which remains for the most part a challenge that still has to be faced.

The chapter concludes by raising the stakes about who should be involved in decision making and how. Hydroinformatics has an ambitious programme to revolutionise the activities of stakeholders in water infrastructure projects.

7.8 KEYWORDS

By the end of the chapter you should have encountered the following terms; ensure that you are familiar with them!

Community needs
Decision makers
Enframing
Governance
Indicators
Integration
Internet
Legislation
Levels of service
Mobile telephony
Monitoring performance
Participatory decision making
Partnerships
Policies
Social justice
Socio-technology
Standards
Standing reserve
Sustainable city

7.9 QUESTIONS

What is the importance of involving stakeholders in decision making?
Explain the use of legislation, regulations, policies, standards, and levels of service in urban water management
What is a sustainable city?

7.10 REFERENCES

Abbott M. B. (1991). *Hydroinformatics: Information Technology and the Aquatic Environment*. Avebury Technical, Aldershot, Brookfield, VT.

Abbott M. B. (2000). The gender issue in hydroinformatics, or Orpheus in the underworld. *J. Hydroinf.*, **2**(2), 87–104.

Abbott M. B. (2002). On definitions. *J. Hydroinf.*, **4**(2), i–xxvii.

Abbott M. B. (2007). Managing the inner world of infrastructure. *Proc. ICE Civ. Eng.*, **160**, 26–32.

Abbott M. B. and Vojinović Z. (2009). Applications of numerical modelling in hydroinformatics. *J. Hydroinf.*, **11**(3), 308–319.

Abbott M. B. and Vojinović Z. (2010). Realising social justice in the water sector: 1. *J. Hydroinf.*, **12**(1), 110–130.

Abbott M. B. and Vojinović Z. (2010). Realising social justice in the water sector: 2. *J. Hydroinf.*, **12**(2), 225–239.

Dijk M. P. V. (2007). Urban management and institutional change: an integrated approach to achieving ecological cities. In: *International Seminar on Sustainable Urbanization*, Tripoli, Libya, Hotel Bab Africa, 30 June–1 July.

Easterly W. (2006). *The White Man's Burden: Why the West's Efforts to Aid the Rest Have Done so Much Ill and so Little Good*. Penguin Group, USA.

Foucault M. and Gordon C. (1980). *Power/Knowledge: Selected Interviews and Other Writings, 1972–1977*. Pantheon, New York.

Heidegger M. (2009). The question concerning technology. In Hanks C (ed.), *Technol. Values: Essential Readings*, (pp. 99–113), Wiley-Blackwell.

Kenworthy J. R. (2006) The eco-city: ten key transport and planning dimensions for sustainable city development. *Environmental and Urbanization* (April, 18), 67–85.

Kurian M. and McCarney P. (eds.) (2010). *Peri-Urban Water and Sanitation Services: Policy, Planning and Method*. Springer, Berlin.

Maslov A. (1943). A theory of human motivation. *Psychol. Rev.*, **50**(1943), 370–396.

Mintzberg H. (1994). The fall and rise of strategic planning. *Harvard Bus. Rev.*, **72**, 107–114.

Rotterdam S. (2008). *1600 Hectares, Creating on the Edge, Five Strategies for Sustainable Development*. Projectbureau Stadshavens Rotterdam, Rotterdam.

Sen A. (1999). *Development as Freedom*. Anchor Books, New York.

Stren R. (1993). Urban management in development assistance, an elusive concept. *Cities*, **10**(2), 125–138.

UNCHS (1999). *Basic Facts on Urbanisation*. UNCHS (Habitat), Nairobi.

van Dijk M. (2006). *Managing Cities in Developing Countries, the Theory and Practice of Urban Management*. Edward Elgar, Cheltenham.

Wiarda H. (2003). *Civil Society: The American Model and Third World Development*. Westview Press, Boulder.

Williams R. (1978/1990/2002). *The Wound of Knowledge*. Darton-Longman-Todd, London.

Chapter 8

Asset Management

8.1 INTRODUCTION

Urban water infrastructure assets represent a vast investment which is made over many generations by both public and private sector organizations in order to support the life of urban communities. The assets are strictly non-stationary systems, but for maintenance purposes they are often regarded as being stationary such that they are maintained to provide certain levels of service (or standards) by the continuing replacement and refurbishment of their components. In the urban water management context, the term 'asset' refers to any physical object of an urban water system which enables the provision of related services (e.g., pipe, manhole, valve, pump, weir, reservoir, detention storage, channel, etc.). The failure of such assets can be very serious for the urban environment and population. Even though most of the actual failures tend to be localised, such as pipe burst, sewer collapse, flooding, or polluted discharges to receiving waters, the solution to such failures may involve an analysis of the whole integrated network. The sustainability of these extensive networks, which frequently, if not continuously, interact with each other (e.g., water distribution network and groundwater, groundwater and wastewater, wastewater and receiving surface water bodies), is an important issue for urban water managers. Therefore, one of the main concerns of urban water utilities is to secure and manage their investments in such a way that they meet appropriate standards and levels of service with an acceptable number and severity of failures while ensuring maximum return on their investments regardless whether they are publicly or privately owned.

Furthermore, besides the sustainability of the operation of the existing networks, urban water utilities are also faced with extensions to the networks with the development of new urban areas, or the alignment of the existing networks when a significant part of the urban area is radically redeveloped. All of these issues highlight the need for a structured and full life-cycle management approach to the management of urban water assets, starting from planning and design, through construction and operation to dismantling and disposal. With such needs most of urban water utilities nowadays are vitally interested in adopting the most appropriate asset management practice for their business, and in particular, the principles and techniques that aid responsible and transparent decision making. A formal approach to the asset management of urban water assets involves number of activities which can be grouped into the following three groups: creation of assets (planning, prioritising and acquiring), operation of assets (including their maintenance and rehabilitation) and asset rationalisation (including their potential reuse, decommissioning and disposal). The efficacy of an asset management practice does not depend only on ability of organisation's resources to carry out such practice but also it depends on the amount and quality of data that such organisation can use in their planning and decision making process. Without adequate data, it is virtually impossible to make cost-effective decisions. In this respect, tremendous efforts and resources have been placed on various spatial and temporal data collection activities, their validation, processing and achieving/storing.

The overall concept of asset management has been described by a number of authors (e.g., ALGENZ, 1998; Water Environment Research Foundation, 2007). This chapter aims to provide an overview of the basic principles behind asset management practice, to give guidance on how to apply such principles to the management of urban water assets, and to indicate how to plan associated data collection and modelling activities.

8.2 ASSET MANAGEMENT CYCLE

Asset management may be defined as a comprehensive and structured approach to the long-term management of assets to serve the needs of urban communities at the lowest possible cost. However, a holistic approach to asset management should also be viewed as a *business discipline* for managing the life cycle of assets to achieve a desired service level while mitigating risks. The approach encompasses management, financial, customer, engineering and other business processes. The reason why asset management is considered a business discipline is that its success is measured using business and financial indicators. It is also important to note that asset management encompasses a large number of diverse processes including engineering, but many are not engineering in nature.

The span of application of asset management extends from identification of desired standards to the daily operation and maintenance of assets that enable provision of related services. Within the entire asset management process (as illustrated in Figure 8.1) it is important to provide the necessary link with desired standards (or levels of service) which should be the driver for the subsequent development of strategies and implementation actions (projects, works programme, operation and maintenance, etc.). Example of such process which contains several phases and runs in a continuous cycle is given in Figure 8.1.

Definition of desired levels of service refers to the selection of a certain quality of a particular service (e.g., 100% compliance with bacteriological requirements of the national water supply standard, water main breaks per 100 km per year, response time for water supply interruptions, frequency of flooding, frequency of CSO spills, etc.) against which service performance can be measured. Therefore, it relates to reliability, quality, quantity, and safety/risk. A clear need should be identified before a particular level of service is adopted. Typically, this is done by referring to appropriate standards and policies/regulations as well as stakeholder expectations. Other inputs that impact on the level of service are the planning, maintenance and renewal. The priority or impact of these inputs varies for each organisation and asset and it depends on the status of the assets and their key corporate objectives or business viability. The key performance indicators or quality assurance issues in relation to level of service are various national industry-related performance, structural and environmental indicators and standards. In that respect, the above mentioned elements of asset management are typically focused around structural integrity of assets, their hydraulic performance and environmental impacts.

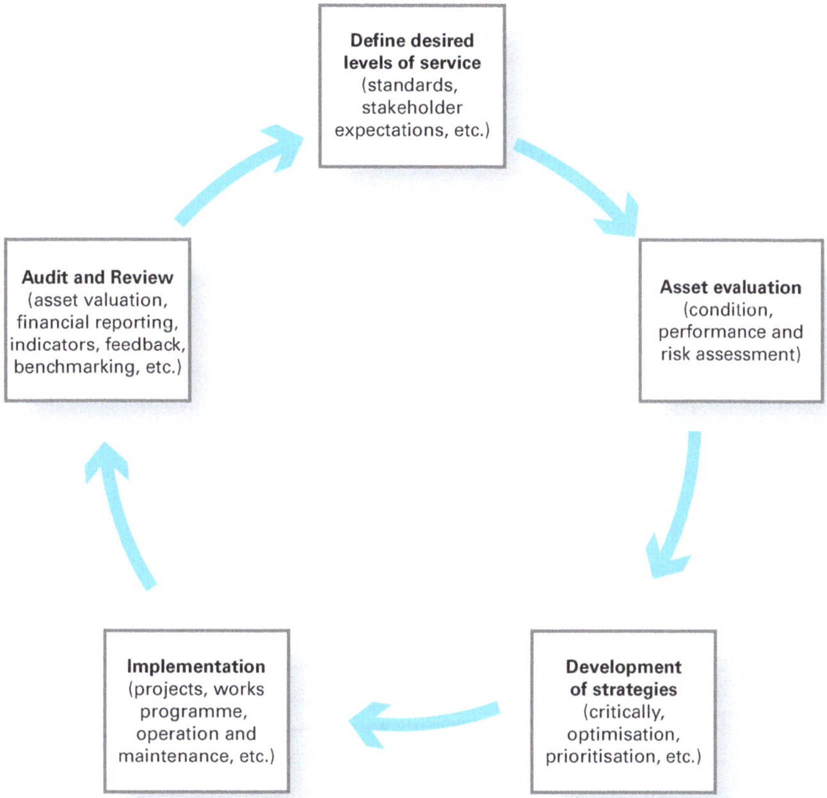

Figure 8.1 Asset management process

Asset evaluation usually refers to the condition, performance and risk aspects. Condition relates to the physical state of assets whereas the performance is closely aligned to the level of service which can be measured in terms of capacity, reliability, availability, etc. Risk relates to analysing and quantifying risks and the development of risk management strategies.

In asset management, development of strategies refers to the process whereby the information gathered from asset evaluation phase is used to optimise and prioritise asset management strategies that relate to operation and maintenance of existing assets, their rehabilitation (renewal and replacement), acquisition of new assets, and dismantling and disposal of surplus assets. In addition to the condition and performance data, it is common to include criticality as yet another aspect in the prioritisation and decision making process. Criticality in this instance refers to the importance of a particular asset with respect to financial, business and service level terms. Typically, assets for which the financial, business or service level consequences of failure are significant are often regarded as critical (e.g., a water distribution main that supplies the water to the hospital or a sewer main that runs under the road of a central business district area), whereas those assets for which such consequences are insignificant are regarded as less critical (e.g., a water distribution pipe that supplies water to a toilet in a peripheral park which is used on rare occasions).

Implementation phase within the asset management process refers to the realisation of the necessary strategies which are needed to satisfy or improve a level of service. This is typically done by undertaking various operational and maintenance activities (i.e., day-to-day running and upkeep of assets), projects, works programme, etc. Also, periodic asset management audit and review are usually undertaken in order to evaluate utility's performance in achieving its asset management objectives, to establish a continuous asset management improvement cycle and to maintain best practice. Usually, this is done by assessing various aspects against a set of indicators, benchmarking and customer feedback. Benchmarking refers to the analysis and comparison of asset management processes by gathering and exchanging the data with other utilities nationally and/or worldwide. Typically, asset evaluation process ends up with the service quality gaps and how well a particular utility performs in meeting desired levels of service. In this respect, poor service relates to inability to provide particular service to the customers (e.g., inability to evacuate wastewater from housing and industry, flooding due to excess rainfall-runoff which is unable to enter sewerage network, etc.).

An important part within the entire asset management process is to enable the ongoing and meaningful engagement of stakeholders. The general purpose of stakeholder participation here is to induce a change in the management of urban water infrastructure in such way that aligns with a positive change in the social environment. For such activity, effective communication of information and knowledge is the first key to ensuring that each phase of asset management is understood by all concerned, and improvement measures are chosen accordingly. Therefore, in order to enable effective communication to take place between all stakeholders it is important to translate the data and technical information into those forms of knowledge and understanding that can be assimilated by the non-technical population so that these may intervene meaningfully and responsibly in the decision-making process. This leads to the development of a framework that employs various spatial and dynamic graphical representation that are best suited to the empowerment of stakeholders, and in promoting their interests. The aim of such asset management framework (as illustrated in Figure 8.2) is to integrate asset performance, condition and criticality with legislative requirements and other constraints defined by regulations, standards and stakeholder interests. The outputs from this framework are the policies, procedures and actions that need to be adopted in order to achieve cost-effective creation, acquisition, maintenance, rehabilitation and disposal of assets that will meet the required level of service and predicted future demands.

Certainly, one of the roles of hydroinformatics in this process would be in bringing the sociotechnological means to weave these together into synergic interactions to provide a balance between technical and social values.

8.3 EVOLUTION OF ASSET MANAGEMENT PRACTICE

Over the past years, the practice of asset management in the urban water sector has evolved through different stages. Initially, most of the asset management practice was reactive and it was mainly based on a 'gut feel' and the local knowledge of maintenance crew, technicians and engineers. In this way, most of the asset replacement or renewal activities were typically undertaken upon the occurrence of an incidence with little anticipation of events. With such old-fashioned asset management decisions that were made have often resulted in ineffective interventions and customer dissatisfaction. Following developments in the asset management theory, data collection and processing capabilities, the focus was placed on a more proactive and structured condition- and performance-based asset management, which requires anticipating failures and planning ahead of them, and even more recently, there has been a further shift towards the service level and risk-based asset management.

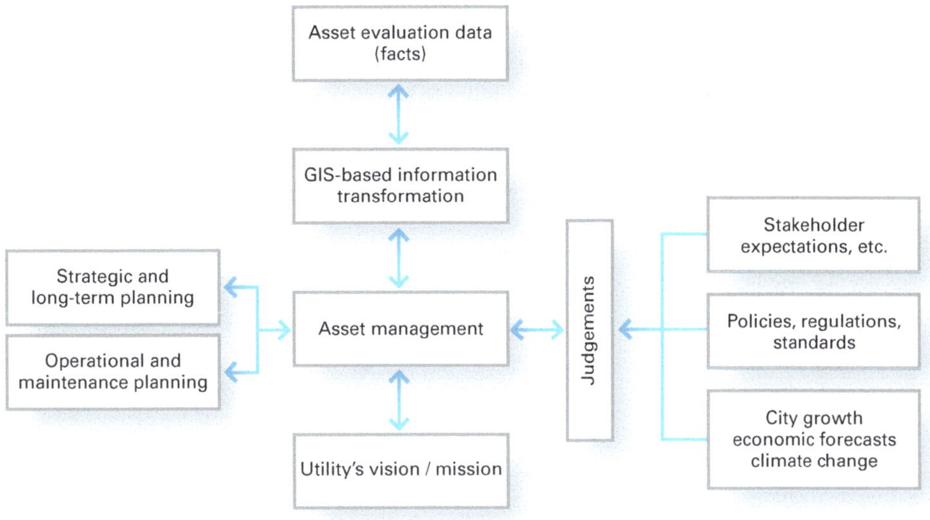

Figure 8.2 A sociotechnical asset management framework

8.4 CONDITION- AND PERFORMANCE-BASED ASSET MANAGEMENT

In condition- and performance-based asset management practice decision making is based on condition and performance assessment for either all assets or for a selected group of assets (e.g., critical assets). In this respect, various approaches on how to identify and rank all assets with respect to their condition and performance characteristics have been developed. Typical asset condition and performance questions considered are:

- How old is a particular asset?
- Where does it stand on its lifecycle curve? In asset management, the term 'lifecycle' refers to the entire life of a particular asset, i.e., from planning and design to decommissioning and disposal.
- What is its estimated residual life?
- Has it been maintained regularly?
- How does it perform with respect to technical, environmental, public health, community standards?

8.5 ASSET CONDITION ASSESSMENT

For asset condition assessment various visual inspection techniques which range from direct visual inspections, acoustic, electrical/electromagnetic to portable Closed Circuit Television (CCTV) technique are used. There are also innovative technologies such as laser profiling, gamma-gamma logging, ground penetrating radar, infrared thermography, micro-defloction and Impact Echo/Spectral Analysis of Surface Waves (SASW) that can be used for the same purpose (see for example EPA, 2009). Such technologies can provide information relevant to the assessment of asset condition. Collection of data such as presence of defects (e.g., cracks, corrosion, displacement of joints, root intrusion, etc.) and their severity collected over a period of time can assist in developing typical asset deterioration curves. In this respect, various decay curves have been established for different urban water assets. On the basis of such curves the potential estimate of an asset condition at different points in time of an asset life can be determined; see Figure 8.3.

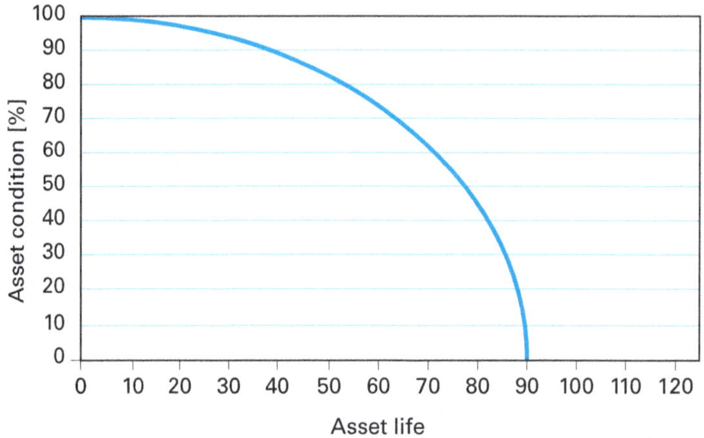

Figure 8.3 The relationship between asset decay/condition and asset life

Some examples of different pipes and their typical manufacturer's claimed life durations are given in Table 8.1. However, the effective (or economic) life of an asset is not necessarily the same as the manufacturer's claimed life. The effective (or economic) life of an asset is the period from its acquisition to the time when it has become unserviceable or obsolete or when the present value of operating and maintenance costs is greater than the cost of replacement. Typically, the effective lives of assets can be estimated from physical inspection, examination of asset service records, discussion wit technical staff and benchmarking against known retirement practices.

Table 8.1 Example of typical manufacturer's claimed life durations for different pipes

Pipe Material	Manufacturer's Claimed Life (years)
Concrete	100
PVC	100
Vitrified Clay	100
HDPE	70+
CMP	30 to 80 depends on soil resistance
Ductile Iron	80
Cast Iron	80

Figure 8.3 provides a typical relationship between the asset condition and asset life. However, the rate of deterioration of assets is very much asset- and site-specific. It is indeed possible that the same type of asset

within a system may have different lives due to different service conditions such as design standards, construction quality, material quality, operational stresses, maintenance history/practice, asset working environment, external stresses. Some examples of pipe failure due to deterioration are given in Figure 8.4 and Figure 8.5.

Figure 8.4 Corrosion of a water distribution cast iron pipe (Photograph: Zoran Vojinović)

Figure 8.5 Corrosion/Erosion of drainage pipes (Photograph: Brigitte Helwigh)

Since the asset deterioration process is rather uncertain (as it is also influenced by random events) it is beneficial to consider the risk or likelihood of its structural failure (e.g., water main burst, sewer collapse, etc.). Hence, if we would add the probability of failure to the asset decay/condition curve, then their typical relationship can be represented as the one given in Figure 8.6.

In order to more precisely estimate the potential residual life of an asset it is important to undertake a form of visual inspection and to identify at which stage of deterioration it is at present. As documented in various literature (see for example, WRc, 1994), the process of pipe collapse can be divided into three stages: initial defect (cracks, leaking joints, etc.), deterioration (erosion of the pipe material under different circumstances) and actual collapse (which can be related to the magnitude of deterioration and/or to some random event). Some of these examples of defects are given in Figure 8.7 and Figure 8.8.

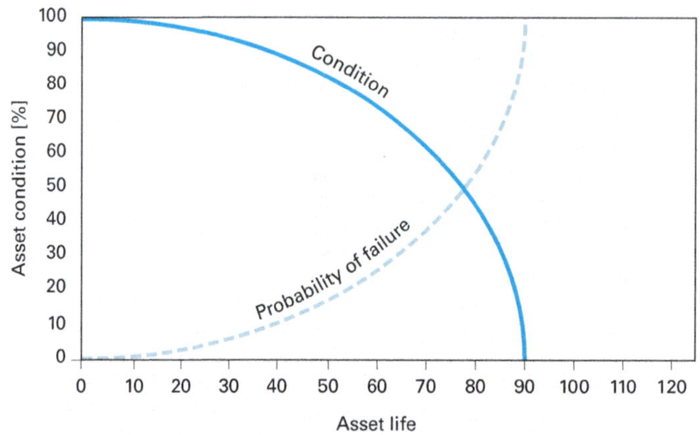

Figure 8.6 The relationship between asset condition and probability of failure (Marlow *et al.*, 2007)

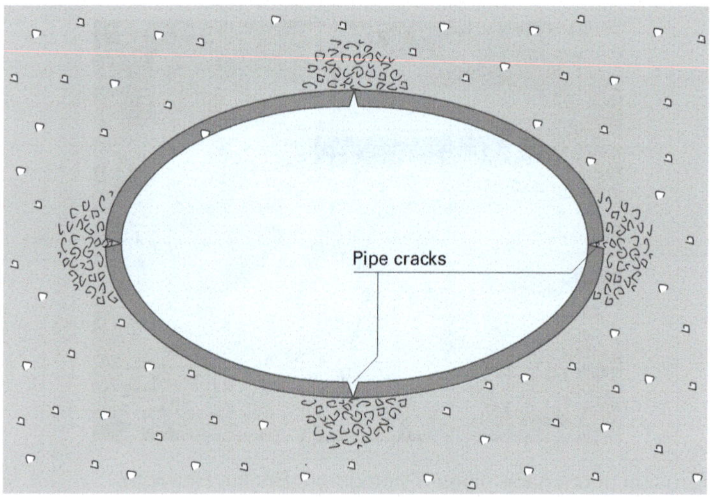

Figure 8.7 Examples of pipe defects at sofit, invert and springing: cracks and fractures (After WRC, Sewer Rehabilitation Manual (WRc, 1994))

Asset condition survey can be very costly and time consuming task. It usually requires employment of significant resources, it is typically carried out in a cost effective way by applying a sampling technique (e.g., stratified sampling, random sampling, systematic sampling, etc.) to only those assets that are assessed to be at higher risk level or those assets that are defined as being critical. The most common technique is the stratified sampling technique in which all assets are initially divided into groups to estimate their proportions and then sampling programmes are designed according to their proportional value and variability.

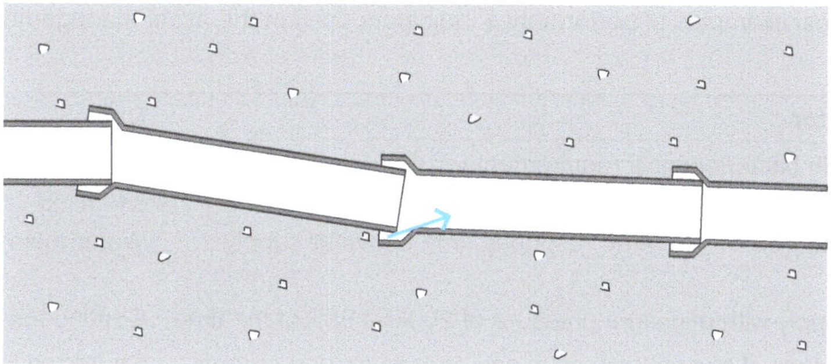

Figure 8.8 Example of pipe defects: open and displaced joints, infiltration/exfiltration and formation of negative gradients (Adopted from Sewerage Rehabilitation Manual (WRc, 1994))

Typical indicators used for the assessment of structural condition of assets are:

- Presence of defects (cracks, fractures, joint displacements, etc.)
- Severity of defects
- Reduction in pipe wall thickness (evidence of erosion)
- Evidence of corrosion
- Evidence of the tree root intrusion.

8.6 ASSET PERFORMANCE ASSESSMENT

In terms of the performance assessment, typically, there are indicators that require different types of data. In some cases regular site observations are adequate whereas in some other cases the use of software modelling tools supported by the short-term and/or long-term data collection is the only way forward. The following Table 8.2 provides some typical examples of performance indicators which are commonly used in the asset management practice of urban water systems.

Performance assessment is an evolving process as it depends on the list of performance indicators chosen and their ever changing standards (customer-related, environmental, legislative, design, etc.). The choice for indicators should represent what all stakeholders regard as important and necessary to meet the required levels of service.

8.7 SERVICE LEVEL AND RISK-BASED ASSET MANAGEMENT

With respect to the improvement of asset management practice over the time, service level and risk-based asset management has recently gained its higher importance over the pure condition- and performance-based asset management. Service level and risk-based asset management seeks to accomplish the practice through consideration of the quality of service delivery and associated risks. Risk is typically defined in this context as a probability of a given number of occurrences in excess of a given magnitude for more than a given duration. It can be further determined by taking into account both probability and consequence of asset failure. For example, a risk associated with the provision of surface water (or stormwater) drainage can be formulated as: *probability of flooding occurring more than once in five years in excess of a depth of 30 cm for a duration of more than 30 minutes (and thus relating to a particular damage cost).*

Certainly, the probability and consequence of asset failure can change over time. The probability of asset failure will typically increase over the life of the asset as it deteriorates and as the city expands. However, the consequence of asset failure may either increase or decrease depending on the system characteristics and its location with respect to the land use and activities of the surrounding area.

Table 8.2 Some typical examples of performance indicators used in the asset management practice of urban water systems

Performance Indicator	Data
100% compliance with bacteriological requirements of the national water supply standard	Continuous water quality sampling
Water main breaks per 100 km per year, response time for water supply interruptions	Water main incidents database
Provision of water supply with minimum pressure of 200kPa 90%of the time	Continuous pressure recording
No more than an average of two shutdowns in any one year per customer due to operational defects	Shutdown incidents database
Interruptions to customer's water supply rectified within four hours	Interruptions database
Speed of response for urgent reactive maintenance works no longer than 4 hours	Maintenance database, customer survey
Speed of response for non-urgent reactive maintenance works no longer than 5 days	Maintenance database, customer survey
All correspondence replied to within 5 working days	Customer relations database, customer survey
No more than 5 unplanned headworks shutdowns with loss of supply to customers up to 4 hours.	Shutdown incidents database
Maximum of 12 unplanned reticulation shutdowns exceeding 4 hours supply loss.	Shutdown incidents database
Unaccounted for leaks no more than 150 litres/connection/day	Operational database
Average time to attend bursts and leaks no longer than 1 hour	Incident database
Not more than ten blockages per 100 km of pipe length	Maintenance database
Response to a pump station failure no longer than 1 hour	Maintenance database
Not more than ten sewage overflows within reticulated networks per 1 in 20 year storm event, due to wet weather infiltration/inflow	Model simulations
Not more than one sewage overflow from a system pump station per fault type per annum.	Pump station database
Not more than two validated odour complaints due to defects in the sewer reticulation	Manual inspections
100% compliance of CSO emissions with resource consent conditions including compliance monitoring.	Model simulations, flow surveys
No flooding for less than 1 in 20 year storm event	Model simulations
The biosolids complies to Grade A National Wastewater Standard	Continuous sampling
Effluent quality complies to Grade A National Wastewater Standard	Continuous sampling
Water quality of receiving waters should meet Grade A National Standards for 90% of time	Model simulations, flow surveys, water quality sampling

Typical risk assessment methodology would contain the following steps:

- Agree expected service and consequences of failure
- Identify probabilities, frequencies and durations of events leading to failure of service
- Determine probability, frequency and duration of each failure
- Determine the extent and potential value of consequences of each failure
- Determine optimal system management strategy to minimise damage (fail-safe)
- Quantify probability and frequency of events consequential on each failure
- Determine business risk by aggregating cost times risk for each outcome

The greatest benefit of asset management is gained when efforts are focused on the more critical assets (i.e., assets with the higher consequence of failure). In this respect, grading of each asset with respect to condition, performance and criticality can aid in the development of effective prioritisation, maintenance and renewal/ replacement (or rehabilitation) strategies.

8.8 PIPE DETERIORATION MODELLING

To evaluate the probability of failure of an asset, such as a collapse of a sewer pipe or a burst of a water distribution pipe, a relationship should be established between asset failures and other parameters of the system. In general, numerous factors such as age, diameter, material, corrosion, traffic, quality of pipe material, installation conditions and operational conditions contribute to deterioration of pipes. Among the parameters that affect water distribution pipe failure, only some of them are measurable such as age, length, diameter, depth and pressure (Berardi *et al.*, 2008 and Tabesh *et al.*, 2009). Pipe deterioration modelling can be performed using either *physically based* or *statistical models*. As described earlier, the former are based on first principles (e.g., physical laws), known variables and known parameters, while the latter are data-driven, predictive models developed by applying statistical tools, e.g., regression, to the data available. Due to the complexity of asset deterioration processes physically-based asset deterioration models are not commonly developed. Contrary to that data-driven models are quite often developed and applied in hydroinformatics practice and their role in asset management will be discussed in more detail here.

The most often applied data-driven methodologies are based on simple regression models on the available pipe failure data from a limited time period. For modelling of water distribution pipe failures, for example, these relationships include a number of influential parameters that affect pipe bursts: diameter (Su *et al.*, 1987 and Goulter *et al.*, 1993), age (Shamir and Howard, 1979), age and diameter (Kettler and Goulter, 1985), climatic conditions (Harada, 1988). Data-driven models are based on the analysis of existing water utility databases which archive data on assets and preserve the historical records of failure events. The scope of such analyses is discovering patterns in asset data for describing pipe failures (e.g., sewer blockages or collapses and water distribution pipe bursts). However, the main drawback of regression-based approaches is that they require the form of the relationship between input and output variables to be assumed before the analysis is performed.

As described in Chapter 5, artificial neural networks (ANN), support vector machines (SVM) and genetic programming (GP) are probably the most well known general-purpose data-driven techniques. An application of a GP-like methodology, namely Evolutionary Polynomial Regression, EPR (Giustolisi and Savic, 2006) to asset deterioration modelling of water distribution pipes is presented next. EPR belongs to the family of GP methodologies and as such can be classified as a grey box technique. Such a technique is based on observed data but allows for the introduction of prior insight on the physical problem at stake, pipe deterioration phenomena. The mathematical structures it returns are symbolic and usually parsimonious. The EPR methodology has two main stages: (1) search for the best model structure using a genetic algorithm (GA) and (2) parameter estimation for the found model structure using the least squares regression method.

During the EPR run, the search for model structures follows user-defined settings such as general polynomial structure, user-defined function type (natural logarithm, exponential, tangent hyperbolic, etc.) and searching strategy parameters. The EPR tool, starts from true polynomials, but allows development of non-polynomial expressions containing user-defined functions:

$$\hat{Y} = a_0 + \sum_{j=1}^{m} a_j \cdot (X_1)^{ES(j,1)} \cdot \ldots \cdot (X_k)^{ES(j,k)} \cdot f\left((X_1)^{ES(j,k+1)} \cdot \ldots \cdot (X_k)^{ES(j,2k)}\right) \tag{8.1}$$

Where $\hat{\mathbf{Y}}$ is the vector of target values; $\mathbf{X}_i$ are the vectors of the k candidate inputs; **ES** is the matrix of exponents; f is a function selected by user; a_j are constant values and m is the length of the expressions.

Berardi *et al.* (2008) describe a case study involving EPR modelling of pipe deterioration in a water quality zone within a water distribution system in the UK. The data in the case study were available at the pipe level for the period 1986–1999 and contain both asset information and bursts recorded. For each individual pipe, data available included pipe diameter, material, year laid, length, number of properties supplied and the total number of bursts recorded during the 14-year monitoring period. As the statistical approach is economically viable for modelling of failures in small pipes, only pipes with the nominal diameter of up to 250 mm have been selected for the analysis. Selected pipes have been grouped into 10 diameter classes with similar classification used to fill-in some existing data gaps. The need for aggregating pipes into homogenous classes is a result of the small number of failures available for the given network. This also makes development of a statistical model for individual pipes difficult to accomplish. Only four fields describing pipe features have been considered for modelling. These are age (*Aep*), diameter (*Dp*), length (*Lp*) and number of properties (*Prp*) supplied, all available at the pipe level. For each diameter-age class the total number of recorded burst events (*Brt*), the sum of pipe lengths (*Lt*), the sum of properties supplied (*Prt*) and the total number of pipes in the class (*Np*) have been computed. Furthermore, to define a significant value of age and diameter for each class, the length weighted mean of relevant variables was computed as shown in Eq. (8.2). The values computed are the equivalent age (*Ae*) and the equivalent diameter (*De*).

$$Ae_{class} = \frac{\sum_{class} Lp \cdot Aep}{Lt} \quad De_{class} = \frac{\sum_{class} Lp \cdot Dp}{Lt} \tag{8.2}$$

Once applied to the aggregated data set, the (single) EPR run returned a multi-objective (Pareto) set of burst prediction models that trade off model's parsimony against model fit (measured as coefficient of determination, CoD) to the observed data. It is interesting to point out that all models demonstrate that burst occurrence depends only on the equivalent pipe class age *Ae*, the equivalent pipe class diameter *De* and the total pipe class length *Lt*, although originally five parameters were considered for modelling. By analysing the form of the equations obtained, their accuracy and using the engineering insight into the problem analysed, the following model is selected:

$$BR = 0.084904 \cdot \frac{Ae\,Lt}{De^{1.5}} \tag{8.3}$$

The model (Figure 8.9) fits the observed data with CoD = 0.822 and provides a good balance between model fit and complexity.

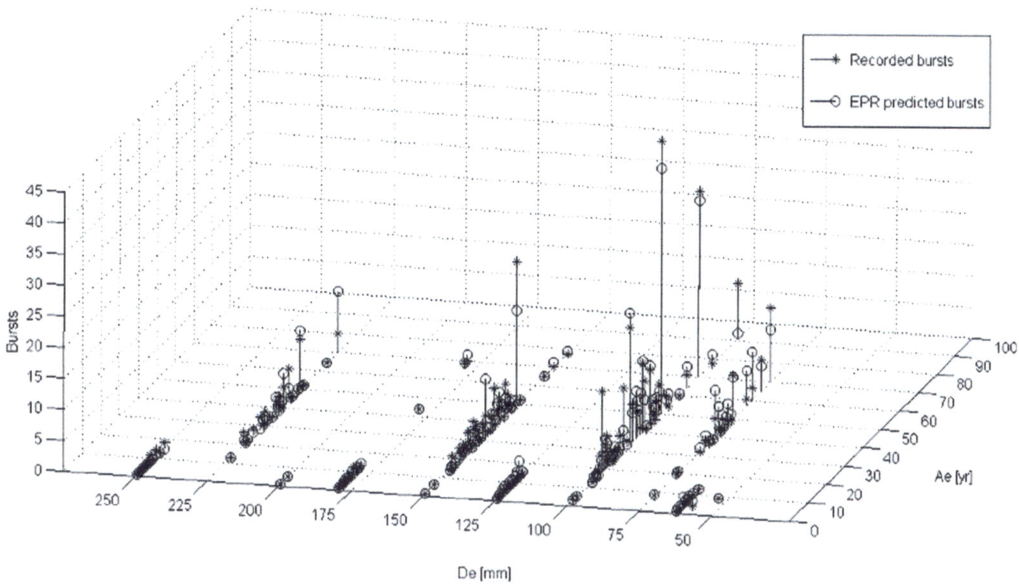

Figure 8.9 EPR model vs recorded pipe bursts (Berardi *et al.*, 2008)

It can be observed that data-driven modelling by EPR produced good, simple and understandable relationships that provide the high level of statistical correlation between the variables and could be regarded as indicators of system performance. The models obtained, can be validated by means of physical knowledge and can be used in the integrated risk assessment methodologies.

8.9 ASSET REHABILITATION

8.9.1 Optimised decision making

As discussed in previous sections, urban water systems face the general infrastructure aging problems and they are subject to continuously refined levels of service. Since the lack of rehabilitation directly increases the failure rate with the time, the need for asset rehabilitation in order to keep them operational is evident. In the past, rehabilitation has often been done on a crisis-based approach. Nowadays, with advances in the theory and practice of asset management the development of rehabilitation plans is based on the optimised decision making process. Within the asset management terminology, the term 'optimised decision making' refers to an optimisation process for considering and prioritising all strategies and options to improve the asset status at the lowest cost. While considering the costs of infrastructure assets as part of the optimisation equation, one should note that there are different costs that occur in each phase of an asset's lifecycle and it is important to reflect the 'true' (or total) economic cost of assets while evaluating different rehabilitation alternatives.

8.9.2 Lifecycle costs

Depending on the standard of service intended it is possible to provide service at different cost levels, and therefore, it is important to be able to attribute the costs to each phase in an asset's lifecycle so that the total cost of each rehabilitation alternative can be established to enable better decision making (Figure 8.10).

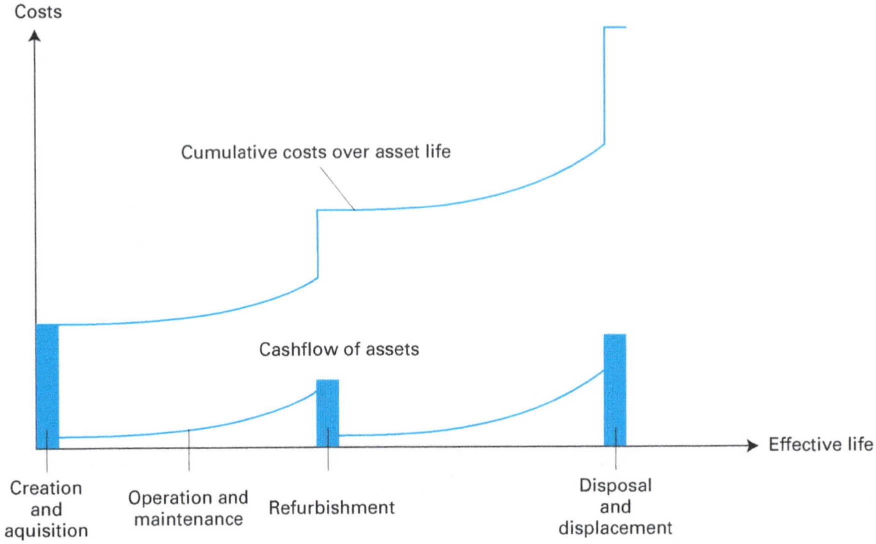

Figure 8.10 Asset lifecycle costs

The full lifecycle asset management process with respect to different functional activities must consider all relevant economic and physical consequences, from initial planning and design through to disposal. Different costs can be identified within a lifecycle of different assets and they can be broadly divided into capital, operation and maintenance costs. Capital investment costs are those costs incurred during the planning, design, construction or acquisition phase of an asset and they generally include all advance expenditure on research, feasibility studies, program planning, land acquisition and improvement, building and site facilities, machinery and equipment, management services, quality control and commissioning, duties and taxes, consulting service fees, cost of raising funds (financial charges), interest charges during the construction period and the construction labour costs.

Operational costs include costs for operations personnel, materials, fuel, chemicals and energy consumption, etc. The operation costs can be greatly affected by factors such as age and condition of the assets (e.g., encrustation of water mains), external environment (e.g., during wet season stormwater discharge may require additional pumping), or maintenance related issues which affect the efficiency of the system.

Maintenance costs include costs of maintenance personnel, plant and equipment, spares required during emergency (unplanned) breakdown, schedulable corrective/predictive planned maintenance or the intermittent maintenance for major mid-life refurbishment, alterations etc. The objective of effective maintenance management is, therefore, to have an optimal maintenance regime where the overall maintenance costs would be minimum whilst providing the required level of service or output (e.g., certain reliability or availability). This requires an appropriate blend of planned and unplanned maintenance.

Figure 8.11 illustrates the typical profile of costs throughout the life of an asset.

As illustrated in Figure 8.11, in an asset lifecycle there are also other costs apart from the capital investment costs and they certainly need to be applied in the process of selecting from among rehabilitation alternatives that would yield the same level of performance or benefits. Life cycle cost assessment enables that the selection of a rehabilitation alternative is not based solely on the lowest initial costs (capital costs), but also considers all the future costs (appropriately discounted) over the project's usable life. Also, it is common that if benefits would vary among the design alternatives (e.g., they would accommodate surface flooding standards), then the alternatives cannot be compared solely on the basis of cost.

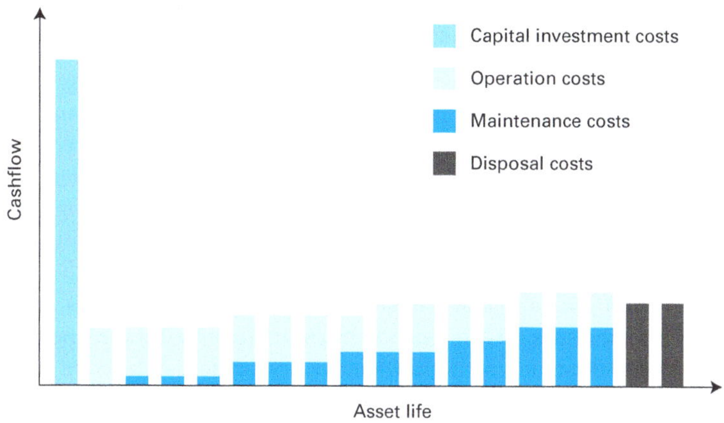

Figure 8.11 Cash flow over the asset lifecycle

Once the performance and costs associated with each alternative have been established, they must be compared over the asset period. This is typically done by calculating net present value (NPV) for each rehabilitation alternative.

8.9.3 Net present value

Net Present Value (NPV) is determined by discounting all alternatives costs to the base, or present, year (usually the present year, year of construction or year of authorisation). Thus the entire project can be expressed as a single base year, or present year, cost. Alternatives are then compared by comparing these base year costs. NPV is a common economic calculation and it can be calculated using Eq (8.4)

$$NPV = Capital\_Costs + \sum_{k=1}^{N} O\ \&\ M\_Costs_k \left[\frac{1}{(1+i)^{n_k}} \right] \tag{8.4}$$

where i is the discount rate

n is the year of expenditure

$\left[\frac{1}{(1+i)^{n_k}} \right]$ is the Present Value (PV) factor

8.9.4 Evaluation of alternatives and the use of optimisation techniques

Typically, as part of the wider asset management process, optimal rehabilitation of urban water assets is implemented by taking into account several structural, hydraulic and environmental indicators that center around the concept of levels of service. With the new standards being introduced the process of identifying optimal rehabilitation strategies has become more and more complex and multidimensional. For example, for optimising sewer system remedial works requirements the methodology should consider the following aspects: full utilisation of the existing systems capacity prior to undertaking improvement works, reduction of pollution affecting local receiving waters as well as main watercourses and groundwater, reduction of infiltration/inflow in the network and its adverse impact on wastewater treatment, prevention of local system surcharges, affecting properties and environment, prevention of structural collapses and damaging other sub-surface infrastructure and the prevention of pollution of water supply systems, due to infiltration of wastewater and minimisation of expenditure. All of these aspects are almost equally important for sustainable urban water asset management and therefore, finding the right balance between the extent of remedial works requirements (i.e., the target level of systems performance) and capital expenditure costs is one of the greater challenges for asset managers.

Even though each of the above mentioned factors has its own part in the planning, design and management of the system and despite their inherent dependence, the process of arriving with an optimal solution (or rehabilitation alternative) is not a trivial task. Initially, this process starts by assessing each and every asset in terms of their physical condition, performance and criticality. As mentioned earlier, the physical condition of assets can be determined though a structural inspection programme, whereas their performance characteristics are evaluated by instantiating and running numerical models. Evaluation of different rehabilitation alternatives within a hydraulic model is normally carried out as a trial-and-error task which implies manual adjustment of model input data followed by running a number of simulations. Due to the limitations of the manual trail-and-error and following the increase in computer power, the research into automatic calibration procedures has led to the use of different automatic optimisation approaches for evaluation of different rehabilitation alternatives. These approaches are based on a general task to optimise (minimise or maximise, as appropriate) the value of objective function, which is a numerical measure expressing the effectiveness of different rehabilitation options. Such problem can be posed as a full-fledge optimisation problem taking into account some of the key factors, Figure 8.12. It deals with determining the pipes diameters for a network with a predetermined layout that needs remediation given the specific objective functions and constraints. Since the objective functions and constraints cannot be expressed as a differentiable analytical function, the methods used for solving this problem belong to the set of multi-extremum (global) multi-objective optimization.

To explain this concept, the objective of selecting an optimal set of remedial alternatives (or measures) of a pipe network system can be described as a process of identifying the vector of some values $x = \{x_1, \ldots, x_n\}$ which are not known a priori (x could be a set of options characterising the possible remedial measures, for example pipe upgrades). This is achieved by feeding the model input data, and subsequently calculating the total cost of works required. Traditionally, following the system analysis a preferred set of system improvements is often selected by engineering judgement. Different from the traditional practices, with the automatic (or automated) selection of system improvements, a set of remedial measures is selected by the use of an optimisation algorithm while satisfying given constraints. With automated practices, depending on the criteria, the remedial measures optimisation problem can be posed in two ways: (1) as a single-criterion optimisation problem, in case of a single output variable (e.g., the total cost of remedial works), and (2) as a multi-criterial optimisation problem – when there are several output variables (e.g., the total cost of remedial works and the total system surcharge/overflow volume, etc.). Example of such system which combines an optimisation tool with a hydrodynamic model for evaluation of system improvement measures is OPTRESS (see the following section).

8.10 ASSET MANAGEMENT DECISION SUPPORT SYSTEMS

There are several asset management decision-support tools currently available. These tools vary in their scope and focus. Figure 8.13 shows some of the available tools and the stages they cover within the infrastructure asset management system. The tools are grouped according to their functionalities and capabilities. The left-hand side of the figure shows the main aspects of an asset management system. The main concepts of each tool are briefly outlined in the following text, see also Ana and Bauwens, 2007.

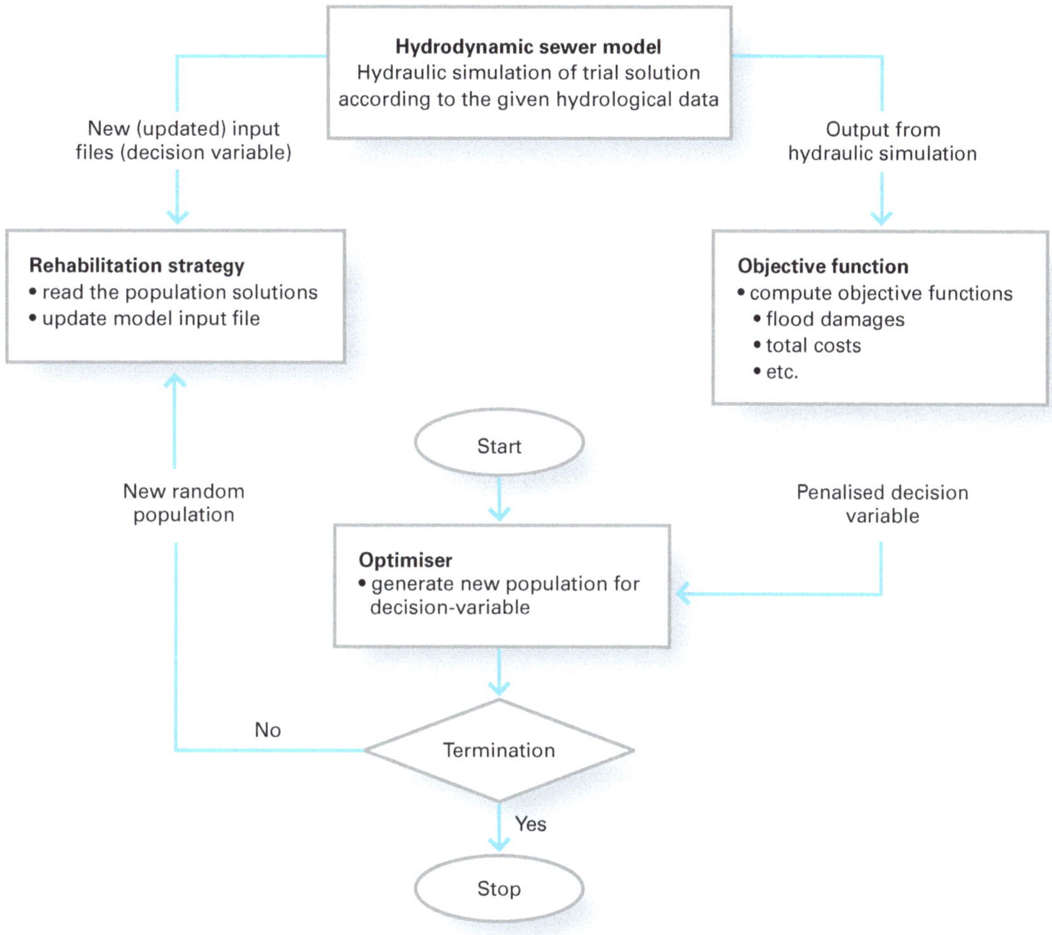

Figure 8.12 Example of an optimisation process for rehabilitation of drainage assets

Group one consists only of performance analysis tools (tools that deal mainly with performance modeling), group two consists of performance and decision analysis tools (they cover the first three asset management stages – from data collection, to performance modelling and decision analysis) and group three consists of the complete asset management tools (they cover all aspects of the asset management stages).

The group one tools are:

- The Baik model (Baik *et al.*, 2006) estimates the transition probabilities of different condition states in Markov chain-based deterioration models for wastewater systems, using an ordered probit model. The idea is to predict the future condition of the sewers so that managers can prepare inspections, rehabilitation and replacement activities in a timely and cost-effective fashion. To estimate transition probabilities, the model requires data from the condition assessments of the existing system: for structural assessment, internal inspection is required and for hydraulic assessment, models are used. Infiltration and inflow are also investigated. The pipe condition rating is calculated based on maintenance and structural points (evaluated using 108 criteria, e.g. deformation, presence of roots) from the inspection.

- Bengassem and Bennis model (Bengassem and Bennis, 2000) is a systematic methodology to evaluate the structural and hydraulic conditions of a sewer system, using fuzzy inference system as an aid in the development of a rehabilitation program. The method encompasses structural inspection and hydraulic simulation to evaluate the condition of the components of the sewer network. Fuzzy theory is then applied at the pipe sections level to integrate all the evaluation factors, to come up with a performance assessment of the sewer network. Three aspects are considered in the evaluation of the structural performance: 1) intrinsic (i.e. pipe defects), 2) extrinsic (i.e. pipe characteristics and environment characteristics affecting pipe degradation, e.g. geotechnical factors, hydro-geological factors and various factors including seismic or tectonic activity), and 3) site vulnerability (i.e. nature of site, soil density,...).
- Hasegawa *et al.* (1999) model estimates the degree of necessity of repairs for existing sewer pipes, based on four viewpoints: 1) decrease in flow capacity, 2) road collapse possibility, 3) sewer overflow and flooding as influenced by inflow/infiltration, and 4) increase in treatment cost due to inflow/infiltration (Hasegawa *et al.*, 1999). The degree of necessity for repair is evaluated at the pipe level.

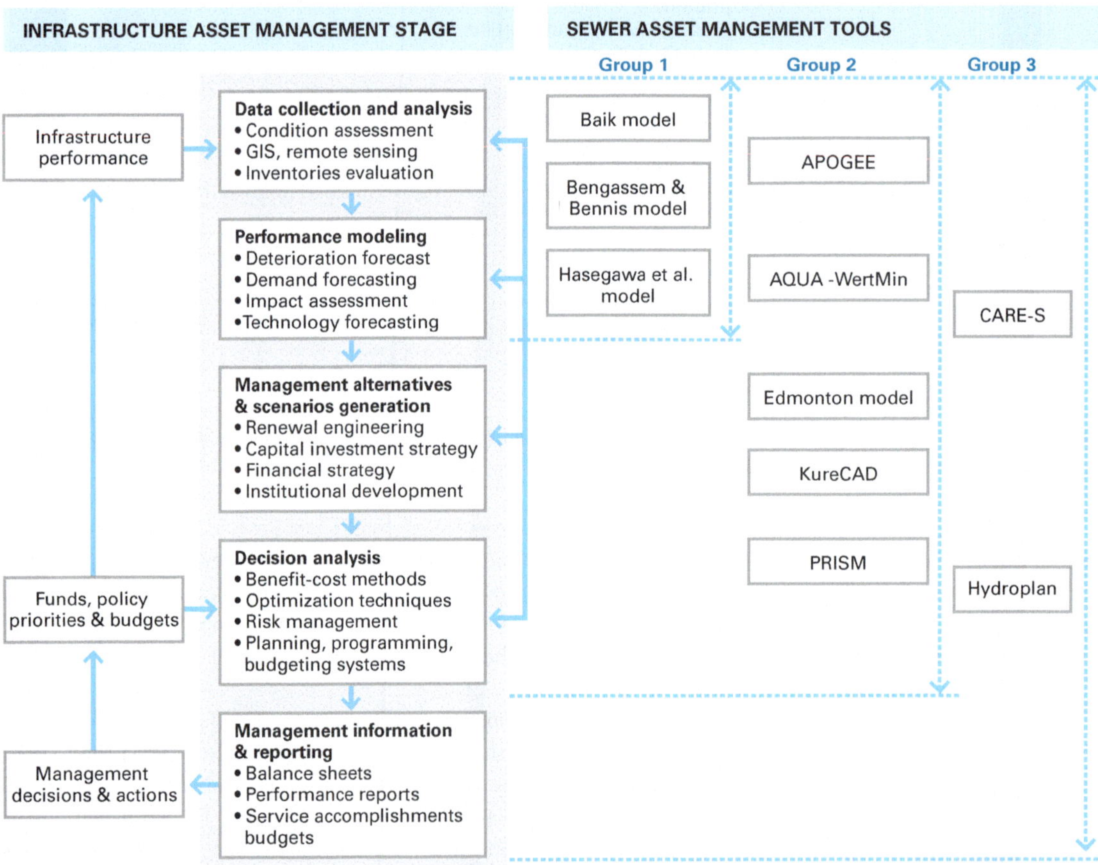

Figure 8.13 The generic infrastructure asset management system (After: Ana and Bauwens, 2007) with the corresponding sewer asset management tools applicable at different stages

The group two tools are:

- APOGEE – it is a decision support system developed with the aim of optimizing the annual planning and rehabilitation of sewer networks (Rommel *et al.*, 1989).
- AQUA-WertMin (Baur and Herz, 2001) is a computer program designed to assist utilities with the planning of TV-inspection, renovation and new construction strategies for wastewater networks.
- Edmonton models – There are three models, each using a combination of rule-based simulation and probability analysis, which were created to assist the city of Edmonton to plan its sewer maintenance expenditure (Ruwanpura *et al.*, 2004). The main aim is to forecast the cost of repair or rehabilitation of existing pipes based solely on pipe condition.
- KureCAD – it is a GIS-based tool for managing sewer pipe rehabilitation with the following functionalities: 1) storage of asset information, 2) sewer pipe rehabilitation prioritization, and 3) provision of documents for the implementation of rehabilitation plans (Stone *et al.*, 2002).
- PRISM (Proactive Rehabilitative Infrastructure Sewer Management) – it is a computerized financial outlay model for prioritizing sewer pipe rehabilitation planning, based on a horizon of budgetary constraints (Ariaratnam and MacLeod, 2002).

The group three tools are:

- CARE-S (Computer-Aided Rehabilitation of Sewer Networks) – it is a decision support system designed to give guidance to municipal engineers to establish and maintain an effective management of the sewer networks (Saegrov, 2006).
- Hydroplan – is an integrated approach to urban water asset management based on structural, hydraulic and environmental risk assessments carried out on the strategic elements of the sewerage network (Vojinović *et al.*, 2005), see also Chapter 10. The Hydroplan risk-based approach capitalizes on access to a vigorous, existing international community of practitioners and their associated information, knowledge and experiences. The procedure (see also Chapter 10) forms a framework of tasks and sub-tasks in sequence and parallel. Each task or sub-task can be regarded as an object with a number of attributes such as input/output to/from other tasks, theoretical/practical background, modelling/analysis tools, expert (rule-based) systems, case studies, acknowledged experts, suppliers of relevant services, etc. In turn this leads to better asset management in terms of safety and reliability. The whole procedure represents a decision support system for the water utilities responsible for urban water asset management. It starts with an inventory of the current situation and historical data, to set-up an initial asset database. Next, a strategic analysis is performed at the pipe level. The pipes that will cause the greatest consequential damage in case of failure are scored based on several factors, which include financial, social and environmental damages. Weighting factors are applied to the scores to designate the global strategic level of a given asset. After estimating the impact of failures, the probabilities of sewer failures are calculated. This is done at structural level (pipe condition) using aging models (Hades) and inspection results, and at the hydraulic and ecological levels using a calibrated hydrodynamic model. To facilitate the evaluation of the large number of possible options in any wastewater or stormwater system scenario an OPTimal REhabilitation of Sewer Systems (OPTRESS) software tool has been developed (see also Vojinović *et al.*, 2006). The methodology embedded within the algorithms of OPTRESS can be used to derive optimal sets of either wastewater or stormwater rehabilitation measures with respect to given levels of service, total life cycle cost of the new assets and the existing assets condition; see Figure 8.14 and Figure 8.15. Scores are attributed for the different aspects of failure, according to fixed criteria, such as probability of collapse (or estimated residual lifetime), frequency of flooding, etc. The results from the strategic analysis and the failure probabilities are combined to come up with the global risk score. This gives the set of critical pipes where proactive rehabilitation investments are most cost efficient. The final outputs of the procedure are various short- and long-term actions and strategies. The long-term investment is determined using a total life cycle model, using Monte Carlo simulations, that integrates all costs by

monetizing the risks and preventive investments. The sewerage database is continuously updated with newly gathered information and the procedure is looped through to ensure the timely detection of new problems and allow better planning of most cost-effective solutions.

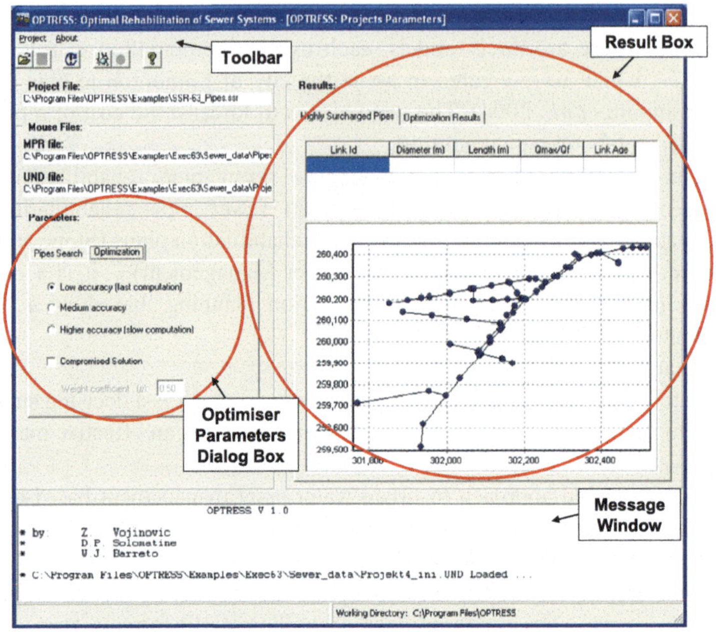

Figure 8.14 The main screen interface of OPTRESS system

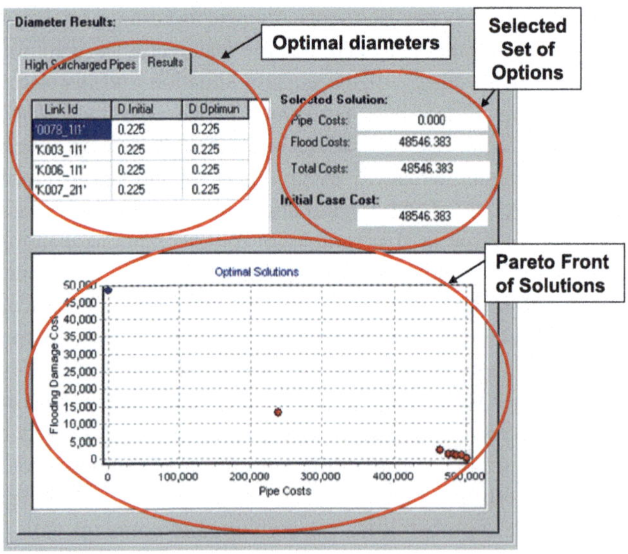

Figure 8.15 The 'Result Box' interface of OPTRESS system

As outlined in this section, there are several asset management decision-support tools currently available and they differ in their capabilities and functionalities. Some tools are integrated, covering data management, performance, optimisation and deterioration analyses to rehabilitation plan developments (e.g. CARE-S, Hydroplan) whereas the others are more specific and cover only part of these requirements. The core of most of these tools is the performance analysis and deterioration prediction using deterioration and hydrodynamic models as well as multi-criteria optimisation or cost analysis techniques.

8.11 CASE STUDY: PROACTIVE ASSET MANAGEMENT STRATEGIES FOR SEVERN TRENT WATER

8.11.1 Introduction

Historically, the UK water industry's approach to asset management has been reactive: 'find and fix' as failures occur, set capital budgets based on the previous year's spending, and favour maintaining the operational status quo over reducing long-term system risk. A key reason for this approach has been the lack the robust tools and technology capable of making more proactive, forward-thinking approaches cost- and time-effective.

One problem with reactive approaches is that they lead enterprises to spend money on the same asset over and over again without eliminating the problem or improving serviceability. Another larger issue is that as the infrastructure ages, its failure rates increase exponentially. Reactive measures simply cannot keep pace with (or outrun) system deterioration and efforts to maintain appropriate serviceability begin to produce extremely large funding gaps. These trends have not gone unnoticed by government regulators.

8.11.2 Background

Since privatisation, water industry funding in the United Kingdom has been based on Asset Management Plans (AMPs). Every five years, all water companies must submit to the UK National Office of Water (Ofwat) their proposed spending plans for the next five years. In the early days, these submissions were largely based on previous spending, particularly in relation to capital maintenance, since it was impossible to forecast the locations and numbers of blockages and collapses. The assumption was that such data would remain roughly the same.

8.11.3 Changing the system

In recent years, water companies began to see that they were spending more and more on maintenance, yet finding it increasingly difficult to maintain serviceability. At the same time, OFWAT was becoming more reluctant to continue increasing budgets without justification. The major change occurred during AMP3 (the third five-year AMP cycle), when the Capital Maintenance Planning Common Framework (CMPCF) was introduced. This guideline was aimed at moving companies away from the 'this is what we've always done' approach toward a risk-based, forward-looking method based on analysis of their historical data.

OFWAT now sets the criteria by which it judges water companies' compliance with the Common Framework, and scores them against each criterion to determine what percentage of their requested capital maintenance budget they receive. Companies assessed as complying with the criteria receive a greater percentage than companies assessed as falling short.

Another major development has been the rewriting of the WRc Sewer Rehabilitation Manual, renamed the Sewerage Risk Management Manual. As the new name suggests, its emphasis has shifted to a risk-based approach in line with the Common Framework. In determining future expenditure companies can no longer simply submit a figure based on previous spending plus inflation.

Customer expectations are also higher, particularly since privatisation. People expect not to experience sewage inside their properties if it rains or a sewer blocks. They expect to be able to commute without falling into a hole caused by a collapsed sewer. Companies therefore have no option but to raise their games. This makes it more critical than ever to know the location of assets and how they are performing.

8.11.4 OFWAT returns

The figures from a recent OFWAT Return show that the water industry in England and Wales owns 374,000 km of sewers, comprising around 7.56 million pipes. Of these, 104,000 are critical sewers on which the water companies must report, including the number surveyed each year. In 2009, just under 2,000 km were surveyed – not an enormous proportion, but a vast amount of data to handle. The majority of these assets are buried. They are also spread over thousands of square kilometres – as are the 4,337 properties that flooded internally in 2009 due to causes such as blockages, collapses and equipment failure. Records show there were 172,500 blockages, nearly 6,000 collapses, and over 1,000 pollution incidents.

Data underpins all of these records, but data alone is not enough. It must be gathered accurately and in a form that can easily be turned into usable information. The key to managing these companies' assets more efficiently is GIS. This not only allows them to see the location of their assets – it reveals where they are failing.

Ten years ago, GIS was more or less a mystery. It was used mainly by major software companies, many of whom did not understand their customers' needs and therefore did not always deliver a useful product. Some companies and councils had such bad experiences with these early systems that they are still wary of GIS.

8.11.5 Severn Trent's DAP programme

Severn Trent's 2000 Drainage Area Plan (DAP) programme included a major CCTV programme which was expected to produce a considerable quantity of data to be managed. This programme proved a turning point in the company's use of GIS. It had GIS details of its sewers, but nothing that could link the CCTV data to them. Fortuitously, MWH Soft had developed InfoNet. Severn Trent had been using the software purely to hold CCTV data. Now, they incorporated it into the data cleansing and validation element of the DAP model build process. This freed up InfoWorks to be used for hydraulic analysis.

Since then, the company has continued to develop its use of InfoNet. It now holds details of nearly 300,000 surveys, all referenced to the pipe surveyed. This data is used to formulate the company's June return, asset inventory and price review submissions. When the company is being audited and the auditor requests details on the pipes surveyed during the year, InfoNet can quickly provide a list, a very impressive capability.

In fact, the OFWAT reporter for Severn Trent's June Return 2008 noted, 'The incorporation of its CCTV into the InfoNet system is commendable. It should enable the company to make use of the benefits of the data produced for Drainage Area Plans.' InfoNet has become more than simply a GIS-based data cleansing tool and CCTV database; it can also be used as a proactive management tool for company assets. At Severn Trent, InfoNet has been used as part of the Periodic Review 2009 (PR 09) process. When shown the data held in InfoNet and the work the firm had undertaken to programme maintenance, one of the reporters said, 'We believe that the company's use of this software and the further applications that they are investigating for its use, for example maintenance scheduling, is industry leading.'

8.11.6 Applications

The first InfoNet application Severn Trent Water used was the CCTV tool. The company has a CCTV framework that uses InfoNet for data management. Once it identifies the pipes to be examined, the company simply emails the contractors an InfoNet Snapshot file of the sewer records and the pipes to be surveyed. The contractors send their survey data back in the same format. The data is validated and stored in a central repository. In AMP 4, the company surveyed over 6,000 km of sewers, with one person managing the entire process in-house.

The company has built on this early work. Over the past two years, it has begun to manage its proactive cleansing and rehabilitation programme in the same way. This effort aims to reduce the number of pollution and internal flooding incidents by reducing the number of blockages and collapses. Sewers with the highest risk of creating pollution or internal flooding are identified and prioritized. Contractors are then given a list of these sewers in InfoNet Snapshot format.

When the contractors survey the sewers, they record the condition, undertake cleansing if necessary, and report the condition after treatment. The company knows the exact condition of its assets before it intervenes as well as the condition after the intervention, with all data recorded in InfoNet. This enables the firm not only to measure the improvements made, but determine which areas need to be re-cleansed. The structural condition of the sewers is also assessed, and a rehabilitation programme is produced and managed in the same way. Finally, InfoNet is central to the process of enabling Severn Trent to keep track of its contractors' work and its cost.

8.11.7 The future

In AMP 5 (the fifth asset management plan period, 2010 to 2015), the company aims to take its activities one step further, using InfoNet to assist with the intervention decision-making process. The software uses SQL to analyze CCTV surveys and chooses the appropriate intervention type, whether a local repair or full-length rehabilitation, based on the defects found.

MWH Soft has ensured that many features the industry will need in the future are already part of InfoNet, particularly scheduling, pipe repair and cleansing tools. Further enhancements are in the pipeline.

One recent development is flow and depth monitoring. InfoNet has always been able to hold Time Varying Data, but the company has not needed it. However, going forward, flow and depth monitoring will be an increasingly important part of pollution strategies. These processes will produce a great deal of data that must be immediately integrated for operational purposes. If loggers report an increase in depth of flow in a sewer, the company needs to know if it was just caused by recent rains, or signals a build-up of debris.

To obtain this data and interpret it, the company is considering using hundreds, if not thousands more loggers. The data they gather will also be invaluable for long-term analysis aimed at determining strategies robust enough to be included in future price review submissions.

InfoNet can also be used to plan monitoring surveys and hold the details of the survey data, linked to the relevant assets and CCTV surveys. This gives the company an overview of the catchment and its developments, not just the monitoring point. The company is currently working on migrating the flow logger and CCTV data, into one massive InfoNet database.

8.11.8 Adoption of private drains and sewers

The impending adoption of private sewers will be a tremendous challenge. There is no accurate data on how many there are or where they are located, let alone their condition. Companies are still deciding how to manage them, but one thing is certain – it will be necessary to store data on them, and the best way to do it will be in a GIS.

Whether pipes are surveyed proactively or only on failure, there is still a need for a significant leap in onsite data capture technology such as MWH Soft InfoNet Mobile. Severn Trent is looking into this promising technology to see if it can be incorporated into its procurement process in the next few months.

These are just a few examples of InfoNet's capabilities. As enhanced data-gathering techniques are developed and the importance of good data is better appreciated, analysis and decision-making will become even easier and more reliable. As managers gain more confidence in the data, greater demands will be placed on analysts and software writers to deliver it.

The water industry has come a long way in the past 30 years, but that growth must continue at a rapid pace. OFWAT expects companies to manage their assets more efficiently and at least maintain serviceability. Failure to do so is not an option. The tools for success are available, and assets can be managed more efficiently than previously possible. It is up to the industry to use these tools to their full potential by working collaboratively to meet the ever-increasing expectations of both business managers and regulators.

8.11.9 Acknowledgement

This case study by John Hateley of Severn Trent Water is based on an original presentation to the MWH Soft User Conference in September 2009.

8.12 KEYWORDS

By the end of the chapter you should have encountered the following terms; ensure that you are familiar with them!

Asset management cycle
Condition assessment
Level of service
Net present value
Performance assessment
Rehabilitation
Risk-based management

The items with an asterisk are dealt with in more detail in later chapters.

8.13 QUESTIONS

What is the asset management cycle?
How should a condition assessment be carried out on drainage or distribution system assets?
How should a performance assessment be carried out on drainage or distribution system assets?
What is the role of risk assessment in asset management?
How and when should assets be rehabilitated?

8.14 REFERENCES

ALGENZ (1998). *Infrastructure Asset Management Manual.* National Asset Management Steering Group, Association of Local Government Engineers, New Zealand.

Ana Jr. E. and Bauwens W. (2007). Sewer network asset management decision-support tools: a review. In: *International Symposium on New Directions in Urban Water Management*, 12–14 September, UNESCO, Paris.

Ariaratnam S. and MacLeod C. (2002). Financial outlay modeling for a local sewer rehabilitation strategy. *J. Constr. Eng. Manag.*, **128**(6), 486–495.

Baik H., Jeong H. and Abraham D. M. (2006). Estimating transition probabilities in Markov chain-based deterioration models for management of wastewater systems. *J. Water Resour. Plann. Manage.*, **132**, 15.

Baur R. and Herz R. (2001). Selective inspection planning with aging forecast for sewer types. In: *Proceedings IWA 2nd World Water Congress: Efficient Water Management – Making It Happen*. 15–19 October, Berlin, Germany, 8.

Bengassem J. and Bennis S. (2000). Fuzzy expert system for sewer networks diagnosis. In: *Proceedings International Conference on Decision Making in Urban and Civil Engineering*. 20–22 November, Lyon, France, 18.

Berardi L., Giustolisi O., Kapelan Z. and Savic D. A. (2008). Development of pipe deterioration models for water distribution systems using EPR. *J. Hydroinf.*, **10**(2), 113–126.

Giustolisi O. and Savic D. A. (2006). A symbolic data-driven technique based on evolutionary polynomial regression. *J. Hydroinf.*, **8**(3), 207–222.
Goulter I. C., Davidson J. and Jacobs P. (1993). Predicting water-main breakage rates. *J. Water Res. Plan. Manage.*, **119**(4), 419–436.
Harada H. (1988). Statistics of the cold wave in the temperate region and prediction of the number of damaged service pipes. *J. AWWA*, **57**(8), 12–15.
Hasegawa K., Wada Y. and Miura H. (1999). New assessment system for premeditated management and maintenance of sewer pipe network. In: *Proceedings 8th International Conference on Urban Storm Drainage*, Sydney, Australia, 30 August–3 September, pp. 586–593.
Kettler A. J. and Goulter L. C. (1985). An analysis of pipe breakage in urban water distribution networks. *Can. J. Civ. Eng.*, **12**(2), 286–293.
Marlow D., Heart S., Burn S., Urquhart A., Gould S., Anderson M., Cook S., Ambrose M., Madin B. and Fitzgerald A. (2007). *Condition Assessment Strategies and Protocols for Water and Wastewater Utility Assets*, WERF Report 03-OCT-20CO.
Ruwanpura J., Ariaratnam S. and El-assaly A. (2004). Prediction models for sewer infrastructure utilizing rule-based simulation. *Civ. Eng. Environ. Syst.*, **21**(3), 169–185.
Saegrov S. (2006). *Care-s: Computer Aided Rehabilitation of Sewer and Storm Water Networks*. International Water Association, London.
Shamir U. and Howard C. D. D. (1979). An analytical approach to scheduling pipe replacement. *J. AWWA*, **71**, 248–258.
Stone S., Dzuray E. J., Meisegeier D., Dahlborg A. S. and Erickson M. (2002). *Decision-Support Tools for Predicting the Performance of Water Distribution and Wastewater Collection Systems*. US EPA/600/R-02/029.
Su Y., May L., Duan N. and Lansey K. E. (1987). Reliability-based optimization model for water distribution systems. *J. Hydraul. Eng. ASCE*, **113**(12), 1539–1556.
Tabesh M., Soltani J., Farmani R. and Savic D. A. (2009). Assessing pipe failure rate and mechanical reliability of water distribution networks using data-driven modeling. *J. Hydroinf.*, **11**(1), 1–17.
Vojinović Z., Bonillo B., Kaushik C. and Price R. K. (2006). Modelling flow transitions at street junctions with 1D and 2D models. In: *7th International Conference on Hydroinformatics*. Acropolis, 4–8 September, Nice, France.
Vojinović Z., Price R. K. and Van den Broek W. (2005). Hydroplan-EU knowledge management framework for urban water asset management. In: *International Conference on Urban Drainage*. 21–26 August, Copenhagen, Denmark.
Water Environment Research Foundation. (2007). *Condition Assessment Strategies and Protocols for Water and Wastewater Utility Assets*. WERF: Alexandria, VA and IWA Publishing: London, United Kingdom.
WRc (1994). *The Sewerage Rehabilitation Manual*. WRc Engineering, Swindon.

Chapter 9

Water Distribution Systems

9.1 INTRODUCTION

The first ancient societies and the great civilizations of the world have evolved around water. Such societies, as those in the Tigress and Euphrates region (Mesopotamia), in the Nile region (Egypt), in the Indus valley (Pakistan and India), in the Yellow River valley (China), on the island of Crete (Greece), and those in Central America, were all founded on their ready access to water. Human settlements normally began where supply of freshwater for drinking and agriculture was readily available from streams or springs. However, because of the variability of the available water or because the settlements tended to grow and with them the demand outgrew the supply, the settlers soon began to develop the means of transporting water from often distant sources to the settlements. For example, it is estimated that over 400 km of aqueducts were supplying water to the ancient city of Rome. However, the Roman aqueducts extended much further beyond Rome, serving large cities of the Empire and many small towns. The Romans also developed an extensive network of drains and sewers that linked to the Cloaca Maxima, which was one of the best known sanitation artefacts from the ancient world. The Roman water supply and sanitation systems represent also one of the word's earliest examples of the so-called urban water cycle in action.

Even today's modern cities rely on the urban water cycle, which is conceptually simple:

1) water is extracted from streams and aquifers and then stored in reservoirs,
2) this water is then processed to drinking (potable) quality,
3) potable water is distributed through an extensive pipe system to consumers,
4) some of the delivered water is used to transport wastes through a network of sewers to treatment plants, and
5) treated wastewater is then discharged into receiving waters, such as rivers, lakes and oceans.

In addition we should include the stormwater component.

A water supply system capable of supplying a sufficient quantity of potable water, together with drainage and wastewater systems, is a necessity for a modern city. This system, typically consisting of a network of interconnected pipes, pumps, valves and reservoirs, is usually referred to as the *clean water system* as opposed to the *dirty water systems* associated with drainage and sewer networks.

Most clean water systems consist of water sources, treatment facilities, pumping stations, transmission mains, storage facilities and distribution lines. *Water sources*, including underground aquifers, rivers, lakes and impounding reservoirs, supply raw water to *treatment plants* where finished or potable water is produced (Figure 9.1). The finished water is then typically pumped into *storage* via the *transmission system* consisting of large diameter pipes (mains). The water is then delivered to individual consumers through the *distribution network*. A distribution network consists of smaller diameter pipes and other network elements, including various types of valves, hydrants, tees, bends, etc (Walski *et al.*, 2001).

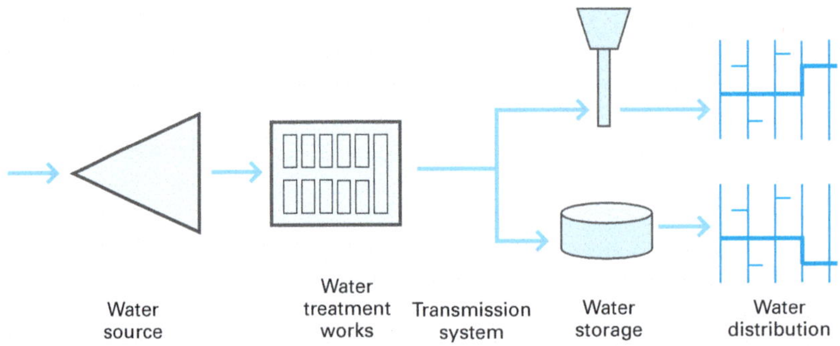

Figure 9.1 Water supply schematic

In the widest sense, a *water distribution system* is considered to include all components of the water supply system as shown in Figure 9.1, apart from the water source and the treatment facilities.

9.1.1 Water sources

Groundwater is water found beneath the ground surface that makes up a substantial proportion of the world's fresh water supply. Groundwater is stored in aquifers, which are layers of porous substrate, such as sand, gravel and fractured rock formation. In areas of undulating terrain, groundwater may appear naturally at a lower elevation in the form of a spring. As the spring water appears on the ground surface, it does not require pumping to get it to the surface, but it is prone to pollution. Abstraction of aquifer water can be achieved through wells and boreholes, which are dug or drilled from the surface to reach the groundwater table. Wells are generally of large diameter, whereas boreholes are small diameter holes (e.g., 400 mm to 500 mm), both requiring pumping to lift water to the surface. Water is normally pumped out of wells and boreholes thus causing a drawdown, i.e., a decrease in groundwater level when water is pumped. Groundwater frequently does not require treatment or such treatment is normally minimal (requiring at most disinfection), but requires protection from pollution, that is, by locating wells and boreholes away from potential hazards.

Direct abstraction from a nearby *river* was one of the easier means of supplying water to early urban populations. Even nowadays, cities and smaller towns depend on river abstractions for their water supply. For example, the Danube River in Europe or the Yellow River in China are sources of drinking water for tens of millions people along their paths. However, due to increased demand and risk of pollution, the construction of intakes upstream of settlements is often necessary.

In situations when a river cannot provide sufficient quantities of water to supply a settlement during dry weather conditions, *impounding reservoirs* may be used to store water. In contrast to *lakes*, which are naturally formed bodies of fresh water surrounded by land, impounding reservoirs are artificial lakes created by a dam (or barrage) across a river or a stream. The main purpose of impounding reservoirs is to balance uneven temporal distribution of streamflows, that is, to store water in times of plenty for use in times of shortage. In addition to water supply, impounding reservoirs can provide many other benefits, such as floodwater control, power generation, irrigation and recreational use. However, there are a number of disadvantages associated with impounding of rivers and streams including loss of high quality agricultural land, water quality problems associated with thermal stratification of reservoirs, displacement of people, changes to downstream flow patterns and impacts on fish migration and microclimates. Many of the adverse environmental impacts associated with impounding reservoirs can bc mitigated, and such measures should be employed wherever appropriate.

9.1.2 Water treatment

All water sources require treatment prior to consumption. After raw water is abstracted from a suitable source it is subjected to a set of water treatment processes to produce a safe product, i.e., potable water. Usually, this process consists of a number of stages, with initial pre-treatment by settling or pre-filtration through coarse media, and then sand filtration followed by chlorination (Figure 9.2). This is called the multiple barrier principle, which provides the basis for effective treatment of raw water and prevents complete breakdown of treatment due to a failure of an individual component.

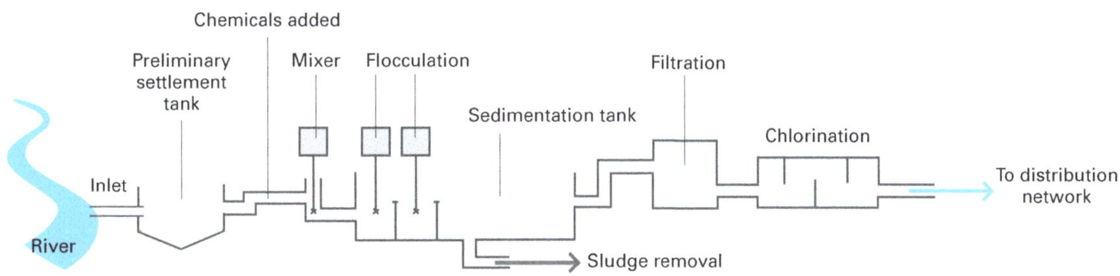

Figure 9.2 Schematic of a typical water treatment process

Preliminary treatment includes *screening* and *sedimentation*. Screening removes large debris, such as leaves, branches and fish. These objects can block up the water treatment system and must be removed before the water enters the treatment plant. For example, treatment facilities that use surface water have large screens covering the site of the water intake. *Sedimentation* is achieved by storing the water free of large debris in a preliminary settlement tank or reservoir, which is used to reduce fluctuations in the water quality and help reduce the suspended solids content. Sedimentation tanks do not require frequent maintenance, apart from cleaning when performance drops off; normally once a year is sufficient.

Sedimentation is also used to remove particles created by *coagulation* and *flocculation*, which is the process of creating larger particles (a precipitate, or floc) that then settle out; otherwise they would be too small to remove through normal sedimentation or filtration. The addition of a chemical coagulant to the water causes the formation of a large particle, which entraps impurities, such as colour, turbidity, algae and other micro-organisms, which then settle quickly to the bottom and form sludge. The advantages of coagulation are that it reduces the time required to settle out suspended solids and it can also be effective in removing protozoa, bacteria and viruses. The most commonly used coagulants are aluminium sulphate and ferric sulphate.

Filtration is another process aimed at removing turbidity, algae and some micro-organisms that are too small to be removed by sedimentation. In this process, the partially treated water is passed through a medium, typically layers of gravel, sand or anthracite, which retains the fine organic and inorganic material and allows the clean water through. Particles are removed from raw water by gravel filters and slow sand or rapid gravity filters. The difference between slow and rapid sand filtration is not only in the speed of filtration, but in the underlying principles employed in the treatment process. Slow sand filtration is essentially a biodegradation process, whereas rapid sand filtration is a physical treatment process. In the former, a few millimetres thick biological layer of algae, plankton and other microscopic plant life, known as the 'schmutzdecke', develops on the surface of the filter and can be effective in removing micro-organisms. The latter have largely replaced slow gravity sand filters, because they are smaller (and can deal with a much higher flow) than slow sand filter, and are hence less expensive. Because of the greater flow, no schmutzdecke is formed and biodegradation does not take place in these filters.

Although sedimentation and filtration can be effective in removing potentially pathogenic micro-organisms, they cannot guarantee the complete removal of extremely small infectious organisms, such as bacteria. Therefore, the partially treated water must be disinfected to ensure it meets potable quality standards. The process that removes these potentially pathogenic micro-organisms, which in many water supplies represent the most serious health threat to water users, is called *disinfection*. Chlorine (Cl_2) is a major disinfecting agent for drinking water, although the use of ozone has recently become more widespread. Disinfection of drinking water is considered to be one of the major public health advances of the 20th century. However, chlorine disinfection could also result in a wide variety of by-products. The most well-known by produces are trihalomethanes (THMs) and haloacetic acids (HAA) that are formed when chlorine reacts with naturally occurring organic and inorganic matter in water. These by-products are suspected carcinogens. See Rodriguez and Sérodes (1998) for a study of residual chlorine evolution.

The primary aim of this section is to introduce the basic concepts of water treatment. For further information on the principles and advanced topics of water treatment see Porteous (2000). There is a number of modelling software products for water treatment plants.

9.1.3 Water transmission

Transmission mains (usually larger diameter pipes) convey water from source, treatment, or storage facilities to the distribution system where the majority of service connections are located. A transmission main (also called a 'trunk' main) does not normally have service connections to it, although it may have a small number of connections that supply isolated users or a group of them. Transmission mains are generally sized to provide flow rates required to serve the anticipated future demand in the vicinity of the system as determined in the long-term water master plan or on the basis of the local land use plan. The size of a transmission main is determined by hydraulic analysis taking into account the network topology and levels of service provided (e.g., required pressure) under various flow conditions.

9.1.4 Service reservoirs and water towers

A water supply system needs to include some form of treated water storage to enable the system to meet peak flow levels and spread the demand placed on the treatment works over an operational period, that is, the whole day or an even longer period. These water service reservoirs also provide a reserve of treated water in the event of planned maintenance or to accommodate the maximum period of interruption of supply or problems with the source or treatment. Additional advantages provided by service reservoirs include a small variation in pressure, reduced on/off cycling of pumps supplying the reservoir, no need to pump and use energy in order to meet peak system demand, and improved reliability.

Ideally, water service reservoirs should be located on high ground allowing water to reach the distribution area by gravity (that is, without pumping); however, they are often situated underground. Sometimes, to reach higher buildings or areas located on elevated ground, they are built as elevated water towers.

9.1.5 Distribution pipes, valves and pumps

The water from service reservoirs is distributed by a network of distribution pipes of various sizes that generally follow town and city streets. Distribution pipes are typically of smaller diameters than the transmission mains and deliver water to customer service lines (i.e., the property boundary of the customer). The majority of distribution mains laid today have sizes 100 mm or 150 mm diameter as standard. These systems of pipes are typically designed as either looped systems or branched systems or a combination of both. Each of these designs has implications in terms of cost, reliability, system control, and other factors (Panguluri *et al.*, 2005).

In a *looped system*, there are extensive loops such that there are generally multiple paths that water can follow from the reservoir to customers. A *branched system* corresponds to a tree-like structure with larger pipes feeding smaller branches. In the extreme case, there are no loops and any disruption of a pipe causes 'outages' to all downstream customers. Branched systems are most commonly found in rural water systems. In the United Kingdom, *district meter areas* (DMAs) have been in use for over 20 years and consist of isolated water distribution zones for which flow volumes can be calculated and thus controlled. This is an important concept in leakage management as the volume of flow into and out of a DMA can be measured, and abnormal usage can signal the presence of leaks or bursts. DMAs generally contain between 1,000 and 5,000 connections and all inflow and outflow from each zone is metered (Panguluri *et al.*, 2005). For an analysis and prediction of short term water consumption see Aafjes *et al.*, 1997.

Although a water distribution system may contain a large number of valves, there are in general only two types: *isolation valves* and *control valves*. Isolation valves are used in the distribution system to isolate sections for maintenance and repair. These are typically located in a network so that the areas isolated will minimise the impact caused to other parts of the system. Control valves are used to regulate the flow or pressure in a distribution system. Typical types of control valves include pressure-reducing valves, pressure-sustaining valves, flow-rate control valves, throttling valves and check valves.

9.1.6 Complexity (or Water distribution labyrinth)

Although the urban water cycle is conceptually simple, this apparent simplicity is deceptive and, even if the physical processes alone are considered, the cycle is difficult to fully comprehend and manage. For example, Colombo and Karney (2003) introduced a labyrinth map (Figure 9.3), which depicts the various technical issues and processes involved in potable water distribution (i.e., step 3) of the urban water cycle. The term 'labyrinth' is particularly suitable, not only because the water distribution system with its looped and interconnected topology resembles the mythical maze, but also because it implies that a successful navigation of the labyrinth's complexity may allow an important goal to be achieved. The water distribution labyrinth contains numerous interconnected processes, sub-processes, causative factors, feedback loops and interrelationships. Practitioners and researchers should appreciate this multiplicity when they model specific processes or limit their analyses to narrowly defined areas.

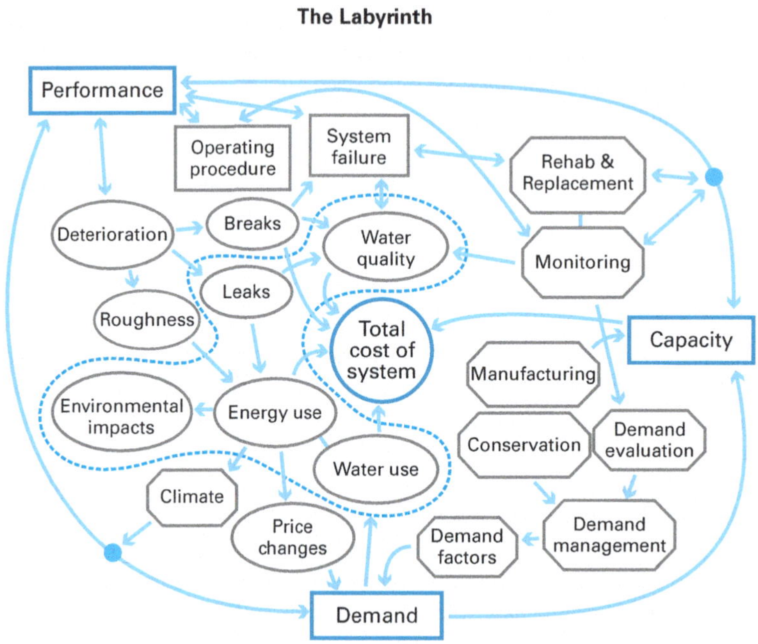

Figure 9.3 Water Distribution Labyrinth (Colombo and Karney, 2003)

There is also further complexity arising from uncertainties associated with water distribution system modelling. These uncertainties come from a number of sources including consumer demands (that change due to stochastic nature of instantaneous demands), pipe characteristics (diameters and energy loss characteristics that change due to the ageing of pipes), water quality reaction coefficients, and others.

9.2 MODELLING WATER DISTRIBUTION SYSTEMS

The past few decades have seen major improvements in our ability to model larger water distribution systems (WDS). Nowadays, it is widely acknowledged that strategic and operational decisions related to WDS depend critically on the efficiency and accuracy of mathematical models utilised to model the systems' behaviour under a variety of conditions. For example, variations of flow and pressure in a system over a 24-hour period could be simulated to check for regulatory compliance.

Chapter 5 describes the mathematical basis of flow in a distribution network. Because the pipes operate under pressure any change in the supply to the network, or adjustment to a valve or pump or a change in the demand at a point, leads to an effectively instantaneous response throughout the network. Although pressure transients exist, they can be ignored so far as the normal modelling of the network is concerned. The primary concern is to conserve mass at nodes and to express the headloss along a pipe in terms of the parameters of the pipe and the discharge through it. The solution of the resulting equations for prescribed inputs (reservoir heads) and outputs (at demand points) gives the heads at all nodes and the discharges in the pipes.

Iterative techniques, such as the Hardy-Cross or Todini-Pilati method, are used to provide the solution to the steady-state, non-linear equations for demand driven conditions. Where there is excessive demand or a pipe burst then a pressure-driven analysis is required (Bhave, 1988 and Todini, 2003). Extended Period

Simulation (EPS) generates a sequence of steady-state analyses at fixed time intervals including mass balance computations at each of the service reservoirs. Where there are rapid changes of the fluid velocity in a pipe line then the transient equations have to be solved for the water hammer phenomenon.

9.2.1 Model instantiation

Techniques and procedures for instantiating an EPS water distribution system model vary from organisation to organisation. The UK Water Research Centre summarises them as follows:

- inspection of supply, distribution and consumer records and maps,
- site inspection of plant and equipment,
- preliminary field measurement,
- field measurement exercise,
- entry of network data for a computer analysis, and
- calibration of the network model.

This sequence of tasks reflects the model instantiation procedure outlined in Chapter 6, and details of the procedure are described there.

Model building consists of particular emphasis on collecting the necessary data from records and flow surveys, simplifying the network and calibrating the pipe roughnesses (and possibly the head losses at valves and other control devices). There are various approaches devised for calibrating a water distribution network model, including manual and automatic methods. The latter involve optimisation techniques. These are detailed in Chapter 6.

9.3 MODELLING APPLICATIONS

Customers, public and regulatory bodies expect water utilities to achieve a number of often conflicting goals when planning for long-term investment and management of day-to-day operations of the water supply system. These goals include: effective and efficient expenditure, environmental protection, serviceability, optimised asset performance, attainment of social goals, minimisation of cost to customer and, in the case of privatised utilities, reasonable shareholder returns (Skipworth, 2002). In doing so, utility planners have to account for both capital and operational costs and expenditures. In the past, network models have been routinely used when utilities considered large capital expenditure, such as long-range master planning, system rehabilitation, strengthening, rezoning, etc., but played little or no part in the day-to-day operation of distribution networks. That has changed with the advent of all-mains models that can adequately represent dynamic behaviour of the system, for example, water quality changes in the network, or reproduce on-line changes in the system behaviour. Such models may be used for a wide variety of problems such as to:

1) assess whether pressures satisfy minimum requirements under different demand scenarios;
2) understand the behaviour of the system;
3) identify and resolve system anomalies (e.g., closed valves);
4) design of expansion or strengthening of the system to meet future demands;
5) consider various rehabilitation options and their effects;
6) develop a leakage management strategy;
7) perform water quality investigations (e.g., water age studies);
8) develop an effective energy management scheme (e.g., by optimised pump scheduling);
9) perform source water coverage studies;
10) train staff without the risk to the system itself.

9.3.1 Modelling for capital investment planning

Water utilities invest significant capital to ensure that their systems meet not only the drinking water requirements, but also any environmental and security standards. This usually means that water is delivered to customers with agreed quality of service, or:

- in sufficient quantity;
- at the required pressure;
- with the required quality ('wholesome and clean', i.e., safe);
- reliably (continuously);
- cost-effectively; and
- sustainably (e.g. without over-abstraction).

Two categories of modelling applications to aid capital investment planning are identified: (1) system design, and (2) asset management planning.

System design

To design a water distribution system to deliver water to customers with an agreed quality of service is a challenging task, which normally involves identifying optimal decisions related to system elements, i.e., pipes, pumps, storage reservoirs and valves. In general terms, system design could be classified into:

1) master planning,
2) system rehabilitation,
3) other design categories (expansion, strengthening, zoning, etc.).

Master planning studies are undertaken to determine major capital improvement projects necessary to ensure the quality of service for the future. Modelling for master planning is concerned with projected system growth, its impact on water usage and system performance over a long-term planning horizon (e.g., from 5 to 50 years). These models are often skeletonised. Master planning involves identifying potential problem areas due to growth in future demand and also sizing and locating of new transmission mains, pumping stations, and storage facilities to ensure that the predicted problems can be avoided.

System rehabilitation is often synonymous for water mains rehabilitation, which encompasses many methods for improving the performance of existing pipe networks by changing their condition or specification. Even without considering future growth, systems age, become less efficient, often experiencing problems with deteriorating infrastructure, water quality and service disruption. Sensitivity toward leaks and breaks increases and accelerates further deterioration, resulting in a loss of water, potential contamination incidents and discoloration through cracked and corroded pipes. Rehabilitation work is undertaken to overcome shortfalls in the level of service provided to customers and would typically include such work as mains cleaning, scraping and relining, slip-lining, various pipe insertion methods, open cut replacement and reinforcement. Examples of reasons for undertaking rehabilitation actions include a need to improve inadequate pressures, water quality, or a need to reduce leakage and maintenance costs. Rehabilitation models generally require a greater level of detail than a master planning model.

Among other categories, design studies are also undertaken for *system expansion* (e.g., for new area being developed), for *system strengthening* (e.g., ensuring better performance of the existing systems) or for *system rezoning* purposes (e.g., creation of district metered areas or pressure zones). These models also require a greater level of detail than a master planning model together with a common requirement that the new design must satisfy operational constraints including pressure constraints under multiple loading conditions.

Developing a water system hydraulic model for design studies begins with evaluating the historical population trends and projected growth patterns. Accurate water demand predictions require knowledge of the magnitude, direction, and characteristics of population growth. In order to derive projected water demand from population projections, the average amount of per-capita water usage in the modelled area is required. Such water usage is commonly measured in litres per capita per day (lpcd), which is estimated using the average day water demand for a year and the population for that year. The calibrated model is then updated by allocating the projected demands to nodes in the model. These demands relate to the ultimate demand condition of the planning horizon as well as to each intermediate year. The proportion and geographical distribution of future commercial and industrial consumption should be reflected in these nodal demands so that concentrations of demand which could stress the system locally can be accommodated by the analysis. Diurnal demand profiles and peaking factors are also needed to design a system that will cope with specific worst-case conditions such as peak demand times, fire protection usage and system component failures.

As an example case study of a master planning exercise and how it could benefit from applying optimisation consider the Regional Municipality of York (Canada). In 1996 the Municipality expected rapid development over the 35-year period, with a doubling of its population (Savic *et al.*, 2000). There were significant implications for the infrastructure required to service this development, with the anticipation of a sizeable capital expenditure. The water supply infrastructure needed to be developed through the construction of additional water mains, pumping stations and storage facilities. The number of options available to achieve a feasible solution was vast and significant potential savings could be achieved by identifying the most cost-effective infrastructure configuration.

An extensive optimisation of the regional infrastructure needed to meet water demands in 2031 was performed using a genetic algorithm (GA) approach, which was applied to a calibrated hydraulic model. A model consisting of all of the regional transmission mains, together with the larger or hydraulically significant distribution mains, was constructed first. Extensive field testing of both the network and the pumping stations generated the data needed to calibrate the model.

Baseline details (for the situation in 1996) and projected future demands (for the planning horizon and intermediate stages) were then added to the calibrated model. Routes along which new pipes to service projected development areas could possibly be built, together with potential locations for pumping stations and storage facilities, were identified and also added to the model. Unit construction costs relating to diameter and ground type (rural, urban or urban congested) were determined. Approximately 300 pipes were included in the model as options for the master plan solution. Each pipe could adopt any of 14 discrete diameters, giving a possible number of solutions of 14^{300}, or 10^{343}. With the additional options for the location and sizing of pumping and storage facilities the total number of possible designs was calculated to be approximately 10^{357}.

The GA algorithm interfaced with the calibrated model to which all of the optional pipes and facilities were added. The optimisation process basically used:

1) a GA with an objective function defined on a set of decision variables, e.g. pipe diameters, reservoir storage requirements, etc.,
2) a calibrated model of the system to simulate its hydraulic behaviour and ensure that continuity and head-loss equations are satisfied at all times (hard constraints), and
3) a penalty term to penalise insufficient level of service (soft constraints), in terms of pressures at nodes, imbalance of reservoir flows, etc.

The GA was programmed to take into account expected operating and maintenance costs as well as capital costs, ensuring that the overall solution reflected cost effectiveness on a whole-life cost basis. Failure to meet a set of performance criteria (for example, acceptable pressure ranges and storage requirements) resulted in cost penalties.

An optimal solution to the final horizon problem was identified first, which proposed the construction of 85 new transmission mains added to the existing 750 mains in the system. Six new pumping stations were proposed and three old pumping stations were identified for expansion with a total of 42 new pumps. The optimised plan also recommended three pumping stations to be decommissioned. Seven new ground reservoirs and elevated tanks were proposed (total volume 78 Ml) with two existing elevated tanks identified for decommissioning. The phased development of the solution to meet progressive demands at intermittent stages during the planning period was conducted later as a desk study to produce a profile of anticipated capital and operational expenditure over the whole planning horizon. The capital cost of these works for the ultimate planning horizon of 2031 was found to be $\$194.42 \times 10^6$ (1997 Canadian Dollars) based on the unit costing data developed for the study. Whilst no directly comparable cost for a manual solution was available for the ultimate planning horizon of 2031, the phased optimal solution for 2011 would cost $\$102 \times 10^6$ instead of the previous manual solution cost of $\$156 \times 10^6$ – a saving of $\$54 \times 10^6$ or 35%. Considering that the size of the problem was estimated to be 10^{357} and that the age of the universe is approximately 14×10^9 years (4.3×10^{17} sec), it is not surprising that the impressive improvement over the manual solution has been achieved by the optimisation approach.

Another, more recent example involves the optimal rehabilitation and expansion of the City of Ottawa (Canada) water distribution system (Randall-Smith *et al.*, 2006). At the time of the analysis, the system served a population of about 780,000 people. The objective was to rehabilitate and expand the system in such a way that it can meet an expected large increase in the future demand over the next 30 years. The rehabilitation and expansion opportunity was formulated as a constrained optimisation problem driven by the minimisation of total costs including capital, operation and maintenance and energy and peak power costs. A 24 hour hydraulic simulation model of the system was used during the optimisation process to check the feasibility of the candidate solutions. Most of the more than 200 decision variables modelled were conventional (replacement/expansion of pipes, tanks, pumps, etc.). Having said this, the novel concept of custom decision variables was introduced to enable considering complex decision variables based on some engineering solutions (e.g. rezoning). The optimisation problem was solved successfully using Genetic Algorithms.

Asset management

The Montgomery Watson Harza's The Asset Group (TAG) defines Asset Management *as 'a business discipline for managing the life cycle of assets to achieve a desired service level while mitigating risks and it encompasses management, financial, customer, engineering and other business processes.'* This definition covers a number of important aspects of asset management:

- Asset management is a *business* discipline and as such, the success is measured using business indicators (e.g. profit).
- Assets need to be managed through the whole *life cycle* of an asset, i.e. from planning and purchasing to installing, maintaining and eventually disposal.
- One of the main asset management goals is to achieve the *target service level* at the lowest reasonable overall cost, whilst mitigating various *risks*.
- Asset management is a complex activity encompassing a large number of *diverse* processes not all of which are of engineering nature (a fact often overlooked by engineers).

An integral part of asset management in a creation and maintenance of the *asset register* where all the information required for the effective management of the company assets and associated delivery of long-term capital programmes is stored. The data is normally stored in a Geographical Information System (GIS) based system, together with a number of associated relational/other databases. One of the main issues to consider is the level of detail to store. The easiest way to tackle this problem is decide at what level assets might be abandoned and replaced by newer assets once their useful life has expired. For example, each pump could be separately listed in the asset register including details of its manufacturer, date of installation, output and power consumption, etc. but there would be no separate listing of pump casing or pump impeller. If the pump impeller needed to be replaced it would be a maintenance item paid from revenue, not capital programme.

Asset management pressures and drivers come from various sources. The main pressures typically come from the relevant regulatory bodies (e.g. OFWAT in the UK, Environment agency, etc.). In addition, water distribution systems are ageing and as such their condition and consequently their performance is deteriorating with time. At the same time, customer expectations are increasing in terms of better service expected. Based on these drivers, a number of different issues exist that have been or are being addressed to improve the quality of the water service. The main asset management issues in water distribution systems are typically as follows:

- Water *supply/demand balance* management including *security of water supply* and *demand efficiency* issues;
- *Leakage management*;
- *Water quality* management including aesthetic, bacteriological and chemical aspects of the problem;
- *Hydraulic capacity* including pressure related issues;
- *Customer service*, including issues related to consumer bills, complaints, provision of information to consumers, etc.;
- *Sustainability* including climate change, renewable energy, carbon footprint and other related issues.

As mentioned above, asset deterioration is one of the key drivers behind large investments required to manage a water company's assets. As a consequence, quantifying and predicting asset deterioration, especially in terms of performance, is critical for any successful asset planning and management. A number of such models exist nowadays that make use of various statistical (Kleiner and Rajani, 2001) and data mining techniques (e.g. Giustolisi and Savic, 2006). An example of the output from a stochastic, Bayesian based pipe deterioration model predicting the cumulative number of pipe bursts is shown in Figure 9.4.

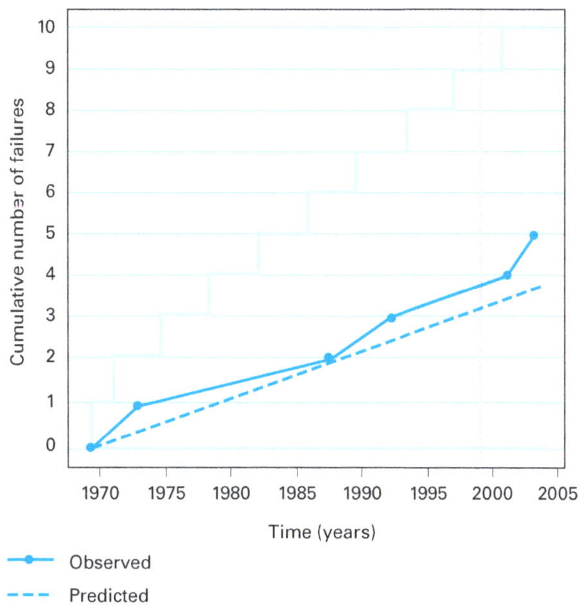

Figure 9.4 Example of the Bayesian-based Zero Inflated Non Homogenous Poisson Process (NHPP) Model Output (Economou *et al.*, 2008)

A wide range of *performance indicators* (PIs) is used to measure the performance of assets and related water distribution systems. Skipworth *et al.* (2002) distinguish between three major types of asset performance indicators:

- Condition PIs: used to measure the physical condition or state of an asset (example: pipe condition grades)
- Performance PIs: used to measure the behaviour of the asset(s) in terms of its performance (example: pipe burst frequency rate)
- Serviceability PIs: used to measure the quality of service received by customers and the ability of the assets (and their operator) to maintain the quality of service (example: percent successful water quality test at tap).

A wide range of *interventions* are available to manage assets effectively. In the water distribution system asset management, interventions are often classified into infrastructure (pipe related) and non-infrastructure interventions (applied to other, typically above ground assets). Another, generic intervention classifications often used are as follows (interventions shown in the order of increasing costs and impacts):

- Asset maintenance (e.g. regular pipe cleaning using jetting, scouring or some other method)
- Asset repair (e.g. a local pipe patch, e.g. a pipe repair to take care of a recent burst)
- Asset refurbishment or rehabilitation (e.g. pipe relining)
- Asset renewal (e.g. pipe replacement).

Another common classification of water distribution system asset interventions is into capital or CAPEX (CApital EXpenditure) and operational or OPEX (OPerational EXpenditure) type interventions. The former are usually one-off interventions occurring over a finite period of time (e.g., pump replacement) whereas the latter denote regular, i.e., recurring interventions over prolonged periods of time (e.g., regular, periodic inspection and maintenance of various assets).

One of the main tasks of asset management is the generation of optimal, long-term asset management plans (also known as capital maintenance programmes). These plans select a number of asset management interventions (form a wide range of options available) to be implemented over some planning horizon (e.g., 5 to 30 years) so that the total net present value of these interventions and other associated costs are minimised while reaching some pre-specified, target level of system performance. The latter is typically defined using a number of different performance indicators and respective target values defined over the analysed planning horizon. When evaluating the costs involved, the *whole life costing approach* is often used accounting for all costs arising from the installation/renovation, use, maintenance and disposal of assets (Skipworth *et al.*, 2002; see Figure 9.5). The optimal asset management plans are nowadays usually generated by specialised software tools. Examples of tools for optimal asset management include Ex-Soft (Skipworth, 2002), CapPlan Water tool (MWH Soft) and a series of tools developed as part of the European CARE-W project (Saegrov, 2005).

9.3.2 Modelling for operational planning

With the ability to create very detailed water distribution models, analysts have moved from using models for capital planning only to support operational decision making. Examples of such analysis include: water quality modelling for chlorine dosing, minimisation of energy costs for pumping of treated water, real-time management of system failures, leakage management, etc.

Water quality analysis

Water quality models rely on the output of hydraulic models (i.e., the temporal and spatial distribution of velocities) to predict time/space distribution of a variety of water quality parameters within a distribution system that could be of interest to a water system analyst. These parameters include (Panguluri *et al.*, 2005):

1) The fraction of water originating from a particular source (e.g., source tracing).
2) The age of water (e.g., the time spent by a specific volume of water in the network).
3) The concentration of a non-reactive (i.e., conservative) constituent or tracer compound either added to or removed from the system (e.g., chloride or fluoride).
4) The concentration of a reactive compound (e.g., chlorine), including the concentration of a secondary disinfectant with additional input of its loss rate (e.g., chlorine or chloramines) and the concentration of disinfection by-products with their growth rate (e.g., THMs).

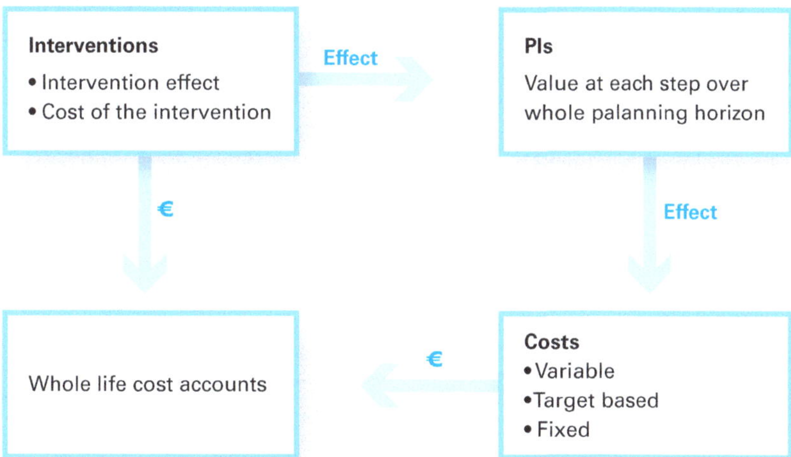

Figure 9.5 WLC based Asset Management Framework (Skipworth, 2002)

When modelling the fraction of water coming from a specific source or the concentration of a tracer, this parameter is modelled as a non-reactive (i.e., conservative) substance that enters the network with the fraction equal to 100 percent. The fractions coming from different sources can then be combined linearly. For example, combining a volume of water that is entirely from the specific source with a similar volume of water from a different source will provide a mixed volume calculated as 50% from the designated source. EPANET (Rossman, 2000) can be used to show to what degree water from a given source blends with water from other sources, and how this blending varies spatially and temporarily.

Due to its simplicity, water age is often used as a surrogate measure for water quality in the distribution system. It is a major factor in water quality deterioration, which occurs due to interactions between the pipe wall and the water and reactions within the bulk water (Vasconcelos *et al.*, 1996). Water age is equivalent to modelling a reactive constituent whose growth follows zero-order kinetics with a rate constant equal to $k = 1$. For example, for every hour that a discrete volume of water spends in a reservoir, its age will increase by one hour. Additionally, discrete volumes of water of different ages are combined linearly. For example, combining a volume of water that is four days old with a similar volume of water that is eight days old will result in an average age of six days.

Existing literature suggests that the key issue for maintaining acceptable water quality is to keep the concentration of chlorine within prescribed limits and simultaneously to control the concentration of disinfection by-products (TTHM and HAA). Chlorine modelling is mainly used to evaluate the spatial distribution of chlorine residuals and locate areas with chlorine concentrations outside the prescribed limits. Chlorine modelling can also serve as an indicator of water age and water main conditions.

Real-time failure management

Water distribution system pipes and various devices occasionally fail. Examples include pipe bursts, blocked valves and pumps that shut down due to power failure. When a failure happens it usually results in partial or full interruption of supply to water system customers and a loss of certain amount of water, together with the indirect loss of energy and chemicals required to treat and pump that water into the distribution system. In addition to all this, the contaminated groundwater or, perhaps, wastewater leaking out of a nearby sewer system may get into the water distribution system and cause serious health problems. Therefore it is desirable to detect failures and mitigate their negative impacts as soon as possible.

To achieve this, a failure management system capable of detecting, diagnosing and mitigating negative impacts of water distribution system failures in (near) real-time can be developed. An example of such a system is the recently developed Neptune Decision Support System (Savic *et al.*, 2008). The management of failures in the Neptune DSS is based on the analysis of real-time information coming in from pressure loggers, flow meters, water tank levels and other sources (e.g., customer complaints) and short-term forecasts of water consumption in the analysed system (see Figure 9.6).

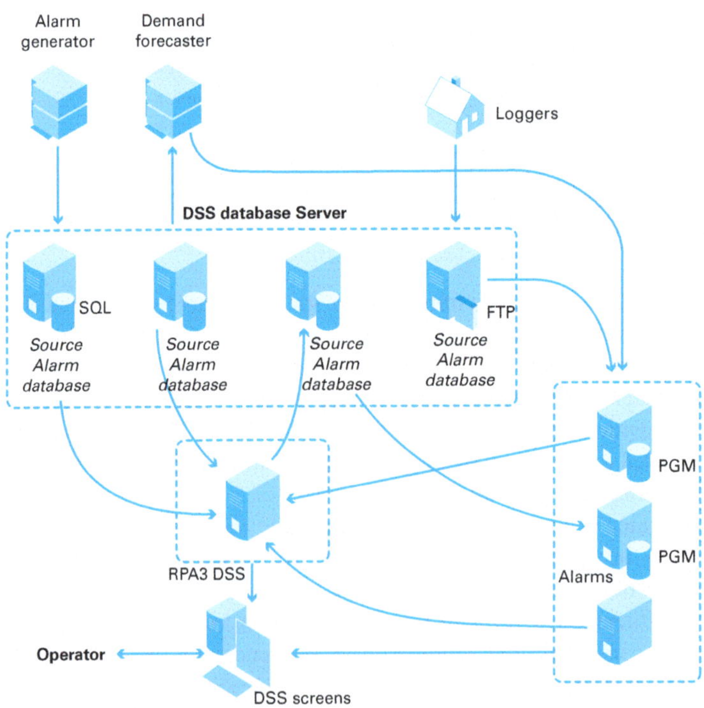

Figure 9.6 Neptune DSS Data Flow

The Neptune DSS failure system starts by identifying the District Metered Area (DMA) where the failure, e.g., a pipe burst/leak, is likely to have occurred. This is done by processing the real-time information coming from flow meters (installed at all DMA entry/import/export points) and pressure loggers (installed at all DMA entry/import/export points and at a critical point within the DMA). The processing involves using several Artificial Intelligence techniques, including Artificial Neural Networks (ANN) for predicting the future pressure/flow signal values (Mounce *et al.*, 2002), followed by the use of a Fuzzy Inference System that is used to classify the discrepancies obtained between predicted and observed signal values (into a normal/failure event). An alternative approach makes use of the Bayesian Network technology to classify events into normal or abnormal. In addition, this system also performs a fully automated de-noising of incoming pressure/flow signals resulting in a significantly improved accuracy of the relevant ANN predictions which, in turn, enables much faster and more reliable detection of failure events (Romano *et al.*, 2009).

Once the pipe burst/leak is located with a particular DMA, a risk-based approach is used to identify the likely pipe within the DMA which failed (Bicik *et al.*, 2009a). Bursts are simulated in each DMA pipe in turn and two risk components are calculated: the likelihood of a failure originating from the analysed pipe, and the consequence, namely the likely impact, of that failure. Dempster-Shafer theory (Shafer, 1976) is used when calculating the likelihood component of the risk, by combining evidence from multiple sources of processed information (see Figure 9.7). At the same time, the overall potential impact caused by the failure of the analysed pipe is estimated using a customised model based on a whole range of related indicators (Bicik *et al.*, 2009b). The estimated risk components associated with each assumed pipe failure are then presented to the decision maker in the form of a risk map; see Figure 9.8. This map is then used to identify the likely area of a pipe burst, that is, an area where the field technicians are sent out to pinpoint the leak (e.g., by using the sounding equipment for leakage detection).

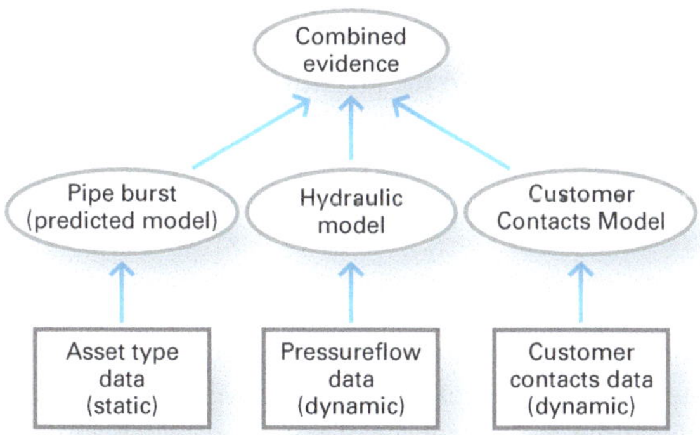

Figure 9.7 Estimating the likelihood of burst location by combining evidence from several information sources

Furthermore, the Neptune DSS enables automated identification of the nearby valves that should be closed to isolate the pipe burst incident (once it is located by the field technicians). The methodology for valve-controlled pipe network segments developed by Jun and Loganathan (2007) is used for this purpose. If some of the valves are found to be inoperable (e.g., due to corrosion) or inaccessible, the system will automatically identify further nearby valves that need to be closed in order to isolate the incident.

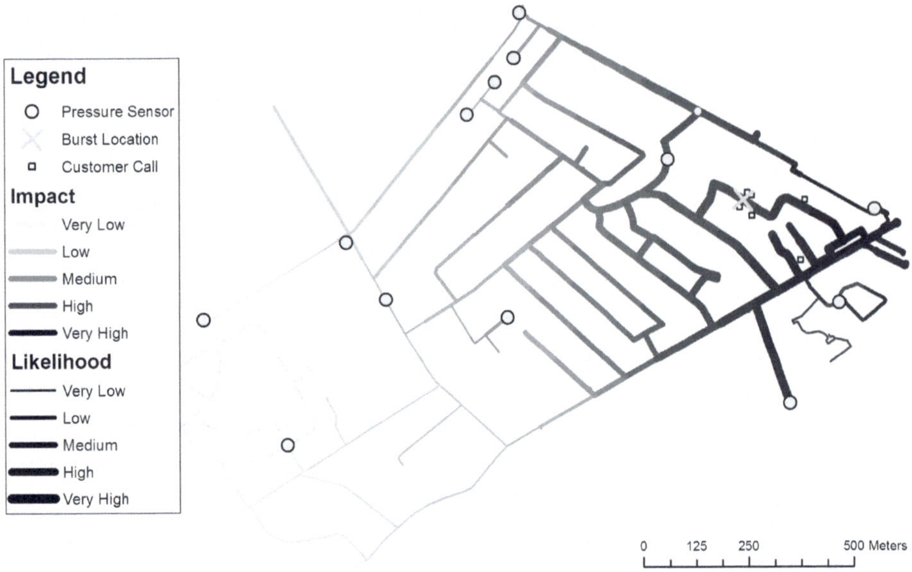

Figure 9.8 A risk map showing likelihood of occurrence and impact of a pipe burst pipe in different locations in a DMA

Once the failed pipe is isolated, a heuristics based method is used to identify the best remediation option to mitigate the negative impact of an incident (Vamvakeridou-Lyroudia *et al.*, 2009). Several possible intervention options are considered including (re)opening/closing some of the DMA boundary valves, changing the settings of the nearby pressure reducing valves and the potential use of overland bypasses (connected to nearby hydrants). The optimal remediation interventions are identified by minimising simultaneously the impact of the incident on customers, and the number of operations (e.g., valve/other manipulations) required.

Pump scheduling

Pumping costs represent a substantial element of operational costs in the majority of water utilities worldwide. Any minor increment in operational efficiency would not only yield significant savings for water utilities, but also ensure compliance with water industry regulations and improved system management among other things (van Zyl *et al.*, 2004). Pump scheduling is the process of choosing which of the available pumps within the system are to be used, the periods when they are to be run, and in the case of variable-speed pumps the pump speeds, while taking into account the different electricity tariff periods in order to minimise costs. Throughout their life pumps also need to be repaired and replaced, resulting in maintenance costs. Therefore, the goal of the pump scheduling problem should include minimisation of the total operational cost, which includes the pumping cost and the pump maintenance cost.

In a normal daily operational cycle water distribution networks are subjected to variable demands, yet they have to be run so as to provide levels of service (e.g., minimum pressure) and satisfy physical constraints (e.g., minimum/maximum reservoir levels). As such, experience and judgement of operational skilled staff is often relied upon while decisions are made on which pumps/valves to operate. However, making such decisions based only on experience becomes difficult because of the multiple constraints, e.g., the minimum water levels in reservoirs that need to be maintained (so that reliability of supply is ensured) and the need to minimize the frequency of turning pumps on/off (in order to lessen wear and tear on pumps). Furthermore, the hydraulic behaviour of water distribution networks is nonlinear making the computer simulation of pump scheduling a complex and time-consuming process and creating difficulties in identifying an optimal operating strategy (van Zyl *et al.*, 2004).

In very general terms, a system that automatically optimises the operation of pumps is made up of three main components:

1) a hydraulic network model,
2) a demand forecast model, and
3) an optimisation model.

A calibrated hydraulic network model is used to calculate the response of the water distribution system to different operational decisions. Another key component for effective pump scheduling is the accurate forecasting of water demand on a daily and hourly basis. For example, if the main aim is to improve water resource management, then a daily water demand forecast is required. However, for operational purposes, such as pump scheduling or alarm management, a forecast based on an hourly or even shorter time step is needed. The optimisation model generates optimal pump schedules by optimising an objective function, usually minimising electricity cost, subject to a number of operational constraints.

Based on how the pump schedules are represented, there are two main approaches to pump scheduling found in the literature:

1) explicit approach, by directly specifying the status of each pump (Mackle *et al.*, 1995), and
2) implicit approach, by linking the operation of pumps to levels in service reservoirs (van Zyl *et al.*, 2004).

The explicit representation divides the scheduling period in smaller time intervals, during which the pump can be on or off. On the other hand, the implicit representation allows the pump status to change when the water level in a service reservoir reaches certain predetermined (trigger) values. The so-called reservoir trigger controls are normally used in pairs, with one control triggering an action (such as switching a pump on) and the other triggering an opposing action (switching the pump off). The implicit approach is widely used in practice, due to its simplicity and proven robustness.

The following constraints should be imposed on the decisions and the operational behaviour of the system to ensure viability of solutions in the optimisation process:

1) Implicit constraints on the network system – the equality constraints defining equilibrium state of the network, which are represented by a set of quasi-linear hydraulic equations that are solved implicitly using the conventional hydraulic simulation model (Rao and Salomons, 2007).
2) A storage reservoir in a water distribution system must also be operated within a minimum and a maximum allowable water level.
3) The initial water levels should be reached in the reservoirs at the end of the optimisation period.

A simplified case study performed by Mackle *et al.* (1995) will demonstrate the basic principles of pump scheduling. The study used a simple system with 4 different pumps delivering water to a single reservoir. Consequently there are 16 different possible pump combinations. The day was divided into 24 time intervals, each one hour long, with the demand drawn from the reservoir varying for each hour. The difference between the rate at which water is pumped to the reservoir and the demand is accommodated by changes in the volume of water stored in the reservoir, subject to maximum and minimum storage volumes. To represent the 16 different pump combinations, a binary string of length 4 was required, and as there are 24 time intervals the heads were pre-calculated (these are summarised in Table 9.1).

Table 9.1 Pump information for the simplified scheduling example

pump 1	pump 2	pump 3	pump 4	V (m^3)	E (KWh)
0	0	0	0	0	0
0	0	0	1	10	12
0	0	1	0	30	30
0	0	1	1	40	42
0	1	0	0	50	44
0	1	0	1	60	56
0	1	1	0	80	74
0	1	1	1	90	86
1	0	0	0	100	80
1	0	0	1	110	92
1	0	1	0	130	111
1	0	1	1	140	124
1	1	0	0	150	127
1	1	0	1	160	141
1	1	1	0	180	165
1	1	1	1	190	182

where V is the volume pumped in an hour,
E is the electricity consumed in an hour and
for the binary coding, 1 means the pump is switched on and zero it is switched off.

In addition, the following data was also needed to generate low cost pump schedules:

- the optimization period and the time interval – 24 hours and 1 hour, respectively;
- the demand data during this optimization period;
- the initial, minimum and maximum water levels of the reservoir;
- the pattern of the electricity tariffs used during this period (it is assumed here that the night-time tariff is 7 money-units per electricity-unit and that the day-time tariff is double that price; the higher daytime tariff has to be paid from 7 a.m. until 8 p.m.).

The best solution was found using a genetic algorithm procedure. As expected, the most of pumping occurs during the cheap night time tariff.

As shown in Figure 9.9, the reservoir storage is consequently as full as possible at the end of the cheap period and as empty as allowed at the end of the expensive period.

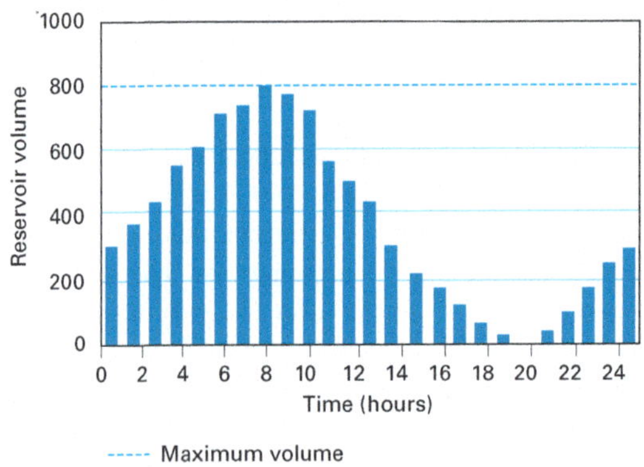

Figure 9.9 Stored volume vs. time

In Figure 9.10 one can see the exact amount which has to be pumped in each time interval in order to achieve the optimum solution. To make the influence of this pump schedule on the reservoir level more visible the demand over the day is also shown in Figure 9.10.

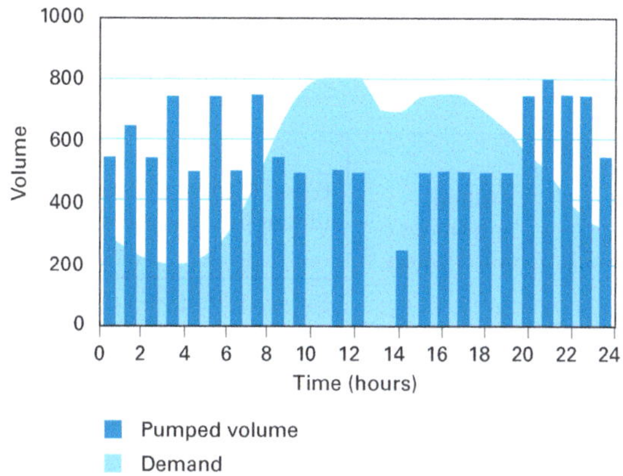

Figure 9.10 Hourly pumped volumes and demand

Leakage management

International figures quoted in traditional terms for 'Non-Revenue Water' (NRW) or 'Unaccounted for Water' (UFW) range from less than 5% to 75% or more of system input volume. However, in the past 30 years, it has been recognised by an increasing number of countries and organisations – including the International Water Association (IWA) – that NRW or UFW expressed as percentage of input volume could be misleading as a performance indicator (PI) for benchmarking operational management performance. The reason for this is that the greater the consumption, the lower will be any particular volume of NRW expressed as a percentage of system input volume.

Based on the initiative by Task Forces of the IWA a new approach to water loss management has been summarised in an International Report on Water Losses Management and Techniques (Lambert, 2002), which includes:

- a best practice standard water balance calculation and terminology;
- a methodology for calculating the Unavoidable (technical minimum) Annual Real Losses (UARL) for well-managed systems with good infrastructure; and
- an improved performance indicator for operational management, the Infrastructure Leakage Index (ILI).

The Report provides the IWA standard water balance and definitions, as the basic but essential first steps in management of water losses. Drawing on the best practice from around the world it advocates a standard approach for water balance calculations; see Figure 9.11, where the following terminology explains each of the water balance constituent (Lambert, 2002):

- System Input Volume is the annual volume input to that part of the water supply system.
- Authorised Consumption is the annual volume of metered and/or non-metered water taken by registered customers, the water supplier and others who are implicitly or explicitly authorised to do so. It includes water exported, and leaks and overflows after the point of customer metering.
- Non-Revenue Water (NRW) is the difference between System Input Volume and Billed Authorised Consumption. NRW consists of:
  - Unbilled Authorised Consumption (usually a minor component of the Water Balance)
  - Water Losses

- Water Losses is the difference between System Input Volume and Authorised Consumption, and consists of Apparent Losses and Real Losses
- Apparent Losses consists of Unauthorised Consumption and all types of metering inaccuracies
- Real Losses are the annual volumes lost through all types of leaks, bursts and overflows on mains, service reservoirs and service connections, up to the point of customer metering.

<table>
<tr><td rowspan="9">System Input Volume</td><td rowspan="4">Authorised Consumption</td><td rowspan="2">Billed Authorised Consumption</td><td>Billed Metered Consumption (including water exported)</td><td rowspan="2">Revenue Water</td></tr>
<tr><td>Billed Unmetered Consumption</td></tr>
<tr><td rowspan="2">Unbilled Authorised Consumption</td><td>Unbilled Metered Consumption</td><td rowspan="7">Non-Revenue Water (NRW)</td></tr>
<tr><td>Unbilled Unmetered Consumption</td></tr>
<tr><td rowspan="5">Water Losses</td><td rowspan="2">Apparent Losses</td><td>Unauthorised Consumpion</td></tr>
<tr><td>Metering Inaccuracies</td></tr>
<tr><td rowspan="3">Real Losses</td><td>Leakage on Transmission and/or Distribution Mains</td></tr>
<tr><td>Leakage and Overflows at Utility's Storage Tanks</td></tr>
<tr><td>Leakage on Service Connections up to point of Customer metering</td></tr>
</table>

Figure 9.11 IWA Standard International Water Balance

The IWA Report concludes that if the term 'Unaccounted for Water' is used, it should be defined and calculated in the same way as 'Non-Revenue Water' in Figure 9.12.

Figure 9.12 shows how the current leakage for a system (the large outer box) can be reduced by an appropriate combination of the four principal leakage management activities. The point at which the cost of reducing leakage is equal to the benefit gained from further leakage reductions is called the economic leakage level, which is a value between the current leakage and the unavoidable real losses.

Active Leakage Control is a proactive strategy aimed at reducing water loss by the detection and pinpointing of non-visible leaks. It uses trained engineers and technicians with specialised equipment followed by a prompt and good quality repair of these leaks. Speed and Quality of Repair is also important as the longer a leak runs the greater volume of water that is lost. The reduction of excess pressures (and excess pressure variation) and surges through Pressure Management will normally reduce:

- The number of new leaks and bursts and repair costs.
- The flow rates of all existing leaks.
- 'Background' (undetectable) leakage (unavoidable losses).
- Pressure-dependent consumption.

Pressure management would also help to defer mains and service pipe replacement (increases asset quality).

The common problem is that some water utilities do not repair small or difficult leaks, do not manage pressures to reduce leak flow rates and new burst frequency, and do not undertake active leakage control. As these three activities are much more cost-effective for leakage management (expressed in volume of water saved per unit cost) than replacing infrastructure only, they should be considered an essential part of any sustainable management policy (together with infrastructure renewal and development of new resources).

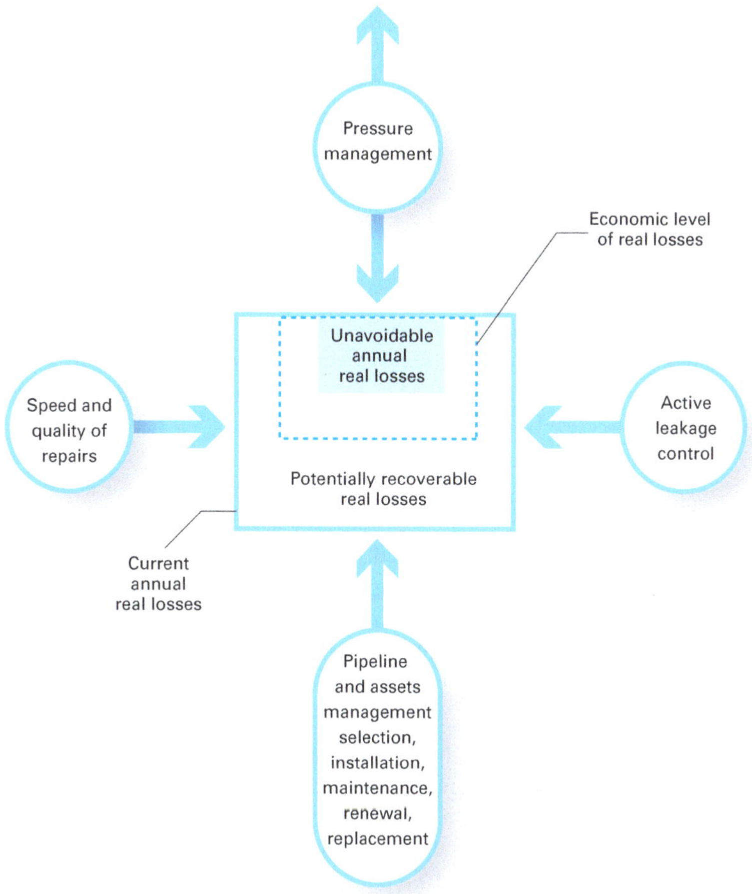

Figure 9.12 The Four Component Approach to Controlling Real Losses (Savic *et al.*, 2005)

Pressure management problem can also be formulated and solved as an optimisation problem. An example of such a methodology is that used in the XpressZone software (Awad *et al.*, 2009). The software determines the locations and settings of different types of PRVs, together with the status of relevant control boundary valves with the aim to maximise the net benefit of pressure management. It uses a novel methodology for the detailed calculation of the locations and settings and other pressure management benefits (Awad *et al.*, 2008). The optimisation problem is solved using a especially customised version of Genetic Algorithms (Awad *et al.*, 2009). The methodology has been applied to a number of real-life systems in the UK and has demonstrated that, even in the case of well maintained water distribution systems, it is possible to identify significant benefits (i.e. financial savings).

The ILI is the ratio of the Current Annual Real Losses (the large box in Figure 9.12) to the Unavoidable Annual Real Losses (the small box). ILIs have been calculated for a large number of systems world-wide (Lambert, 2004), and typically range from close to 1.0 (for the world best utilities) to 20 or more; the highest yet identified is over 200, in a system with grossly wasteful consumption, and a UFW of 20%.

9.4 CASE STUDY: APPLICATION OF HYDRAULIC MODELLING FOR LEAKAGE MANAGEMENT IN THE BANGKOK WATER SUPPLY SYSTEM

9.4.1 Introduction

High levels of unaccounted-for water (UFW) in water supply systems has been a major concern in many urban centres in developing countries as this implies substantial financial losses and a low quality of services. One of the main causes for such high UFW is leakage in the pipe network. In Bangkok, Thailand, leakages due to broken pipes and fittings account for 81% of UFW (Babel *et al.*, 2010).

Pressure management has been found to be a very effective means of reducing background leakage in water distribution systems (Babel and Islam, 2009). This case study demonstrates the application of hydraulic modelling in managing pressures to curb UFW in water supply systems.

9.4.2 Leakage management study in Bangkok

Given the high rate of UFW in Bangkok, there is a need to evaluate the potential of pressure management to minimize leakages in the distribution system. Accordingly, a study (Babel *et al.*, 2009) was conducted in a selected area within the service area of the Metropolitan Waterworks Authority, a state enterprise responsible for piped water supply in Bangkok and the two neighbouring provinces of Samut Prakan and Nonthaburi.

The study area

The study was carried out in one of the District Metering Areas (DMAs) of the Bangkoknoi Branch, one of the 15 branch offices of MWA (Figure 9.13). In particular, the DMA-0144-01 has a service area coverage of 2.5 km^2, an estimated served population of 3,750, a total pipe length of 17.5 km, and non-revenue water (NRW) of 38% in 2004. The distribution network (Figure 9.14) is composed of an inlet (entry of water from the large network), 60 pipes connected by 77 junctions, and 19 types of valves.

Model development

EPANET 2.0 was used to develop the network hydraulic model for the selected DMA. This model, developed by the US Environmental Protection Agency (US EPA), performs an extended period simulation of hydraulic and water quality behaviour within pressurized pipe networks consisting of pipes, nodes or pipe junctions, pumps, valves, and storage tanks or reservoirs (Babel *et al.*, 2009). The model computes junction heads and link flows for a fixed set of reservoir levels, tank levels, and water demands over a succession of points in time (Rossman, 2000). The developed model was applied for Extended Period Simulation (EPS).

In the model, two types of demands are assigned to each node: consumer demand or base demand, and leakage demand. Consumer demand is assigned by counting the number of properties supplied with water from the node, which in this case was estimated based on billed water (Islam, 2005). In EPANET 2.0, the leakage demand or pressure-dependent demand is modelled using emitters with assigned appropriate emitter coefficients, which are the power exponent of the head loss-pressure relationship.

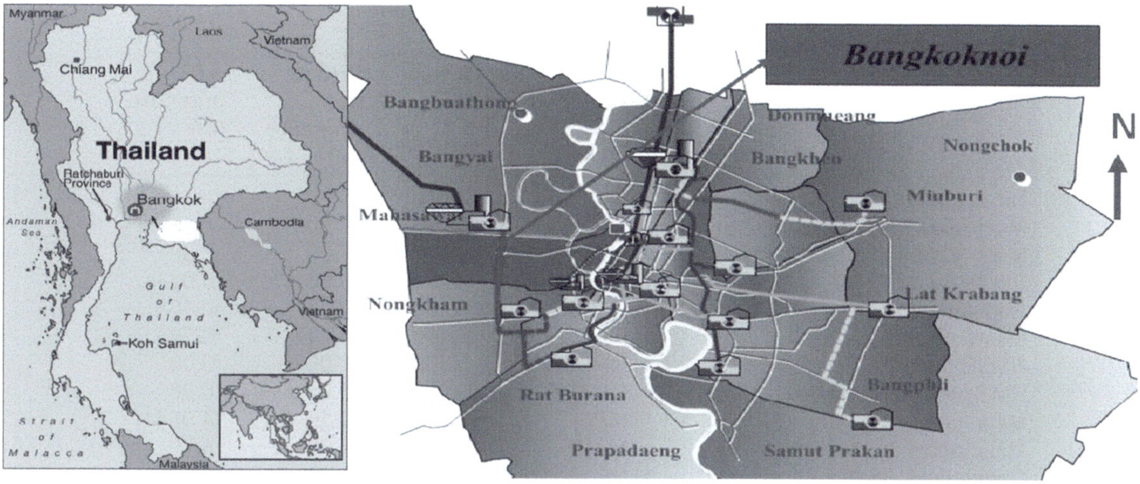

Figure 9.13 The MWA service area (Babel *et al.*, 2009)

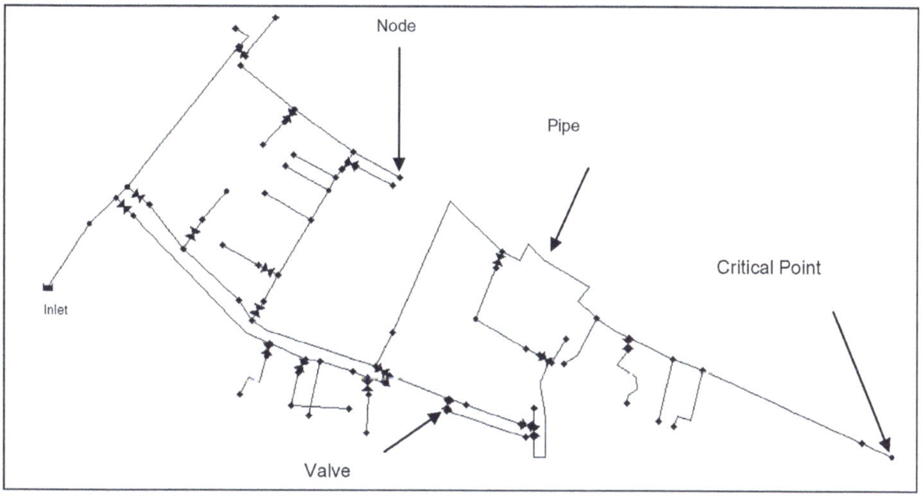

Figure 9.14 EPANET 2.0 network hydraulic model of the study area. Source: Babel *et al.*, 2009

Model calibration

Using emitter coefficients and the roughness of pipe materials as calibration parameters, the model was calibrated and verified for extended period simulation (EPS). Results of the calibration show that the calculated and observed inflow at the inlet point, pressure at the average zone point (AZP), and the pressure at critical points are comparable, indicating the acceptable performance of the model (Table 9.2). The testing for leakage was also conducted by simulating night flow as it is assumed that during late nights (1:00–2:00 am), the total flow into the network is composed only of the customer's night use and the leakage in the network. The simulated leakage matched reasonably well the estimated leakage based on the night use survey.

Table 9.2 Model performance statistics Babel (Babel *et al.*, 2009)

		Inlet flow (m^3/h)	Critical point pressure (m)	AZP pressure (m)
No. of observations	N	24	24	24
Mean value, simulated*		1,702	7.86	8.17
Mean value, observed*		1,696	7.89	8.09
Efficiency index	EI	0.97	0.96	0.93
Root mean square error	RMSE	3.82	0.01	0.48
Mean absolute error	MAE	2.80	0.44	0.59
RMSE mean	RMSEM	0.05	0.06	0.08
RMS over standard deviation	RMSES	0.16	0.01	0.20
Mean percentage error, %	MPE	−2.08	0.16	4.96
Mean absolute percentage error, %	MAPE	4.78	4.78	4.78
Correlations coefficient	R	0.99	1.00	0.99

*Note: *in m^3/day.*

Model verification

Model verification results for different sets of observed inflows and critical point pressures, at five different times, show that the errors are small, ranging from −2.44 to 2.32%, i.e., the model has 97.5% or higher accuracy in simulating the hydraulic behaviour in the study network.

Model application

To determine the effects of pressure management on leakage, the developed network hydraulic model was applied to analyze various inlet pressure profile alternatives. The alternatives were developed considering surrounding DMAs, available pressure and flow, control of the pressure etc. to optimize pressure in the study area without affecting other areas. To ensure the minimum level of service (LoS), the pressure at the average zone point (AZP) and the pressure at the critical point (CP) were considered, with more emphasis given to the latter as the study area is relatively flat and small. The pressure optimization procedure is shown in Figure 9.15.

Apart from the current MWA operation profile, three pressure profiles were analyzed with due consideration to the LoS (6 m) of the MWA (Figure 9.16). Results of the analysis indicate that if Pressure Profile 2 is implemented, about 17.5% of water can be saved in the DMA. An important assumption is that the decrease in the inflow into the system is mainly due to the reduction in pressure dependent leakages and the pressure dependent customer demand remains constant.

Implementation results

In the absence of pressure reducing valves (PRV) in the selected DMA and no individual control in any of the DMAs of the network, it was not possible to operate the network following exactly the suggested inlet pressure profiles. The whole network is fed and controlled from a pumping station. Thus, the MWA was requested to reduce the pressure for a period of seven days without significantly affecting the other DMAs in the network. The MWA controlled the pressure at the pumping station and the actual reduced profile implemented in the study DMA is shown in Figure 9.17. The observed average pressure and flow into the system before (January 2005) and after (March 2005) the implementation of the reduced pressure profile are presented in Table 9.3. The average pressure reduction is 2.4 m and the water saved during this period was 12.5%, which is equivalent to 1,768 m^3 for 7 days or 92,000 m^3/year.

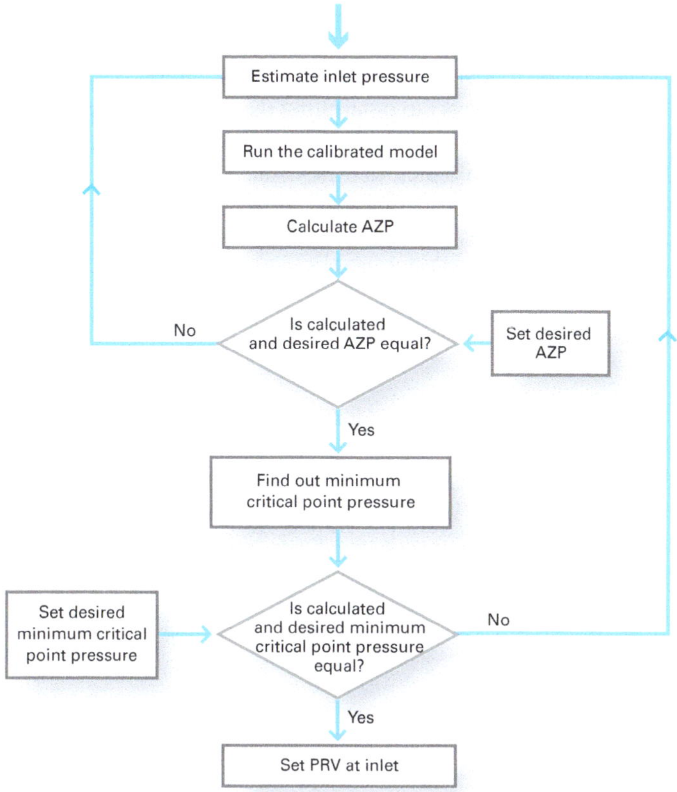

Figure 9.15 Framework for pressure optimization (Babel *et al.*, 2009)

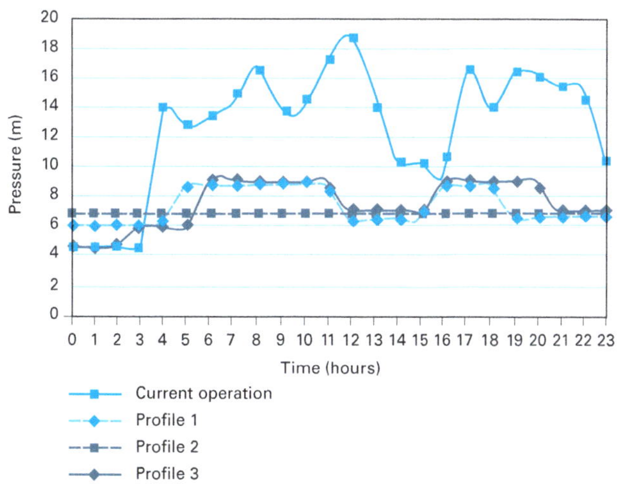

Figure 9.16 Alternative inlet pressure profiles analysed (Babel *et al.*, 2009)

The developed model was run with the reduced pressure profile. The calculated inflow into the network during the period was 12,632 m^3, which is very close to the actual observed inflow of 12,372 m^3. These results clearly indicate the usefulness of the model in managing the water supply system and in improving the overall hydraulic performance, especially with respect to controlling the leakages in the network.

A field survey carried out for 30 households close to the critical pressure point immediately after the implementation of the reduced pressure revealed that more than 73% of the households did not even notice any change in the pressure, whereas the rest complained of low pressure at night time. The latter arises due to limitations on ensuring the desired pressure at the inlet point of the study network. Given the absence of a pressure control device at the inlet of the DMA, it was not possible to estimate the pressure required at the control point (i.e., pumping station).

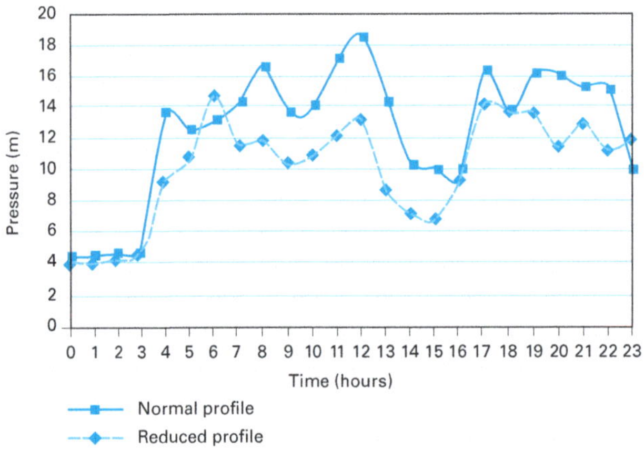

Figure 9.17 Comparison of current and implemented inlet pressure profiles (Babel *et al.*, 2009)

Table 9.3 Implementation results with reduced pressure profile (Babel *et al.*, 2009)

Day	Average pressure (m)		Average pressure reduction (m)	Flow (m^3/d)		Savings	
	Before change (Jan 2005)	After change (Mar 2005)		Before pressure change	After pressure change	(m^3/d)	(%)
1	12.95	10.86	2.09	2,013	1,740	273	13.56
2	12.28	10.68	1.60	2,044	1,777	267	13.06
3	13.78	09.92	3.87	2,001	1,773	228	11.39
4	11.68	10.09	1.60	2,033	1,782	251	12.35
5	11.80	10.14	1.65	2,016	1,751	265	13.14
6	12.30	09.53	2.77	1,956	1,778	178	09.10
7	13.42	10.17	3.26	2,078	1,772	306	14.73
	Average/ Total		2.42	14,140	12,372	1,768	12.50
	Flow calculated from the model (m^3/7d)				12,632	1,508	10.67

9.4.3 Conclusion

This case study has shown the usefulness of hydraulic modelling in assessing the applicability of pressure control and management in reducing leakages in a water supply distribution network. In the case of Bangkok, a small reduction in pressure was seen to decrease the leakage significantly without compromising the level of service.

9.4.4 Acknowledgement

This case study was prepared by Mukand S. Babel, Md. S. Islam and Aldrin A. Rivas.

9.5 KEYWORDS

By the end of the chapter you should have encountered the following terms; ensure that you are familiar with them!

Asset management
Branched system
Clean water systems
Control valves
Dirty water systems
Distribution network
Filtration
Flocculation
Impounding reservoir
Isolation valves
Leakage
Looped system
Master planning
Performance indicators
Pipe burst
Pump scheduling
Sedimentation
System rehabilitation
Treatment plant
Transmission system
Urban water cycle

9.6 QUESTIONS

How would you go about managing leakage in a distribution network?
What is the value of performance indicators, and how should they be used?
How should a hydraulic simulation model be instantiated for hydraulic analysis of a distribution network?
What is the value of optimisation techniques in network design and rehabilitation?
How can an existing distribution network be extended to serve a newly urbanized area?

9.7 REFERENCES

Aafjes J., Hendriks L. J. W., Verberne A. J. P. and Vingerhoeds R. A (1997). Voorspellen van het drinkwaterverbruik met een neuraal netwerk. *H2O (in Dutch)* (Predicting drinking water usage with a neural network), **30**(15), 484–487.

Awad H., Kapelan Z., Walters G. and Savic D. A. (2008). Analysis of pressure management economics in water distribution systems. In: *Proceedings 10th International Water Distribution System Analysis Conference*, 17–20 August, Kruger National Park, South Africa.

Awad H., Kapelan Z. and Savic D. A. (2009). Optimal setting of time-modulated pressure reducing valves in water distribution networks using genetic algorithms. In: *Proceedings 10th International Conference on Computing and Control for the Water Industry (CCWI)*, 1–3 September, Sheffield, UK.

Babel M. S., Islam M. S. and Kallidaikurichi S. (2009). Leakage management in a low-pressure water distribution network of Bangkok. *Water Sci. Technol. Water Suppl.*, **9**(2), 141–147.

Babel M. S., Rivas A. A. and Kallidaikurichi S. (2010). Municipal water supply management in Bangkok: achievements and lessons. *Int. J. Water Resour. Dev.*, **26**(2), 181–205.

Bhave P. (1988). Calibrating water distribution network models. *J. Environ. Eng.*, **114**, 120.

Bicik J., Kapelan Z. and Savic D. A. (2009a). Operational perspective of the impact of failures in water distribution systems. In: *Proceedings World Environmental & Water Resources Congress*, 17–21 May, Kansas City, Missouri.

Bicik J., Makropoulos C., Kapelan Z. and Savic D. A. (2009b). The application of evidence theory in decision support for water distribution system operations. In: *Proceedings 8th International Conference on Hydroinformatics*, 2–16 January, Concepcion, Chile.

Colombo A. and Karney B. (2003). The labyrinth of water distribution systems: demand, energy and climate change. In: *Pumps, Electromechanical Devices and Systems Applied to Urban Water Management*, Taylor and Francis, Leiden, p. 239.

Economou T., Kapelan Z. and Bailey T. (2008). A zero-inflated Bayesian model for the prediction of water pipe bursts. In: *Proceedings 10th International Water Distribution System Analysis Conference*, Kruger National Park, South Africa.

Giustolisi O. and Savic D. A. (2006). A symbolic data-driven technique based on evolutionary polynomial regression. *J. Hydroinf.*, **8**(3), 207–222.

Islam M. S. (2005). *Leakage Analysis and Management in the Water Distribution Network in a Selected Area of Bangkok.* Master of Engineering Thesis, Asian Institute of Technology, Bangkok, Thailand.

Jun H. and Loganathan G. (2007). Valve-controlled segments in water distribution systems. *J. Water Resour. Plann. Manage.*, **133**, 145.

Kleiner Y. and Rajani B. (2001). Comprehensive review of structural deterioration of water mains: statistical models. *Urban Water*, **3**(3), 131–150.

Lambert A. O. (2002). International report on water loss management and techniques. *Water Sci. Technol. Water Supply*, **2**(4), 1–20.

Lambert A. O. (2004). ILIs over four continents. *Environ. Agency Demand Manage. Bull.*, 67, www.environmentagency.gov.uk/savewater (accessed 4 October 2010).

Mackle G., Savic D. A. and Walters G. A. (1995). Application of genetic algorithms to pump scheduling for water supply. In: *Genetic Algorithms in Engineering Systems: Innovations and Applications, GALESIA '95*, IEE Conference Publication No. 414, 12–14 September, Sheffield, UK, pp. 400–405.

Mounce S., Day A., Wood A. S., Khan A., Widdop P. D. and Machell, J. (2002). A neural network approach to burst detection. *Water Sci. Technol. J. Int. Assoc. Water Poll. Res.*, **45**(4–5), 237.

Panguluri S., Grayman W. M. and Clark R. M. (2005). *Water Distribution System Analysis: Field Studies, Modeling and Management.* EPA Report, EPA/600/R-06/028, Cincinnati, OH, USA.

Porteous A. (2000). *Dictionary of Environmental Science and Technology*. John Wiley & Sons, New York.

Randall-Smith M., Rogers C., Keedwell E., Diduch R. and Kapelan Z. (2006). Optimized design of the City of Ottawa water network: a genetic algorithm case study. In: *Proceedings 8th Annual Water Distribution System Analysis Symposium*, 27–30 August, Cincinnati, Ohio.

Rao Z. and Salomons E. (2007). Development of a real-time, near-optimal control process for water-distribution networks. *J. Hydroinf.*, **9**(1), 25–37.

Rodriguez M. and Sérodes J. (1998). Assessing empirical linear and non-linear modelling of residual chlorine in urban drinking water systems. *Environ. Modell. Software*, **14**(1), 93–102.

Romano M., Kapelan Z. and Savic D. A. (2009). Bayesian-based online burst detection in water distribution systems. In: *Proceedings 10th International Conference on Computing and Control for the Water Industry (CCWI)*, 1–3 September, Sheffield, UK.

Rossman L. (2000). *EPANET 2 Users Manual.* US Environmental Protection Agency, Cincinnati, Ohio.

Rossman L. A. (2000). *EPANET 2 Users Manual.* US Environmental Protection Agency, Cincinnati, Ohio.

Saegrov S. (2005). *Care-w: Computer Aided Rehabilitation for Water Networks.* International Water Association, London.

Savic D. A., Boxall J. B., Vlanicki B. and Kapelan, Z. (2008). Project neptune: improved operation of water distribution networks. In: *Proceedings of the 10th Annual Water Distribution Systems Analysis Conference (WDSA 2008)*, 17–20 August, Kruger National Park, South Africa, pp. 543–558.

Savic D., Lambert A. and Kapelan Z. (2005). Water losses management and leakage detection techniques. *Water Sewerage J.*, **2**, 25.

Savic D. A., Walters G. A., Randall-Smith M. and Atkinson B. M. (2000). Large water distribution systems design through genetic algorithm optimisation. In: *ASCE 2000 Joint Conference on Water Resources Engineering and Water Resources Planning and Management*, R. H. Hotchkiss and M. Glade (eds.), Minneapolis, Minnesota.

Shafer G. (1976). *Mathematical Theory of Evidence*. Princeton University Press, Princeton, New Jersey.

Skipworth P. (2002). *Whole Life Costing for Water Distribution Network Management.* Thomas Telford Services Ltd, London.

Todini E. (ed.) (2003). A more realistic approach to the "extended period simulation" of water distribution networks. In: *Advances in Water Supply Management.* Lisse, Swets and Zeillinger, Lisse.

Vamvakeridou-Lyroudia L., Bicik J., Awad H., Savic D. A. and Kapelan Z. (2009). Developing and implementing a real-time intervention management model for water distribution systems. In: *Proceedings 10th International Conference on Computing and Control for the Water Industry (CCWI)*, 1–3 September, Sheffield, UK.

van Zyl J., Savic D. and Walters G. A. (2004). Operational optimization of water distribution systems using a hybrid genetic algorithm. *J. Water Resour. Plann. Manage.*, **130**, 160.

Vasconcelos J., Boulos P., Grayman W. M., Kiene I., Wable O., Biswas P., Bhari A., Rossman I. A., Clark R. M. and Goodrich J. A. (1996). *Characterization and Modeling of Chlorine Decay in Distribution Systems.* AwwaRF, Denver, Colorado.

Walski T. M., Chase D. V., Savic D. A., Grayman W., Beckwith S. and Koelle E. (2001). *Advanced Water Distribution Modeling and Management.* Haestad Press, Waterbury, Connecticut.

Chapter 10
Collection Systems

10.1 INTRODUCTION

As discussed in Chapter 2, there are different elements of the urban water cycle (Figure 10.1) which together provide the necessary services to the community. In particular, the collection systems play a vital role in collecting and draining (or transporting) stormwater and wastewater (or sewage) from urban areas. The systems consist of open channels or closed conduits (that is, sewers). Here, the term 'sewers' is restricted to closed conduits linked together in a network, which conveys sewage from a certain area, whereas the term 'sewerage' refers to the entire wastewater/stormwater system infrastructure in a particular city.

The supply and use of potable water in urban areas leads to the contamination of the water with significant materials and substances (that is, pollutants) that are dissolved or in suspension. The resulting wastewater needs to be returned to the environment and should be disposed of safely, preferably by removing at least some of the pollutants. In contrast, the extensive paving of the ground surface in an urban area means that infiltration of rainfall is limited compared to the former natural catchment, and therefore excess rainfall-runoff, which has the tendency to pond on the ground surface, has to be removed in order to limit the frequency of surface flooding. In both cases there is a need to collect the unwanted water and to convey it for processing or discharging to an appropriate receptor.

Whereas in some cities around the world open drains along streets have been and continue to be a primary means of evacuating wastewater and stormwater from the urban area, the health risks posed to the community are considerable. Therefore, where the community is affluent enough, at least the wastewater (or dry weather flow) is collected from domestic, public and industrial sources via underground pipes, and conveyed under gravity either to a suitable receiving water, such as a stream, river or coastal waters, or to a treatment works.

Excess rainfall-runoff forming the stormwater component may similarly be collected from roofs, roads and other paved areas and conveyed through surface channels which may discharge to underground pipe networks. The objective is to dispose of the stormwater from the more frequent rainfall events to the nearest appropriate

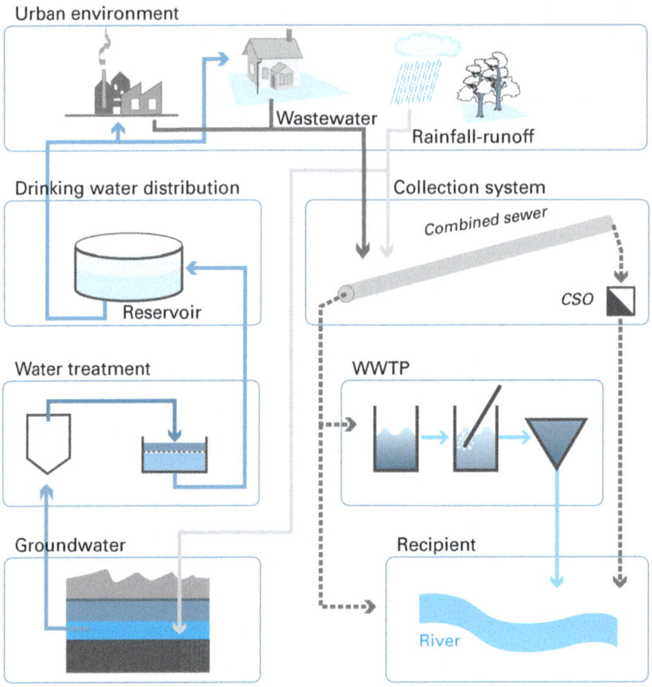

Figure 10.1 Overview of key elements of urban water cycle

stream, river or coastal water under gravity. Of course, this will not necessarily be possible for rarer events. Recognizing this, some countries make provision for these events by designing over ground routes through an urban area for the excess flows.

Normally, the stormwater on its own has not been regarded as needing treatment before being discharged to the environment. However, the cost of providing two collection systems for the same area: one for the wastewater and another for the stormwater with much larger conduits, lead readily to the notion of combining them. The result is a combined sewerage system that is designed largely to convey the stormwater component while transporting the wastewater to the treatment works or the outfall to the receiving waters. Therefore there are two main types of sewerage collection systems: combined and separate (in addition, there are small collection systems which operate under pressure or vacuum; these are known as vacuum systems).

The key elements sewerage collection systems can be summarised as follows:

- reticulation or collection networks (pipes, channels, culverts, manholes, etc.),
- gully inlets (that is, catchpits),
- ventilation shafts,
- outfalls,
- controls (orifices, gates, penstocks, vortex regulators, flap valves, etc.),
- pumping stations, bifurcations, weirs, inverted siphons, treatment and other ancillary structures,
- storage units (offline, online, open, closed, above ground, below ground),
- watercourses,
- overland flow paths and designated floodplains,
- dams.

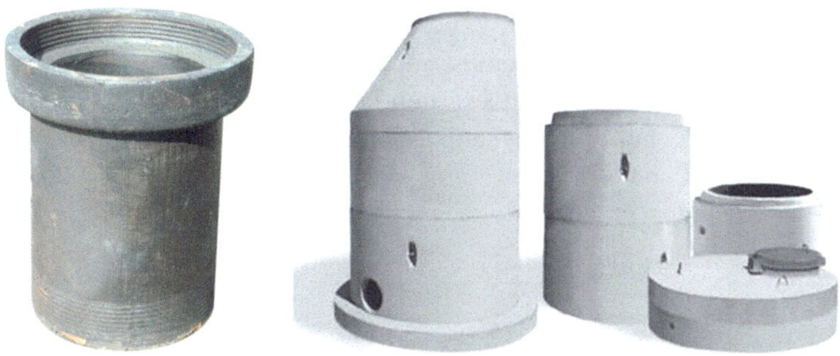
(a) (b)

Figure 10.2 Key elements of urban drainage systems. (a) Pipe (b) Manhole

Figure 10.3 Combined sewer overflow (Pattaya, Thailand) (Photograph: Sutat Weesakul)

Most collection systems are designed to operate under gravity. However, in some cases it is not feasible to achieve efficient operation of the system only by gravity, and the use of pumps is inevitable. Such facilities normally operate on a stop-start basis with respect to the water level in the wet well. The hydraulic performance of a pump is determined by its characteristic curve which is expressed as a head-flow relationship.

10.2 COMBINED VERSUS SEPARATE COLLECTION SYSTEMS

Vigorous debate has existed for a long time over the advantages and disadvantages of combined and separate collection systems. An advantage of a combined sewerage system is the cost saving of having only one network of pipes that needs to be constructed, operated and maintained. Also, sediments that tend to be deposited in the combined sewers during dry weather periods may be flushed during wet weather periods due to the bed shears stresses generated in the pipes by larger flow velocities. Moderate amounts of stormwater runoff that may be polluted before entering the sewerage network may also pass through the treatment works. This is important for cities that need to protect valuable surface water resources from stormwater pollution, that is, where the stormwater runoff from the urban area is heavily polluted, particularly from nutrients and heavy metals, so that the surface resource, such as a lake or river, experiences the long term build up of pollution attached to the sediment. To protect these resources, major cities, such as Chicago and St Louis in the USA or Sydney in Australia, have sought relief by building large underground storage conduits to retain the stormwater flows for treatment or for diversion to a safer environment (such as another less sensitive river).

Figure 10.4 Combiner sewer pumping station, Pattaya beach, Thailand (Photograph: Zoran Vojinović)

A summary of the advantages and disadvantages of separate and combined sewer systems is given in Table 10.1.

The construction of separate or even combined sewer systems is a luxury that some developing countries cannot afford. What is more, those countries that can afford the capital investment to build separate or combined underground sewerage systems often find that they do not have the income or the culture to maintain the systems in a serviceable condition. For example, inlets to closed drains readily become blocked; it is difficult to access such drains, to monitor their state and to clean and maintain them, and like any artificial asset which is subject to such a corrosive environment it degrades and decays. Therefore other ways of dealing with wastewater are needed, particularly for cities in developing countries. These are discussed further in Chapter 14.

10.3 WASTEWATER SYSTEMS

The main sources of wastewater can be grouped into: domestic, non-domestic (commercial and industrial) and infiltration and inflow. The domestic component of wastewater has a strong link with water supply in terms of its pattern as well as its quantity. Usually, the percentage of potable water discharged as wastewater is between 70% and 95% depending on the local climate conditions, the availability of water, the culture, etc. The pattern of the daily use of water depends on whether it is a working or non-working day and what season it is. As a result the cumulative wastewater flow is continuous in practice with a repeatable (diurnal) pattern.

The wastewater includes dissolved substances and particulate matter with various pollutants attached. Much of the particulate matter is carried in suspension, but coarser, denser material may be moved as a bed-load along the bottom of the channels or pipes. Chemicals in wastewater include nitrites, phosphates, dissolved oxygen, carbon, heavy metals, as well as materials that have chemical and biological oxygen demand, bacteria, and a sophisticated range of trace elements originating in the various modern substances that are incorporated as ingredients in modern washing and other domestic products. Water is a very good solvent and dissolves most substances, but there are certain chemicals that it is unable to dissolve, such as various oils. These are carried along in wastewater as colloidal globules in suspension or on the surface.

Table 10.1 Advantages and disadvantages of separate and combined systems

SEPARATE SYSTEM		COMBINED SYSTEM	
Advantages	**Disadvantages**	**Advantages**	**Disadvantages**
Lower frequency and smaller volume of overflow spills.	Additional costs due to the construction of two pipe systems.	Lower construction costs.	Greater frequency and volume of overflow spills.
Less inflow to the WWTP.	Possibility for illegal- or cross-connections.	Occupation of a smaller public space.	Larger inflow into the WWTP.
Smaller variation in flows.	The two pipe systems together with their other elements occupy a larger public space.	Smaller maintenance costs.	Greater variation in flows.
Apart from the first flush, the stormwater is relatively less polluted.	Greater maintenance costs (including possible additional pumping).	The pipe slope may be lower.	In the case of flooding, flood water will also contain raw sewerage.
Flooding is by stormwater only and as such it can be used for irrigation purposes.	Smaller flows may require that the wastewater pipes are constructed with higher slopes in order to maintain self-cleansing characteristics.	Some portion of the stormwater will be treated at the WWTP.	Design and construction difficulties associated by including the two constituent flows in one pipe system.
The two systems can be laid separately following their own optimal route and depth.	Difficulties due to the need to construct pipes with higher slopes.	Less planning and asset management activity due to the production of a single asset management plan.	Stormwater cannot be used for irrigation purposes.

In most urban areas the aim is to provide wastewater collection from every significant building where people live or work. Pipes with small diameter run from kitchen sinks, baths and toilets, and possibly surface drains in back-yards to sewers along the adjacent street. The sewers are sized on estimates of the total wastewater flow draining to each pipe. This is done on the basis of the number of people living in an area times the volume of generated wastewater per person per day plus an allowance for infiltration and any industrial effluent discharges. Furthermore, a peaking factor is included to account for the peak flows during the diurnal pattern. Another factor may take into account the standard of living in the area, as this is known to affect the amount of water used, and therefore the amount of wastewater produced. Where the time variation of the wastewater flow is important, such as planning the daily operation of a treatment works, then a diurnal unit hydrograph is used with peaks in the morning and evening when people are getting up or having their evening meal. Typically, the hydrograph will have small flows during the night.

Traditionally, the sewer systems are designed to carry sediments by specifying the minimum 'self-cleansing' velocity (typically, 0.6 m/s). This approach has proved to be only partially effective due to the fact that such a velocity does not take into account the characteristics and concentration of sediments or the hydraulic ability of sewer flows with different depths to transport sediments. More recent design practice is based on methods that determine both the local bed shear stress values (based on a logarithmic velocity profile and the Reynolds shear stress distribution) and the mean bed shear stress values (determined from the energy slope).

Generally a treatment works is designed to take, say, 6 to 8 times the dry weather flow, which enables it to accommodate the daily flow variations comfortably. Because the wastewater draining from housing and industry is reasonably consistent over a long period of time some storage upstream of a treatment works is normally sufficient to balance any extreme wastewater flows due to unusual events.

However, there are other factors at work which can make even storage in the sewerage system unable to prevent flooding from the network before the wastewater reaches the treatment works. Often the most important factor is the illegal connection of stormwater drains to the sanitary sewerage system, called *inflow* by drainage engineers. Developers sometimes make mistakes about which is the separate sanitary sewer, or which is the separate stormwater drain, and connect the collectors from buildings to the wrong service. Typically, in such situations, smoke surveys are carried out to detect the location of such connections.

Another factor is the *infiltration* of groundwater from the ground around the pipes (or leakage from the water supply/distribution system), such that the sewer network is acting like a land drain. The extent of infiltration usually depends on the quality of design and construction works, material and age of the system, surrounding soil characteristics and frequency of surcharge. This is particularly difficult to manage because, although grouting and pipe lining can improve the integrity of the main pipe network, it is not so easy to repair the small pipes taking the wastewater from buildings to the main sewers.

Due to both inflow and infiltration (which are often abbreviated as I and I, or I/I) the wastewater flow can be augmented to be several times the normal dry weather discharges such that the pipe network is as much or more of a storm drain than a wastewater collector. If the treatment works cannot cope with the peak flows that occur, and the storage facility upstream of the works is inadequate, then provision has to made for excess flow to be discharged directly from the sewerage network to the environment, whether to urban streams or rivers or to coastal waters. This can of course generate difficult water quality conditions in the receiving water.

In addition to I and I the reverse process where the wastewater leaks out from pipes, also known as *exfiltration*, is yet another concern as it creates the potential for groundwater contamination. The extent of exfiltration usually depends on similar factors to those facilitating infiltration.

The key to planning and implementing effective strategies for designing or rehabilitating wastewater networks are dynamic models, and not just the steady flow models that are traditionally used for sizing the pipes in such networks. A dynamic model takes into account the short term responses of the flow in a network due to the stormwater component. A number of authorities around the world have developed procedures based on the use of a dynamic model for the wastewater network to identify and manage illegal (stormwater) inflows and unwanted groundwater infiltration. Such procedures depend on the strategic use of short term flow surveys that take advantage of technological developments in velocity and depth monitors in pipes. This sort of procedure is considered in more detail below in Section 10.23.

10.4 STORMWATER SYSTEMS

As described above an urban area is characterized by an extensive sealing of the land surface using concrete, asphalt, stone and other construction materials in order to provide buildings, parking lots, roads, side walks and cycle paths. Even where some of these surfaces are reasonably porous their existence means that rainfall (or snow melt) which would normally infiltrate to the subsurface is limited or prevented from doing so. As a consequence the runoff over the paved surface tends to collect in surface depressions giving rise to the need for special provision to be made in order to remove the collected water. This is done through adjusting or adapting to the surface topography by providing a network of carefully positioned and sized surface channels or through the construction of a network of underground pipes that accept the stormwater through specially designed inlets (e.g. road gullies). In both cases, stormwater is conveyed to a convenient urban stream, pond or lake often following the natural drainage paths, or for urban areas adjacent to the sea to a coastal outfall. Most of the time, separate stormwater systems will remain empty during dry weather periods and they will only operate during and immediately after the wet weather periods. Care has to be taken to ensure that the conduits, including the receiving waters are large enough to convey a certain flow. There is a limit to the flow that can be conveyed due to the cost of constructing the channel or pipe network. This cost has to be balanced against the damage that would ensue if the generating storm created a runoff with a potential discharge in excess of the capacity of the conduits, leading to unacceptable surface flooding. In other words, the drainage network has a limiting capacity which defines a return period of flow beyond which damage due to flooding may occur. A consequence is that any storm drainage system can only limit the frequency of surface flooding and cannot prevent the phenomenon.

Flooding in drainage networks of channels or pipes can occur in one of two ways: water can emerge from the network to flow overland, or the runoff locally may be too large to enter the network and so ponds up on the surface, possibly flowing away from any inlet or channel. Where the management of excessive surface flooding has been recognised as important, separate provision has been made for overland flood flows as well as the traditional above ground channel or below ground pipe networks. In such cases, urban drainage systems are divided into *minor* (or primary) and *major* (or secondary) systems; see Figure 10.5. Minor systems therefore refer to the underground pipes, watercourses and open drains, whereas major systems refer to the above ground drainage and consist of overland flow paths and the floodplains associated with urban streams. In some cases, the major system utilises the roads as part of the drainage system. In such cases, the multiple use of roads needs special design and maintenance considerations.

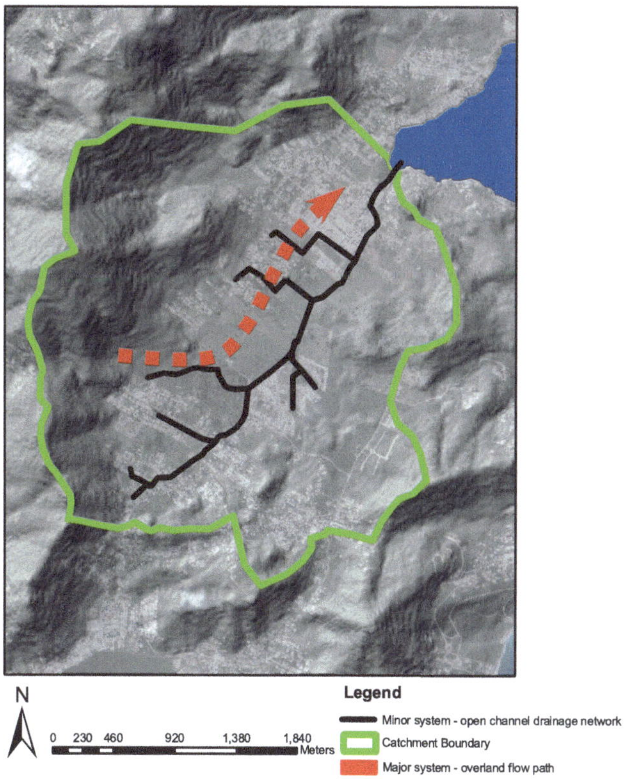

Figure 10.5 Example of minor and major systems

The stormwater runoff over the urban surface tends to collect and transport detritus and sediment deposited by human activities, such as vehicle transport and construction sites, and by the wind. Street cleaning is done to limit the accumulation of sediment and detritus, but each significant storm event carries with it a certain quantity of pollutants, including rubbers and heavy metals, and decaying vegetation. This wash-off of pollutants from the urban surface can lead to a degradation of water quality in the receiving waters for the storm runoff.

As with wastewater systems, strategies for improving or extending stormwater drainage can benefit considerably from simulation modelling. However, in this case the modelling is more complex than for water distribution networks in that a range of different phases in the drainage process have to be identified and modelled separately, including rainfall, runoff from urban catchments and possibly rural catchments, loading and wash-off from urban surfaces, groundwater, flow and quality in receiving waters, and above ground-overland flow. All of these phases are also included in the case of combined sewerage, and therefore discussion of their modelling is given below.

10.5 COMBINED SEWERAGE SYSTEMS

As described above, there are two types of minor urban drainage systems currently in use: a combined system in which stormwater and wastewater (i.e. sewage) flow together in the same pipe, and separate systems in which stormwater and wastewater are conveyed in separate pipe networks. Figure 10.6 shows typical layouts of these systems. Both types of system may have at certain points an overflow connection, which serves as a relief point for the separate wastewater systems (that is, sanitary or separate sewer overflows – SSOs) or a means for conveying excessive flows from combined pipes (that is, combined sewer overflows – CSOs) into the receiving waters or the separate stormwater system. In dry weather conditions, the dry weather flow component within the combined system is transported to the wastewater treatment plant (WWTP) for treatment. When it rains and the surface runoff inflow increases, any flows in excess of the capacity of the WWTP are either stored or allowed to spill into the receiving waters via overflows. The hydraulic design of overflow structures requires special care in order to avoid frequent (or premature) spills (such as in the case of too low a weir crest level) or excessive surcharge upstream (such as in the case of too high a weir crest level). Vortex separators storages and screens are often constructed in order to minimise the impacts from sewer overflows, associated structures (or facilities) such as stilling ponds.

The service provided by a combined system can fail for a number of reasons. For example, wastewater does not drain away from the generating source because of blockages in the system downstream caused by sediment, congealed fats or a sewer collapse or because the hydraulic capacity of the system downstream is inadequate due to increased loading of the system from developments upstream. Such blockages or lack of capacity may increase the frequency of surface flooding upstream. Alternatively, despite a treatment works at the end of the combined system there may be too frequent or too polluted discharges to receiving waters from combined sewer overflows. This leads to a general deterioration of the water quality in the receiving waters and the possibility of short term fish kills.

Obviously, the management of a combined system with its integration of a variety of different processes and potential problems presents the greatest challenge for urban hydroinformatics. For this reason we consider the implications for simulation modelling in some detail. In particular, we identify the following phases for such modelling: rainfall (precipitation), rainfall-runoff for urban catchments, surface pollutant loading and wash-off, dry weather flow, groundwater, pipe network flow, above ground-overland flow, pipe network flow, treatment works, rainfall-runoff for rural catchments, receiving waters (urban streams, lakes, coastal waters); see Figure 10.7.

10.6 SIMULATION MODELLING

The modelling of combined sewerage system is done by deconstructing the flow processes into a number of distinct phases, some of which are in sequence and some in parallel. The model space consists of the network of pipes or channels that are assumed to be uniform along their lengths, and each conduit has wastewater and stormwater inflows that are input at the 'upstream' manhole rather than distributed along the length of each conduit. Here 'upstream' denotes that end of the conduit where the bed is higher than the other end. The stormwater component is divided into surface runoff and groundwater infiltration. This implies that a sub-catchment is attached to a conduit. Rainfall is normally considered to apply uniformly and simultaneously to all the urban sub-catchments that contribute rainfall-runoff to the pipe network. Where there is a spatial distribution of rainfall, each sub-catchment can receive a different amount of rain. Similarly the surface loading of pollutants is assumed to be distributed over the sub-catchments, and the washoff is directly related to the rainfall and rainfall-runoff from each sub-catchment.

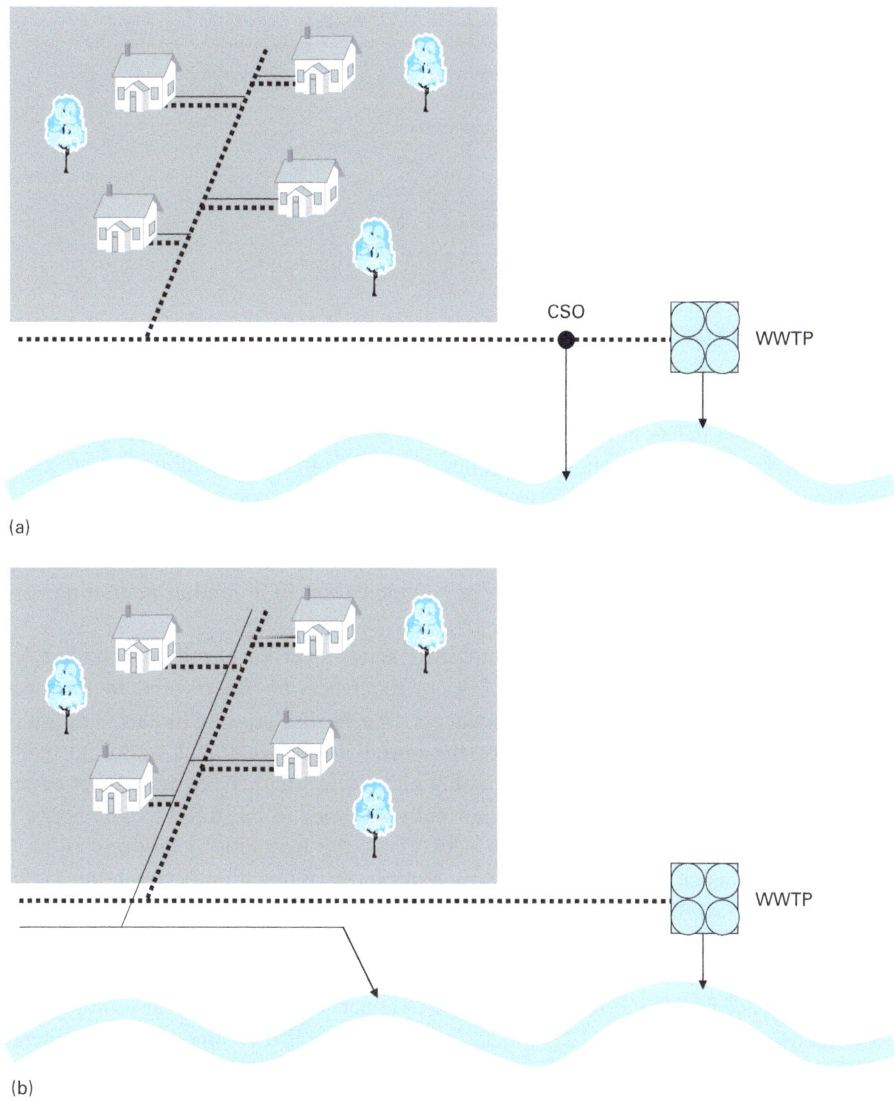

Figure 10.6 Combined (a) and Separate (b) Drainage Systems

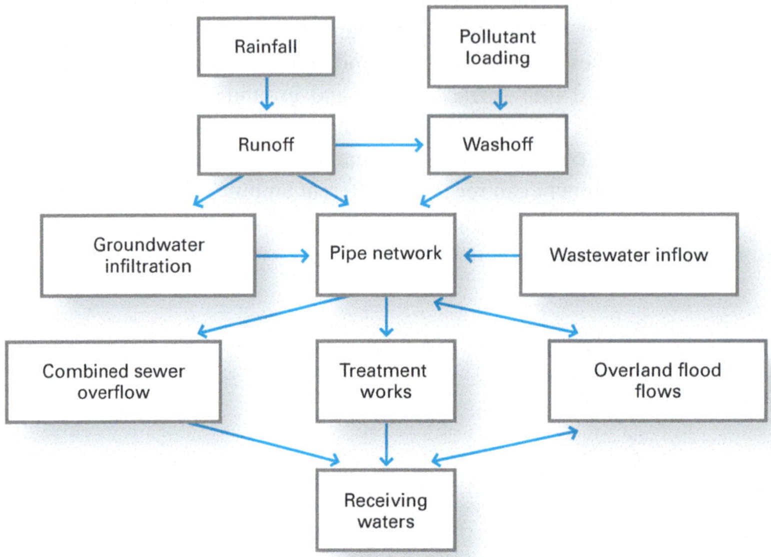

Figure 10.7 Phases for urban drainage

The sub-catchments all act independently in parallel, and feed forward rainfall-runoff to the corresponding conduits in the network. Obviously the conduits feed flows to and from those conduits that are adjacent to them. They are so interdependent that it makes sense to model the flows in the conduits together simultaneously. The transport of pollutants from wastewater and stormwater runoff is calculated on the basis of the flows in the conduits. Generally, it is assumed that there is no feedback of information from the water quality model to the calculation of the flow in the conduits, such as sedimentation in a conduit restricting its capacity and therefore the flow.

The outfalls from the network, namely the combined sewer overflows and the conduit leading to the treatment works, discharge flows out of the network, to the receiving waters and to the treatment works respectively. It is normally assumed that the overflows are free to discharge without restriction generated by high water levels in the receiving waters. Similarly, the outfall to the treatment works normally consists of a pump with a prescribed discharge (which can possibly change in time). Even water that emerges from the network onto the urban surface causing flooding is normally assumed to be stored locally above the manholes only to be lost to the system or to return into the network as the hydraulic conditions allow. This concept however, has been modified as engineers now regularly explore the interaction between the 1D network flows and the possibility of a 2D overland flow due to rainfall-runoff being unable to enter the network or water emerging from the network onto the above-ground surface. Modelling this interaction is not trivial since information on the flows is being fed to and from the 1D model and the 2D model. Like the conduits in the 1D network, the interdependence of the overland flow with the 1D network would appear to imply that they should be modelled together. However, the 1D and 2D models are normally self-contained. So their interaction is done explicitly at each of the larger time step of the two models (usually that for the 1D model); see Figure 10.8. In modelling the interaction between the network and overland flows it is important to distinguish between the open channel and closed conduit networks. This is because flood flows from the open channel network can occur overbank at any point along the channel, whereas flows from the closed conduit network are assumed only to emerge at the manholes.

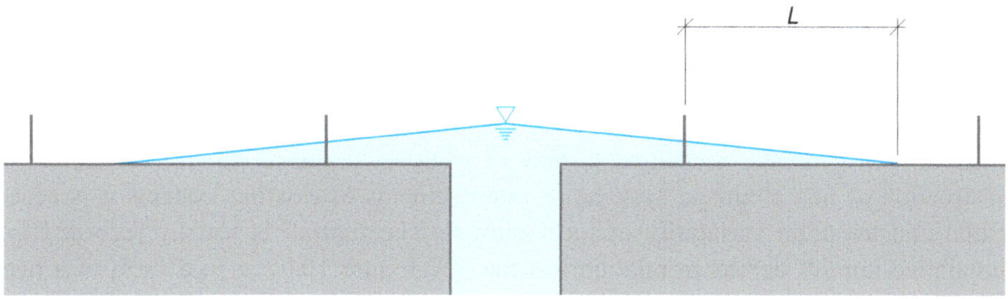

Figure 10.8 1D-2D interaction

We examine models for each phase in turn.

10.7 MODELLING RAINFALL

Light or moderate rainfall has a positive value for an urban area. It provides water necessary for the ecology, recharges groundwater, refreshes and to some extent cleanses the urban surface. Heavy rainfall, however, can cause damage, especially where the paved surface limits infiltration and surface water collects unacceptably in streets, affecting buildings and other structures adversely. Water will tend to collect at low points and there will always be a greater rainfall (with a lower frequency); therefore, the objective in collecting and disposing of excess rainfall-runoff is always to limit the frequency of surface flooding. In this respect rainfall is a stochastic event which varies in time and space over the urban catchment. We can similarly acknowledge precipitation in other forms than rain (liquid water), such as snow. In this case, we have to identify a loading of water on the urban surface followed by a gradual release of liquid water as the snow melts according to the prevailing air temperature; see Liston, 1999.

Rain falls as drops of water that have condensed on dust particles from cloud vapour. Rather than treat each drop separately, it is customary to regard the rain as a continuum and to measure it as a depth per unit time (typically in mm/hr). The rainfall is normally measured over a small area ($0.25\,m^2$) determined by the instrument. In effect the area is so small compared with the length scales of the urban surface that the gauge can be said to measure the rainfall at a point. Although the rainfall gauge may represent the rainfall over a sizeable area the actual rainfall varies both in space and time. A finite time is needed to accumulate rainfall for measurement. Typically one or two minutes is needed for rapid variations in time, though much larger times of 15 minutes to an hour are often used. The shorter time steps are important for calculations with short lag times between rainfall and runoff in the drainage system. Spatially, rainfall can vary significantly over a kilometre during a summer thunderstorm, which is the sort of event that can cause problems for urban runoff. Such spatial variations can be accommodated by a sufficiently dense network of point rainfall gauges. If this is not available an area reduction factor (ARF) can be applied to the point rainfall. This factor makes use of the fact that on average the total rainfall over a large area is less than the corresponding total of rainfall at a point. Alternatively, weather radar data can be used if it is available. As yet the rainfall determined from cloud temperatures by remote sensing satellites is considerably uncertain; although significant advances are being made in improving the reliability of the rainfall recorded using this technology.

Whether the rainfall is given as a data time series or a calculated forecast from weather radar or global circulation of the atmosphere model, the rainfall can be said to be modelled. In the latter case the interest is in extrapolating rainfall from weather radar records, or in working with ensemble forecasts from global circulation of the atmosphere models. Much work is being done in this area, but the interested reader should consult Molteni *et al.* (2006) for further information.

We distinguish between design (or synthetic) rainfall, time series rainfall and forecast rainfall.

10.7.1 Characterisation of rainfall

As already explained, rainfall generates considerable flow in an urban drainage network. Therefore, in order to plan for the construction of new drainage systems or improvements of existing systems it is necessary to understand the spatial and temporal variability of local rainfall. The rainfall is usually recorded in a form of mass curve (cumulative rainfall depths in relation to time; see Figure 10.9) or in a form of a hyetograph (rainfall depths in relation to time; see Figure 10.10). From the meteorological point of view the rainfall is convective, orographic, frontal, cyclonic or convergent. A statistical characterisation of rainfall refers to 'interior' statistics (i.e., time and spatial distribution of rainfall rate within storms) and 'exterior' statistics (i.e., rainfall depth, duration, intensity and time between the storms).

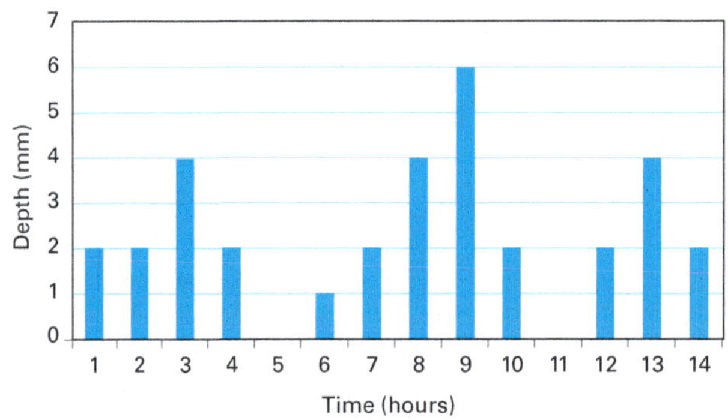

Figure 10.9 Hyetograph (rainfall presented in a histogram form)

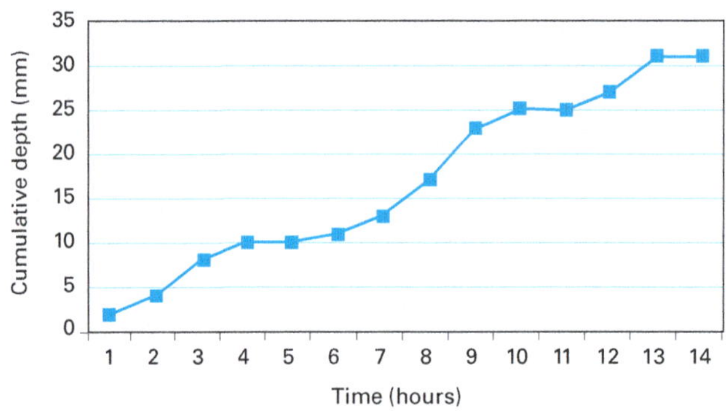

Figure 10.10 Mass curve

In terms of the 'interior' statistics, some rainfall events may exhibit similar characteristics. Convective rainfalls have more or less triangular hyetographs with the highest intensity towards the start of the event, whereas tropical cyclones have a bell-shaped hyetograph with the largest intensity towards the middle of the event. Furthermore, tropical cyclones have a much longer duration and a much larger rainfall depth than thunderstorms; see Figure 10.11.

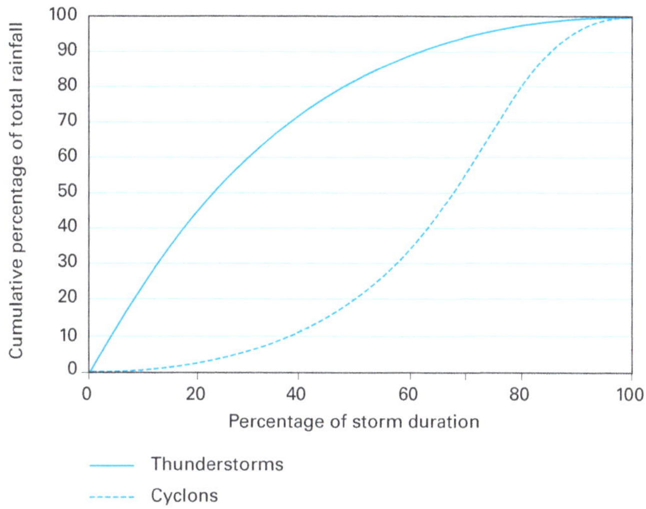

Figure 10.11 Typical percentage mass curves of rainfall for thunderstorms and cyclones (based on de Laat and Savenije, 1992)

Another important 'interior' characteristic is the spatial distribution. The shape, timing and peak of a rainfall are significantly influenced by spatial and temporal variability across the catchment. An illustration of spatial variability of rainfall for an urban catchment is given in Figure 10.12.

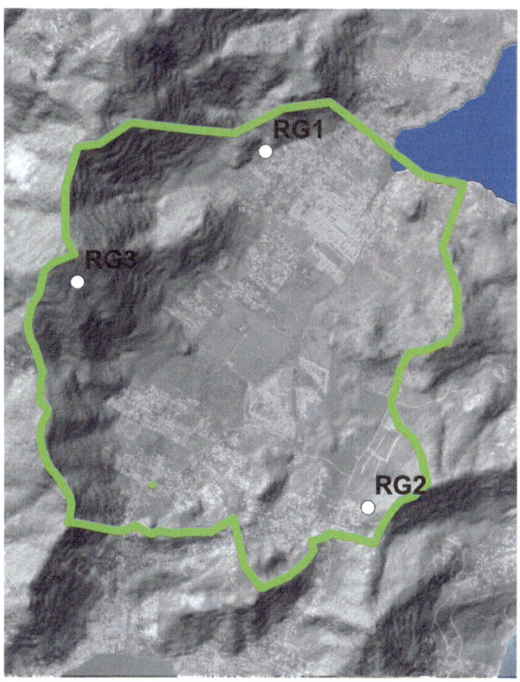

Figure 10.12 Location of raingauges in a catchment area

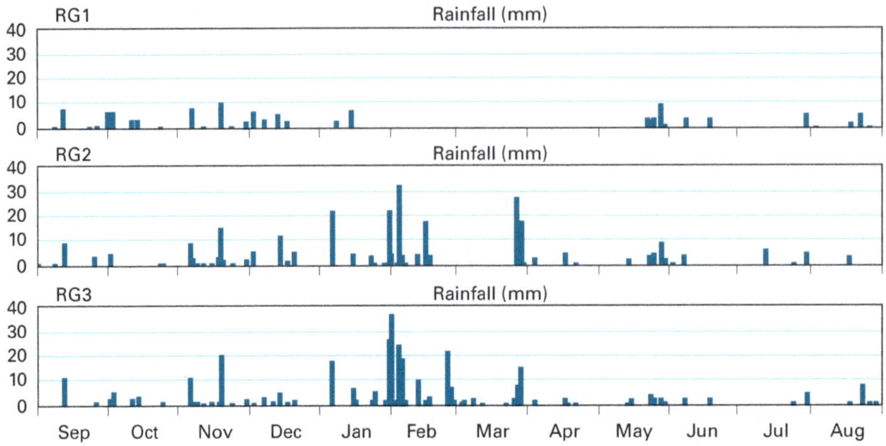

Figure 10.13 Rainfall records from a catchment area given in Figure 10.12. The raingauge station 5CH01 has recorded significantly different rainfall pattern from the other two raingauges

There are various methods that can be used to estimate the average rainfall over an area (i.e., areal rainfall) from point measurements. Some of the commonly used methods are: average depth method, Thiessen method, Isohyetal method, Kriging, reciprocal distance method and IDW (see also the section on Interpolation in Chapter 4).

The average depth method is a method based on the arithmetic mean of the rainfall amounts measured across an area. The arithmetic method uses only those gauging stations within the catchment area:

$$\bar{P} = \sum_{i=1}^{n} \frac{P_i}{n} \tag{10.1}$$

where P_i is the average precipitation over cell (mm/hr)

n is the number of gauges within the catchment.

For example, if the measured precipitation by gauge 1 is 10 mm, gauge 2 is 12 mm and gauge 3 is 16 mm then the average precipitation is: $(10+12+16)/3 = 12.67$ mm.

The Thiessen method adjusts for the non-uniform location of gauging stations by attempting to determine the areas of influence for each gauge. The procedure consists of connecting each precipitation station with straight lines and constructing perpendicular bisectors of the connecting lines to form polygons with these bisectors. In effect, the precipitation surface is assumed to be constant and equal to the gauge value throughout its polygon area of influence (see also Figure 10.14). Therefore, the average precipitation over a cell can be calculated as:

$$\bar{P} = \sum_{i=1}^{n} W_i P_i \tag{10.2}$$

where $W_i = A_p/A$,

A_p is area of the polygon (ha)

A is the total catchment area (ha)

n is the number of polygons

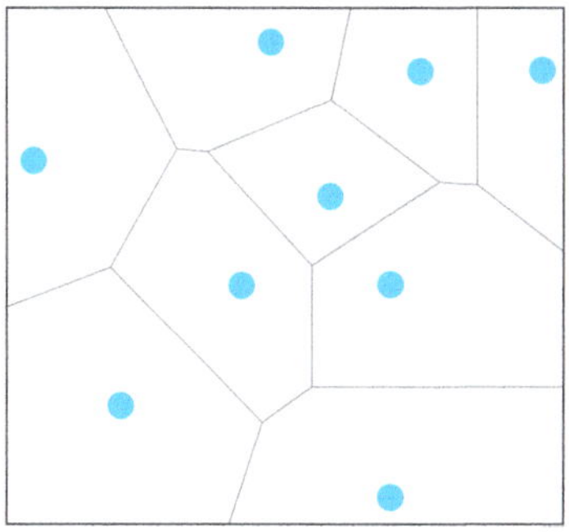

Figure 10.14 Thiessen polygons

The Isohyetal method is also regarded an accurate method to estimate average precipitation within a catchment area; however, this average is difficult to reproduce because of the complexity of drawing the isohyets; see Figure 10.15. The following formula is used to calculate the average values:

$$\bar{P} = \sum_{i=1}^{n} W_i P_i \tag{10.3}$$

where $W_i = A_p/A$,
A_p is area of the polygon (ha)
A is the total catchment area (ha)
n is the number of cells.

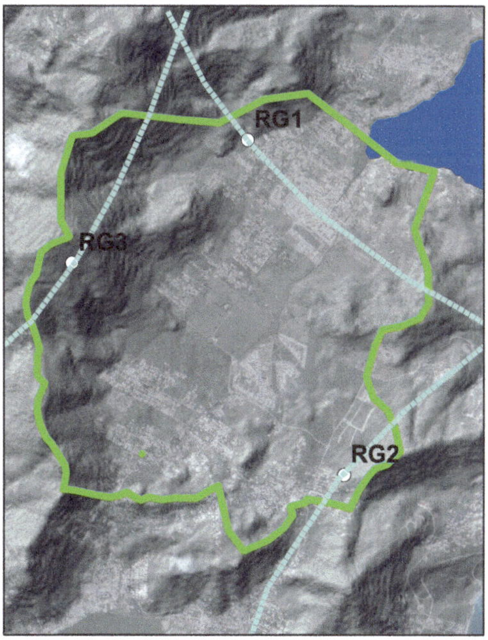

Figure 10.15 The Isohyetal method

Typical parameters that define the 'exterior' statistics for the rainfall are:

- Intensity – the rate of rainfall (usually expressed in mm/hr),
- Duration – a period of rainfall event (usually expressed in mins or hrs),
- Depth – thickness of water layer on the surface (usually expressed in mm),
- Area – geographic extent of the rainfall (usually expressed in km^2), and
- Frequency – the number of repetitions of a particular rainfall event per unit time (usually expressed in terms of 'return period' values).

The Average Recurrence Interval (ARI) and the Annual Exceedance Probability (AEP) are usually used as a measure of the frequency of a rainfall event. ARI can be defined as the average, or expected, value of the periods between the exceedances of a given rainfall total accumulated over a given duration (e.g., 10-year ARI). AEP can be defined as the probability that a given total rainfall accumulated over a given duration will be exceeded in any one year (e.g., 10% AEP). The relationship between the ARI and AEP can be expressed as:

$$AEP = 1 - \exp\left(\frac{-1}{ARI}\right) \tag{10.4}$$

This relationship yields the values given in Table 10.2:

Table 10.2 Annual Exceedance Probability

ARI (years)	AEP	AEP (%)
1	0.632	63.2
2	0.393	39.3
5	0.181	18.1
10	0.095	9.5
20	0.049	4.9
50	0.02	2
100	0.01	1

When dealing with urban drainage systems (that is, the design of a new system or the evaluation of the performance of an existing system), extreme rainfall events are often used for the analysis. For example, if a new drainage system is to be designed for a 10 year return period with, for example, a rainfall of 62 mm/hr, on average, 62 mm/hr rainfall occurs in a particular area once every 10 years. It is important to remember that this is based on statistics and there is no guarantee that once this much rain has occurred it will not happen again for another 10 years. In a 100 years of record there should be ten such rainfall events (an average of one every 10 years). These events could be in consecutive years or even multiple events in a single year! There is also no guarantee that a nearby catchment will not experience such an event in, say, consecutive years.

10.7.2 Design rainfall

Much of the concern of urban drainage engineers is with limiting the frequency of flooding. This leads to the analysis of observed rainfall data in order to derive synthetic (or design) rainfall data in terms of frequency of event. There are different forms of design rainfall that need to be considered:

- Intensity-duration-frequency curves, derived specifically for return periods greater than a year, and developed separately for return periods less than a year

- The rainfall event profile (hyetograph)
- Time series of critical events
- Stochastic time series

Design storm events are particularly useful for analysing the frequency of a range of phenomena in a drainage system, such as surface flooding, surcharging in pipes, filling of storm tanks, operation of overflows and other diversion structures, and impact of effluent from sea outfalls on bathing beaches. In particular, design storms are used for the prescriptive sizing of pipes in a new drainage system. A design storm is normally derived with a rainfall depth for the event, a duration and a profile. Each of these parameters may have their own statistical distributions.

Intensity-duration-frequency curves

Three parameters are needed to describe the statistics of rainfall:

- The *duration* of the period in which the rainfall occurs
- The rainfall *intensity* or depth during that period
- The *frequency* with which the rainfall intensity over the particular period is likely to occur, or the probability of it occurring in any particular year. (Note that the inverse of the probability is the return period).

These three parameters are linked in an intensity-duration-frequency (IDF) relationship. Generally, as the rainfall duration increases the average intensity reduces, and the rarer the event the greater the intensity; see Figure 10.16. Note that an actual storm rainfall event can be analysed for a range of durations and each duration will have its own (average) intensity and frequency.

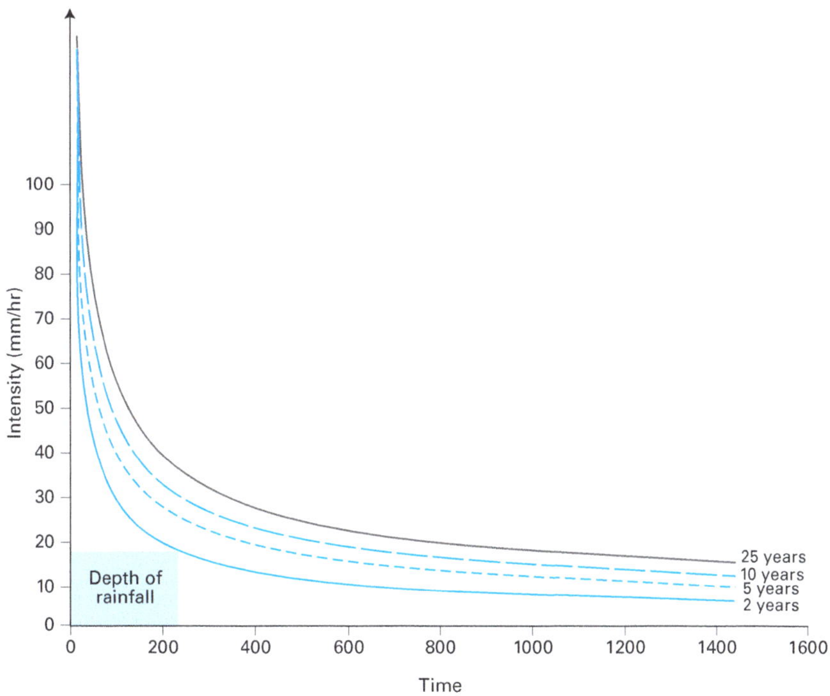

Figure 10.16 Intensity – Duration – Frequency curves (after Kellagher, 1993)

IDF relationships are formally derived by analysing a long set of rainfall records for a large range of durations to determine the largest depth of rain for each duration in each year of the record. These depths form an annual maxima series for each duration from which the depth of rain for various return periods can be derived.

Bilham (1936) carried out such an analysis for 10 years of data from 12 stations situated in lowland England and Wales with an average annual rainfall of 889 mm. He came up with the formula:

$$n = \frac{1.25D}{(I / 25.4 + 0.1)^{-3.55}} \tag{10.5}$$

where n is the number of times the rainfall occurs

I is the rainfall depth (mm)

D is the duration (hrs)

The analysis is valid for durations between 5 min to 2 h.

Storm profile

The rainfall intensity during an event is not uniform in time, and it is therefore necessary to define a storm profile to describe the variation. The profile can be characterised by two parameters, although the details of the shapes used require further parameters. The two basic parameters are:

- The *peakedness* of the storm, which defines the relationship between the peak intensity and the average intensity of the design storm relative to recorded events.
- The *skew* that defines the position of the peak within the duration of the storm.

These are illustrated in Figure 10.17 and Figure 10.18.

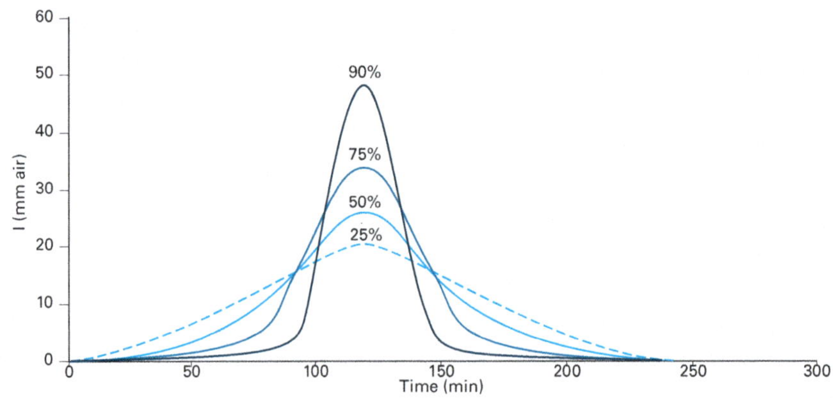

Figure 10.17 Storm peakedness profiles (after Kellagher, 1993)

Design rainfall events (or hyetographs) for use in simulation modelling can be derived from IDF data using the Chicago method or analysing an extended data set of recorded autographic rainfall. The Chicago method derives a time varying storm that has the same frequency for each duration contained in the storm; see Figure 10.19. Obviously, this does not necessarily happen in practice, but it simplifies the analysis of the return period of flooding in an urban drainage system by running a single event defined by the Chicago method. (It happens that the critical storm duration at any point in the drainage network is approximately equal to the time of concentration to the point; see below). Whole events are therefore derived for specific return periods. They can be assumed to be symmetric or skewed. Different countries or states have adopted alternative ways of defining such profiles, but their aim is to provide a synthetic storm that has an appropriate peak intensity and also the correct rainfall depth. Where possible the locally recommended design hyetograph should be used to preserve consistency in the calculations.

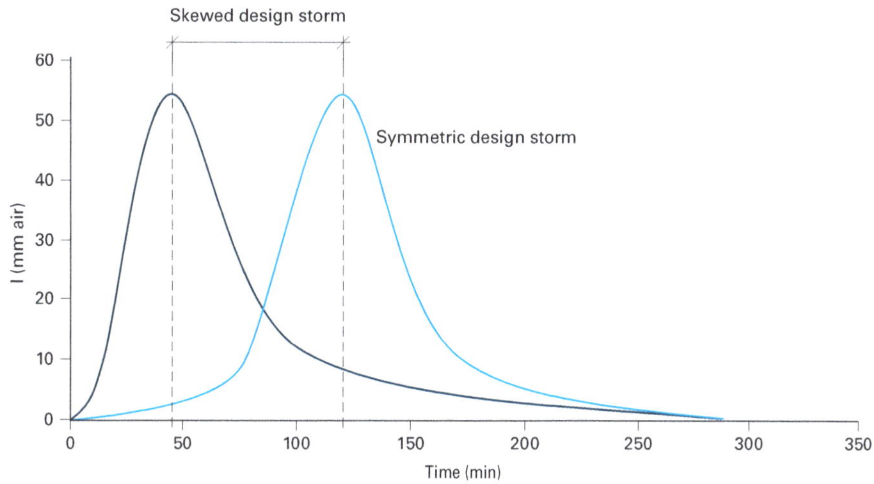

Figure 10.18 Storm skewness profiles (after Kellagher, 1993)

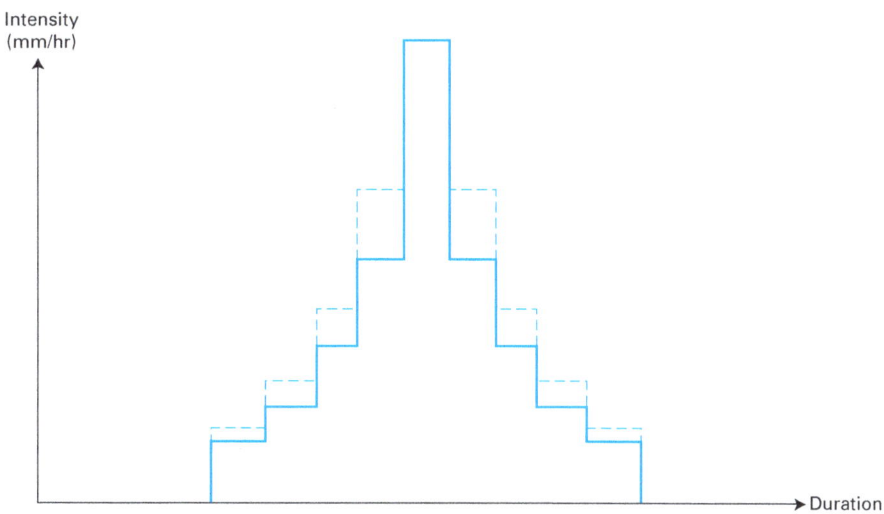

Figure 10.19 Chicago profile

10.7.3 Example of large amount of rainfall data (UK rainfall)

When enough data exist it is possible to carry out an extensive analysis that takes into account the IDF curves adjusted for location and includes suitable rainfall profiles. As an illustration of what can be done with sufficient rainfall consider the analysis of UK rainfall data carried out by the Meteorological Office originally in the early 1970s for the first edition of the Flood Studies Report (FSR) (NERC, 1975) and then for the Wallingford Procedure for the Hydraulic Design of Storm Drainage, released in 1981 (Wallingford, 1981). Subsequently the analysis has been updated for further editions of the FSR. For detailed information on design rainfall for urban drainage practice, see Volume 1 of the Wallingford Procedure (Wallingford, 1981).

10.7.4 UK synthetic design storms

A large number of rainfall records at a wide distribution of autographic rainfall stations were analysed by the UK Meteorological Office to determine depth-duration-frequency relationships for the whole of the UK in terms of:

- *M5*-60; the depth of rain in 60 minutes for a return period of 5 years.
- *r*; the ratio between *M5*-60 and *M5*-2 day.

Here *M5*-60 refers to a 60 minute period starting in any minute rather than a one clock hour rainfall for a 1 hour period starting on the hour. In contrast, *M5*-2 day refers to a period of 2 calendar days and normally starts at 9 am GMT. *M5* refers to a return period of 5 years. The variations of these two parameters spatially are given on maps covering the whole of the UK. See, for example, Figure 10.20 and Figure 10.21. The method first calculates the depth of rain for a 5 year return period for the required duration (M5-D):

$$\ln M5 - D = \ln D + \ln\left(1.06\frac{M5 - 60}{48r}\right) + \frac{\ln\left(\dfrac{721}{1+15D}\right)\ln\left(\dfrac{48r}{1.06}\right)}{\ln\left(\dfrac{721}{16}\right)} \tag{10.6}$$

The subsequent calculation of the depth of rainfall (mm) for duration D (hrs) and return period T (*MT-D*) (years) depends on the location in one of two geographic regions, namely England and Wales or Scotland and Northern Ireland. For return periods greater than 5 years:

$$\ln\left(\frac{MT - D}{M5 - D}\right) = C_r(\ln(T) - 1.5) \tag{10.7}$$

where C_r is a coefficient varying with geographical location and with the value of *M5-D*. For return periods less than five years no general equation has been derived, and so values for C_r given in tables are used with linear interpolation where necessary.

Besides the IDF curves the data was analysed to provide scaled rainfall profiles for summer and winter periods.

Areal reduction factors

Rainfall will always vary across a catchment, and the rainfall measured at a rain gauge in the catchment will not be the same as the average falling across the whole catchment. The IDF analysis described above is biased towards storms that are particularly intense at the rain gauge by selecting the largest events from the rainfall record. The average rainfall over the catchment during these events will therefore be less than that measured at the rain gauge. A reduction factor is needed to reduce the depth of rain in these synthetic storms to represent the average across the catchment.

Figure 10.20 UK rainfall data

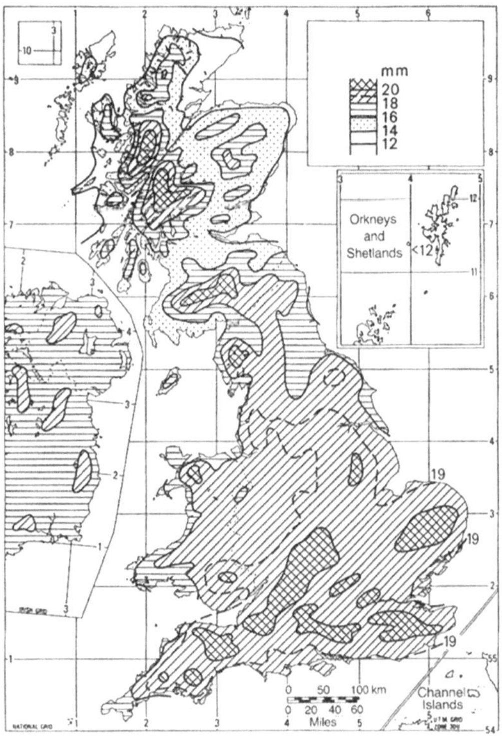

Figure 10.21 UK data

Strictly, a different areal reduction factor should be applied to different durations within the storm so that it has a greater effect on the peak intensities than on the average. However, this is not normally done, and a single factor is used for the total duration of the storm.

The reduction factor varies with catchment area, and for simplicity the total catchment area is used for this calculation. For storms of short duration the area of influence of the storm is only that over which flow can travel during the duration of the storm. However this problem can be largely ignored for catchments less than $5\,\text{km}^2$.

The UK-FSR (NERC, 1975) also includes information on areal reduction factors to represent the average rainfall over large catchment areas. One relationship is used for the whole of the UK. The relationship was derived for return periods of between two and three years, but as the areal reduction factor varies only slightly with return period, these values are used for all return periods. For most urban situations the areal reduction factor will be greater than 0.9 indicating a reduction in rainfall depth of less than 10%.

The area reduction factor is given by:

$$ARF = 1.0 - 0.0394A^{0.354}D^{-(0.40-0.0208\ln(4.6-\ln A))} \tag{10.8}$$

where D is the duration (hrs)

A is the catchment area $< 20\,\text{km}^2$

There is currently some concern that large urban areas in summer should have greater reduction factors. The derivation of ARF was calculated from assessments of river catchments for which critical periods occur in winter and where winter storms have much less spatial variation.

It is important to be aware of the modelling approximations that are used in representing and manipulating rainfall inputs in models. There are basically two categories of rainfall we need to consider: recorded (real) rainfall and synthetic, design or forecast rainfall. Recorded rainfall can come from point rainfall gauges, weather radar, or satellites. Design rainfall is derived by analysing recorded data, whereas forecast rainfall may be generated directly from recorded data or from global circulation of the atmosphere models, in which case forecasts are increasingly given in terms of ensembles; see Molteni *et al.*, 2006.

10.7.5 Selection of design storm

The return period of the storm is not necessarily the same as the return period of the flow. This is because there are several independent statistical variables contributing to the flow. These include: rainfall depth, storm profile, duration and the antecedent wetness of the catchment. Given the rainfall depth of given return period, the duration equal to the time of concentration to the point in question, and the 50% summer profile then the antecedent wetness can be determined as that which leads to the generation of a peak flow of the same return period as the rainfall depth. The summer profile was chosen because in urban areas the critical storms usually occur in the summer rather than the winter, and the rarer events are typically thunderstorms. This is discussed further in the section below on rainfall-runoff.

For design storm analysis it is necessary to identify the critical duration of the rain (i.e., *critical rainfall*) that generates the largest discharge at a point in the system. Sometimes, depending on the rainfall profile and the system characteristics, rainfalls of a shorter duration (and higher intensity) may cause flooding upstream, whereas, longer storms will not produce a flow in excess of the maximum peak flow of the critical duration storm.

10.7.6 Annual time series

The IDF analysis is done using an annual maxima series for which only the single largest event is taken from each year of the rainfall data. This means that the analysis cannot be used to define storms with a return period of less than one year. However, such storms can be very important when analysing the frequency with which overflows operate. The IDF relationship needs therefore to be complemented by an alternative approach to deal with frequencies of event more than once a year. Such an approach can be provided by an annual rainfall time series, which includes events in a typical year with frequencies greater than once a year. As an example, the UK-WRc has derived three regional time series covering the UK, namely for the South East, the South West, and the East Midlands & Yorkshire.

10.7.7 Synthetic time series

A criticism of the annual time series is that it is biased towards short storms, and tends to be catchment specific. Also, its use in model simulation is time consuming and is not suitable for the assessment of drainage solutions to meet the bathing beaches consent criteria of 3 spills a season, which requires a simulation of a 10 or 20 years time series. To overcome these problems a stochastic rainfall model has been produced by Cowpertwait (1991) generating a continuous rainfall series for a period of up to 30 years, which is representative of the rainfall climate for the catchment. The rainfall generator is based upon the Neyman-Scott Rectangular Pulses rainfall model; see also Entekhabi *et al.*, 1989. The model was developed using five main stations, each located in one of the five regions.

The accuracy of the rainfall generator depends to some degree on the location in which it is being applied, but the generally held view is that the accuracy is adequate for most project activities and it reasonably reflects the rainfall character of the area.

10.8 DELINEATION OF CATCHMENTS AND SUB-CATCHMENTS

An urban catchment area can be defined as an area where the water is collected by the natural landscape. In a catchment area, the surface run-off eventually flows to the receiving environment or into the groundwater via the gullies (or catchpits), soakholes and pipe network system. For modelling purposes, after defining the overall catchment boundary it is important to determine the relevant contributing areas of each individual subcatchment covering the key features that are present in the catchment. Figure 10.22 and Figure 10.23 illustrate the schematics of a delineation process for the stormwater catchment and individual subcatchments boundaries.

In the case of combined sewer system modelling, the wastewater catchment boundary (and also related subcatchments) may not necessarily follow the ridgeline of a natural landscape and as such may not coincide with the stormwater catchment boundary (and its subcatchments). Usually, the wastewater subcatchments follow the property lines (as per the household connections) and not the natural landscape; see Figure 10.24. From the modelling point of view, this does not pose a problem since the catchments can be separated with rainfall applied to the stormwater subcatchments and the equivalent population (for residential) or area (for commercial and industrial) applied to the wastewater subcatchments.

The process of catchment and subcatchment delineation is normally initiated using built-in GIS functions and is further refined manually. After this process, inflow loading points (i.e., a manhole of a pipe network or a branch of an open channel network) must be defined. These loading points then effectively define the upstream extents of the model. In principle, the subcatchment loading point should be defined at least one manhole upstream from the known problem areas or ancillaries (i.e., overflows, pumping stations, bifurcations, control devices and inlet structures). Loading points should be defined at an appropriate manhole (or a branch) within the subcatchment such that the resultant flows entering the model result in representative flows and surcharge levels in the system. This is particularly important in the vicinity of flow monitors, bifurcations and overflows. Special attention should be paid to the constructed overflow locations to make sure that subcatchment boundaries do not include flows downstream of the overflow.

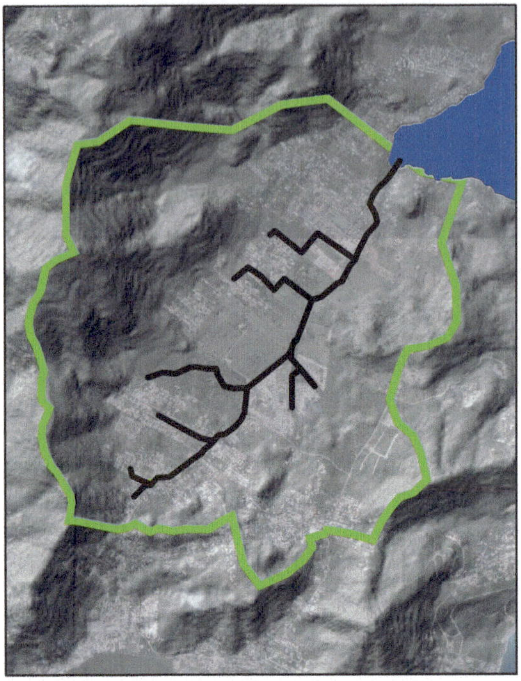

Figure 10.22 Delineated catchment boundary, 2D-Hillshade view

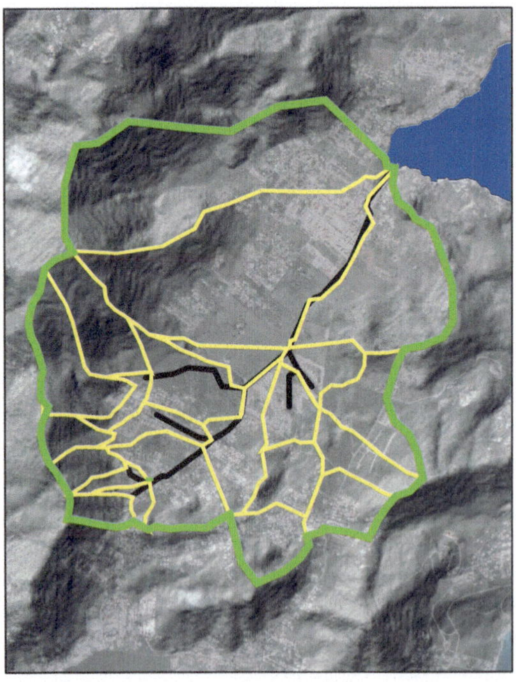

Figure 10.23 Delineated subcatchment boundaries, 2D-Hillshade view

Figure 10.24 Separate wastewater subcatchment schematisation/delineation. The closed rectangles show the subcatchment boundaries. The lines and dots represent pipes and manholes

Subcatchment definition should also take into account the location of flow monitors to ensure that subcatchment boundaries coincide with monitored (sub)catchments. Large areas that exist within an urban catchment such as playing fields, parkland and major roadways which provide no dry weather sanitary flow contribution to the drainage system should be differentiated from land use maps during the model subcatchment delineation. The land use maps should clearly identify these areas such that wet weather flow parameters can be assigned appropriately. Certainly, the entire area within the catchment boundary needs to be divided into subcatchments without omitted areas. This also includes areas that do not drain directly to the drainage network. Furthermore, consideration should be given to the location of major problem sites identified from the incident/complaint records.

10.9 MODELLING RAINFALL-RUNOFF FROM URBAN CATCHMENTS

Once rainfall reaches the urban surface the water making up the rain undergoes a far-reaching transformation. Some of the rain falls on buildings and vegetation and is therefore intercepted before it can reach the ground surface or level. Other rain reaches the ground surface. In both cases a small amount of water is used to wet the different surfaces. Some water begins to infiltrate the ground surface, whether it is through cracks and voids in a concrete or asphalt layer or directly into the soil. If the rainfall intensity exceeds the rate at which water infiltrates the local surface, or the absorption capacity is filled, water begins flowing over the surface and fills any remaining small voids or depressions in the surface. Once a void or depression is full, water moves under gravity to an adjacent depression, and therefore moves further and further, creating streams and sheet flow over the two dimensional surface. Necessarily the urban surface is extremely complex, consisting of roads with their camber and drainage gullies or channels each side, side walks, paved parking areas, buildings with roof drainage, trees, and pervious areas with shrubs and grass. Consequently the shallow flows over the urban surface are controlled by very small differences in topography, and especially steps formed by curbs and low walls.

Modelling the detailed flow over this complex urban surface is too time consuming for practical purposes at present, and it would be difficult to guarantee a level of accuracy in the results for, say, the depths of flow over the surface. Therefore the modelling of rainfall-runoff over an urban catchment follows the compartmental approach and treats the subcatchment contributing to a pipe or surface channel as a whole, forming deterministic equations to relate the volume of runoff to the volume of rainfall taking into account the characteristics of the subcatchment, and similarly the rate at which the excess rainfall-runoff flows from the subcatchment.

Note that the rainfall-runoff from an urban catchment is very different from the runoff from a natural rural catchment. The runoff from the urban catchment is usually much faster. In addition, there is (or should be) very little infiltration into the conduits from groundwater in the urban catchment, so that almost all runoff occurs over the surface, whereas the groundwater component in the natural catchment feeding the stream network can be considerable. Any groundwater inflow can be estimated through the analysis of flow gauging data. A groundwater model may be needed to determine areas where the stormwater or wastewater systems intersect with elevated groundwater levels during different seasons.

There are two major concepts associated with rainfall-runoff from an urban subcatchment draining to a node of a drainage network. The first is the notion of a rainfall 'loss' that fails to reach the outfall of the urban subcatchment. A basic assumption is that the rainfall event has a certain volume, which is reduced by an initial amount before the water starts flowing over the urban surface, and thereafter it has a constant or variable continuing loss depending on the nature of the losses model used. The losses model can however, take one of several different forms. Given that the initial loss is removed from the total rainfall volume to leave an *effective rainfall* volume, the simplest approach is one of using a runoff coefficient that determines the factor of the total effective rainfall that runs off. However, runoff can be said to occur from three or more types of surface, including roofs, paved areas and pervious areas. A different continuing losses model can therefore be assumed for each surface type. Whereas runoff from a paved area is subject to a small constant continuing loss, the continuing losses on the pervious area might be subject to the variation in the infiltration rate. Yet another losses model lumps the continuing losses from all surface types and then reallocates them to each surface type according to some rule.

The second major concept associated with the rainfall-runoff concerns the rate at which the remaining surface flow runs off the subcatchment to the outfall or node. This is a matter of routing the runoff to the outfall.

10.9.1 Runoff coefficient model

The continuing runoff losses after initial losses can be defined as being at a fixed rate and therefore independent of antecedent conditions and changes taking place during an event. This type of representation is only advised for use with impervious areas: the runoff coefficient from pervious areas is known to vary significantly with antecedent conditions.

Typical values for runoff coefficients are given in Table 10.3. These are conservative values. It is advised that the use of these coefficients is supported either by field observations or by expert guidance.

10.9.2 The Horton infiltration model

Whereas the runoff coefficient model deals with runoff from paved areas, a separate model may be needed for runoff from permeable surfaces, such as the Horton model, which assumes that runoff from permeable surfaces increases as an event takes place; see Horton, 1933. The paved runoff factor is assumed to be constant while the permeable runoff is a function of the conceptual wetting and infiltration processes, and takes the form:

$$f = f_c + (f_0 - f_c)e^{-kt} \tag{10.9}$$

where f_o is the initial infiltration rate (mm/hr)
f_c is the final (limiting) infiltration rate (mm/hr)
k is the exponential term coefficient (1/hr)
t is the time (hr)

Table 10.3 Typical values of the runoff coefficient

Surface type	Description	Coefficient
Paved	High quality paved roads with gullies <100 m apart	1.00
Paved	High quality paved roads with gullies >100 m apart	0.90
Paved	Medium quality paved roads	0.85
Paved	Poor quality paved roads	0.80
Permeable	High density housing	0.55
Permeable	Medium density housing	0.45
Permeable	Low density housing or industrial areas	0.35
Permeable	Open areas	0.0–0.25

10.9.3 Conceptual framework for rainfall-runoff models (UK)

The rainfall-runoff models in the UK Wallingford Procedure consist of a set of models which enable engineers to represent the transformation of rainfall into runoff for both urban and rural catchment areas contributing to piped or channelled drainage systems. Irrespective of the model option selected, there is a basic conceptual framework underpinning the use of the model. First, the overall catchment model is divided into a series of sub-catchments. Each sub-catchment can then be subdivided into three surface types. It is normally assumed that the first two surface types represent impervious areas, whereas the remainder of the sub-catchment represents pervious surfaces. The surface runoff model is divided into two major process models, representing rainfall losses and runoff routing, each of which can act independently on the surfaces in each sub-catchment; see Figure 10.25.

The runoff model for the UK Wallingford Procedure (Wallingford, 1981) is dependent on experimental studies carried out largely in Europe but including some studies from North America. First, 34 studies were carried out in the UK, Denmark, France, Italy, Norway, Sweden, Yugoslavia, USA, Canada and Australia where measurements were taken in the sewer, monitoring the response of a collection of individual areas and surfaces. The resulting experimental data enabled a percentage runoff to be determined for each catchment, and subsequently a generic regression equation to be derived for general application. Second, 20 experimental studies in the UK, The Netherlands and Sweden based on runoff from individual urban surfaces (such as roads and roofs) measured at the inlets into the sewerage system were carried out in order to calibrate a generic rate of runoff model.

10.9.4 Rainfall-losses models (UK)

There are a number of processes that have been explicitly represented in the Wallingford Procedure runoff models for the losses:

(1) Depression Storage Model for the initial losses
(2) Percentage Runoff Model including an Infiltration Model for pervious areas

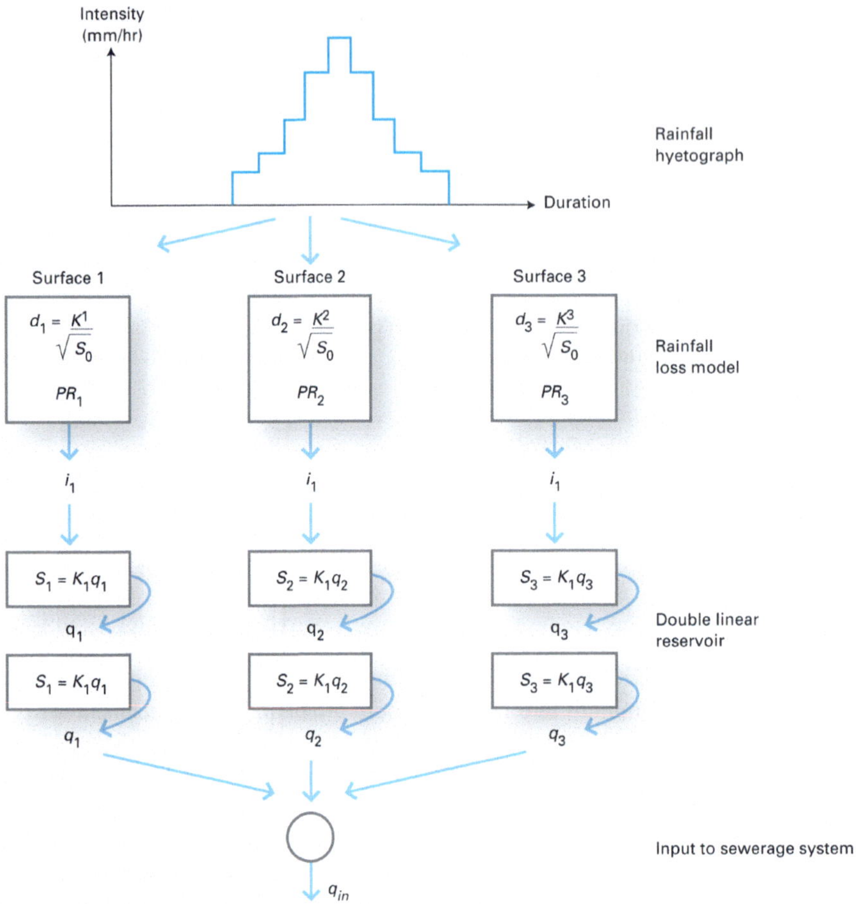

Figure 10.25 Wallingford Procedure rainfall-runoff model (after Kellagher, 1993)

Depression storage model (UK)

Although on paved surfaces this is dominated by depression storage, other losses such as surface wetting, infiltration and evapotranspiration all have an impact on the value of the initial loss. On pervious surfaces interception is a further complicating process. Initial losses are defined as the quantity of rainfall required just before overland runoff begins. The experimental data shows that

$$D = \frac{k}{\sqrt{s}} \tag{10.10}$$

where D is the average depth of initial losses (m)

s is the ground slope (m/m)

k is a coefficient dependent on the surface type (0.000071 for impervious surfaces and 0.00028 for pervious surfaces)

Percentage runoff model (UK)

A better model for the continuing losses is provided by an equation that relates the percentage runoff PR equation to the more important parameters describing the characteristics of an urban catchment. From data collected for a number of small urban catchments it was deduced that the more important catchment characteristics are the soil type and the antecedent wetness. At first a regression equation for PR was derived from the data for discrete storms using the percentage impervious area (PIMP) for the catchment, the soil index defined by the Flood Studies Report (NERC, 1975) and an urban catchment antecedent wetness index (UCWI). UCWI was based on a 5-day antecedent precipitation index (API5) and the soil moisture deficit (SMD). The equation was found to under predict the runoff volume for catchments with relatively low PIMP and with light soils, and for long duration storms. The equation was not easy to use with time series rainfall. Consequently, an alternative equation was developed, which calculates the runoff from the paved and pervious surfaces separately; see Eq (10.11).

$$PR = IF * PIMP + (100 - IF * PIMP) * \frac{NAPI}{PF} \tag{10.11}$$

where *PIMP* is the percentage impervious area

IF is the effective impervious area factor; see Table 10.4

PF is the moisture depth parameter (mm) calibrated to be 200 mm

NAPI is the 30 day antecedent precipitation index (*API*30) derived from the net rainfall after the subtraction of running depression storage

Table 10.4 Recommended values of IF

Surface condition	Effective impervious area factor, *IF*
POOR	0.45
FAIR	0.60
GOOD	0.75

The losses on pervious surfaces and non-effective impervious areas are represented by the second term in the equation. *NAPI* is defined as a 30-day *API* with evapotranspiration and initial losses subtracted from rainfall. API_{30} is given by

$$API_{30} = \sum_{n=1,30} P_{-n} C_p^{n-0.5} \tag{10.12}$$

The constant value *C* of the *API* has been made dependent on the soil type to reflect the faster reduction of soil moisture on lighter soils; see Table 10.5.

Table 10.5 Relationship between C and soil type

Soil type	C
1	0.1
2	0.5
3	0.7
4	0.9
5	0.99

10.9.5 The US soil conservation method SCS model

The SCS method is a widely accepted model for predicting storm flows from rural catchments. It can therefore also be used as the permeable component in a semi-urban environment. This is a simple runoff model which allows for variation in runoff depending on catchment wetness. The model is described by two parameters: soil storage depth and catchment wetness index. The basis of the method is the continuity equation

$$P = I_a + F + Q \tag{10.13}$$

where I_a is the initial abstractions (initial losses) (mm)
F is the cumulative actual retention (mm)
Q is the total runoff (mm)
P is the total rainfall (mm)
together with a relationship between rainfall, runoff and actual retention of the form

$$\frac{F}{S} = \frac{Q}{P - I_a} \tag{10.14}$$

where S is the maximum potential retention (storage) (mm)

These equations combine to give the SCS model:

$$Q = \frac{(P - I_a)^2}{P - I_a + S} \tag{10.15}$$

for $P > I_a$. It is normally assumed that S does not include I_a (Hjelmfelt, 1990).

The form of this model can be adapted to be suitable for continuous simulation. The differential form of Eq. (10.15) is given as

$$\frac{dQ}{dt} = \frac{(P - I_a)(P - I_a + 2S)}{(P - I_a + S)^2} \frac{dP}{dt} \tag{10.16}$$

Numerical representation of the derivative terms leads to

$$\frac{q}{p} = \frac{(P - I_a)(P - I_a + 2S)}{(P - I_a + S)^2} \tag{10.17}$$

where q is the runoff over time interval (mm)
p is the rainfall in time interval (mm)

This equation is used incrementally enabling the runoff coefficient (q/p) to change during the event. The two parameters S and I_a are related by

$$I_a = kS \tag{10.18}$$

where $0 < k < 0.2$. The original SCS approach recommended is $k = 0.2$. However other studies suggest k values between 0.05 and 0.1 may be more appropriate.

The storage parameter itself is related to an index known as the runoff curve number (CN) representing the combined influence of soil type, land management practices, vegetation cover, urban development and antecedent moisture conditions on hydrological response. i varies between 0 and 100, 0 giving no runoff and 100 giving 100 per cent runoff. The storage parameter S is related to CN by

$$S = \frac{25400}{CN} - 254 \tag{10.19}$$

The CN value can either be defined globally for the catchment model or can be associated with specific surface types. CN values for different conditions are available from various sources.

10.10 RAINFALL-RUNOFF ROUTING MODELS

The rainfall losses model is a compartmental model that converts the rainfall into and effective rainfall for the runoff. Similarly, the runoff routing model is another compartmental model that transforms the effective rainfall into the runoff. Necessary features of this latter model are that it should provide a time delay and attenuation between the peak effective rainfall (times the catchment area) and the surface runoff. There are two approaches that can be readily adopted to model the surface runoff: (design) unit hydrograph, time-area diagram, linear reservoir or kinematic wave. All of these methods have been introduced in Chapter 5; see Section 5.12.

10.10.1 Design unit hydrograph

The design unit hydrograph of Harmes and Verworn (1984) for the HYSTEM model takes the form shown in Figure 10.26 such that

$$\begin{aligned} Q &= \frac{t}{t_p} Q_p \qquad 0 < t < t_p \\ Q &= Q_p e^{-\frac{t-t_p}{k}} \qquad t \geq t_p \end{aligned} \tag{10.20}$$

where Q is the flow rate (m^3/s)

Q_p is the peak flow rate (m^3/s)

t is the time (s)

t_p is thc time to peak (s)

k is an exponential decay constant (s^{-1})

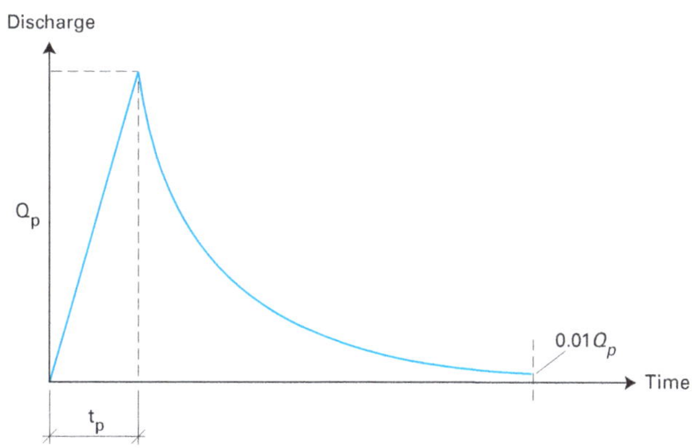

Figure 10.26 HYSTEM design unit hydrograph

The three parameters are related to catchment characteristics by

$$
\begin{aligned}
Q_p &= \frac{0.96A}{t_L} \\
t_p &= 0.49t_L \\
k &= \frac{\frac{A}{Q_p} - \frac{t_p}{2}}{0.99}
\end{aligned}
\tag{10.21}
$$

where t_L is a lag time (hr) calculated from catchment characteristics
A is the catchment area (ha)

The USA-SCS method uses a design unit hydrograph defined by

$$
\begin{aligned}
Q_p &= \frac{484A}{t_p} \\
B &= 1.67t_p \\
t_p &= t_r/2 + t_L \\
t_L &= \frac{L^{0.8}(S+1)^{0.7}}{1900y^{0.5}} \\
S &= 1000/CN - 10
\end{aligned}
\tag{10.22}
$$

Here Q_p is the peak discharge (cfs) for one inch rainfall excess
t_p is the time to peak (hr)
t_L is the lag time (hr)
A is the catchment area (sq.mi)
L is the main stream length (ft)
y is the average basin slope (%)
CN is the curve number

See Figure 10.27.

10.10.2 Time-area method

Typically, in time-area method the runoff is controlled by the initial loss, size of the contributing area and continuous hydrological loss. The shape of the runoff hydrograph is controlled by the concentration time and by the time area (T-A) curve; see Figure 10.28.

If the rainfall is given by the sequence of rainfall intensities $\{P_n\}$ then the flow at the outfall is given by

$$
Q(N\Delta t) = \sum_{j=1}^{n} \Delta A_j P_{N-j} \tag{10.23}
$$

where Δt is the time increment (s)
ΔA_j is the area contributing to the flow at the outfall after j time increments (m^2)

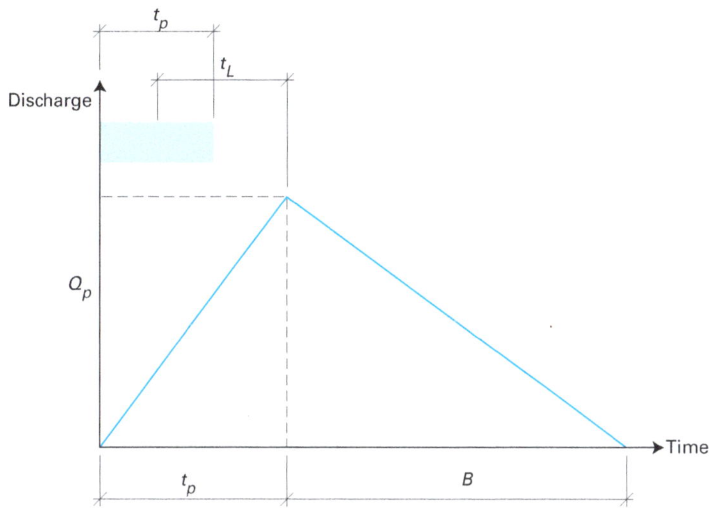

Figure 10.27 Design unit hydrograph for the SCS method

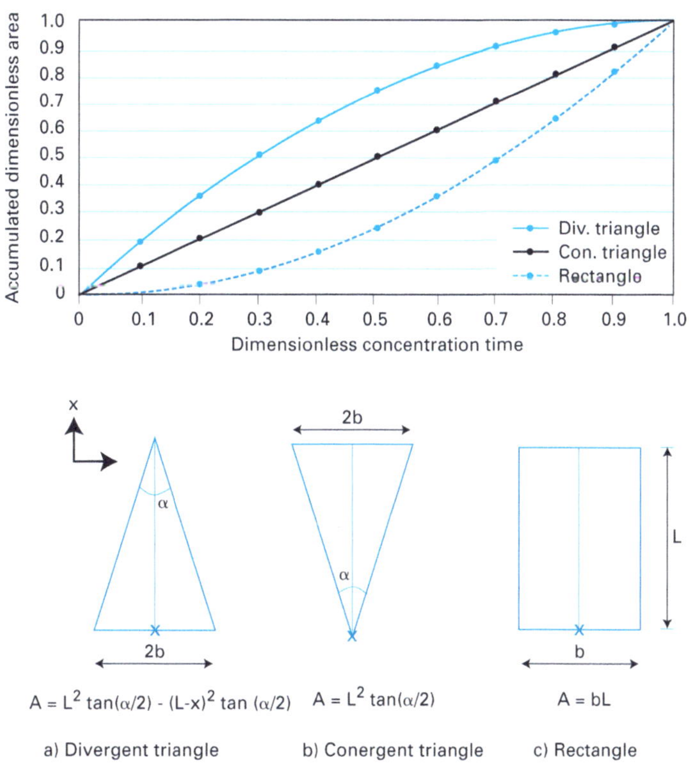

Figure 10.28 Time-area method for different shaped catchments

10.10.3 Kinematic wave (Nonlinear reservoir)

In this method the surface runoff is computed as flow in open channel, taking into account only the gravitational and friction forces and ignoring the contribution from the water surface gradient; see Section 5.2.3. The runoff amount is controlled by the size of the contributing area as well as the hydrological losses. The shape of the runoff hydrograph is controlled by the catchment parameters such as length, slope and roughness of the catchment surface. The surface runoff at time t is calculated as:

$$Q(t) = MBI^{\frac{1}{2}} y_r(t)^{\frac{5}{3}} \tag{10.24}$$

The depth $y_r(t)$ is calculated from the continuity equation:

$$I_{eff} A - Q(t) = \frac{dy_r}{dt} A \tag{10.25}$$

where M is Manning's number,

B is the flow channel width (m), calculated as: $B = A/L$
I is the surface slope,
$y_r(t)$ is the runoff depth at time t (m)
I_{eff} is the effective rainfall (mm/hr)
A is the contributing catchment surface area (m^2)
dt is the time step (s)
dy_r is the change in runoff depth (m)

10.10.4 Runoff routing models (UK)

The overland flow routing model in the UK Wallingford Procedure (Wallingford, 1981) is based on the use of a double quasi-linear reservoir applied in series for each surface type with the same storage-output relationship for each reservoir:

$$S = kq \tag{10.26}$$

where

$$k = C\bar{i}^{-0.39} \tag{10.27}$$

in which

$$\bar{i} = 0.5(1 + i_{10}) \tag{10.28}$$

where i_{10} is the running ten minute average of rainfall intensity.

Combination of the two storage relationships with the continuity equation leads to a second-order ordinary differential equation of the form:

$$k^2 \frac{d^2q}{dt^2} + 2k\frac{dq}{dt} + q = \bar{i} \tag{10.29}$$

C was derived by optimisation for each of the experimental catchments referred to in Section 10.8.3. Then a generic functional form for C was derived from the characteristics of the catchments: An overall relationship relating this value to catchment characteristics was then developed and the final relationship is:

$$C = 0.117S^{-0.13}A^{0.24} \tag{10.30}$$

where S is the slope (m/m)
A is the area (m^2)
The parameters are subject to the following limits:
if ($S < 0.002$) $S = 0.002$
if ($A < 1000$) $A = 1000$
if ($A > 10000$) $A = 10000$

10.10.5 Extension for large sub-catchments

The routing model is developed for small catchments contributing to a pipe. For larger, possibly sewered catchments an alternative is needed. When systems are in surcharge, the value and use of the big catchment model is much reduced.

$$K' = 0.03A^{-0.022}S^{-0.228}L^{0.46} \tag{10.31}$$

$$T' = 4.334A^{-0.009}S^{-0.173}L^{0.46} \tag{10.32}$$

These terms are applied individually to each subcatchment. K' is the multiplier of k and must be greater than 1.0, whereas T' is a time shift applied to the nearest whole time step used in the rainfall file and effectively applies a shift to the rainfall file so that it reads T-T' at each time step. Where figures for length and slope are not provided, these are assumed to be:

$L = 2A^{0.5}$
$S = 1.5$ times the slope of the downstream sewer link.

10.11 DRY WEATHER FLOWS

In case of modelling separate or combined sewer system, in addition to the surface runoff component (which is generated from the hydrological model computation) the dry weather flow component needs to be applied to each modelled subcatchment as well. A typical wastewater production per person per day should be defined together with the diurnal dimensionless profile to represent the 24-hour cycle (Figure 10.29). The wastewater production values can vary significantly due to various reasons (e.g., climate, culture, etc.).

The average daily water consumption for each residential property should be summed for each modelled subcatchment and as such it should be used as a comparison and a reality check of the final per capita discharge rates used in the calibrated model and it may assist in identifying areas with significant exfiltration.

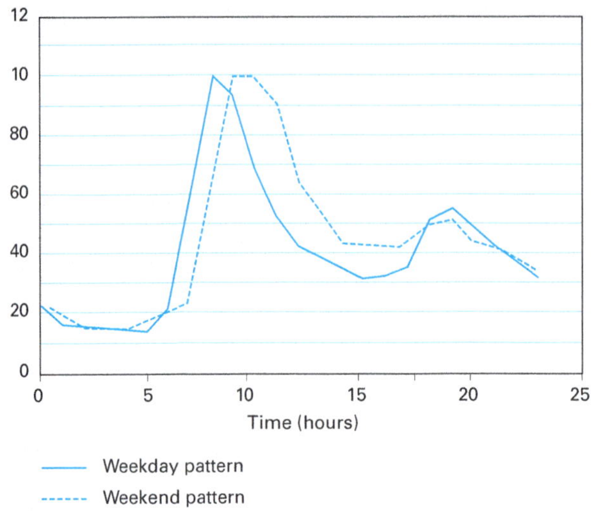

Figure 10.29 Example of a typical weekday and weekend diurnal pattern

The average daily water consumption for each commercial and industrially designated property should be also estimated. However, if available, detailed information regarding the discharge pattern from any major commercial or industrial facilities can be used to assign the most appropriate profile from the standard profile list or in exceptional circumstances create a new dimensionless curve following discussion with the system operator.

10.12 POLLUTANT LOADING AND WASHOFF

In addition to the modelling of flows in collection systems, the pollution-related aspects can be also incorporated. Typically, the following aspects are modelled:

- wash-off and build-up of sediments and attached pollutants (inputs into the model are: size, density and porosity of fine and coarse sediments);
- sediment transport (transport of cohesive and non-cohesive sediments and deposition/erosion processes);
- advection dispersion (transport of dissolved and suspended material; inputs into the model are: dispersion coefficients, initial conditions, boundary conditions and decay);
- water quality (processes in the pipe network system considering chemical compounds and biological constituents; examples are: BOD, COD, H_2S, etc.).

The modelling of water quality in a sewerage system assumes various sources of pollution: domestic wastewater, surface washoff, industrial inflows, and erosion of sewer sediments. Both dissolved and particulate pollutants are transported through a sewerage system. The dissolved pollutants normally include ammoniacal nitrogen NH4+ (total ammonia) and oxygen demanding substances represented by Biological Oxygen Demand (BOD). These are normally treated as being conservative, that is, they do not degrade with time. Sediments with different characteristics (size and density) are also transported through the system. These are classified into sediment fractions. Typically, a minimum of two sediment fractions are used, one coarse high-density material (grit) and one fine (organics). In addition to being dissolved, pollutants such as BOD and COD are also assumed to be permanently attached to sediments. (Ammonia is normally only present in the dissolved phase).

Each source of sediment is categorised by the same sediment fractions (i.e. size and density). However the pollutants associated with a given sediment fraction may be different for each source. As the sediments are routed through the system those from different sources are mixed together, and therefore the concentrations of associated pollutants change during a simulation. The sediments from each source are assumed to have a characteristic amount of pollutants attached to them. This is assumed to be fixed and the ratio is referred to as the potency factor of the pollutant, expressed as kg of pollutant per kg of sediment.

All surface sediments (sometimes described as suspended solids) are defined by one value of size (d_{50}) and one value of density. These values apply to sediments derived from any source, including household and industrial inflows, and sediments eroded from and around pipes.

The build up of sediment on the catchment surface is a function of time and an accumulation rate coefficient. Sediment 'build up' on the catchment surface during periods of dry weather starts from an initial quantity present at the end of the previous event. The basic hypothesis is one of a linear accumulation of pollution on the ground, which is related to the land use of that catchment. There is a limiting value for surface mass (B_s/K_s) as the number of days tends to infinity.

$$\frac{dM_r}{dt} = B_r - K_r M_r \tag{10.33}$$

where M_r is the mass of deposit per surface unit (kg/ha)
B_r is the build up factor (kg/ha.day)
K_r is the decay factor (/day)

Therefore the mass of sediment built up at the start of an event is:

$$M_0 = M_i e^{K_r N \Delta t} + \frac{B_r}{K_r}(1 - e^{K_r N \Delta t}) \tag{10.34}$$

where M_0 is the mass prior to the event (kg)
M_i is the initial mass of sediment (kg)
N is the duration of dry weather (days)

The rate of sediment washoff is dependent on an erosion coefficient and the quantity of available sediment. As the event proceeds and sediment is removed from the catchment, the quantity of available sediment, and hence the rate of sediment washoff, reduces.

The washoff model can be linked, for example, to a single linear reservoir coefficient. The mass of sediment or dissolved pollutant mobilised is described by:

$$M_s(t) = K.f_s(t) \tag{10.35}$$

where $M_s(t)$ is the mass of pollutant mobilised (kg/ha)
$f_s(t)$ is the pollutant flow (kg/s)
K is the linear reservoir coefficient.

The mass of sediment eroded from the bed is a function of the rainfall intensity based on a 5 minute moving average:

$$\frac{dM_r}{dt} = -K_a.M_r(t) \tag{10.36}$$

where

$$K_a(t) = c_1 i_5(t)^{c_2} - c_3 i_5(t) \tag{10.37}$$

Here K_a is rainfall coefficient

c_1, c_2, c_3 are coefficients

$i_5(t)$ is the effective 5 minute rainfall

The mass of sediment is carried from the catchment in suspension

$$\frac{dM_s}{dt} = K_a.M_r(t) - f_s(t) \tag{10.38}$$

The outflow of sediment mass for each time increment is:

$$f_s^{n+1} = f_s^n e^{-\Delta t/K} + (e^{-\Delta t/K_a} - e^{-\Delta t/K})\frac{K_a^2}{K_a - K} M_r^n \tag{10.39}$$

where $f_s(t)$ is the TSS flow per unit of active surface (kg/s/ha)

This leads to an effective flow of sediment from the catchment of

$$F_s(t) = C.A.f_s(t) \tag{10.40}$$

where C is the proportion of the catchment that is impermeable

A is the catchment area (ha)

F_s is the sediment flow (kg/s)

10.12.1 Attached pollutants

The amount of pollutants attached to sediments is determined by means of potency factors. A potency factor describes the strength of a pollutant attached to a particular sediment, and is defined as the ratio of the volume of pollutant to the volume of sediment. The potency factor does not change during the rainfall event. However, when sediments with different potency factors mix, the overall potency alters.

The relationship between total suspended solids (TSS) and attached pollutants is defined as:

$$P_{k,n} = a_1(i_5 - a_2)^{-b_1} + a_3 \tag{10.41}$$

where $P_{k,n}$ is the potency factor

i_5 is the 5-minute rainfall intensity (mm/hr)

a_1, a_2, a_3 and b_1 are coefficients

Thus the calculation of washoff for the particular attached pollutant is:

$$F_{s,n}(t) = P_{k,n}(t).CA.f_s(t) \tag{10.42}$$

where $F_{s,n}(t)$ is the pollutant flow (kg/s)

10.12.2 Dissolved pollutants

Dissolved pollutants build up and are removed from each subcatchment via the 'gully pot' model. With regard to build up there are assumed to be no dissolved pollutants in the gully pot at the end of the previous event (this does not necessarily mean that all dissolved pollutants are removed during a rainfall event). Starting from zero, dissolved pollutants build-up in a linear manner according the equation

$$P = \frac{(C + MN)V_g}{1000} \tag{10.43}$$

where C, M are parameters
N is the number of days since previous event
V_g is the volume of gully pot calculated from depth of gully pot and catchment area (m^3)

The rate of washoff of dissolved pollutants is a function of the time for which it has been raining and the quantity of available pollutant.

In calibration of the washoff model, the only term found to be significant is the one for erosion caused by rainfall impact.

A major input of the UK-UPM Research Programme has been the assimilation of data describing the nature and behaviour of sediments found in combined sewers and storage tanks. Crabtree (2007) presented a 5 category classification of the types of combined sewer sediments; see Table 10.6.

Table 10.6 Combined sewer sediments

TYPE A	Coarse, predominantly granular mineral material found in the invert of pipes.
TYPE B	Similar in composition to Type A but concreted by the addition of mineral cements, fats and tars.
TYPE C	Mobile, fine grained, largely organic sludges found in quiescent zones.
TYPE D	Organic pipe wall slimes and biofilms.
TYPE E	Fine grained organic and mineral deposits found in storage tanks.

As explained by Butler and Davies (2000) the pollution processes take place in and between the wastewater, the atmosphere, the biofilm on the sewer wall and the sediment both in suspension and on the bed. The biodegradation of the organic material is caused by the micro-organisms in the biofilm and the wastewater, and is dependent on the presence of dissolved oxygen (DO) to facilitate these aerobic processes. Often water quality models for sewer networks assume that the pollutants are conservative, but degradation modelled by simple decay laws are appropriate in systems with long retention times.

10.13 MODELLING FLOW IN NETWORKS OF CHANNELS AND/OR PIPES

Modelling flow in networks of pipes and channels is described in Chapter 5. What is important to recall is that generally the full 1D Saint Venant are solved, rather than the non-inertial or kinematic wave equation approximations. Note that the latter approximation assumes no downstream boundary condition and therefore cannot take backwater effects into account. Such effects are very important in most sewerage network simulations. Care has to be taken with the dry bed condition, the hysteresis in the flow at the top of a pipe where there is a transition between the free surface and pressurised flow, the inclusion of the imaginary Preissmann slot to preserve the same numerical algorithm for pressurised flow as for free surface flow, the specification of discrete head-losses at the upstream and downstream ends of each pipe, and the preservation of the sub-critical flow algorithm to solve for super-critical flow. In addition, a range of ancillary structures should be anticipated as possibly being needed for incorporation in the model, such as various weirs (such as for combined sewer overflows), gates, sluices, different pumps, outfall flaps and orifices. Allowance should be made for the emergence of sewage from a manhole. (All surface and wastewater inflows and outflows are assumed to be at the upstream manhole of a pipe rather than being distributed irregularly along its length). The emerging effluent from an underground pipe network or and above ground channel network can interact with an overland 2D flow model.

Traditionally, collection systems have been modelled using the 1D approach which was limited to the modelling of flows within channels and/or pipes. A significant improvement has been the dual drainage concept, where above ground flows over the urban surface are treated as a network of open channels and ponds (major system) connected to the channel/pipe system (minor system), commonly referred to as the 1D/1D approach. More recently, coupled 1D/2D models have been developed in which a channel/pipe network flow model is fully coupled with an above-ground model that treats the urban surface as a two-dimensional flow domain. In both approaches, complex interactions take place through surface/sub-surface links, which are explicitly taken into account using appropriate equations. A majority of processes within the collection systems can be adequately represented by the use of 1D, 1D/1D or 1D/2D modelling approaches. However, care must be taken, particularly in the case of overflow structures, where three dimensional effects of flow can significantly reduce or increase flows over weirs compared to the values predicted by standard weir equations. In addition, the use of low side weirs, where supercritical flows and hydraulic jumps can take place, is difficult to represent explicitly in standard 1D models. The ever-increasing power of computers is now allowing a 3D fluid dynamic analysis of such structures, and in time these may be integrated into 1D models where such locations require this kind of analysis.

Water quality is simulated using the advection-diffusion equation for the different concentrations of the primary pollutants with particular source terms identified for the pollutant and associated processes. The advection speed is the mean velocity calculated from the hydraulic model. Note that the full water quality model is dependent on having a safe and reliable hydraulic model. Again, see Chapter 5 for details of the techniques that can be used.

The modelling of the interaction of infiltration to a pipe network from groundwater and infiltration from the surface rainfall and flood flows to groundwater can be done using a two layer groundwater flow model, with the upper layer representing the unsaturated zone and the lower layer the saturated zone.

10.14 1D MODELLING APPROACH

As discussed earlier, the hydraulic models used for the modelling of collection systems have a sound physical basis in the shallow-water equations and their reliability increases with the amount of measurement data that can be used for their instantiation and calibration (or validation; see Chapter 5, and Cunge, 2003). For these and other reasons, one or another form of hydraulic modelling, such as the one-dimensional (1D) approach has been standard industry practice for some time. Consequently there are still many researchers and practitioners who favour the use of 1D modelling approach over other approaches. A reason for this is that it is relatively easy to set up and calibrate these models and to defend their results.

10.15 SIMPLIFICATION OF 1D MODELS

The 1D pipe flow (or open channel) models are either developed on a skeletal (or simplified) level or on a more detailed level depending on the number of pipes and nodes (or branches and cross-sections). The choice for the appropriate extent of the model (and the level of its complexity) is normally driven by the modelling objectives, the availability of data and the financial capacity of the modelling project.

The purpose for the development of a skeletal (or simplified, strategic planning) model is to gain the basic understanding of system functioning and to undertake simple planning/management analysis. This can be also regarded as a cost-effective way of approaching the modelling task. The detailed model is usually developed to understand the detailed relationship between problems in the system and the hydraulics of the drainage infrastructure. Therefore, due to the complexity and capacity to describe various characteristics of an urban drainage system and its functioning, detailed models are also used in operational and system maintenance activities.

Very often, studies using a collection system model would begin by commissioning a skeletal (or simplified) model and then continue with a more detailed model in the critical (or problem) areas only. Typically, as a minimum, pipes (or branches of the open channel network) of a simplified network are selected on the basis of their dimension (i.e., trunk lines), proximity to the known problem area, and the location of key ancillary structures (e.g., overflows, pumping stations, etc.). This process should take into consideration also the location of flow/level gauges.

The simplification of a model requires that the head losses and storage units from manholes/pipes/channels which are not included explicitly in the simplified model network are also incorporated in some manner. This is usually done in such a way that additional storage (due to the omitted pipes and ancillary structures) is added to the modelled manhole downstream, and additional head losses are added to the modelled manhole upstream. In doing so it is important to take into consideration any other particular local circumstances such as low terrain elevations and significant slope changes.

Computing the additional storage can be done in a GIS package by cross-referencing the ground level of the highest upstream node in the simplified network and the invert levels of all nodes upstream of that node (i.e., beyond the simplified network). Their values need to be subtracted in order to compute the additional storage volume; see Figure 10.30. Thereafter, the additional storage volume needs to be calculated for every pipe and node upstream.

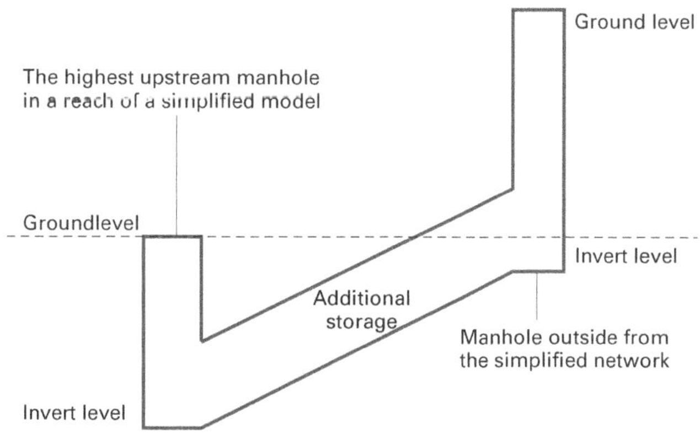

Figure 10.30 Schematisation of the additional storage area/volume

Furthermore, there could also be intermediate nodes which can be omitted from the simplified model network. Their storage volumes must also be summed and added to the next model node downstream. For converging junctions the storage should be summed along each incoming branch and added to the junction node. An example of a simplified network layout compared with the total pipe network is given in Figure 10.31.

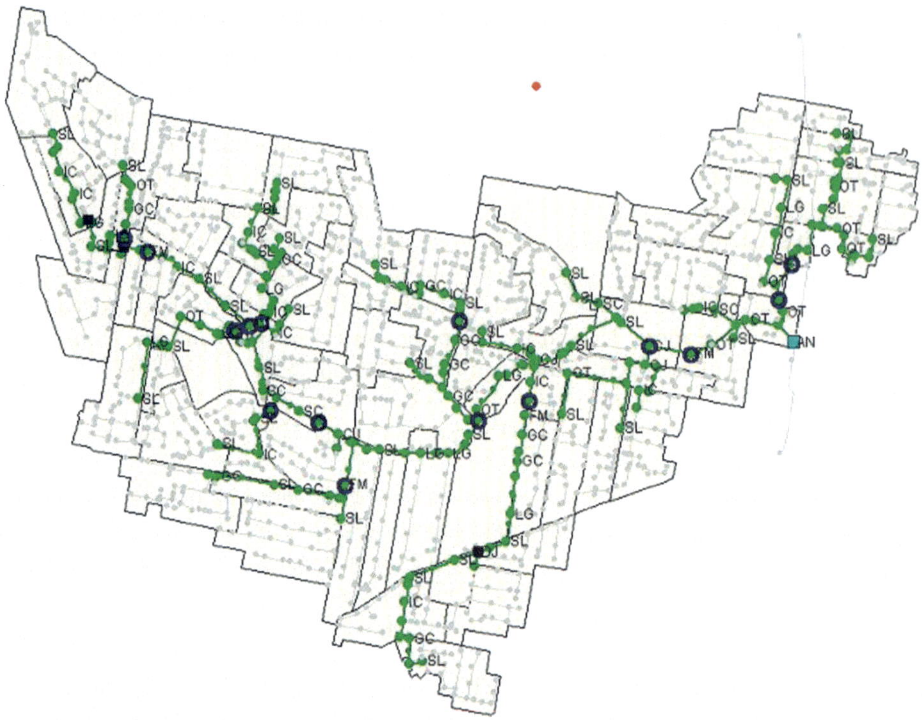

Figure 10.31 An example of a simplified model layout against the total drainage network

10.16 1D/1D MODELLING APPROACH

The *dual drainage* concept, where the urban surface is treated as a network of open channels and ponds (major system) connected to the pipe (or channel) network system (minor system), is commonly referred to as the 1D/1D approach.

10.17 1D/2D MODELLING APPROACH

The appropriate level of modelling for the assessment or particular issues depends crucially on the nature of the physical situation and on the availability of data. Urban topography, drainage network layouts and even detailed geometry of the networks can be readily surveyed using current technology. Where flood flows are confined to well-defined conduits, a robust 1D model can usually be well instantiated and its results considered safe for decision-making. However, the flows generated along urban floodplains are normally highly complex because the morphology of the urban surface is eminently artificial with highly irregular geometries, and the flows may run contrary to natural flow paths. Modelling of flows in such complex geometrical situations is difficult. Small geometric 'discontinuities' such as road or pavement curbs can play a significant role in diverting the shallow flows that are generated along roads, through fences and around buildings. Head losses due to flows over or around such structures are particularly difficult to represent and accommodate. Such issues necessitate the coupling of simulations using 1D and 2D modelling systems; see, for example, Hsu *et al.*, 2000; Mark *et al.*, 2004; Djordjevic *et al.*, 2005; Chen *et al.*, 2005; Vojinović *et al.*, 2006 and Vojinović and Tutulic, 2009.

The simulation process in the case of coupled 1D/2D modelling is rather complex and it can become computationally very expensive (i.e., time consuming). It is based on complex numerical solution schemes for computation of water levels, discharges and velocities. The surface model (i.e., 2D model) simulates vertically-integrated two-dimensional unsteady flow given the relevant boundary conditions (e.g., resistance coefficients, etc.) and bathymetry (as provided by a digital terrain model of the catchment area). In this regard, the use of parallel computing seems to be one of the promising ways of reducing the computational time. Furthermore, an approach based on the adjusted conveyance and storage characteristics to account for fine scale flow obstructions (i.e., change in storage and lateral fluxes) is also giving promising results.

The interactions between the below-ground (or minor) and above-ground (or major) system are determined according to the type of link. For example, discharges generated by pumping stations, weirs or orifices are regarded as the lateral inflow to the 2D model. Also, if the pipe flow exceeds the ground level or bank levels (for open channels) then the discharge is computed by the weir (or orifice) discharge equation and it is considered as an inflow to the 2D model (that is, the source term F_s); see also Chapter 5.

10.18 DETERMINING PEAK FLOWS IN A DENDRITIC NETWORK

The prescriptive design of stormwater pipes or channels in a dendritic network is normally carried out by determining a peak runoff for rainfall of a given return period and duration. The most convenient way of doing this is to use the Rational Method. This uses the simple formula:

$$Q = 0.00278CAi \tag{10.44}$$

where Q is the peak discharge (m^3/s)
A is the total area (ha)
C is a runoff coefficient
i is the rainfall intensity corresponding to a given return period and duration (mm/hr)

Normally the return period is prescribed, and the duration is the same as the time of concentration. This is the time taken for water particles to travel from the furthest point of the catchment to the location in the pipe or channel network where the peak discharge is being calculated. Typically, the time of concentration is equal to a 'time of entry' for water to enter the pipe or channel over the ground surface plus the longest time of travel along the network to the location. The time of entry is typically 2 minutes for small contributing areas, but can be longer for larger areas, up to 5 minutes. The time of travel along the pipes or channels is determined by the length divided by the pipe full velocity or the flow velocity at some given depth in the channel.

The runoff coefficient can be determined as described above in Section 10.8.1. Note that for the UK a (percentage) runoff coefficient which takes into account a design value of the antecedent wetness of the catchment is used.

When pipes or channels are being sized it is important to ensure that each pipe or channels upstream of a given pipe or channel has been sized so that the time of concentration can be properly defined. The most efficient way of doing this is through ordering the sizing of the pipes or channels in 'end order'. This requires starting with the pipe furthest away (in terms of number of pipes) from the outfall. The ordering continues with the pipes downstream of the current pipe, unless a new branch of pipes joins the current pipe at its downstream end. In this case the next pipe is the pipe on the new branch furthest away from the junction. This procedure is followed strictly until the outfall of the system is reached; see Figure 10.32.

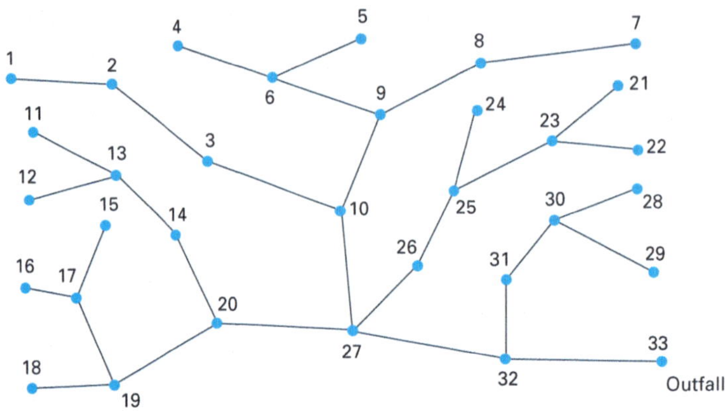

Figure 10.32 Pipes in a dendritic network with the nodes identified in end order

10.19 USING EVENT-BASED AND TIME SERIES RAINFALL WITH PIPE NETWORKS

There are two fundamental types of model simulation which can be applied to assess the performance of an urban drainage system, namely, event-based and time series simulation.

An event-based simulation is used when a prescribed frequency-based standard has to be met. For example, a 1 in 100 year event might be defined as the acceptable house floor level for protection against flooding. These events are typically of short duration and can be analysed either on the basis of a frequency analysis of measured data over a long period, or by applying rainfall patterns developed after an analysis of long term rainfall records. The major assumption in the event-based approach is normally that the effect (i.e. the flood level) is assumed to have the same frequency of occurrence as the cause (the rainfall). This approach is applicable when the frequency of the effect is closely linked to the frequency of the cause, as is the case in urban flooding.

The time-series approach is normally used for the analysing the combined sewer system. The approach follows long-term (typically, 10-year plus 1 year for initial conditions) model simulations using historical or synthetic rainfall and evaporation time series as input in addition to the wastewater dry weather flow. The long-term time series are simulated using the rainfall-run-off model (calibrated or built using regionalised parameters) and the overflow performance is then statistically evaluated. In the analysis of the combined sewer system performance, the long-term statistical analysis is considered a vast improvement on the traditional steady-state approach. It provides a better understanding of the system and leads to more appropriate and cost-effective solutions.

The use of design rainfall events in modelling is appreciated because of the identification of the return period of flow with the return period of the rainfall depth, and the economy of running what can be a sophisticated and time consuming simulation model. Strictly however, there is no guarantee that the return period of the flow is the same as the return period of the rainfall. What is more, it can be difficult to be confident that the return period of the rainfall is well defined for return periods less than a year, especially in areas with low annual rainfall. For these reasons, modellers tend to make increasing use of time series rainfall when they can accommodate the lengthy computer run times involved.

Design rainfall events are adopted for prescriptive design of pipes and channels in dendritic networks. In particular, the Wallingford Procedure advocates the use of rainfall events (rainfall depth, duration, profile and antecedent wetness) for sizing pipes. The concept is that a pipe network has all of its pipes sized for each run of a sequence of rainfall events with the given design return period, 50 percentile profile, a design Urban Catchment Wetness Index (UCWI) of 120, and with increasing duration (for example, 5 min, 10 min, 20 min 30 min 60 min 120 min etc). Every pipe is sized to convey the peak discharge from the pipes immediately above unless the pipe is already large enough. In this way the pipes can be designed consistently.

10.20 MODELLING TREATMENT WORKS

A fuller description of the treatment works simulation model is given in Chapter 11. The important point made here is that the outfall of the collection network discharges sewage to the treatment works via a pump, which is often an Archimedes screw pump. Upstream of the pump will be an overflow to divert excess flows to the adjacent receiving waters. Combined with the overflow is some form of storage, which is designed to capture the first foul flush or its equivalent, and to delay as much polluted wastewater for treatment over a longer period than the duration of the storm.

10.21 MODELLING RECEIVING WATERS

Rivers, watercourses coastal influences in urban areas can be modelled with 1D, 2D or even 3D models. Rivers and watercourses are typically modelled with 1D models and floodplains are represented by either 1D or 2D models. The open channel geometry is represented as a series of irregularly spaced cross sections perpendicular to the 1D flow. Boundary conditions typically contain an inflow hydrograph at the upstream cross section and a stage hydrograph at the downstream end. If necessary, rivers and watercourses can be modelled in a common framework together with the pipe network models. This is usually done in cases where there is a potential for greater influence of rivers/watercourse on hydraulic conditions at stormwater pipe network outlets. Also, the same approach can be applied in cases where the impacts of sewer networks are evaluated.

In some cases, the elevation of the groundwater table during wet weather can cause direct (above ground surcharge) or indirect (infiltration and overload of the pipe network system) flooding in urban areas and as such it might be of importance for modelling purposes. Depending on the complexity of issues associated with groundwater influences different modelling approaches may be chosen (e.g., conceptual, detailed, etc.).

Coastal models (2D or even 3D) can also be used in directly with pipe (or open channel) network models to assess drainage and flooding problems. However, in many cases, the use of historical tidal variation water level data as a downstream boundary condition for pipe (or open channel) network models is adequate. Certainly, if there is a risk of coastal inundation and more substantial influence on a drainage pipe (or open channel) network then the use of a fully-fledged coastal and/or estuarial model might be required.

The discharges diverted by the overflows and the effluent from the treatment works enter the corresponding receiving waters, whether a lake, urban stream or river, or coastal waters. The pollutants in the discharges will adversely affect the receiving waters, and the associated damage should strictly be taken into account when considering the performance of the collection system. Modelling is used to quantify the impact of the pollution on the receiving waters, and thereby to deduce the economic consequences of the damage incurred. Again, a fuller description is given in Chapter 12, where the impact of sewage from CSOs and from treatment works on the stream or coastal waters is discussed and followed up.

10.22 INSTANTIATING AN URBAN DRAINAGE SIMULATION MODEL

The previous sections have described the concepts and assumptions that are made in isolating the essential features of urban water collection systems for modelling purposes. These features are included generically in a number of simulation modelling software products such that the application of the software to a particular collection system requires the addition of specific data describing the geometry of the model space (sub-catchments, pipes, channels, manholes, ancillary structures, urban surface topography, receiving water channels or 2D coastal bed topography, etc) and the appropriate boundary inputs (rainfall, wastewater inflows, groundwater levels, river discharges and water levels, ancillary structure settings, astronomic tides, etc). Considerable expertise is needed to instantiate such a model, that is, to construct and confirm the integrity of the structural data for the model space, and to calibrate and verify the model parameters. The model is a significant part of the virtual world which parallels particular aspects of the real world. As such it needs to be 'fit for purpose', that is, it should generate information that is safe and reliable to use in formulating interventions in the real world, whether directly affecting the structural aspects of the collection network or the non-structural consequences for the community benefiting from the network. For this reason, experience gained in instantiating models for collection networks is accumulated in engineering procedures such as the Wallingford Procedure (Wallingford, 1981), the WaPUG Code of Practice for the Hydraulic Modelling of Sewers (WaPUG, 2002), the UK Sewerage Rehabilitation Manual (WRc, 1994) and the UK Urban Pollution Management Procedure (FWR, 1998); see Chapter 6. In addition, many water companies have produced their own in-house manuals on model building requirements, consistent with internal QA and standards. HarmoniQuA is an EU project that has focussed on a general procedure for computational model instantiation. In that there are thousands of urban drainage networks world-wide such procedures provide quality assurance guidelines for model instantiation and application. These procedures are part of the procedural world in which professionals operate on behalf of society. The key elements of an instantiation procedure for a collection model are given in outline in Chapter 6.

10.23 MODEL APPLICATION

A number of applications of simulation modelling in urban drainage can be identified:

- Conduit design
- Hydraulic analysis (including flooding)
- Infiltration and inflow analysis.
- CSO analysis
- Storage facility analysis
- Real time control and warning
- Rehabilitation of sewerage systems
- Urban pollution management

For the above applications, either event-based or time series simulations are used. A general procedure for applying hydraulic and water quality models within the context of these applications is given in Figure 10.33. Each application is now considered in turn.

10.23.1 Design of systems

The sizing of pipes in reticulated sewerage systems has depended largely on the use of the Rational Method with local IDF curves. The basic notion is that the pipe layout as a dendritic collection network is given, with the pipes largely following the road network for ease of construction and maintenance. The pipes are laid such that their gradients induce flow under gravity down the main (collecting) branches of the pipe network. A proper engineering practice is that the sizes of the pipes increase downstream. The outfall of the network discharges to a treatment works or directly to receiving waters. During the sizing of the pipes adjustments may be made to the gradients of the pipes so that self cleansing velocities for reasonable flows are maintained to limit sediment deposition.

The use of the Rational Method for sizing the pipes is described in Section 10.12 above. The pipes are sized such that they convey the Rational design discharges flowing just full (or at a certain proportion of pipe full).

For details of the design process see Butler and Davies, 2000. See also the Wallingford Procedure (Wallingford, 1981).

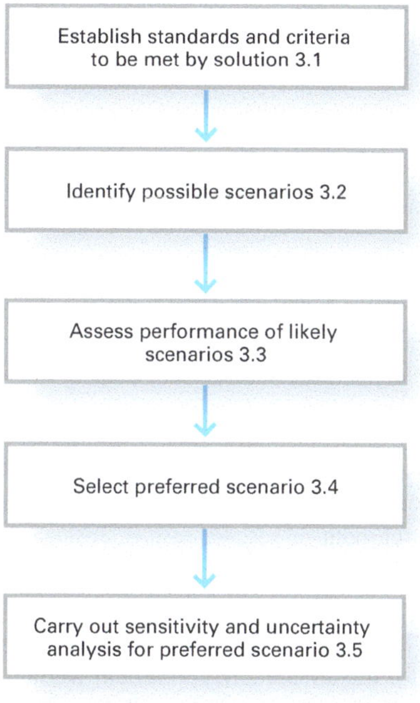

Figure 10.33 Procedure for application of models

The shortcoming of this approach to designing the pipes is that it says nothing about the frequency of surface flooding, which the pipe network is designed to limit. However, because the pipes are laid at a depth such that the cover over the pipe is typically more than a metre, then the frequency of surface flooding might be considerably less than the design frequency for the pipes. The frequency of significant surface flooding at nodes throughout the network defines the level of service provided by the network in the drainage catchment, and this will vary from place to place depending on the topography of the catchment and the pipe layout. Obviously, the actual levels of service in the catchment are an important factor in justifying the design, but to obtain these a simulation of the designed system for events rarer than the design frequency is needed. For this reason the Wallingford Procedure for the design of urban storm drainage was developed in the late 1970s as a computer-based method which gives the options of using the Rational method or a hydrograph method for the design of the pipe network followed by the assessment of the performance of the designed network using a simulation method; see Wallingford, 1981. Where the simulation model gives depths of flooding at nodes in the network, there is the potential to estimate the damage costs as a function of the flooding depth. Given a designed pipe network and the cumulative damage costs associated with a set of rainfall events rarer than the design event, it is possible to trade off the cumulative damage cost against the construction cost of the pipe network; see Figure 10.34. That is, the pipe network can be redesigned for a larger design rainfall, and the corresponding cumulative damage cost, which should be lower than the dame cost for the first design, can again be compared with the new construction cost. The damage costs should decrease as the design rainfall for the sizing of the pipe network is increased, while the construction cost of the pipe network increases. This gives a standard problem in economics that leads to optimal sizes for the pipes.

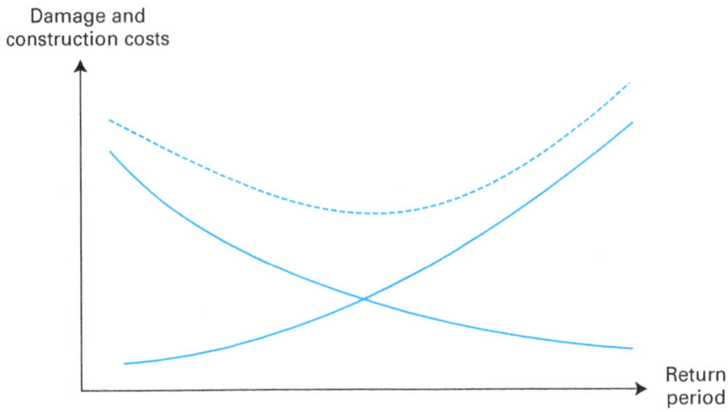

Figure 10.34 Deriving the minimum cost

The Wallingford Procedure includes a number of radical innovations in modelling rainfall-runoff:

- A (UK) rainfall depth for a prescribed duration and a design frequency (or return period) specifically for the geographic location of the catchment
- A (UK) specific symmetric rainfall profile (a summer 50 percentile profile)
- The (UK) volume of runoff from a catchment given by a regression equation based on the percentage paved area, the soil type and the antecedent catchment wetness
- A (UK) calibrated design value for the antecedent catchment wetness
- A rate of runoff calculated through one or two calibrated linear reservoirs (in series)
- A sequence of design rainfall events with increasing duration
- A simulation model consistent with the hydrograph design method
- (Actually, the simulation model is limited and was subsequently replaced by a more complete solution of the full Saint Venant equations)

10.23.2 Hydraulic analysis

Once simulation modelling of sewerage networks became feasible, its potential to analyse the performance of a system led to an explosion in methods of hydraulic analysis. Firstly, simulation modelling is used to assess the performance of a designed system for events rarer than the design event. This leads to the determination of the frequency of surcharging in each pipe and surface flooding at all nodes in the network, and therefore the level of service (in terms of flooding) throughout the catchment. For a combined sewerage system the frequency of CSO operation can be determined by running the model for, say, an annual time series of rainfall, or a series of rainfall events with higher frequency than the design event for the CSO. By running the model for the complete range of higher and lower frequency events it is possible to study the frequency of filling the tanks, operating the pumps at maximum and so on. Hydraulic analysis therefore, is the assessment of the performance of sewerage or drainage networks in terms of the service provided. The aim is to formulate upgrading scenarios.

When carrying out a hydraulic analysis of a drainage network, the main problems that arise include such items as:

- Shortage of network data
- Unknown dry weather flows

- Unknown infiltration
- Uncertainty about pipe blockages, sediment deposits etc
- Shortage of temporal and spatial rainfall data
- Shortage of measured flows in the network
- Unknown headlosses at structures

Figure 10.35 shows the procedure that can be followed when applying a model to hydraulic analysis.

Normally, for a green field site a discharge limit is provided by the Authority in the form of l/s/ha. Typically this is based in the UK on an analysis of the site in its green-field state for a return period anywhere between 1 and 100 years according to the Flood Studies Report (NERC, 1975). Storage rates are usually prescribed in the range of 1 to 10 l/s/ha, though higher rates up to 30 l/s/ha have been known. A Flood Studies analysis may also be required to demonstrate what the 'green field' site discharge is to determine the current limit of peak flow. The return period specified ranges widely and occasionally a 2 year 'green field' runoff will be applied to a 100 year development state. This, in effect, reduces the risk of flooding in the receiving stream by a significant degree from the pre-development situation.

Criteria applied by water companies tend to be less onerous with specifications requiring return periods in the region of 20 or 30 years. Sometimes a developer is asked to provide a model of both his network and the receiving sewer system (often a large system) to prove that his proposals can be served by the system or determine the improvements needed to enable the additional are to be connected.

Task 3.1 Establish the standards and criteria to be met by the solution

Description	Define the standards and criteria that the solution should satisfy
Input	Project objectives and scope, standards, policy statements, levels of service
Output	Audit documentation, review report
Sub-tasks	Check all levels of service
Tools	DSS

So far as flooding standards are concerned, the UK Department of the Environment stated in 1989 that 'the sewerage system will be considered unsatisfactory if a flooding incident due to hydraulic overloading of public sewers occurs at a statistical frequency of more than ten years'. Sewers for Adoption (WRc, 2006) identifies that for sites with ground slopes of greater than 1%, the design return period should be one year. For sites with ground slopes less than 1%, the design period should be two years, and for sites where the consequences of flooding are severe the recommended return period is five years.

CIRIA (1992) recommends that there should be no surface flooding of car parks and local access roads in a five year return period, of adopted roads in a ten year return period; residential buildings should be protected from flooding up to the design return period of fifty years.

General guidelines on good practice exist, such as CIRIA, 2006a and CIRIA, 2006b. Other useful reports include

- Infiltration; (White *et al.*, 1997)
- Use of detention tanks for flow attenuation; (WRc, 1995)
- Code of practice for building drainage; (BSI, 1985)
- Drain and sewer systems outside buildings. Performance requirements; (BSI, 2008)
- Sewers for Adoption, 6th edition; (WRc 2004; WRc, 2006)

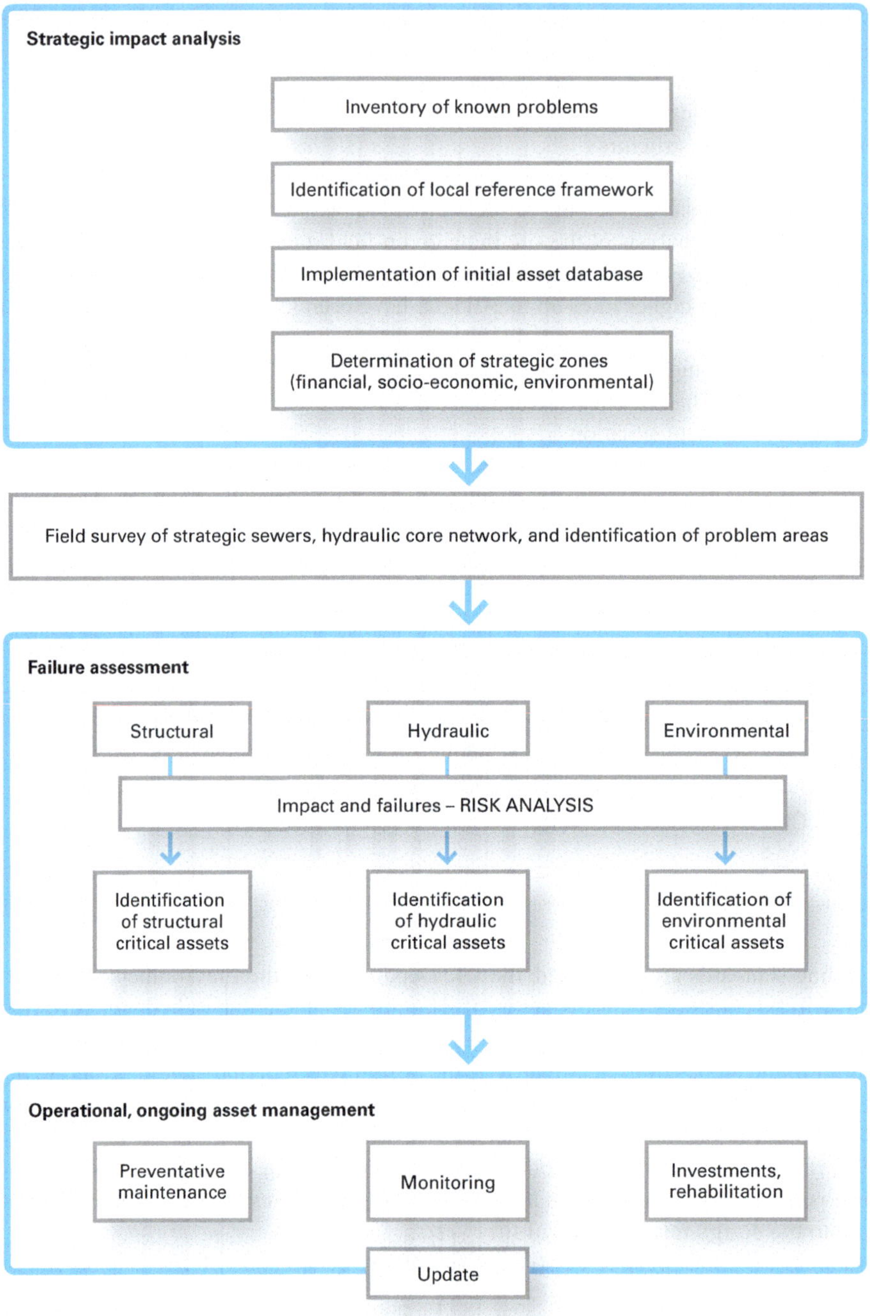

Figure 10.35 Hydroplan – defining optimal cost-efficient investment and operation and maintenance scenario (Vojinović *et al.*, 2005)

More recently, the EC Urban Waste Water Treatment Directive (CEC, 1991) has placed the responsibility of member states to decide on measures to limit the pollution discharged from sewer systems and indicated that measures could be based on three techniques:

- A dilution ratio between the quality of the sewer flow and that in the receiving stream.
- The capacity of the sewer system in relation to the dry weather flow.
- A specified number of overflows per year.

Resource questions:

- Have the objectives for the study been clearly identified?
- Have the standards for the performance of the network been clearly enunciated?
- Has the regulator (if required) been briefed on the objectives?

Task 3.2 Identify possible scenarios

Description	Identify structural and non-structural approaches to solving the problems and addressing the issues raised by the project.
Input	Project objectives and scope, standards, levels of service and criteria
Output	Audit documentation, possible scenarios
Sub-tasks	Consult stakeholders and regulator
Tools	DSS

Ingenuity provoked, say, by brainstorming should lead to a range of possible scenarios being tested under different conditions. Case studies elsewhere should be reviewed to generate different structural and non-structural options for inclusion in the scenarios. Preferably at least three scenarios should be tested, two representing more extreme conditions around a likely scenario. The modellers should be prepared to be flexible in adjusting the scenarios to meet the standards, levels of service or criteria agreed beforehand.

Resource questions:

- Do the scenarios cover the main aspects of the project objectives?
- Is there a realistic estimate of a likely scenario?

Task 3.3 Assess the performance of likely scenarios

Description	Assess the performance of the likely scenarios using the instantiated model
Input	Project objectives and scope, instantiated model, set of scenarios for analysis
Output	Audit documentation, report on the performance of each scenario
Sub-tasks	Modify structural and non-structural measures to improve scenarios, use optimisation to minimise cost
Tools	DSS, optimizer, instantiated model

Each likely scenario is tested for a range of conditions arising from the input data. The input data should include the data used when defining the base performance of the original/existing system. This enables an objective assessment of the performance of a particular scenario and eases the comparison made during the following task. Where the performance is demonstrably poor and can be improved, the modeller should seek a better design of the proposed structural or non-structural solution that is characteristic of the scenario. It is possible to use optimisation if the instantiated model runs sufficiently quickly on the relevant computer. Otherwise, it may be possible to replicate the essential features of the physically based instantiated model using a data driven model (for example, an Artificial Neural Network) and to benefit from the significant gain in computational speed. Alternatively, networks of PCs can be linked together to run the instantiated (physically-based) model on separate units and to address the optimisation problem in this way; see Barreto *et al.*, 2010.

Resource questions:

- Has the assessment been as objective as possible?
- Has each scenario received appropriate attention?

Task 3.4 Select the preferred scenario

Description	Select the preferred scenario by comparing benefit-cost ratios with other considerations (including social and political)
Input	Project objectives and scope, performance of each scenario
Output	Audit documentation, selected scenario
Sub-tasks	Determine benefit-cost ratio or other similar indicators for each scenario, optimise
Tools	DSS

The selection of a preferred scenario depends on having a criterion for the comparison. A suitable indicator could be the benefit-cost ratio, where the benefits might be generated by having better levels of service for flooding in the catchment, or less frequent overflows discharges to the receiving waters, while the costs could be made up of additional storage tanks associated with or without overflows, replaced or relined conduits, or new diversion structures. The scenario with the largest benefit-cost ratio may be selected, provided other social and political factors are comparatively unimportant. Again, optimisation may be used if there are parameters pertaining to each scenario that can be varied.

Resource questions:

- Is the selected scenario clearly preferable to other scenarios?
- Are the benefits of the preferred scenario clearly enunciated?

Task 3.5 Carry out a sensitivity and uncertainty analysis for the preferred scenario

Description	Conduct a sensitivity and uncertainty analyses on the preferred scenario to link the output of the instantiated model to input data and parameters
Input	Project objectives and scope, preferred scenario, uncertainty bounds on input data and parameters
Output	Audit documentation, report on sensitivity and uncertainty analyses
Sub-tasks	Terms of reference and conduct of additional data collection, recalibration and validation of model
Tools	DSS, instantiated model, sensitivity and uncertainty analysers

To round off the selection of the preferred scenario, it is helpful to establish the sensitivity of the instantiated model output results to the input data and parameters so that the decision maker knows that all reasonable effort has been put into refining the data values for those in put data and parameters for which the model is sensitive. Similarly, it is also helpful for the decision maker to know the uncertainty bounds for the model output in so far as they affect the choice of scenario and the benefit cost ratio for the preferred scenario.

If a model predicts both the location and the approximate return period correctly, it can be used with confidence to propose solutions to the problem.

Resource questions:

- What are the conclusions of the sensitivity and uncertainty analysis?
- Can the model be used now with confidence?

10.23.3 Infiltration and inflow analysis for wastewater systems

Infiltration and inflow in wastewater systems are not intended by the design of the networks but occur because of deficiencies in the structural integrity of the pipes or because of illegal or wrongly connected storm flows from streams, gullies draining yards, roof downpipes or manhole covers respectively. Infiltration can arise because of cracks and fissures in pipes and manholes or inadequate joints between pipes. There may be many reasons why the integrity of the elements of the network are compromised, including poor construction, adverse surface loadings due to passing traffic, aging of pipes, thinning of pipe walls due to high shear velocities, high groundwater levels, aggressive chemicals in the soil, frost damage, ground subsidence, penetration by tree roots, human interference, and so on. On the other hand, inflow is usually due to some form of human intervention. Both infiltration and inflow can lead to an inadequate sewer capacity to convey the designed wastewater flow, leading to surcharging and possibly flooding, overloading of pumping stations and treatment works, higher frequency of operation of sewer overflows installed to protect the treatment works, and possible higher sediment loads due to the ingress of soil leading to a greater risk of blockages and the requirement of increased maintenance as well as possible ground subsidence and the associated risk of collapse of the affected sewers.

There is value in using simulation modelling for separate wastewater systems. Here the dry weather flow should strictly follow a slowly time varying profile dependent on the largely diurnal cycle of the domestic use of potable water. But it is well established that separate wastewater systems deteriorate over time, especially from illegal stormwater connections and infiltration from groundwater. In the latter case wastewater systems can in some cases act as land drains. Consequently there can be a considerable stormwater component in the flow, so much so that overflows have to be constructed to cream off excess flows that the wastewater treatment works cannot cope with. This is harmful for the environment and therefore expensive for the community. This leads to rehabilitation techniques applied to the wastewater network, involving the identification of stormwater inflows and the sealing of pipes, sometimes with relining, to reduce the infiltration to a tolerable level. It may be difficult to completely eliminate infiltration in house connection pipes.

Simulation modelling is frequently used to investigate the performance of a separate wastewater collection system and to identify the presence, flow rate, and type of infiltration/inflow conditions which exist in the sewer system. The investigation focuses on describing the existing wastewater collection system using estimates of average residential, industrial, commercial and institution wastewater flows. Continuous flow monitoring and in some cases flow isolation monitoring, is used to determine infiltration and inflow rates, and to reproduce these in the model. On the basis of the results for the existing situation a detailed survey of the system may be needed. Obviously prior to any modelling or survey all available data and information on the performance of the system will have been collected and archived.

10.23.4 CSO analysis

As described above, combined (sanitary or separate) sewer overflows are included in a drainage system to divert excess flows directly to receiving waters in order to limit the overloading of a treatment works. Their performance is usually evaluated by statistical analyses which is used to:

- Assess the frequency of occurrence of surcharging and percentage of deficiency with respect to existing capacity,
- Assess the flow, duration and volume of overflow frequencies (Figure 10.36),

- Assess the average annual overflow volumes,
- Perform various statistical tests and produce the charts to assist in conceptualisation of remedial/ mitigation options,
- Tabulate combined sewer overflow (CSO) spill volumes, peak and average daily dry weather flows,
- Summarise system performance based on the above analyses.

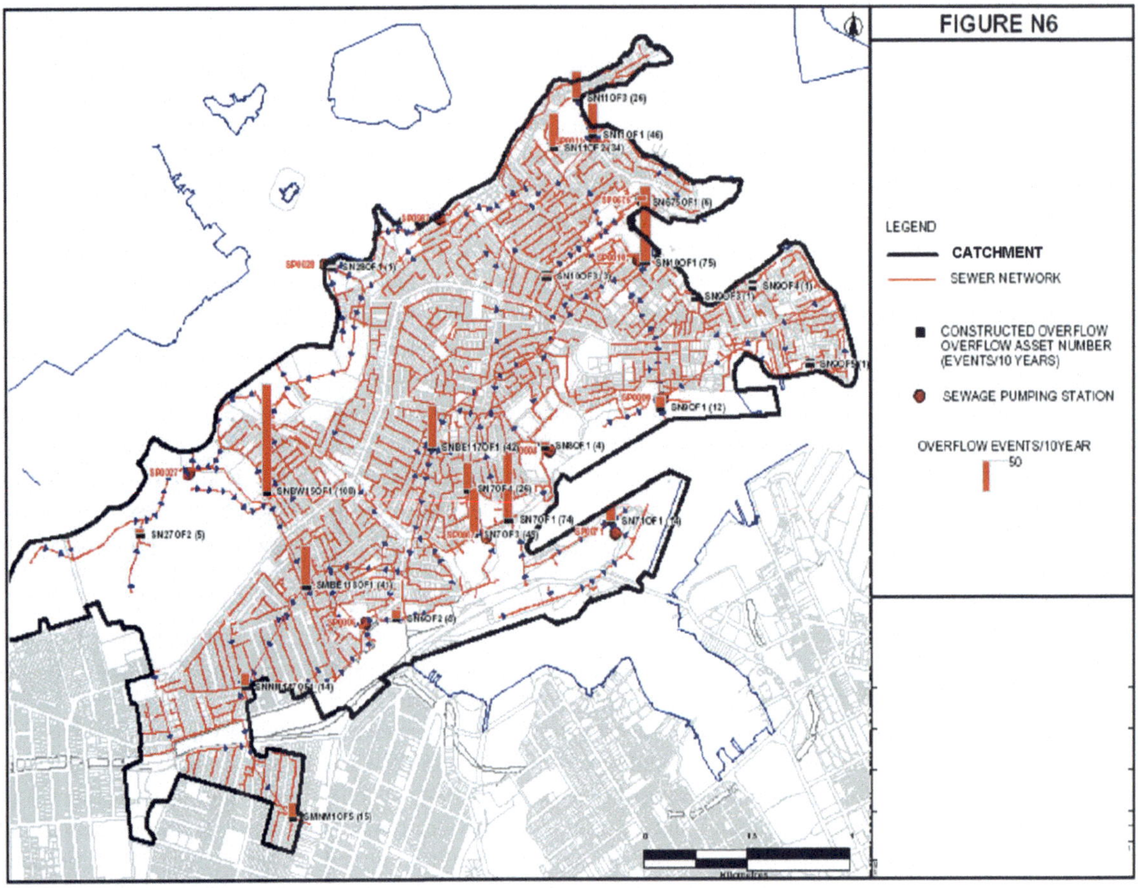

Figure 10.36 A typical GIS map showing combined sewer system performance in terms of the overflow frequencies

Unfortunately the CSOs (or SSOs) discharge dilute sewage that can be heavily polluted. Many of the problems of pollution from sewerage systems are due to inadequate overflows that do not meet current design guidelines. This causes them to spill too much and too frequently and to have poor retention of gross solids and of finer polluting solids. Drainage area plans will therefore frequently involve the re-construction and rationalisation of overflows to reduce these problems. However these changes can have side effects on the operation of the rest of the system.

Some systems exhibit what is called a first foul flush. This is a phenomenon that occurs early on in a storm, when washoff of deposited solids from the catchment, the flushing of gully pots, heavily polluted wastewater and near-bed organic solids in pipes, and the erosion of deposited solids also in pipes produce a high concentration of pollutants at the beginning of the runoff hydrograph for the event. A combined sewer overflow would be designed in conjunction with storage to minimise the discharge of any highly polluted first foul flush and to retain the polluted sewage until the pollutant concentration is acceptable.

An effective CSO should provide adequate hydraulic relief of the sewerage system so as to meet the targets for flood control in the catchment, and control the continuation flow to ensure there is no overloading of the downstream sewerage system and wastewater treatment works. The regulatory requirements for the intermittent discharges to the receiving water must be met, while minimising operational requirements and whole life costs.

Reconstruction of overflows to spill less volume and less frequently may require the construction of new sewers downstream of the overflow to takc the increased continuation flow. Note that the increase in flows through the treatment works may reduce the quality of the effluent.

Rationalisation of overflows to replace several old overflows with one new one can have the side effect of causing a larger impact on the receiving water at the new discharge point than at the old discharge points by discharging the entire spill into one reach of river. The use of a river model will identify this problem.

New overflows based on proven designs, such as stilling pond, high side weir and vortex overflow, have much better retention of gross solids within the sewer than old hole-in-wall or low side-weir overflows. As the emission of gross solids is one of the most obvious forms of pollution caused by overflows this is a significant benefit. However, even these more recent overflows are not very effective at retaining the fine sediment particles which carry much of the polluting matter. They are of course also unable to retain dissolved pollutants from the spill. The concentration of fine sediment in the spill should be assumed to be equal to that in the continuation flow.

Before 1970 in UK it was customary to define the CSO setting as 6 times the dry weather flow. Subsequently the UK Technical Committee on Storm Overflows and the Disposal of Storm Sewage (HMSO, 1970) reported that it would be better to make the setting more dependent on the population and its consumption of water and industrial use of water. They proposed 'Formula A':

$$Q_0 = DWF + 1360P + 2E \tag{10.45}$$

where Q_0 is the overflow setting (litres/day)

$DWF = PG + I + E$

P is the population

G is the water consumption per person (litres/day)

I is the pipe infiltration rate (litres/day)

E is the average industrial effluent (litres/day)

Eq (10.44) does not include any allowance for the flow in the receiving stream. The Scottish Development Department recommended in a report from the Working Party on Storm Sewage (Scotland), Storm Sewage: Separation and Disposal (HMSO, 1977) that for a dilution of more than 7 to 1 Formula A is satisfactory, but for a dilution of 6 to 1 or less then Q_o should be increased by $455P$ and storage should be introduced with the CSO.

In 1993 the National Rivers Authority in the UK introduced the AMP2 Guidelines (NRA, 1993); see Table 10.7. These were designed to classify existing CSOs in terms of their significance and subsequently to identify CSOs that were not performing well and needed to be improved. For bathing beaches, the guideline was that CSOs should not spill more than 3 times per bathing season.

Further to these approaches the Urban Pollution Management Manual (FWR, 1998) was introduced. The Manual provides a procedure for assessing not only the pollution runoff from a sewerage system but also the impact of CSOs and effluent form treatment works on the receiving waters. In addition the Manual provides quantitative standards for intermittent discharges; see Table 10.8.

Table 10.7 AMP2 guidelines

Significance	Dilution	Interactions	Population equivalent	Fisheries	Assessment criteria
Low	>8:1	None	–	–	Emission control (Formula A)
Medium	<8:1	Limited	>2000	Cyprimid	Simple models (QUALSOC, CARP + sewer hydraulic model)
High	<2:1	Significant	>10000	Cyprimid or salmonid	Complex models (sewer and river quality model)

Table 10.8 Intermittent standards for dissolved oxygen and ammonia concentration/duration thresholds for sustaining cyprinid fisheries (after FWR, 1998)

Return period (months)	DO concentration (mg/l)*		
	1 h	**6 h**	**24 h**
1	4.0	5.0	5.5
3	3.5	4.5	5.0
12	3.0	4.0	4.5
	NH_3-N concentration (mg/l)**		
1	0.150	0.075	0.030
3	0.225	0.125	0.050
12	0.250	0.150	0.065

**Applicable when NH_3-N < 0.02 mg/l*
***Applicable when DO $>$ 5 mg/l, pH $>$ 7 and T $>$ 5°C*

There are several aspects of the CSO performance that should be examined during an upgrading of the structure. One is that care should be taken over the design of the hydraulic storage volume associated with a storm overflow. In order to correctly deal with the volumes as well as the peak flows during storm events the analysis of storage is normally carried out using a rainfall time series approach, such as produced by StormPac™ or the synthetic design storm time series developed at HR Wallingford. The storage tank is normally designed to provide some settling of sediment such that a greater concentration of sediment goes on to treatment than is conveyed in the overflow discharge. A scumboard will be used to deflect gross floating solids from the overflow.

Another aspect is the ability of the CSO to remove gross solids from the effluent. There are two standards: the first is the separation from the effluent of a 'significant quantity' of solids with a size of 6 mm in any two dimensions, and a screen that removes 10 mm solids.

Resource questions:

- Has the design of the CSO been though through in terms of storage and removal of sediment and gross solids?
- Have the standards for intermittent operation of the CSO been agreed with the regulator?
- Is the rainfall input capable of properly defining the frequency of operation of the CSO?

10.23.5 Performance analysis of pipes and channels

The pipe (or channel) capacities (i.e., performance) in relation to design standards can be established and presented for the current and future development scenarios in colour-coded GIS maps; see Figure 10.37.

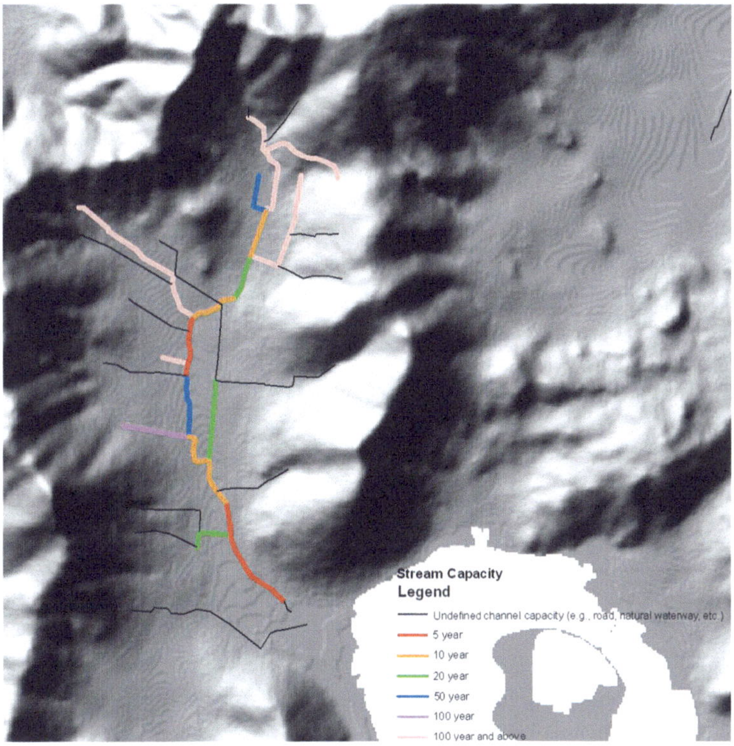

Figure 10.37 Example of a GIS map displaying channel capacities against design standards

Furthermore, in the case of sediment transport analysis, the calculated maximum bed shear stresses in the drainage network system can be displayed in a similar way; see Figure 10.38.

10.23.6 Storage facility analysis

A detention tank can be used to limit the peak discharge from a catchment. It does this by storing excess flow in an on- or off-line storage tank or pond until the storm has passed, and then the stored water is released from the tank to the pipes downstream or directly to the sewage treatment works. The volume continuing on downstream can be reduced through source control techniques (as introduced through the UK Sustainable Urban Drainage Systems (SUDS) initiative; see Woods Ballard *et al.*, 2007) or through diverting excess flows through an overflow to the nearest receiving water. In both cases the sizes of the pipes downstream of the tank or the overflow can be reduced. The introduction of storage is therefore a means of upgrading existing sewerage systems and of reducing the pollution impact on receiving waters downstream of new sewerage systems.

A storage tank is also needed to deal with a possibly increased first foul flush of pollutants at the start of storms as described above. Sediment will be transported into a tank where fine sediment will tend to be deposited as the tank drains down at the end of a storm and flushed out at the start of the next storm. If there is another overflow further downstream it may spill a more polluted flow, or the treatment works may be unable to cope with the pollutant loading especially the final clarifiers of activated sludge plants. The long detention times in the tank can also lead to the generation of methane and hydrogen sulphide in septic sewage causing both pollution and operational problems. The use of a sewerage water quality model in conjunction with a treatment works model can help with the investigation of these problems and with the design of the tank.

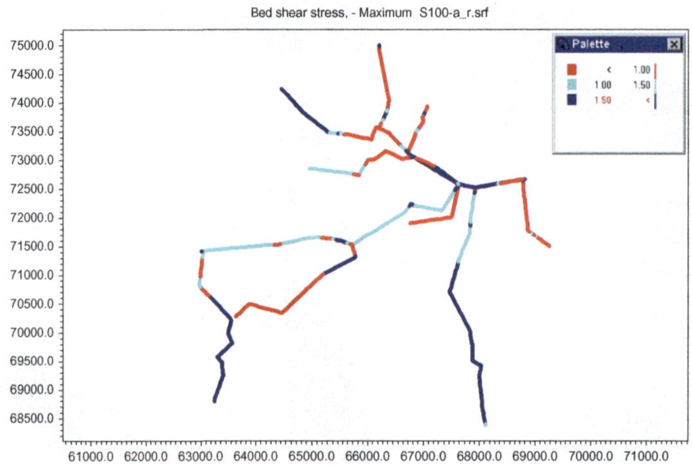

Figure 10.38 Example of a map displaying maximum bed shear stresses in the drainage network

The sizing of a tank is based on standard storms of the desired return period, whereas the verification of the model is done with long events, typically those that include runoff from permeable areas, (such as are generated during the winter in the northern hemisphere). So adjustments may be needed to the designed tank size.

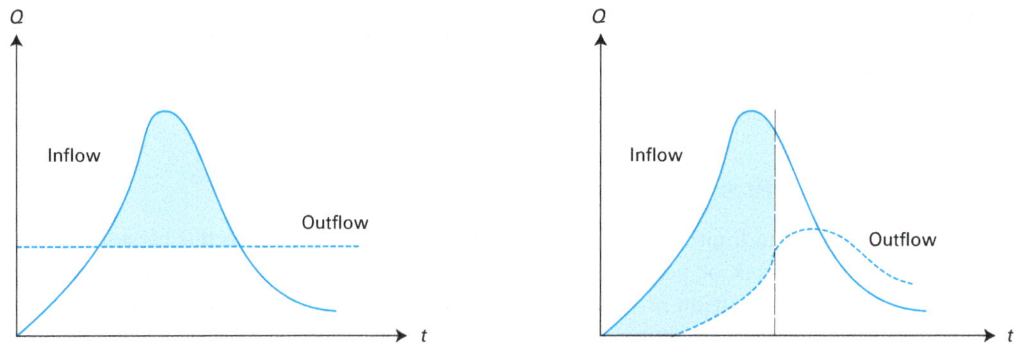

Figure 10.39 Use of storage

On-line tanks

On-line tanks form an integral part of the sewer system, consisting of enlarged flow sections that fill when the inflow exceeds the required continuation flow to the downstream sewer. The flow into the downstream sewer is usually controlled using an orifice plate, penstock, throttle pipe or vortex flow regulator. Pumps are rarely used in on-line tanks.

Off-line tanks

Off-line tanks are physically separate from the sewer system, with the flow entering the tank via an overflow device incorporated into a diversion structure. The flow to the downstream sewer is again controlled in a similar way to that of the on-line tank. The stored flows are returned to the system either by gravity, where the site allows, or by pumping.

Pumped storage tanks

Pumped storage tanks may be used to prevent flooding, to balance hydraulic loading, to reduce the frequency of spill, to reduce pollution spilled from the sewer system to the receiving watercourse, and to allow response time in the event of pump or power failure at pumping stations.

The UPM manual states that 'detention tanks should be designed to be, as far as possible, self cleansing whilst in operation'. In pumped storage tanks, which are designed to meet such criteria, it is preferable that the weir is located at the upstream end of the tank so that the operation of the tank results in the retention of the heavily polluted first flush in the tank downstream of the weir. This means that the spilled flow to the watercourse is comprised of the more dilute stormwater that enters the tank during the later stages of the storm event.

In contrast, pumped storage tanks which are designed to settle out the sediments rely on the creation of a tranquil flow pattern within the tank that is conducive to the settlement of the gross and finely suspended solids. In this case the tank should be so dimensioned to create conditions of ideal settlement with all sediments deposited to the bed of the tank as they are discharged through the tank. Hence, in this type of tank, the position of the overflow weir should be located at the downstream end of the tank.

The retention time of effluents in the pumped storage tank is a particularly important parameter in respect of the deterioration of the quality of the effluents retained in the tank. Long retention times, in excess of 12 hours, may lead to problems of septicity and odour. Microbiological activity in the effluent and in the sediment deposits on the bed of the tank results in an uptake of oxygen from the effluent in the tank with a corresponding reduction in the concentration of dissolved oxygen. The rate of oxygen uptake is a function of many parameters which include temperature, the initial concentration of DO, the constituents and pollutant strength of the effluent, the characteristics of the bed sediment deposits and the re-entrainment of oxygen via the water surface. Equations have been developed by Pomeroy and Parkhurst (1972) to predict the uptake of DO and the potential for the generation of hydrogen sulphide in pumped rising main systems. Their applicability to large pumped storage tank systems needs further research.

Designing for sediments

Flow velocities are greatly reduced on entry to a storage structure, and the hydraulic conditions favour sediment deposition. Deposits of sediments in storage chambers may however, lead to blockages, surcharging, flooding and premature spill operation. In addition, the ageing of deposited sediments, and their subsequent re suspension in later storms, can lead to poor quality continuation flow and overflow spilled discharges, and may also lead to odour and septicity problems. It is therefore important that sediments are not allowed to build up in storage tanks. Saul and Ellis (1992) recommended criteria for the design of on-line tanks, taking sediment into account.

Levels of service for receiving waters

Where a first flush occurs, pumped storage tanks can be designed to capture the flush, thus allowing the storm flow with lower pollutant concentration to enter the river during the later stages of the storm event. Hedley and

King (1971) observed that the major part of the pollution load in storm sewage was carried either in solution, or as a very fine suspension, and could not, therefore, be retained within the sewerage system by the settling processes occurring at a conventional overflow structure. They suggested therefore that the storage volume should be equivalent to the excess volume of the rising limb of a design storm hydrograph above that of the continuation flow. They recommended that the return frequency of the design storm is a function of the size and quality of the receiving stream, ranging from 6 months for a small stream to an overflow frequency of 10 times per year for large rivers.

Where a storage tank is required because river amenity is of special concern, the Scottish Development Department, (HMSO, 1977), suggested a simple method to estimate the desired storage capacity based on available dilution in the receiving watercourse. The guidelines are presented in Table 10.9 below. The dilution is based on the dry weather flow of sewage and on the minimum flow in the stream at the discharge point, that is, the mean daily discharge which is exceeded in the long term for 95% of the time. The UK Environment Agency normally asks for 1 to 2 hours storage at 3 DWF at pumping stations before spills can be made to a water course.

Table 10.9 SDD (HMSO, 1977) Design guidelines for storm overflows

Dilution*	Percentage Pollution**	Type or storm overflow
8	100	Formula A, no tank
6	75	Formula A + 455P, no tank, if sewer capacity available. Alternatively adopt next recommendation.
4	50	Formula A + tank capacity 40 l/hd.
2	25	Formula A + tank of capacity 80 1/hd.
1	12	Formula A + tank of capacity 120 1/hd.

** Minimum dilution in terms of 95% exceedence mean daily discharge in receiving watercourse and sewage DWF*
*** Approximate percentage of total polluting load of storm sewage spilled from sewer that may be discharged to stream (%)*

Sizing of tanks: European practice

Historically, in Europe, it has been common practice to design the size of the tank based on the storage of a rainfall amount falling on the impervious area of the catchment. Typically 1.5 mm to 4.0 mm may be used for design purposes, which gives a volume of storage equivalent to 15–40 m^3/hectare of impervious area. This design philosophy was based on the recommendations of Krauth which have subsequently been written into the (German) ATV A128 design guidelines for storage tanks (ATV, 1992).

Resource questions:

- Have the right standards been taken into account when designing storage?
- Has the design taken sediment and gross solids into account, and is the performance of the designed system adequate?

10.23.7 Real time control options

An urban drainage system is controlled in 'real time' if flows within the system are governed by moveable regulators which operate based on information of the behaviour of the system at one or more sites at each moment in time. The reason for using real time control (RTC) is to make optimum use of the sewers and facilities, especially the storage available in them. This allows operating and maintenance costs to be minimised, flooding and environmental impacts to be reduced and capital expenditure to be deferred or limited.

In contrast to passive systems, RTC enables the control structures to manage the flows actively. The degree to which flows can be managed is largely related to the extent and amount that available storage can be utilised.

With RTC, operators have a degree of freedom that does not exist in passive systems. Control rules are adopted that improve the performance of a system, and these rules are normally the subject of an agreement with the Regulator such that he can grant a consent for the structure. Then it is important that monitoring procedures are in place to log the RTC actions and demonstrate that the control rules are implemented in the event of a failure in the system performance. Such a failure may be due to a malfunction of the monitoring system providing the information for the RTC decisions. Safeguards are therefore needed for these situations to ensure that the system does not perform worse than an equivalent passive system.

It is also important to ensure that the effects of any RTC action on the whole drainage system (collection network, treatment works and receiving waters) are properly understood such that the effects of any action can be predicted and can be shown to be of benefit. Modelling therefore becomes strategically important in RTC in producing safe and efficient rules for controlling the flows, and in training operators to run the system reliably.

The essential feature of RTC is that, during a rainfall event, parts of the sewerage system may be overloaded while other parts are under utilised, and the parts affected change dynamically during the event. The flow control structures are operated such that the stress on the overloaded parts of the network is alleviated by storing sewage or moving it to the under utilised parts. In this way RTC can benefit the following:

- Deciding when to divide flows
- Accounting for spatial rainfall
- Flushing sediment
- Controlling diurnal dry weather flow
- Minimising pumping station costs
- Controlling spill to rivers
- Minimising the transient impact on receiving waters
- Providing improved flood protection

An RTC system consists of regulators, sensors and controllers; see Table 10.10. Rules are used to operate the regulators controlling the flows and these rules depend on the inflow conditions. The controllers receive information from sensors and this information is used to control the regulators in order to produce the optimum performance of the network under different conditions.

It is generally acknowledged that there are three levels of RTC, namely, local, area and global; see Table 10.11. Obviously, the complexity increases according to the area affected by the RTC. The strategy adopted in each case depends on the peculiarities of the sewerage system. If rainfall is uniform across a catchment then the best strategy to minimise the volume and frequency of spills is a uniform filling of the available storage elements, though this does not consider the quality of the discharges or the sensitivity of receiving watercourses. In theory, all storage should be utilised before any spills from CSOs start to occur, and this condition has the advantage of meeting the current requirements of overflow consents more closely.

There are two basic ways in which the rules of an RTC system can be developed. The first way is to deduce rules for the regulators and then to demonstrate their effectiveness by analysing historical events. This is off-line control. Alternatively, on-line control is about continuously revising and updating the rules based a prediction of future system behaviour, and finding the best way of minimising the failure or damage. For guidelines about implementing RTC see CIWEM, 1996.

One of the principles of RTC is that pass-forward rates from overflows can be varied. In fact, under certain circumstances auto-penstocks can be fully closed, with all the flow being retained in a certain part of the catchment. Sensible design criteria should be developed in consultation with the appropriate regulatory body.

Table 10.10 The elements of RTC

Item	Description	Examples
Regulators	Physical structures such as auto penstocks that move or store flow.	Auto penstocks Variable level weir crests Pumps
Sensors	Equipment such as level sensors that record aspects of system behaviour.	Flow monitor Depth monitor Velocity probe Raingauge Timer / clock pH meter Water quality meters
Controllers	Decision making units which communicate the required actions to the regulators	Manual – An operative observes the current state of the system and physically alters the setting of a regulator(s) based on this information Remote – An operative observes the current state of the system and directs the setting on a regulator remotely via a telemetry link Automatic – Direct communication between sensors and regulators. Supervisory – The system normally operates under automatic control but can be over-ridden by an operative.

Table 10.11 The levels of RTC

Level	Description
Local	One regulator controlled with information from one sensor
Area	Parts of a system are controlled in a coherent manner
Global	Multiple regulators across the network are controlled with information from multiple sensors

In developing scenarios for RTC care should be taken over the selection of the appropriate rainfall:

- Synthetic (design) rainfall data
- Annual time series rainfall data
- Stochastic rainfall generator (such as StormPac™)
- Local historical time series data

Design storms may be of value for certain types of analysis, but a series of events should normally be used.

It is important to ensure that during the failure of any element of RTC, the sewerage system does not perform any worse than it would have done if it was operated under passive control. It is therefore necessary to identify the 'safe' condition that the system should revert to in the event of a breakdown of the RTC system.

Figure 10.40 presents the generalised procedure that can be followed when designing and implementing real time control.

Resource questions:

- Has the rainfall used in the design and assessment of the RTC rules been selected properly and justified?
- What has been done to minimise the risk of failure of the system?

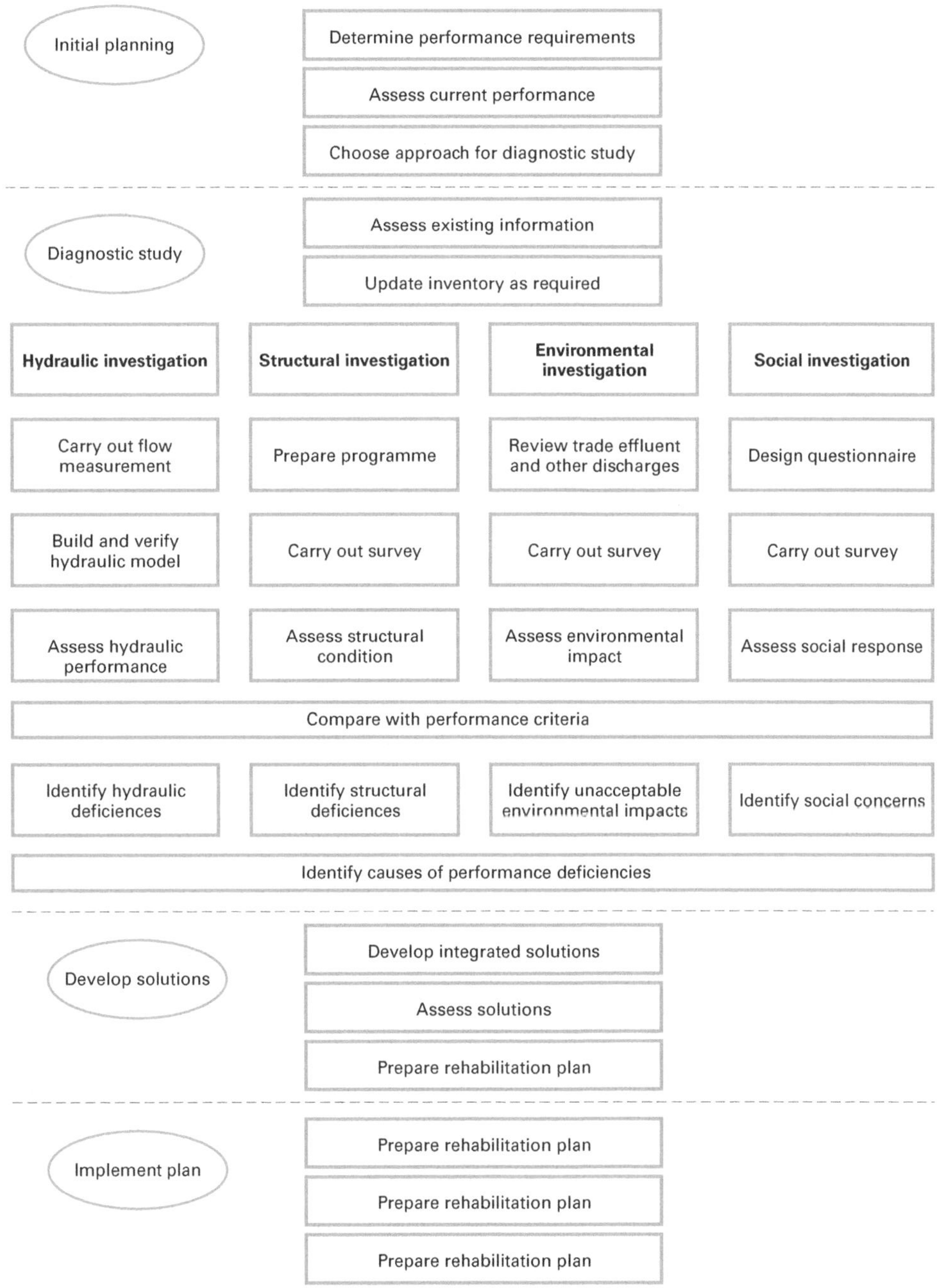

Figure 10.40 Sewerage rehabilitation (after WRc, 1994)

10.23.8 Sewerage rehabilitation

Rehabilitation of an urban water utility network is a major aspect of its management. Any sewerage network deteriorates in time for a variety of reasons: pipes degenerate with age, structural stresses and strains lead to pipe fracture and collapses, there is unacceptable infiltration, the levels of service against flooding in the catchment become unacceptable in time due to increased loading from rainfall-runoff, overflows operate with increased frequency, and so on. In view of the large number of sewerage networks around the world, considerable effort has been put into deriving an appropriate procedural approach to sewerage rehabilitation (see also Chapter 8). Note that the word 'rehabilitation' is not strictly correct for the task in hand. 'Rehabilitation' conveys the notion of restoring the system to its *original* performance. But this is not what the sewerage manager is normally attempting to do. His concern is to improve the performance of a system that has inevitably changed its structure and loading. The population being served by the system may well have changed significantly, and the drainage network may have been considerably extended. Also, community expectations may have been raised such that they seek less frequent flooding in a combined system. In other words, levels of service may have increased and therefore standards have been raised. Consequently the sewerage manager is not necessarily seeking to restore the performance of the system to its original level of service, but to improve the service provided.

A number of procedures for sewerage rehabilitation have been put forward. The original version in the UK was the Sewerage Rehabilitation Manual promoted by WRc in 1983. It is currently in its 4th edition (WRc, 1994). Subsequently, extensions of this procedure have been advocated by the Guide for the Rehabilitation of Sewerage Systems (Kellagher, 2002) and the HydroPlan procedure (Vojinović *et al.*, 2005), among others; see Chapter 8.

The Guide

Stage 1 (Initial Planning) of the Guide procedure concerns the definition of the extent of the rehabilitation project; see Figure 10.40. It may be initiated by changes in legislation, poor performance of the network, evidence of a physical deterioration of the system, or the threat of additional loading on the network. Tangible and intangible indicators are used to determine the performance of the system. A 'trigger' is defined for each indicator as a criterion for giving attention to the particular aspect of performance. A 'target' criterion is then defined as an objective for the performance as measured by the indicator. In both cases, trigger and target levels will normally be specified by local, regional or national government. Criteria for the added value of the improvements in system performance are then specified, along with criteria that take into account the intangible benefits.

The performance requirements for a rehabilitated system are similar to those for a new system. BSI (2008) EN752-2 is entitled 'Performance Requirements', and in Clause 6, it lists the following 'Basic performance requirements':

a) The pipework operates without blocking.
b) The flooding frequencies shall be limited to prescribed values.
c) Public health and life shall be safeguarded.
d) The sewer surcharge frequencies shall be limited to prescribed values.
e) The health and safety of operator personnel shall be safeguarded.
f) Receiving waters shall be protected from pollution within prescribed limits.
g) Drains and sewers shall not endanger existing adjacent structures and utility services.
h) The required design life and structural integrity shall be achieved.
i) Drains and sewers shall be watertight in accordance with testing requirements.

j) Odour nuisance and toxicity do not arise.
k) Appropriate access shall be provided for maintenance.

Indicators therefore relate to hydraulic performance, structural condition, environmental impact and social response, and are selected to minimize overlap. (Note that the social response has been introduced here in recognition of the importance of consulting stakeholders; it is not in the original Guide). Performance criteria for the system must be selected, along with objectives. For example, the number of spills for overflows, the life expectancy of the rehabilitated system, and the flooding frequency at specified locations. Basic hydraulic, structural and environmental performance requirements are specified in Part 2 of EN 752 (BSI, 2008), whilst recommendations for design criteria and procedures are given in Part 4; (basic design recommendations for structural performance are given in EN 1295 (BSI, 1998).

Given that the trigger levels indicate that a full diagnostic study is required then the objective is to assess the hydraulic, structural, environmental and social performance of the sewerage system against defined performance criteria, and to identify the causes of performance deficiencies. For this a comprehensive sewer system inventory should be developed and maintained using a GIS.

In large sewerage network it is likely that 80% of the problems are caused by 20% of the network. Therefore there is clearly a need where possible to focus the rehabilitation on a limited set of pipes, and to identify three categories of pipe: critical, strategic and normal. Every pipe is normal (and acceptable in its current state) or critical or strategic. A strategic pipe is one that is key to the performance of the system whereas a critical pipe is one that might be difficult and expensive to rehabilitate, such as a pipe under a main railway line or major road; see HydroPlan (Vojinović *et al.*, 2005) below.

A hydraulic simulation model is an essential tool for the assessment of both the hydraulic and environmental performance; see Table 10.12. The integrity of the structural data for the model should be confirmed together with any calibration and verification of model parameters, including parameters relevant to pollutant loading and transport. The hydraulic model may then be used to assess the service levels of flooding throughout the catchment and the pollutant concentrations reaching the treatment works or being discharged through the overflows. If necessary, the modelling can be extended to include the impact on the receiving waters (lake, river, estuary or coastal waters) and the interaction with groundwater; see Chapter 12.

Table 10.12 Impacts

State	Environmental impact
Hydraulic	Sewage spills to rivers Aesthetic (litter) Dissolved and organic solids STW discharges Stormwater discharges Flooding
Structural	Groundwater pollution
Other	Odour Noise Operational and maintenance activities Illicit connections to surface water systems Industrial effluent Domestic effluent

In order for the pipe network to fulfil its purpose the pipes must not collapse or leak, contaminating the environment, and infiltration should be minimal. Excessive infiltration reduces the hydraulic performance of the network and the efficiency of the treatment works, and can create voids in the backfill and soil around the pipes which can put unacceptable stresses on the pipe structures leading to their collapse. A structural study using man-entry or CCTV should therefore be carried out to ensure that the integrity of the pipe system. A scoring system may be used to highlight those pipes that are failing structurally, or could fail in the future.

A cost-benefit approach can be applied to the development and selection of appropriate rehabilitation proposals in order to improve the performance of a sewer network. Account is taken of the costs of phased solutions, maintenance and operating costs and also the estimated costs of the consequences of failure. Other intangible costs such as public perception of the service provision may also be included. In parallel with costing, a model of the network would be used to evaluate the level of service of the future, preferably taking into account deterioration of the sewer fabric, the operation and maintenance proposed, future development and other influences on system performance. The model would be used to ensure trigger levels are complied with for the duration of the planning period. The main limitations of this approach are in determining the deterioration rate of the sewer and assessing the risks and consequence costs of failures. The full analytical process involved in assessing the best option for rehabilitation can be summarised as follows:

a) Understand the 'Trigger' failures of the current system
b) Establish Client criteria 'Targets' for performance indicators
c) Develop Performance scoring criteria for assessment
d) Develop intangible indicators for Value Engineering
e) Score the baseline network for comparative purposes
f) Develop engineering solutions for rehabilitation that generally meet Client Target levels for indicators and ensure all Trigger failures are removed
g) Cost the engineering solutions
h) Score the engineering solutions to determine Performance scores;
i) Score Value Engineering indicators
j) For each engineering solution, evaluate costs against the system performance score and the Value Engineering scores and compare with the baseline network performance score
k) Eliminate high cost, low value options
l) Further develop high value options
m) Choose the Best Engineering solution.

The sewerage rehabilitation strategy to find a solution can consider the following mix of options:

a) Rehabilitating individual sewers
b) Introducing additional storage or sewerage ancillary structures
c) Improving performance of CSOs or other existing structures
d) Introducing RTC to make better use of existing (or new) storage

The choice on how to rehabilitate an individual sewer will depend on a variety of *local* factors. However, solutions cannot just be local as every change to the system will have an impact downstream and possibly a short distance upstream as well. This means that there are also important *global* considerations for nearly all solutions. Therefore, the design of a rehabilitation scheme to meet certain levels of service in terms of frequency of (significant) surface flooding and pollution impact must consider the whole system. In addition, it should be understood that a system is not 'frozen' at the time the scheme is implemented. In fact, the system continues to change: pipes age, the population changes, development takes place within the catchment. The design life of the system is therefore also important. Not only might it be better to have a shorter design life, but it might also be better to phase the rehabilitation into the future. We have therefore a multi-criteria optimisation problem.

It should also be noted that predicted failure of a sewer is a probablistic problem due to hydraulic failure being related to the frequency of rainfall events. It is impossible to take all issues into account because the future, by definition, is uncertain, but these issues illustrate the importance of deriving solutions which take into account whole life costs, the future of the system and the risks and consequences of delaying rehabilitation activities.

A catchment that has a range of problems that need to be addressed over a period of time will require a range of different rehabilitation options. Due to the interactive nature of networks, each solution may affect the performance of the system at other locations as well. This will influence the requirements for other rehabilitation solutions. It is therefore clear that some proposals, particularly for complex systems, will benefit from comparing the rehabilitation costs with an objective evaluation of a measurement of the improvement in system performance. Other proposed solutions may have very little influence on other parts of the network and the choice of option is dictated by the cost of the proposals.

The Guide (Kellagher, 2002) therefore requires a decision to be made as to whether it is worthwhile making an assessment of the options using a Performance Enhancement and Value Engineering evaluation of the network in guiding the selection of the most appropriate solution. It may be sufficiently obvious that this analysis need not be made.

The rehabilitation plan should be part of a continuous procedure of monitoring the system performance, taking account of changes in the required levels of service and minimising operating and maintenance costs.

The performance score enables a continuous evaluation of the system performance, drawing attention to any reduction in performance over time and also flagging up 'Trigger' failures.

One of the problems faced by operators is that of treating such an exercise as an end in itself and not keeping an adequate record of the changes made: a rolling programme of phased rehabilitation changes requires a continuously updated model and the related GIS database needed to provide all the relevant data.

Stage 4 of the Guide provides guidance on the activities relating to the needs during the construction process, which will ensure a satisfactory outcome to the construction, and also on the various testing procedures available to confirm that the renovated sewers meet the specified performance criteria.

Hydroplan

The Hydroplan procedure takes a risk-based view of the on-going management of the sewerage asset (Vojinović *et al.*, 2005). In effect, it aims at a comprehensive management of the sewerage assets while incorporating the insights of rehabilitation. A strategic analysis is first carried out to assess the impact of a possible failure of each sewer. The impact includes the costs of reconstructing the pipe, delays to traffic, environmental damage, and so on. Separate scores are defined for financial, socio-economic and environmental aspects. A hydraulic model is used to determine the probabilities of different forms of hydraulic and environmental failure. A separate aging model using CCTV data is used to estimate the probability of the structural failure of the sewer. Then impact and probability information for each pipe are combined to provide a corresponding risk score. By comparing scores the most critical sewers are highlighted for each of the three main failure modes (hydraulic, structural and environmental). This information then feeds a short to medium term action plan. But the on-going aging of the sewers prompts the need for a long term operation and maintenance strategy. A new total life cycle model, HADES, runs Monte Carlo simulations and integrates the monetary values of the risk with preventative investments. This way, alternative strategies can be identified and compared in order to select the optimal strategy for each pipe. HADES can indicate which sewers can be neglected in favour of more frequent inspection of strategic sewers.

10.23.9 Urban pollution management

The discharges from combined sewer overflows will normally contain significant concentrations of sewage pollutants. Even the effluent from wastewater treatment works will include residual quantities of different pollutants. These pollutants have an adverse effect on the receiving waters, whether a river or stream, or coastal waters. In each case the receiving waters have a self-treatment capacity, but in terms of normal runoff from urban areas this capacity is limited and the impact of the pollutants can have serious short term and long term consequences for the water quality of the receiving waters. The impact on different rivers and coastal waters has become so significant with serious consequences for environmental and public health that many governments have had to bring in legislation to control urban discharges to receiving waters through the setting of environmental quality standards (EQOs) to meet agreed environmental quality objectives (EQOs) for the various target uses of the receiving waters. So, for example, in 1994 the UK advocated an Urban Pollution Management (UPM) programme which adopts such an approach to enable the control of CSO discharges such that levels of impact are compatible with the desired uses of the receiving water. Water quality criteria are used to define pollutant concentration profiles in terms of magnitude (pollutant concentration), duration, and frequency, against which output from a river model can be evaluated in order to assess the performance of the sewerage system in respect of the EQO's.

The UPM approach is to use a variety of models in an integrated methodology for management of pollution on the river catchment scale. The tools include water quality models to simulate the behaviour of pollutants and sediments in sewer systems, sewage treatment works, and rivers. The UPM procedure is not one specific methodology, but a set of methods selected on the basis of appropriateness. In general, the method uses detailed water quality models (including river, sewage treatment works and sewers) though simplified approaches are also recommended.

The UPM procedure has four key stages

1. Initial planning
2. Data collection and modelling
3. Development of solutions and testing for compliance
4. Obtaining approval and detailed design

These are discussed in detail in Chapter 12. The WaPUG guide to Quality Modelling of Sewerage Systems (WaPUG, 2006) outlines a procedure for building, calibrating and operating a model of water quality in a sewerage system.

10.24 KEYWORDS

By the end of the chapter you should have encountered the following terms; ensure that you are familiar with them!

Combined sewerage system
Combined Sewer Overflow
Design rainfall
Horton infiltration
Hydraulic analysis
Infiltration
Instantiating a model
Infiltration and inflow analysis
Pollutant loading
Pollutant washoff
Rainfall
Rainfall losses
Rainfall-runoff
Rainfall-runoff routing
Rational method
Real time control
Receiving waters
Separate sewerage system
Sewage
Sewerage
Sewerage rehabilitation
Simulation modelling
Storage facility
Stormwater system
Time-area diagram
Urban pollution management
Unit hydrograph
Wastewater system
Wastewater treatment plant

10.25 QUESTIONS

What design rainfall should be used when modelling the operation of CSOs?
What role can optimisation play in sewerage rehabilitation?
What are the advantages and disadvantages of combined and separate sewerage systems?

What does hydraulic analysis of collection systems provide to the engineer?
How should a combined sewer overflow be assessed?
How can real time control be included in the design of an urban drainage system?
What is the role of storage in the successful operation of a combined sewer overflow?

10.26 REFERENCES

ATV (1992). *Standards for the Dimensioning and Design of Stormwater Structures in Combined Sewers*. ATV–A 128E. Sankt Augustin Gesellschaft zur Förderung der Abwassertechnik e.V. (GFA).

Barreto W., Vojinović Z., Price R. K. and Solomatine D. P. (2010). A multi-objective evolutionary approach for rehabilitation of urban drainage systems. *J. Water Resour. Plann. Manage. ASCE*, **136**(5), 547–554.

Bilham E. G. (1936). Classification of heavy falls in short periods. In: *British Rainfall 1935*. HMSO, London.

BSI (1985). *Code of Practice for Building Drainage*. BS 8301:1985: 74. British Standards Institute, London.

BSI (1998). *Structural Design of Buried Pipelines Under Various Conditions of Loading. General Requirements*. BS EN 1295-1:1998. British Standards Institute, London.

BSI (2008). *Drain and Sewer Systems Outside Buildings. Performance Requirements*. BS EN 752-2. British Standards Institute, London.

Butler D. and Davies J. (2000). *Urban Drainage*. Spon Press, London.

CEC (1991). *Directive Concerning Urban Waste Water Treatment*. 91/271/EEC. CEC, Brussels.

Centre W. R. (1994). *Sewerage Rehabilitation Manual*, 3rd edn. WRc Engineering, Swindon, UK.

Chen A. S., Hsu M. H., Teng W. H., Huang C. J., Yeh S. H. and Lien W. Y. (2005). Establishing the database of inundation potential in taiwan. *Natural Hazards Review*, **37**(1–2), 107–132.

CIRIA (1992). *Scope for Control of Urban Runoff*. Report 124. CIRIA, London.

CIRIA (2006). *Designing for Exceedance in Urban Drainage – Good Practice*. Report C635. CIRIA, London.

CIRIA (2006). *The SUDS Manual*. FR/CP/110. CIRIA, London.

CIWEM (1996). *Evaluating Real Time Control for Waste Water Systems, A Managerial and Technical Guide*. Report of RTC Steering Group. CIWEM, London.

Cowpertwait P. (1991). Further developments of the Neyman-Scott clustered point process for modelling rainfall. *Water Resour. Res.*, **27**(7), 1431–1438.

Crabtree R. (2007). Sediments in sewers. *Water Environ. J.*, **3**(6), 569–578.

Cunge J. A. (2003). Of data and models. *J. Hydroinf.*, **5**(2), 75–98.

Djordjevic S., Prodanovic D., Maksimovic C., Ivetic M., Savic D., Koegst T. and Krebs P. (2005). SIPSON – simulation of interaction between pipe flow and surface overland flow in networks. *Water Sci. Technol.*, **52**(5), 275–283.

Entekhabi D., Rodriguez-Iturbe I. and Eagleson P. (1989). Probabilistic representation of the temporal rainfall process by a modified Neyman-Scott rectangular pulses model: Parameter estimation and validation. *Water Resour. Res.*, **25**(2), 295–302.

FWR (1998). *Urban Pollution Management Manual*. Foundation for Water Research, Marlow.

Harms R. W. and Verworn H. R. (1984). HYSTEM – a hydrological urban drainage model: Part 1. Model description. *Korrespondenz Abwasser*, **31**, 1112–1117.

Hedley G. and King M. (1971). *Suggested Correlation Between Storm Sewage Characteristics and Storm Outflow Performance*. Thomas Telford-ICE Virtual Library, London.

Hjelmfelt M. (1990). Numerical study of the influence of environmental conditions on lake-effect snowstorms over Lake Michigan. *Mon. Wea. Rev.*, **118**, 138–150.

HMSO (1970). *Technical Committee on Storm Overflows and the Disposal of Storm Sewage*. Final Report. Ministry of Housing and Local Government, London.

HMSO (1977). *Storm Sewage: Separation and Disposal*. Report of the Working Party on Storm Sewage (Scotland). Scottish Development Department, Edinburgh.

Horton R. (1933). The role of infiltration in the hydrologic cycle. *Trans. Am. Geophys. Union*, 14, 446–460.

Hsu K.-L., Gupta H. V., Gao X. and Sorooshian S. (eds.) (2000). Rainfall Estimation from Satellite Imagery. In: *Artificial Neural Networks in Hydrology*, Kluwer Academic Publishers, Dordrecht.

Kellagher R. (1993). *Sewer Modelling for Rehabilitation*. EU SPRINT Project Reports. HR Wallingford, Wallingford UK.

Kellagher R. (2002). *Development of Guidelines on the Structural, Hydraulic and Environmental Rehabilitation of Sewers*. HR Wallingford, Wallingford.

Laat P. d. and Savenije H. H. G. (1992). Principles of hydrology. In: *UNESCO-IHE Lecture Note Series*. LN0201/05/1. UNESCO-IHE, Delft.

Liston G. (1999). Interrelationships among snow distribution, snowmelt, and snow cover depletion: implications for atmospheric, hydrologic, and ecologic modelling. *J. Appl. Meteorol.*, **38**(10), 1474–1487.

Mark O., Weesakul S., Apirumanekul C., Aroonnet S. and Djordjevic S. (2004). Potential and limitations of 1D modelling of urban flooding. *J. Hydrol.*, **299**(3–4), 284–299.

Molteni F., Buizza R., Palmer T. and Petroliagis T. (2006). The ECMWF ensemble prediction system: methodology and validation. *Quart. J. Roy. Meteorol. Soc.*, **122**(529), 73–119.

NERC (1975). *Flood Studies Report* (in 5 volumes). Natural Environment Research Council, London.

NRA (1993). *Guidelines for AMP(2). Periodic Review, Volume 2, Water Resources*. National Rivers Authority, London.

Pomeroy R. and Parkhurst J. (1972). Self purification in sewers. In: Jenkins S. H. (ed.) *Proc 6th Joint Conf. on Advances in Water Pollution Research*, 18–23 June, Pergamon Press Oxford, Jerusalem.

Saul A. and Ellis D. (1992). Sediment deposition in storage tanks. *Water Sci. Technol. WSTED 4*, **25**(8), 189–198.

Vojinović Z., Bonillo B., Chitranjan K. and Price R. (2006). Modelling flow transitions at street junctions with 1D and 2D models. In: *7th International Conference on Hydroinformatics*, Acropolis, 4–8 September, Nice, France.

Vojinović Z., Price R. K. and van den Broek W. (2005). Hydroplan-EU knowledge management framework for urban water asset management. In: *International Conference on Urban Drainage*, 21–26 August, Copenhagen, Denmark.

Vojinović Z., Seyoum S. D., Mwalwaka J. M. and Price R. K. (2009). Effects of model schematization, geometry and parameter values on urban flood modelling. In: *8th International Conference on Urban Drainage Modelling*, 7–11 September, Tokyo, Japan.

Wallingford (1981). *Design and Analysis of Urban Storm Drainage – The Wallingford Procedure*. Department of Envrionment, Natioinal Water Council, Standing Technical Committee Reports 28. Hydraulics Research Limited, Wallingford.

Wallingford H. (1981). *The Wallingford Procedure for the Design of Urban Storm Drainage*. HR Wallingford, Wallingford.

WaPUG (2002). *Code of Practice for the Hydraulic Modelling of Sewers*. CIWEM, London.

WaPUG (2006). *Guide to the Quality Modelling of Sewer Systems*. WaPUG (CIWEM), London.

White M., Johnson H. and Anderson G. (1997). *Control of Infiltration to Sewers*. R175: 72. London, CIRIA.

WRc (1994). *The Sewerage Rehabilitation Manual*. WRc Engineering, Swindon, UK.

WRc (1995). *Use of Detention Tanks for Flow Attenuation*. ER 171E. WRc Engineering, Swindon, UK.

WRc (2004). *Sewers for Adoption*. WRc Engineering, Swindon, UK.

WRc (2006). *Sewers for Adoption*, 6th edn. WRc Engineering, Swindon, UK.

Chapter 11

Wastewater Treatment

11.1 INTRODUCTION

It is estimated that more than 1.2 billion people have no access to safe drinking water and over 2.4 billion lack basic sanitation. The need for proper sanitation was made explicit in the United Nations Millennium Development Goals (UNESCO, 2009). Goal number 7 urges for the reduction by half of the population living without proper sanitation.

The protection of our clean water resources and the sustainability of our water systems as such, is one of the biggest challenges of the 21st century. Water is essential for most living organisms on this planet. However, the growth of the earth's population and industrialisation has increased the need for clean water enormously, putting a large pressure on our water resources. Apart from this, the environment is also exposed to a growing amount of (untreated) wastewater.

The European Union recognised this problem for the first time in the 1970s and introduced several water directives in its environmental policy. Among them are the EU Urban Wastewater Directive (CEC, 1991), and the EU Water Framework Directive (CEC, 2000), which has among its goals 'water management based on river basins' and 'getting the citizens more involved'.

On the other side of the Atlantic, the US Environmental Protection Agency (EPA) adopted the National Pollution Discharge Elimination System (NPDES) as part of the Clean Water Act (CWA) in 1972 (USEPA, 2004). In their latest regulation, the EPA requires states to develop prioritised lists of polluted or threatened water bodies and to establish the maximum amount of pollutant that a water body can receive while still meeting water quality standards, the so-called total maximum daily load (TMDL).

11.1.1 Short history of wastewater treatment

Sewage has for a long time been considered a potential health risk and nuisance in urban agglomerations. The Ancient Greeks (300BC to 500AD) used public latrines which drained into sewers conveying the sewage and stormwater to a collection basin outside the city. The Romans took the system further and constructed the Cloaca Maxima. The system worked well, but collapsed along with the Roman Empire. The period between 450 and 1750 AD is therefore known as the 'Sanitary Dark Ages'. During this period the main form of waste disposal was simply to dispose of it on the streets using buckets. Around 1800, collection systems appeared in many cities since people did not want to put up with the smell and farmers welcomed this 'humanure'. The collected sewage did not undergo treatment. Instead, it was spread out over the land as fertiliser. However, water-logging became a major problem, as well as finding sufficient land due to the expansion of the cities. At that stage, the idea of using organisms within the treatment process started to grow; biofilms growing on rocks in the river bed were applied as biofilters. The activated sludge process was discovered in the UK by Ardern and Lockett in 1914. At that time, rivers were considered as part of the treatment process. But at some point, the increased load of sanitary waste due to city growth could no longer be accommodated by the rivers, which meant that higher requirements were set for the wastewater treatments plants to reach better removal efficiencies. Moreover, the toxic effect of ammonia on aquatic species introduced awareness of nitrification. This process did not always work well, especially in winter. Nitrate was not considered as being undesirable at that time. This viewpoint changed in the second half of the 20th century when eutrophication became a severe issue; it led to an explosive growth of algae through the fertilising effect of nitrate. As a consequence there was further understanding of the processes and denitrification was introduced with particular bacteria using nitrate as an electron acceptor. This led to the development of pre- and post denitrification systems and the introduction of recycled flows. As well as nitrogen, phosphorus also poses a eutrophication problem. Unlike nitrogen, however, phosphorus needs to be removed by converting it to a solid phase. This resulted in the development of chemical precipitation followed by tertiary filtration in the 1970s. Later, biological phosphorus removal was discovered by accident. However, it led to the established EBPR system we know today. The energy crisis of the 1970s associated with an increased demand for industrial wastewater treatment, shifted attention from aerobic to anaerobic treatment, which was especially important for concentrated and warmer industrial influents. After a century of urban development and expansion, the problems generated by rapid urbanisation became apparent. This raised the urgent need for the development of a whole range of new treatment processes. Also the demands of the increasing effluent triggered the need for upgrading wastewater treatment plants as well as the treatment of sludge effluents containing high loads of nitrogen and phosphorus. This led to the development of SHARON-ANAMMOX and the BABE-process. The control of plants became yet another point of attention, and this was a driver for development of mathematical models. More recently, aspects such as control of sludge properties, disinfection and micropollutants have received growing attention.

Today's municipal wastewater treatment (WWTP) in developed countries is commonly achieved through the activated sludge process. This process consists of

- mineralisation and conversion of the biodegradable components by a broad variety of micro-organisms
- nutrient removal (nitrification, denitrification, phosphorus removal)
- separation of the grown activated sludge from the purified water through either gravitational settling or membrane filtration

The efficiency of the entire process depends on the efficiency of the processes separately. The first two biological steps have found considerable attention in research, which has resulted in a sufficient understanding of the various biological processes and conversions that take place. This knowledge has been translated into mathematical models which are able to describe this degradation process adequately (Henze *et al.*, 1987; Henze *et al.*, 1995a; Henze *et al.*, 2000a and Henze *et al.*, 2000b). As for the second, separation step, the processes involved are not yet fully understood. However, models have been developed for these as well, but are not as advanced and conclusive compared to the biochemical models.

In some cases, technologies other than activated sludge are used, mainly to save costs and due their easier implementation. These are often beneficial for developing countries. Note that some of these cheap alternatives are not feasible in densely populated areas due to lack of space. The need for space is even sometimes problematic when upgrading existing facilities which have space limitations, resulting in having to use expensive solutions like membrane bioreactors. Examples of alternative treatment solutions are waste stabilisation ponds and reed beds. These are not discussed here.

This chapter has two major parts. The first focuses on wastewater characterisation and the different treatment steps. The second part discusses the most commonly used mathematical models for the different processes and the description of the wastewater treatment plant hydraulics.

11.2 WASTEWATER CHARACTERISATION

Wastewater can characterised by both its **quantity** and **quality**. The former refers to its flow rate, whereas the second is concerned with its composition. Wastewater can originate from municipal use, i.e. households, or from industry, or even a mixture of both if industry is allowed to discharge untreated wastewater directly into the sewers. All these resulting wastewaters have different specifications with respect to flow rate and composition, which have implications for their treatment. This chapter focuses mainly on municipal wastewater.

11.2.1 Wastewater quantity or flow

The flow of wastewater coming to a wastewater treatment plant is typically collected in a sewer system (see Chapter 10). The origins of the wastewater are households, industry, and rain (whether indirectly through infiltration to sewers or directly through rainfall-runoff to combined sewers). The actual flow coming to the treatment plant is also influenced by exfiltration taking place in the sewer.

A very important feature of wastewater flow or hydraulic load is that it is highly dynamic, which poses some serious challenges for the design and operation of wastewater treatment plants. First, a diurnal pattern can be recognised because of the typical water consumption pattern of people (Figure 11.1). A first peak can be observed in the morning when they wake up and take a shower or bath. However, the hydraulic load drops when they go off to work. After returning home, the flow increases again giving an evening peak. Overnight, the flow drops again. The occurrence of the peaks in time can be shifted slightly depending on the length and structure of the sewer system. Moreover, the maximum peak flow can be as high as 5–10 times the minimum flow.

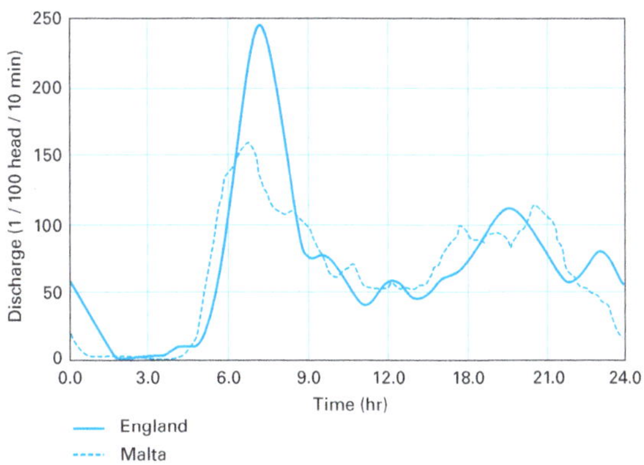

Figure 11.1 Examples of diurnal patterns of influent flows of different treatment plants (Henze *et al.*, 1995a)

Second, a differentiation is made between dry and wet weather flow (DWF vs WWF). This is strictly only applicable for combined sewer systems. WWF occurs when a rainfall event takes place in the catchment of the sewer system. It results in peaks to the treatment plant as much as 6 to 10 times the average DWF (being the average of a diurnal dry weather day). This is illustrated in Figure 11.2.

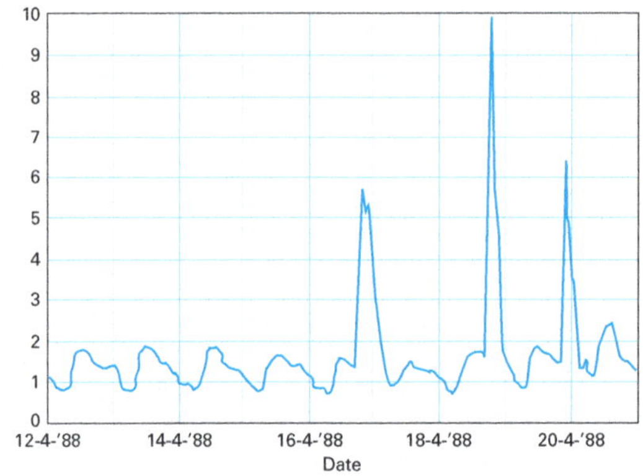

Figure 11.2 Examples of WWF diurnal patterns (Henze *et al.*, 1995a)

Sewer systems are designed in such a way that they can convey as much wastewater as possible to the treatment facility, but the maximum capacity is sometimes reached in combined sewers, leading to the need for combined sewer overflows (CSO). In order to limit discharges from CSOs to receiving waters, excess water is stored in retention tanks installed in the sewer system. The installation of such tanks increases the load to a treatment plant due to the need to empty the tanks as quickly as possible after a rain event to be ready for the next event. This may make a treatment plant subject to prolonged maximum hydraulic loads that would affect its performance. Finding the optimal compromise between the frequency of operation of combined sewer overflows and the treatment plant overload is the subject of considerable research (Vanrolleghem *et al.*, 2005).

In order to design a WWTP, it is important to have a good knowledge of the flow dynamics. Cumulative flow distribution diagrams (CFDD) can be used to summarise both diurnal and wet weather peak variations; see Figure 11.3. The CFDDs indicate the amount of time (expressed as % of days) that a certain flow rate is exceeded. They can be obtained from long-term historical measurements of sewer flows (for new plants) or from the WWTP influent flow (for plant extensions). As a rule of thumb, the 60percentile is used to represent the daily average flow, while the 85percentile serves as the maximum daily dry weather flow. The latter assumes that 15% of the days are actually wet weather days. These numbers can, of course, vary geographically.

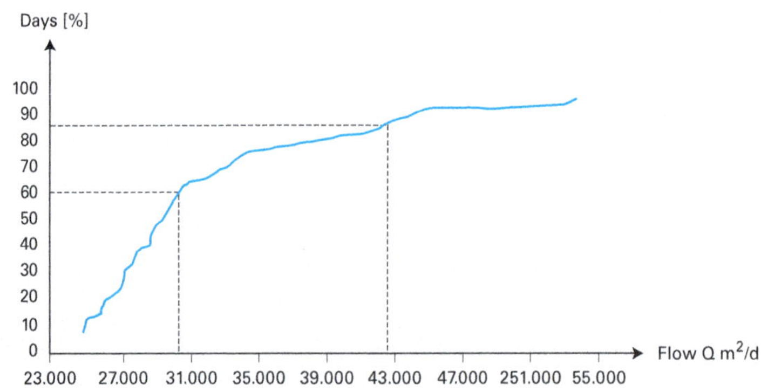

Figure 11.3 Example of a cumulative flow distribution diagram (Henze *et al.*, 1995a)

11.2.2 Wastewater constituents

An overview of different components that are likely to be present in wastewater along with their special interest and environmental effects is summarised in Table 11.1. This clearly illustrates why raw wastewater should not be discharged as such into receiving waters such as a river, lake or even coastal waters.

Table 11.1 Components in wastewater and their environmental effects (after Henze *et al.*, 2008)

Component	Of special interest	Environmental effect
Micro-organisms	Pathogenic bacteria, virus and worms eggs	Risk when bathing and eating shellfish
Biodegradable organic materials	Oxygen depletion in rivers, lakes and fjords	
Other organic materials	Detergents, pesticides, fat, oil and grease, colouring, solvents, phenol, cyanide	Toxic effect, aesthetic inconvenien-ces, bio accumulation
Nutrients	Nitrogen, phosphorus, ammonia	Eutrophication, oxygen depletion, toxic effect
Metals	Hg, Pb, Cd, Cr, Cu, Ni	Toxic effect, bio accumulation
Other inorganic materials	Acids, e.g. hydrogen sulphide bases	Corrosion, toxic effect
Thermal effects	Hot water	Changing living conditions for flora and fauna
Odour (and taste)	Hydrogen sulphide	Aesthetic inconveniences, toxic effect
Radioactivity		Toxic effect, accumulation

In order to quantify the composition of wastewater, several determinants have been proposed. They can be grouped in terms of the type of pollution.

- ***Organic pollution***:

Biochemical Oxygen Demand (BOD) represents the amount of oxygen required by a small amount of micro-organisms to degrade a wastewater sample's organic pollution in a small (litre-scale) batch experiment equipped with a Dissolved Oxygen (DO) sensor according to:

$$\text{Sewage} + \text{few micro-organisms} \rightarrow \text{growth} + \text{oxygen consumption} \tag{11.1}$$

BOD is measured over time, resulting in a cumulative curve that is highly temperature dependent; see Figure 11.4. BOD values can be determined after different time instants. An often used value is BOD_5, which is the reading of the consumed DO after 5 days. 5 days was chosen to speed up the measurement, assuming that most of the readily biodegradable organic pollution is degraded. BOD_7 uses 7 days, which was regarded as the hydraulic residence time of most rivers to their mouth. If the total BOD (BODt) is to be determined, then typically BOD_{28} is used. In Figure 11.4, there is an intermediate plateau after which a further increase occurs (beyond the dashed lines) due to nitrification, which is brought about by slower growing organisms and, hence, occurs at a later time. This is often referred to as NBOD (nitrogenous BOD) as opposed to CBOD (carbonaceous BOD).

Chemical Oxygen Demand (COD) represents the amount of oxygen required to completely chemically oxidise the organic pollution in a wastewater sample into carbon dioxide, water, ammonium and sulphate.

The chemical reaction is conducted at a high temperature and in a very acid environment. As not all organic pollution is biodegradable (some part is inert), the COD of a wastewater is always somewhat higher than the BODt. A typical value observed for municipal wastewaters is 1.33.

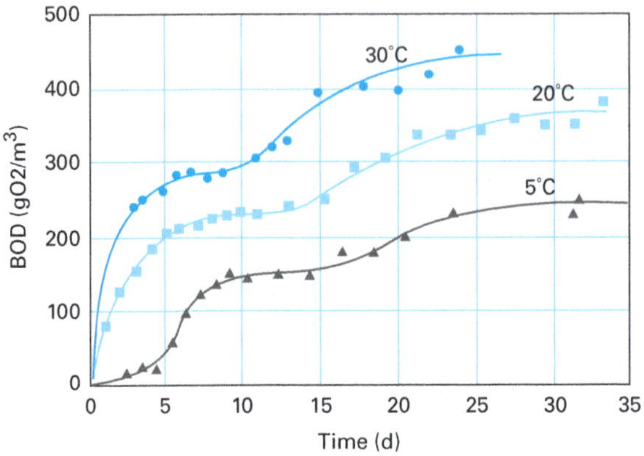

Figure 11.4 Illustration of typical BOD curves monitored at different temperatures. The oxygen consumption beyond the dashed lines represents the NBOD. Temperature significantly slows down the biodegradation processes

Unlike BOD, which always needs to be measured, COD can be calculated for specific components based on the elemental composition of the sewage components and the oxidation reaction. The former enables the computation of the molar mass of the component, whereas the latter yields the amount of oxygen molecules needed to perform the complete oxidation. Normalising to 1 g of a component yields the mass of oxygen that this 1 g of organic pollution represents. For acetate:

$$\text{Oxidation reaction: } CH_3COOH + 2O_2 \rightarrow 2\,CO_2 + 2\,H_2O \tag{11.2}$$

60 g of acetate (= $2 \times 12 + 2 \times 16 + 4 \times 1$) requires 2 molecules of oxygen or 64 g of oxygen. Hence, 1 g of acetate represents 64/60 g or 1.07 g of COD.

Exercise: compute the amount of COD represented by ethanol (CH_3CH_2OH), sewage ($C_{18}H_{19}O_9N$), biomass ($C_5H_7O_2N$), methane (CH_4), sulphide (H_2S).

As can be seen above, COD can only be computed when the elemental composition is available. If this is not available then an actual measurement is needed. Since the original standard measurement is time consuming and rather dangerous (due to the harsh environment) and typically many COD measurements are required, fast test kits with high reliability are now commercially available.

COD has the advantage over BOD in that it enables the mass to be balanced (closed) in the system. This is very important when including organic pollution in a mathematical model where it the mass has to be balanced (conserved). This is developed further below.

Total Organic Carbon (TOC) aims at measuring the total amount of carbon atoms present in the organic pollution. The technique uses a catalytic reaction at high temperature. Here also the TOC can be readily calculated when the elemental composition is known. For acetate this yields: per mole of acetate, 2 moles of carbon dioxide are produced. This means that 60 g of acetate produces 88 g of CO_2 or 24 g of carbon (because 44 g of CO_2 contains only 12 g of C; 88 × 12/44 = 24).

Obviously, TOC is only a measure for the organics that contain carbon and cannot be used for other organic components. Also, the measurement is more difficult and not easy, though fast test kits exist.

Exercise: compute the amount of TOC present in: ethanol (CH_3CH_2OH), sewage ($C_{18}H_{19}O_9N$), biomass ($C_5H_7O_2N$), methane (CH_4).

Total Suspended Solids (TSS) determines the dry solids present in the sample after drying at 105°C for 24 h. TSS includes all suspended solids and also mineral salts.

Suspended solids (SS) determines the dry solids in a filtered (0.24 micron) sample after drying at 105°C for 24 h. The difference with TSS is that it does not include the mineral salts.

Settleable solids are determined from a wastewater sample that has been presettled for 2 hours, and are the difference between the SS and the settleable solids. They are discussed in the sludge-water separation step below.

- ***Nitrogen content of wastewater***:

This form of pollution can be present as ammonium (NH4), NO_3. Organic nitrogen can be determined as total Kjeldahl-N or total-N. Standard methods are available for these measurements.

- ***Phosphorous content of wastewater***:

This form can be present as ortho-phosphate. Organic phosphorus can be determined as Total-P. Again, standard methods are available for all of these measurements.

- ***Heavy metals***:

Standard methods are also available for measuring Hg, Ag, Cd, Zn, Cu, Ni, Pb, As and Cr.

11.2.3 Wastewater composition

The wastewater from domestic use is often expressed in the unit Population Equivalent (PE), which can either be expressed in water volume or BOD. General worldwide accepted definitions are:

- 1PE = 0.2 m^3/d
- 1PE = 60 g BOD/d

These two definitions are based on fixed non-changeable values. The actual wastewater contribution from a person living in a sewer catchment, the so-called Person Load (PL), can vary considerably as shown in Table 11.2. The reasons for the variation can be: working place outside the catchment, socio-economic factors, lifestyle, type of household installation, etc. Hence, values vary regionally. PE and PL are both based on average contributions, and are used to give an impression of the loading to the wastewater treatment processes that are needed to deal with them. They should be calculated on long term time intervals (months-years).

Table 11.2 Variations in person load

Parameter	Unit	Range
COD	g/cap.d	25–200
BOD	g/cap.d	15–80
Nitrogen	g/cap.d	2–15
Phosphorus	g/cap.d	1–3
Wastewater	m^3/cap.d	0.05–0.4

The concentrations found in wastewater are a combination of pollutant load and the amount of water with which the pollutant is mixed. The daily or yearly polluting load may thus form a good basis for an evaluation of the composition of wastewater. The composition varies significantly from one location to another. At a given location the composition varies with time. This is partly due to variations in the discharged amounts of the various pollutants. However, the main reasons are the variations in water consumption by households and infiltration and exfiltration during the transport of the wastewater in the sewage system.

The composition of typical domestic/municipal wastewater is shown in Table 11.3 where concentrated wastewater (high) represents cases with low water consumption and/or infiltration. Diluted wastewater (low) represents high water consumption and/or infiltration. Stormwater further dilutes the wastewater as most stormwater components have lower concentrations compared to very diluted wastewater.

Table 11.3 Typical composition of raw municipal wastewater with minor contributions of industrial wastewater (in g/m^3) (after Henze *et al.*, 2008)

Parameter	High	Medium	Low
COD total	1,200	750	500
COD soluble	480	300	200
COD suspended	720	450	300
BOD	560	350	230
VFA (as acetate)	80	30	10
N total	100	60	30
Ammonia-N	75	45	20
P total	25	15	6
Ortho-P	15	10	4
TSS	600	400	250
VSS	480	320	200

The fractionation of nitrogen and phosphorus in wastewater has an influence on the possible treatment options for the wastewater. Since most of the nutrients are soluble, they cannot be removed by solid-liquid separation (e.g. settling, filtration). The distribution between soluble and suspended matter is given in Table 11.4.

Table 11.4 Distribution of soluble and suspended matter for medium concentrated municipal wastewater (in g/m^3) (after Henze *et al.*, 2008)

Parameter	Soluble	Suspended	Total
COD	300	450	750
BOD	140	210	350
N total	50	10	60
P total	11	4	15

Many treatment processes are based on biodegradation. Hence, the degradable fraction is another important characteristic of wastewater. Typical values are given on Table 11.5.

Table 11.5 Degradability of medium concentrated municipal wastewater (in g/m^3) (after Henze *et al.*, 2008)

Parameter	Biodegradable	Inert	Total
COD total	570	180	750
COD soluble	270	30	300
COD particulate	300	150	450
BOD	350	0	350
N total	43	2	45
Organic N	13	2	15
P total	14.7	0.3	15

During WWF, two phenomena typically occur (in WWTP connected to combined sewer systems):

- ***First flush***: This can be observed when the storm event occurs after a long period without rain, which has allowed the sedimentation of suspended solids in the sewer system. A first flush event results in a peak load of TSS to the plant; see Figure 11.5. This figure shows the same period as the second part of Figure 11.2. The peak brings about ten times the normal TSS load to the plant. This number is, however, dependent on the nature of the sewer system and the DWF period preceding the event: Figure 11.5 shows that a TSS peak does not occur during the second storm.

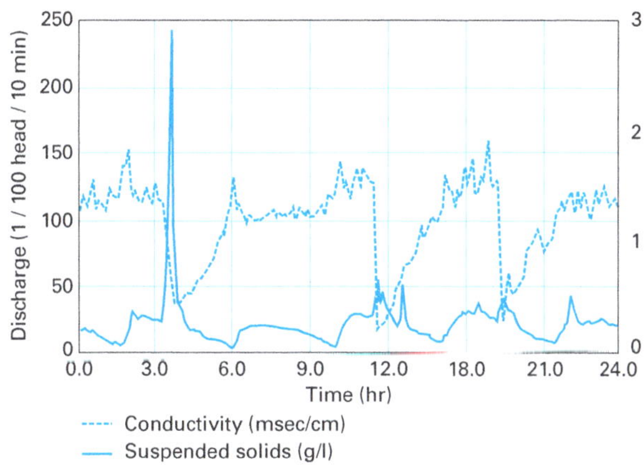

Figure 11.5 Illustration of the effect of wet weather flow on conductivity (dilution) and suspended solids (first flush effect) (Henze *et al.*, 1995a)

- ***Dilution***: A diluting effect occurs when there are large amounts of rainwater. Figure 11.5 illustrates this from conductivity measurements in a sewer. Unlike the first flush, dilution occurs with all storm events. Dilution is not beneficial for WWTP as the food to micro-organism ratio can be great reduced. One way to avoid this is to construct separate sewer systems for wastewater and rainwater. This is, however, very expensive, especially if a combined system is already in place.

11.3 TREATMENT STEPS

An overview of the different treatment steps is given in Figure 11.6.

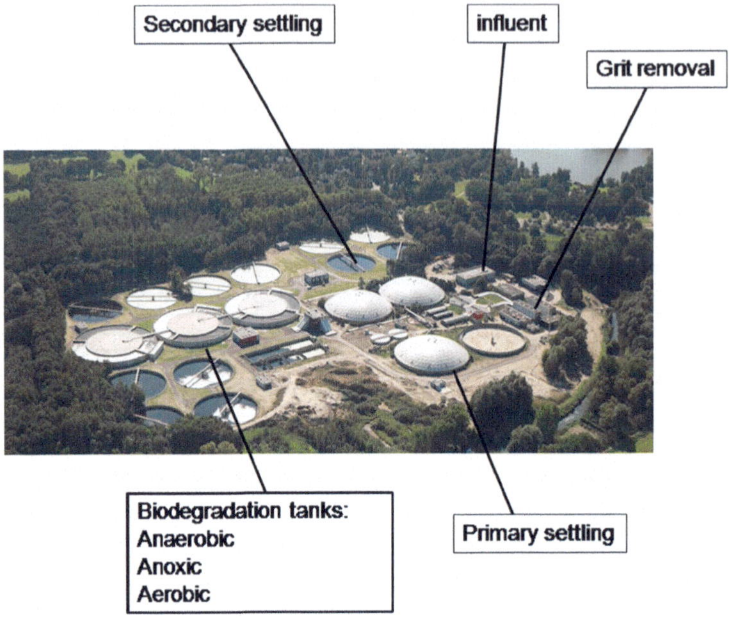

Figure 11.6 Illustration of the different treatment steps in biological wastewater treatment

11.3.1 Primary treatment

The primary treatment step consists of physical-chemical processes with the aim to remove:

- Grit and large material: this is achieved by filtration screens with certain pore sizes (typically 6 mm followed by 1 mm). This process is essential for membrane based systems to avoid clogging.
- Inert material: this material is not biologically degradable and accumulates in the reactors requiring considerable maintenance (i.e. costs). A typical example is sand. These units are referred to as sand traps and exploit the higher density of sand compared to other TSS components in wastewater.
- Difficult to degrade components: examples are oil and grease. Due to bad solubility and lower density, fat traps can be based on surface separation.
- Excess BOD: BOD is needed for nitrogen and phosphorus removal; see below. However, if the BOD load is too high compared to TN and TP (the indicative ratio is 100/10/1), it makes no sense to allow this excess BOD into the plant as it increases the aeration cost and sludge production. Therefore, this excess (that is, the slowly biodegradable BOD) is removed in primary settlers, which are sometimes preceded by a flocculation step to enhance the primary settling. The resulting 'primary' sludge is high in energy content and is typically sent to a sludge digester where energy can be recuperated in the form of biogas.

11.3.2 Secondary treatment

The secondary treatment step consists of two main steps: biological treatment processes and sludge-water separation. The latter produces water purified to a certain level that is either discharged to the receiving water body or further polished in a tertiary treatment step where effluent limits are not met or if the water is intended for reuse (industry, agriculture) or drinking water production. The sludge enters the sludge treatment step, also called the 'sludge line'.

The biological treatment is performed by the activated sludge process. Activated sludge is a mixed culture of micro-organisms that typically occur in suspended flocs. An illustration is given in Figure 11.7. Different micro-organism species take care of different specific biochemical reactions. The energy gained from these reactions is used for growth and cell maintenance.

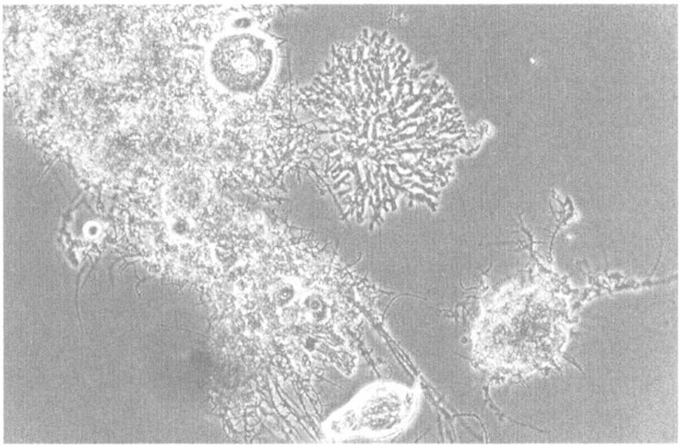

Figure 11.7 Microscopic photograph of an activated sludge floc

The growth of micro-organisms results in the multiplication of organisms. Next to an electron donor (C-source) and an electron acceptor (O_2 or NO_3), nutrients (N, P, S) and favourable environmental conditions are required. The basic reaction can be expressed in a simplified way as follows:

$$\begin{aligned}&\text{C-source} + NH_4 + PO_4 + H^+ + \text{electron acceptor } (O_2 \text{ or } NO_3)\\&\quad\rightarrow \text{Biomass } (C_5H_7O_2N) + \text{by-products } (H_2O, CO_2, N_2, NO_3)\end{aligned} \tag{11.3}$$

Since the biomass wants to grow, a number of compounds are converted:

- Organic pollutants $\rightarrow CO_2$ + biomass
- $NH_4 \rightarrow NO_3$
- $PO_4 \rightarrow$ Poly-P stored in biomass
- Organic pollutants (primary + secondary sludge) $\rightarrow$ biogas (CH_4, CO_2)

The compounds that are converted are those present in the wastewater as described above. The next important question that arises is how much of these components are converted. Process rates are typically described by the product of the *process kinetics* and the *process stoichiometry*. The former describes the rate of the conversion reaction, the latter describes the ratio of conversion of the different compounds.

Suppose the following reaction taking place (i.e. aerobic sewage degradation):

$$C_{18}H_{19}O_9N + O_2 + H^+ \rightarrow C_5H_7O_2N + CO_2 + H_2O \tag{11.4}$$

For each molecule of sewage degraded, a proportional amount of other products is used (all those components to the left of the arrow in Eq (11.4)) or produced (all components to the right of the arrow).

Adding this information can be done by including proportionality constants, yield or stoichiometric coefficients:

$$a\,C_{18}H_{19}O_9N + b\,O_2 + c\,H^+ \rightarrow d\,C_5H_7O_2N + e\,CO_2 + f\,H_2O \tag{11.5}$$

Typically, one of the coefficients is chosen to be equal to 1 and the others are normalised.

A reaction does not occur (i.e. the reaction rate is 0) when its sources or substrates (components to the left of the arrow in any process equation) are absent. On the other hand, when all components are available in excess, reaction rates are a maximum. The intermediate presence of components yields reaction rates between 0 and the maximum rate. This behaviour is typically described by the well-known Monod equation:

$$\mu(S) = \mu_{max} \frac{S}{K_S + S} \tag{11.6}$$

where μ is the reaction rate (1/s)
μ_{max} is the maximum reaction rate (1/s)
S is a substrate that the micro organisms can degrade for energy and growth
K_S is the half saturation constant or affinity constant.

K_S refers to that substrate concentration where the reaction rate equals half of the maximum reaction rate, i.e. $\mu_{max}/2$. This is illustrated in Figure 11.8 (left).

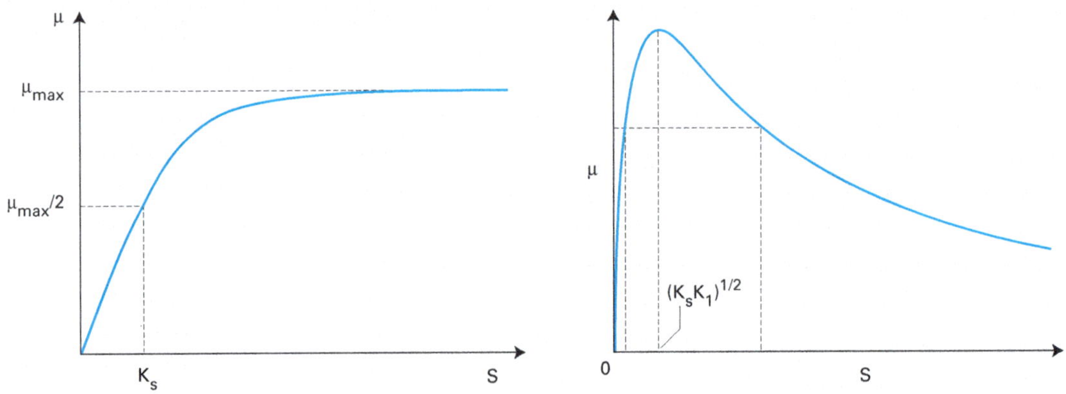

Figure 11.8 Illustration of Monod (left) and Andrews (right) kinetics

Sometimes substrates in excess can affect the maximum reaction rate negatively. In this case Andrews (or Haldane) kinetics can be used:

$$\mu(S) = \mu_{max} \frac{S}{K_S + S + S^2/K_I} \tag{11.7}$$

where K_I represents the inhibition constant.

The shape of this type of kinetics is shown in Figure 11.8 (right).

If the conversion rates for the different compounds are computed from the normalised reaction:

$$\mathrm{a\,C_{18}H_{19}O_9N + b\,O_2 + c\,H^+ \rightarrow 1\,C_5H_7O_2N + d\,CO_2 + e\,H_2O} \tag{11.8}$$

assuming Monod-type kinetics in the substrate and first order kinetics in the biomass concentration X (kg/m^3)

$$\rho(S) = \mu_{\max}\frac{S}{K_S + S}X \tag{11.9}$$

then:

$$\begin{array}{llll} \mathrm{C_{18}H_{19}O_9N:} & -\,\mathrm{a}\,\rho(\mathrm{S}) & \mathrm{C_5H_7O_2N:} & +\,1\,\rho(\mathrm{S}) \\ \mathrm{O_2:} & -\,\mathrm{b}\,\rho(\mathrm{S}) & \mathrm{CO_2:} & +\,\mathrm{d}\,\rho(\mathrm{S}) \\ \mathrm{H^+:} & -\,\mathrm{c}\,\rho(\mathrm{S}) & \mathrm{H_2O:} & +\,\mathrm{e}\,\rho(\mathrm{S}) \end{array} \tag{11.10}$$

This means that the conversion rate of a component consists of 3 parts:

- a sign (+/−) dependent on whether it is used or produced
- a stoichiometric coefficient in the reaction
- a rate of the reaction $\rho(S)$

When a compound appears in more than one reaction, the conversion rate takes one term per reaction. Suppose the following reactions are taking place with respective reaction rates $\rho_1(S)$ and $\rho_2(S)$:

$$\begin{array}{l} \mathrm{a\,C_{18}H_{19}O_9N + b\,O_2 + c\,H^+ \rightarrow 1\,C_5H_7O_2N + d\,CO_2 + e\,H_2O} \\ \mathrm{f\,CO_2 + g\,O_2 + h\,NH_4{}^+ \rightarrow 1\,C_5H_7O_2N + i\,NO_3 + j\,H_2O + k\,H^+} \end{array} \tag{11.11}$$

The conversion rates of those compounds occurring in both reactions are then given by

$$\begin{array}{llll} \mathrm{C_5H_7O_2N:} & +\,1\,\rho_1(\mathrm{S}) + 1\,\rho_2(\mathrm{S}) & \mathrm{O_2:} & -\,\mathrm{b}\,\rho_1(\mathrm{S}) - \mathrm{g}\,\rho_2(\mathrm{S}) \\ \mathrm{CO_2:} & +\,\mathrm{d}\,\rho_1(\mathrm{S}) - \mathrm{f}\,\rho_2(\mathrm{S}) & \mathrm{H^+:} & -\,\mathrm{c}\,\rho_1(\mathrm{S}) + \mathrm{k}\,\rho_2(\mathrm{S}) \end{array} \tag{11.12}$$

The latter can be generalised for compound S_i taking part in j reactions:

$$r(S_i) = \sum_j \mathrm{sign}(ji)v_{ji}\rho_j \tag{11.13}$$

where ρ_j is the rate of the jth reaction in which S_i participates (1/s)
v_{ji} the stoichiometric coefficient for S_i in the jth reaction
sign(ji) is the sign (−/+) indicating whether S_i is consumed or produced in the jth reaction.

As mentioned above, biological processes are impacted by the environmental conditions. Typical examples of influencing factors are temperature and pH. The former changes diurnally and seasonally, whereas the latter can be influenced by a spill in an industry, but can also be impacted by biological processes like (de)nitrification, consumption of fatty acids (e.g. acetic acid) and CO_2-stripping through aeration. Since the impact on conversion rates can be rather large, they need to be accounted for.

After the measurement of the conversion rates at different temperatures in a large number of sludges, the solid black line along with a variability band in Figure 11.9 was found.

It can be clearly observed that in the lower temperature range the conversion rate increases with temperature. As a rule of thumb, the conversion rate doubles for an increase of 10°C. The exponential increasing behaviour is not valid anymore above 25°C. On the contrary, above 30°C, a decrease can be observed, due to the denaturation of enzymes (note that this does not apply for mesophilic and thermophilic organisms present in digestion processes). However, most WWTP are operated in the exponential increasing region. Hence, reaction rates are typically corrected using an exponential equation:

$$\mu(T) - \mu(20) \,.\, e^{k(T-20)} \tag{11.14}$$

Here $\mu(20)$ is the conversion rate at 20°C (1/s)
T is the current temperature (°C)

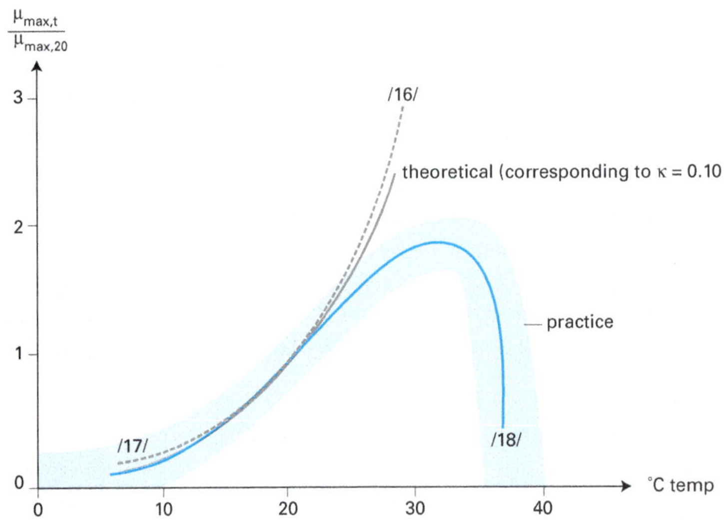

Figure 11.9 Illustration of the impact of temperature on maximum specific growth rate of micro-organisms

Knowing this temperature dependency, it can be deduced that the WWTP suffers when exposed to cold temperatures, i.e. in winter time in moderate and cold climates, when conversion rates are lower. This means that sludge residence times should be sufficiently high to avoid a sludge washout. This is especially true for slow growing organisms. As shown below nitrification, or the biological process responsible for NH_4-removal, is governed by slow growing organisms and, hence, it is vulnerable during winter. Temperature is, therefore, a parameter that must be accounted for during system design.

A similar vulnerability holds for pH variations. Micro-organisms have an optimal pH-range in which their maximum conversion rate is a maximum, roughly ranging from 6.5 to 8.5. Outside this range, a deterioration in the conversion rate occurs, as illustrated in Figure 11.10.

pH changes may occur due to processes that either produce or consume H^+ or when *pH* shocks occur in the influent (e.g. due to an illegal discharge from industry). *pH* inhibition can be included in models through

$$\mu(pH) = \mu(pH_{opt})\frac{K_{pH}}{K_{pH} - 1 + 10^{|pH - pH_{opt}|}} \tag{11.15}$$

It is often discarded when the sludge has enough buffer capacity (alkalinity) to avoid *pH* upsets. It is recommended to account for this phenomenon when modelling industrial WWTP.

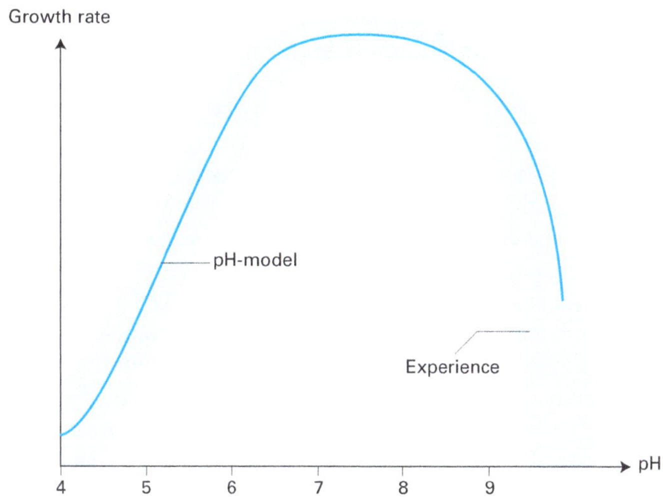

Figure 11.10 Illustration of the impact of pH on maximum specific growth rate of micro-organisms

Several specific **biological treatment processes** can be distinguished in an activated sludge system:

- **Aerobic organic substrate removal**: This process is conducted by a consortium of heterotrophic organisms, that is, organisms that use an organic C-source. Dissolved oxygen is required as electron acceptor resulting in

$$C_{18}H_{19}O_9N + O_2\,(+\,H^+) + NH_4 \rightarrow C_5H_7O_2N + CO_2 + H_2O \qquad (11.16)$$

  As can be seen, the biomass requires both sewage and nutrients for respiration and cell growth. The amount of substrate COD being built into biomass COD is referred to as the yield. Heterotrophic organisms exhibit a high yield of 0.4 g biomass COD/g substrate COD.

- **Biological nitrogen removal**:
  An overview of the different biological nitrogen removal processes is shown in Figure 11.11.
  - *Nitrification*: This process is conducted by a consortium of autotrophic organisms, that is, organisms that use an inorganic C-source such as CO_2. Dissolved oxygen is required as an electron acceptor resulting in

$$NH_4 + CO_2 + O_2 + \rightarrow C_5H_7O_2N + NO_3 + H_2O + H^+ \qquad (11.17)$$

    As can be seen, NH_4 is oxidised to NO_3 and the reaction has an acidifying effect. Autotrophic organisms have a lower yield: 0.24 g biomass COD/g N, and also exhibit a lower growth rate. The latter makes them vulnerable to process disturbances such as pH, T and toxic components. Therefore, nitrification is often the first process that is lost from an upset system. It is, however, highly unwanted given the toxic nature of NH_4 when discharged to the surface water.

    The process is actually a 2-step process ($NH_4 \rightarrow NO_2 \rightarrow NO_3$) carried out by different organisms (AOB and NOB). When the first step is hampered, this can lead to a build up of NO_2, which is toxic to many biological processes in the activated sludge system. This should be avoided. Complete nitrification consists of steps 1 and 2.
  - *Denitrification*: This process is mostly conducted by heterotrophic organisms (DEN), although some steps can be performed by AOB organisms as well. Unlike aerobic organic carbon removal,

denitrification only occurs in an anoxic environment, that is, when dissolved oxygen is absent and NO_3 is used as electron acceptor. The resulting reaction is

$$C_{18}H_{19}O_9N + NO_3 + H^+ + NH_4 \rightarrow C_5H_7O_2N + CO_2 + H_2O + N_2 \tag{11.18}$$

which indicates the usefulness of the reaction, i.e. the reduction of NO_3 to N_2 which volatilises from the system. Therefore, nitrogen is removed from the wastewater. Note that some portion of the N is built into biomass. The reaction has a relatively high yield of 0.3 g biomass COD / g substrate COD and actually performs both N and COD removal. Note that the dissolved oxygen used in the nitrification step is reused here for the oxidation of COD via NO_3. Complete denitrification consists of steps 3 to 6. Recent research has revealed that incomplete denitrification can lead to the production of nitrous oxide which is a strong greenhouse gas and should be avoided.

- *Anammox*: This recently discovered process is performed by special organisms that are capable of oxidising NH_4 using nitrite. It is especially useful as it saves both aeration and carbon addition costs. However, it only works if no nitrate is formed, which requires a special environment to avoid step 2 in the scheme. This process is especially promising for high N-loaded wastewater streams (e.g. reject water from digester supernatant).

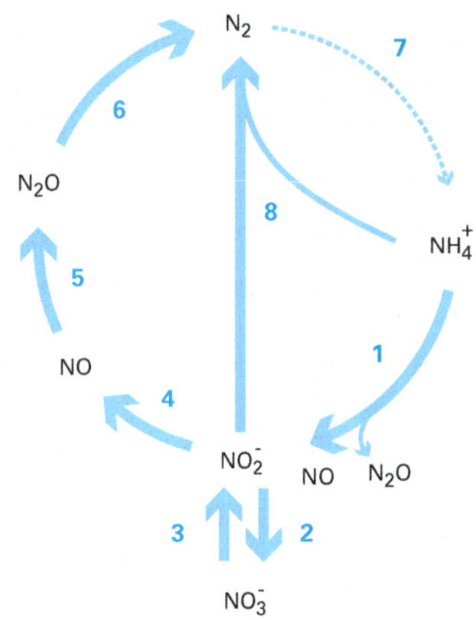

Figure 11.11 Overview of the different processes involved in biological nitrogen removal (after Kampschreur *et al.*, 2009) (1) Aerobic ammonia oxidation (Ammonia Oxidizing Bacteria – AOB), (2) aerobic nitrite oxidation (Nitrite Oxidizing Bacteria – NOB), (3) nitrate reduction to nitrite (denitrifiers – DEN), (4) nitrite reduction to nitric oxide (AOB and DEN), (5) nitric oxide reduction to nitrous oxide (AOB and DEN), (6) nitrous oxide reduction to N_2 (DEN), (7) nitrogen fixation (not relevant in most WWTPs), (8) ammonium oxidation with nitrite to N_2 (Anammox)

- **Biological P-removal**: This process is also conducted by heterotrophic organisms (called Phosphorous Accumulating Organisms or PAO) and is rather complex as it requires a sequence of anaerobic (absence of O_2 and NO_3) and anoxic/aerobic conditions. The mechanism is illustrated in Figure 11.12. During anaerobic conditions, PAOs take up easily degradable organic substrate (volatile fatty acids or VFA) and store them in the form of poly hydroxyl alkanoates (PHA) inside the cell. The energy required by the PAOs to take up the VFA is obtained by releasing phosphate from their internal poly-phosphate

pool. To convert the VFA into PHA reducing equivalents are needed that are provided by degradation of glycogen, another storage product that is essential for biological P-removal. During the next phase, i.e. during anoxic or aerobic conditions, PAOs oxidise the internal PHA pool and use the energy to grow, store phosphate as polyphosphate and regenerate the glycogen pool that they require in a next passage under anaerobic conditions. Thanks to the growth of the PAOs, more P is stored than was originally released, resulting in a net uptake of P from the wastewater.
The net reaction is given by

$$VFA + PO_4 + NH_4 + O_2\ (+NO_3) \rightarrow C_5H_7O_2N + CO_2 + H_2O + \text{Poly-P}\ (+\ N_2) \tag{11.19}$$

Also, this process exhibits a fairly high yield of 0.3 g biomass COD/g substrate COD. As can be noted, the process performs N, P and COD removal. However, due to the complexity of the process, it is difficult to operate in a stable way in practice.

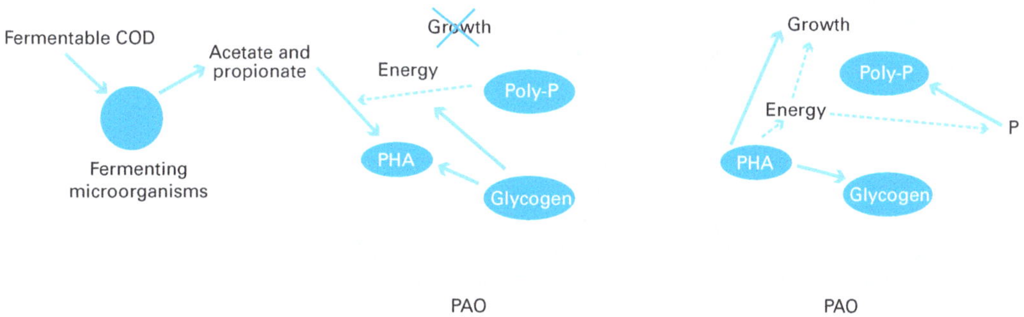

Figure 11.12 Illustration of the behaviour of PAOs under anaerobic (left) and aerobic/anoxic (right) conditions (after Henze *et al.*, 2008)

- **Anaerobic digestion**: This process is part of the sludge treatment process. It takes settled sludges from both the primary and secondary settler as input. The process is also complex and takes place at an elevated temperature of 35°C where specialised organisms (acidogens, methanogens) exhibit their optimal growth conditions. In a first step large dissolved molecules and particles are extracellularly hydrolysed into small dissolved molecules by the acidogens, thereby producing acid molecules. In a second step, a second consortium of organisms, the methanogens, use the acids and hydrogen to produce biogas, i.e. CH_4 and CO_2. The process can be summarised as

$$C_{18}H_{19}O_9N + NH_4 \rightarrow C_5H_7O_2N + CO_2 + H_2O + CH_4 \tag{11.20}$$

The organisms are slow growers, however, necessitating rather long sludge ages of about 20 days. Moreover, H_2 is inhibitory and, hence, a delicate balance needs to be established to keep the process running. A more detailed overview of the 2 steps is illustrated in Figure 11.13.

The different required environmental conditions (anaerobic, anoxic, aerobic) make the plant configuration rather complex as the different conditions are established based on a spatial separation of the different processes.

A first typical configuration that was built merely for COD-removal and nitrification (if the sludge age is high enough) consisted of a single or sequence of aerated reactors (Figure 11.14a). This was common practice in Europe back in the 1980s, when nutrient removal was not obligatory, and is still widely observed in the USA. However, in the 1990s European legislation with regard to nutrients became more stringent and existing plants were retrofitted to achieve nutrient removal as well, and new plants were built already containing nutrient removal. This led to a second typical configuration including either pre- or post-denitrification, requiring an additional anoxic tank either in front of or after the aerobic tank (Figure 11.14b) along with a recirculation flow. The latter only achieves N-removal, and, hence, to include bio P-removal, yet another configuration has to be used. The latter is often referred to the UCT-process as it was designed at the University of Cape Town (South Africa); it adds an anaerobic tank (Figure 11.14c). This also requires several recirculation flows.

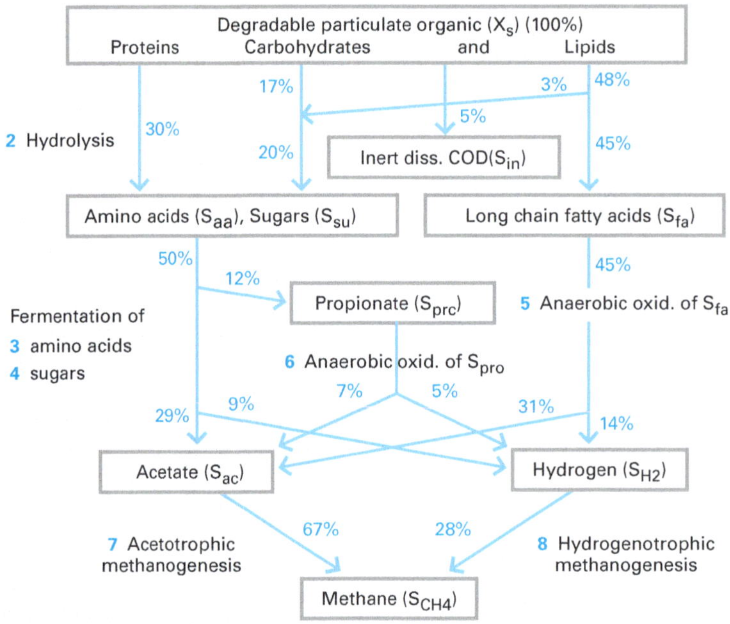

Figure 11.13 The different degradation steps in anaerobic digestion of sludge

Next to these 3 configurations, a number of modified configurations have been developed in the course of the last decades. They cannot be discussed here in detail and, therefore, the reader is referred to literature for more detailed information on the configurations (Tchobanoglous *et al.*, 2003 and Henze and Comeau, 2008).

To overcome some of the shortcomings of the conventional activated sludge process, including the sensitivity of some processes to sludge age and the spatially complex configurations needed, an alternative technology has been developed, namely, biofilm reactors. If the residence time of organisms in the process is smaller than the inverse of their growth rate, the organisms are washed out of the system. Organisms try to counteract this by forming either flocs that settle in the secondary clarifier or by growing as a biofilm attached to a surface. This allows them to stay sufficiently long in the system to survive.

An old technology that uses this biofilm principle is the trickling filter. It uses plastic carrier material that serves as surface for the biofilm to grow on. Water is then trickled over the carriers from the top, and oxygen is provided through natural ventilation. This has the advantage of being simple, performs reasonably and exhibits cheap aeration. The latter cuts costs, making this an appealing alternative for developing countries. Disadvantages of the technology are clogging and the attraction of flies that feed on the biomass.

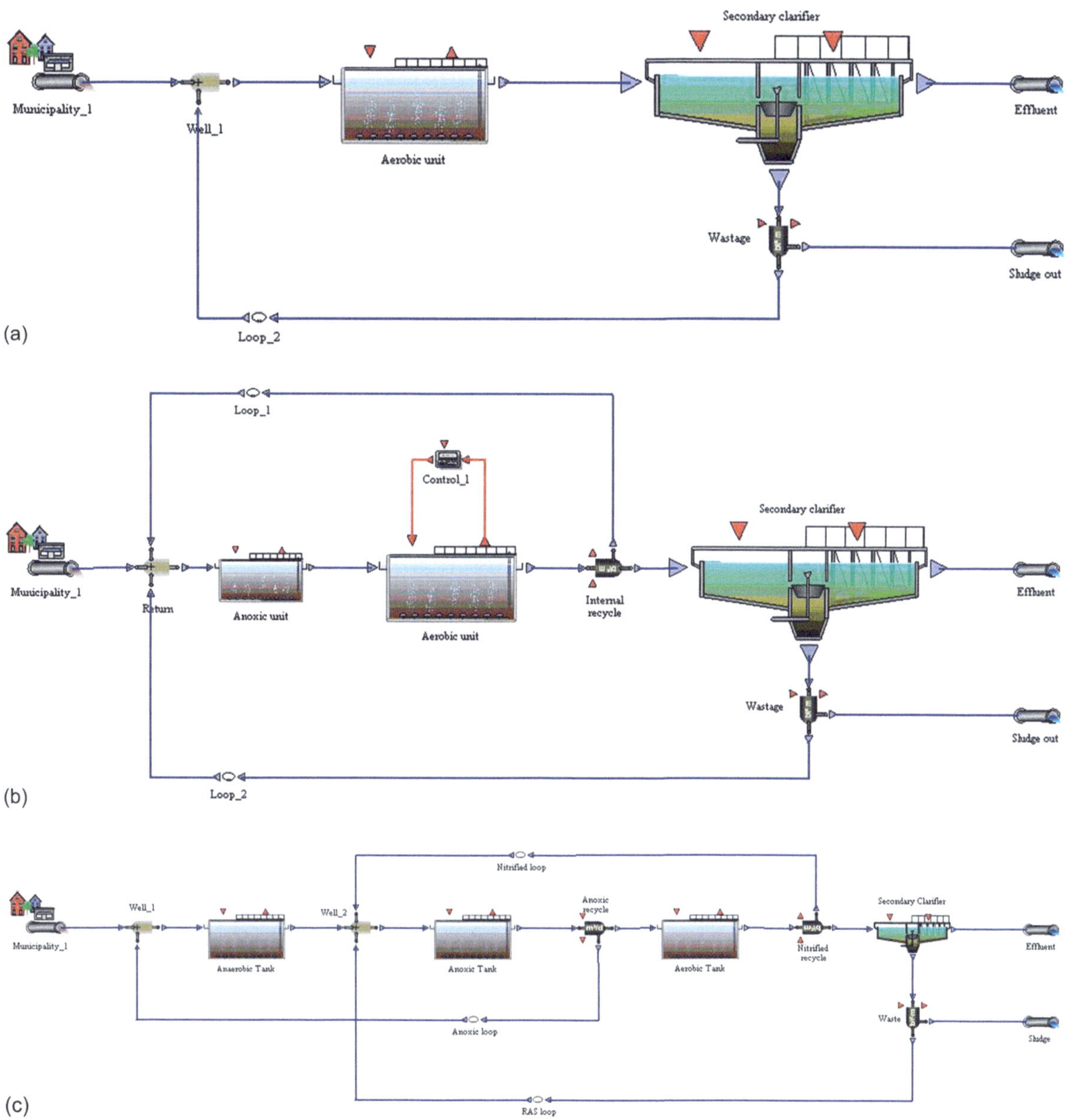

Figure 11.14 Illustration of different typically used configurations of activated sludge systems: (a) carbon removal and nitrification; (b) predenitrification system and (c) UCT process (screenshots courtesy of WEST®; www.mostforwater.com)

More recent processes based on biofilm technology are Moving Bed Bio-Reactors (MBBR), integrated Fixed-film Activated Sludge (IFAS) and (an)aerobic granule reactors.

Biofilm processes govern microbial conversion in a thin film of micro-organisms sitting on a carrier (in the case of granule reactors, the biomass acts as carrier itself). An important feature of biofilms is the diffusion of bulk products into the biofilm, giving rise to gradients along the depth of the biofilm. These gradients obviously affect the actual processes taking place locally and their conversion rates. However, an advantage is the fact that also electron acceptors like dissolved oxygen and nitrate need to diffuse. This leads to zones with different conditions, allowing several of the previously described processes to take place in one location (see Figure 11.15) compared to having to construct separate tanks for all of them and use expensive recirculation flows. Instead, different layers in the biofilm perform different processes simultaneously. As indicated in Figure 11.15, dissolved oxygen only diffuses to a certain depth providing only a small aerobic top layer of the biofilm where aerobic carbon degradation and nitrification can take place (reference the bump in NH_4 in this layer). Since NO_3 is produced in this aerobic layer, it diffuses deeper into the biofilm, resulting in an anoxic zone. With the presence of readily biodegradable BOD, NO_3 is depleted. This results in a third anaerobic zone where sulphate reduction and methane formation take place (note that in the outer aerobic and anoxic layers SO_4 is produced and methane is oxidised).

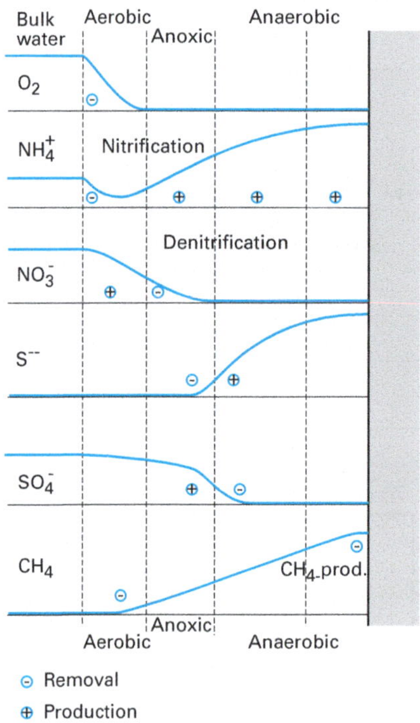

Figure 11.15 Illustration of the build-up of different zones due to diffusion limitations in a biofilm

Seemingly elegant from a theoretical point of view, making this biofilm technology work requires good control of the biofilm thickness which is a tedious task, influenced by both biological growth and physical forces (e.g. shear).

Water-sludge separation

As a final step in the secondary treatment, the purified water and the sludge need to be separated. This can be achieved through either gravitational settling or membrane filtration.

So-called secondary settlers are used with **gravitational settling**. Figure 11.16 shows an example of a circular clarifier (rectangular clarifiers also exist, but the principle is the same). Sludge enters the clarifier in the centre energy dissipation well and then flows into the flocculation well where large flocs are formed that settle faster. The sludge then flows in radial direction towards the peripheral launder or weir. In the mean time, the sludge flocs can settle out and form a sludge blanket. The performance of the settling is dependant on the sludge settling properties. As the settling velocity is decreasing with increasing sludge concentration, the design sludge concentration in the biological tanks is actually limited to 3–4 g/l by the settling step.

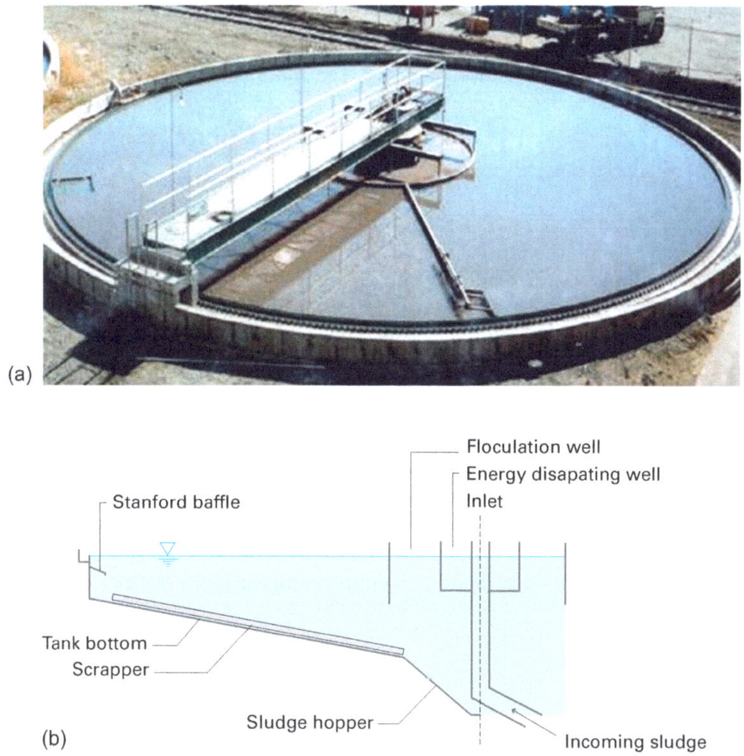

Figure 11.16 Illustration of a circular secondary clarifier (a) and an axisymmetric sideview (b)

The clarified water flows over the weir and forms the effluent of the treatment plant. It typically contains a non-settleable fraction which is in the order of 5–50 mg SS/l. The sludge moves towards the sludge hopper (the combination of the floor slope and a sludge scraper that is attached to the bridge that rotates constantly around the clarifier at a slow velocity). The collected sludge exhibits an elevated sludge concentration of about 8–10 g/l and is either directly disposed of or fed to the sludge treatment line.

More detailed information on the working principles of secondary clarifiers can be found in the IWA Scientific and Technical report (Ekama *et al.*, 1997).

An alternative technology that has been used in the last decade is membrane filtration in so-called membrane bioreactors (MBR). Instead of gravity, this uses a pressure drop over a membrane with pore sizes smaller than the size of a microbial cell (1 μm). Therefore, the process is often referred to as microfiltration. The process has some clear advantages: the effluent quality is superior (0 mg SS/l), biomass in the biological reactors can be operated at higher concentrations, which also implies a smaller footprint as less volume is required. However, the process has the major disadvantage of the fouling of the membrane surface. The latter is cured using air scour and liquid movement. However, these make MBRs very energy consuming. Even though the unit price of membranes has dropped significantly over the last decade, the technology is still twice as expensive from an operational point of view (€1 vs €0.5/m^3 of treated water). This has kept the technology from conquering the wastewater treatment market. MBRs are designed in two configurations, either submerged or side-stream (Figure 11.17). The former is the most widely used as it does not require an additional cross-flow over the membrane unit. However, the sidestream configuration can be operated with significantly higher fluxes (50 vs 20l/m^2/h), which reduces the installed membrane surface to treat the same volume of water. Also, a variety of membrane products exist. Submerged systems either use flat sheet or hollow fibre membranes. Both operate in the outside-in mode. Sidestream MBR use tubular membranes. For more detailed information on MBR, see Judd, 2006 or Henze *et al.*, 2008.

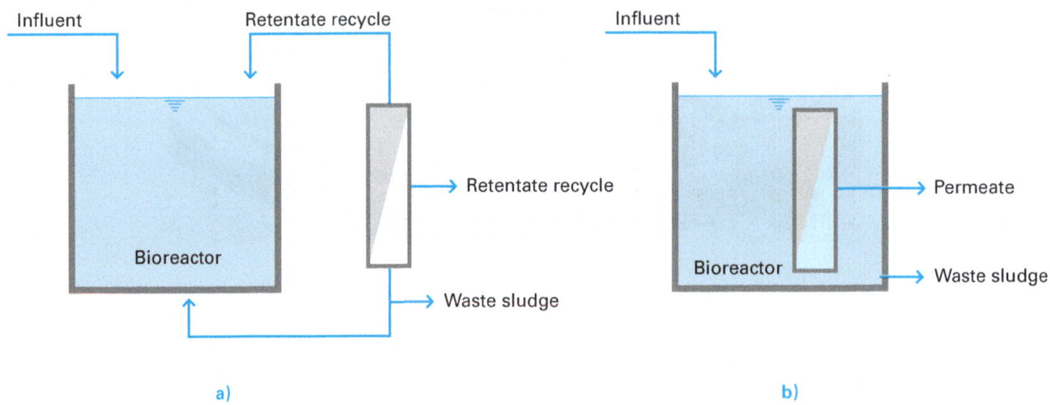

Figure 11.17 Illustration of the working principle of membrane bioreactors (MBR): sidestream configuration (a) and submerged configuration (b)

Sludge line

In the early days of the activated sludge process, the excess sludge produced was directly disposed of by incineration, landfill or as an agricultural fertiliser. However, due to stricter legislation, the options of disposing sludge have decreased drastically in Europe. The main options remaining are incineration or sludge digestion in which biogas and, hence, energy is recovered from this waste product.

The classic way of sludge treatment includes a sludge thickening step in order to increase the dry solids concentration drastically. This is then followed by a sludge dewatering step, which is typically brought about by using a coagulation step (using polymers) and a belt press or centrifuge. The removed water is recycled to the entrance of the biological tank.

A more recent practice of sludge treatment is to use digestion. The digestion step is included in between the thickening and dewatering step; see Figure 11.18. Since the digestion step releases both N and P into the liquid, the return flow from the digester to the entrance of the plant ('reject water stream') contains high loads of NH_4 and PO_4. Either, this extra load should be accounted for at the plant design stage, or it should be reduced through an additional treatment step (e.g. Sharon-Anammox). Recently, efforts have also been made to precipitate the nutrients as struvite ($(NH_4)MgPO_4{\cdot}6H_2O$) which allows their recovery and separation. However, this requires additional crystallisation and separation steps.

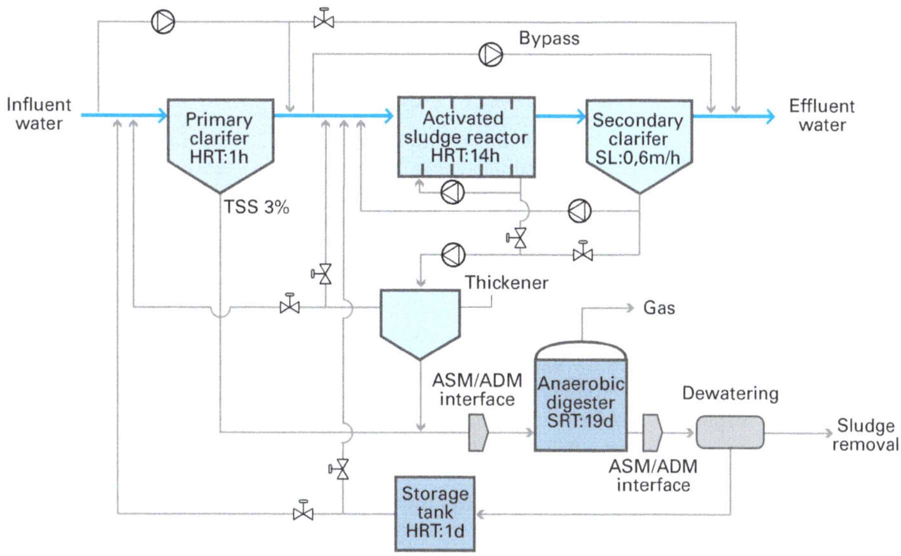

Figure 11.18 Illustration of the layout of the Benchmark Simulation Model no. 2 (BSM2), an example of a plant-wide model (after Nopens *et al.*, 2009)

A major difference in the dewatering effluent between the classic and digestion approach is the fact that nutrients are again released during the digestion process. Recycling them to the entrance of the plant results in an extra nutrient-load to the plant. This needs to be accounted for during design. Recently, efforts are being made in the field of nutrient recovery, by trying to precipitate the nutrients into struvite, which can be directly reused as fertiliser.

11.3.3 Tertiary treatment

Depending on the effluent requirements and the technology used, an additional treatment step might be required. MBR technology typically delivers a superior effluent quality (with all TSS removed), however, at a significantly higher cost. Other technologies might suffer from some remaining TSS that contributes to the effluent quality. Typical ways of dealing with this is to add a tertiary polishing step consisting of sand filtration (TSS removal and biofilm-based (de)nitrification), and micro- or nanofiltration. To remove micro-pollutants ozonation or the addition of Powdered Activated Carbon (PAC) may be adopted, and in some situations even reverse osmosis is applied to generate high purity water.

For more information on this matter the reader is referred to Tchobanoglous *et al.*, 2003.

11.4 MODELLING OF WASTEWATER TREATMENT PLANTS

In general, mathematical process modelling can be used for two purposes:

- Increase understanding of processes by testing certain hypotheses and confronting them with experimental data. This is referred to as 'model use for process analysis'
- Use well-established models for the design of the system and optimise its operation.

Hug *et al.* (2009) provide a more extensive overview of objectives for modelling, the type of modellers and the environment in which the modelling work takes place (Table 11.6).

Table 11.6 Categories that structure the demands on modellers for different jobs with wastewater treatment models (after Hug *et al.*, 2009)

Objective of model work:	Type and level of model work:	Environment for model work:
• Academic research (as graduate student, research staff or faculty) • Industrial research and development of processes and equipment • Plant design and optimization • Process control • Model-based support for plant operation • Model-based support for strategic and political decisions • Communication of a complex or dynamic system behaviour • Software development • Approving modeling studies (e.g. as supervising agency) • Teaching and training with different targets at different levels	• *Basic model users* (performing simple simulations without need for in-depth understanding, possibly advised by advanced model user) • *Advanced model users* (performing complex simulation studies requiring sound knowledge of possibilities and limitations) • *Model developers* (developing or adjusting model equations and structure) • *Software programmers* (implementing models into software, might need to deal with user interfaces or with numerical solutions) • *Instructors* (training one of the above listed groups of modellers)	• Universities (learning, teaching, research) • Consulting companies and equipment suppliers of different specialty and size • Wastewater treatment plants • Software developers and suppliers • Supervising agencies • Professional organizations

In this chapter the focus is on the description of models established through system analysis. When modelling a WWTP, models have to be selected to describe:

- Hydraulics
- Mixing
- Biological processes
- Aeration
- Sludge-water separation: settling or filtration

For each process a variety of models exist ranging from very simple to very complex. The selection of the model complexity depends on the objective of the modelling study. Hence, the problem definition is the first issues that the modeller should resolve. The art of selecting the model complexity can be based on personal expertise or on dedicated model selection techniques. The latter is, however, beyond the scope of this chapter and can be found elsewhere (Dochain and Vanrolleghem, 2001). For further discussion on these issues see Chapter 6.

11.4.1 Modelling hydraulics

In Chapter 5 mass balance leads to an equation consisting of a transport term and a rate term. The former needs to reflect the hydraulic behaviour of the system. In a WWTP completely stirred tank reactors (CSTRs) are often used. The hydraulics may not be accounted for in detail, implying that incoming and outgoing flow rates are the same and the reactor volume is therefore constant. This is a valid assumption when steady state system behaviour is modelled. However, when one is interested in accurate dynamic behaviour, more detailed modelling is required to correctly capture the related phenomena. This is illustrated by Figure 11.19, which is taken from de Clercq *et al.* (1999). Here, a step is imposed on the inflow at t = 0 and an additional step in the recycled flow after t = 0.5 h. The results show that a flow measurement at the back of the WWTP does not continue to contain steps, but rather a smooth transition. This cannot be modelled by a fixed volume tank because it would predict a step down for the decrease in inflow rate and not even notice the step down in the recycled flow. If this is to be accounted for, there is a need to use either a detailed hydraulic model (e.g. InfoWorks or CFD), or a variable volume approach where

$$\frac{dV}{dt} = Q_{out} - Q_{in}$$
$$Q_{out} = f(V) \quad \wedge \quad f(V) = \alpha(V - V_{\min})^{\beta} \tag{11.21}$$

where V is the volume (m^3)
V_{min} is the minimum volume (m^3)
Q_{out} is the outflow (m^3/s)
Q_{in} is the inflow (m^3/s)
β is a constant

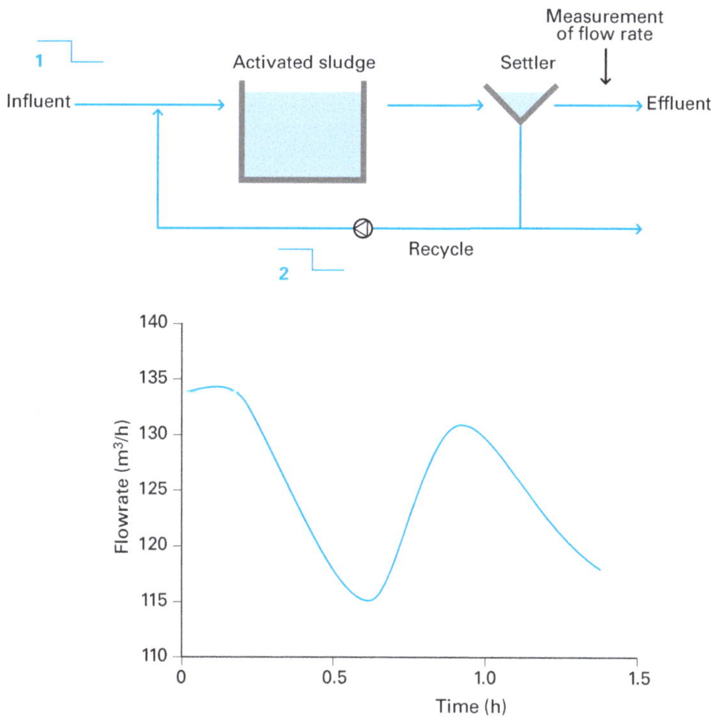

Figure 11.19 Illustration of the measured and simulated dynamic response on a step down change in the influent flow rate (at t = 0) and the recycle flow rate (at t = 0.5 h) (after De Clercq *et al.*, 1999)

The actual outgoing flow rate is computed based on a power law depending on the degree with which the current volume V exceeds the minimum volume of the tank (V_{min}). In the latter case, Q_{out} is zero. The parameter β is determined by the shape of the outflow works and takes values between 1 and 2. When applying this model to the example in Figure 11.19, a good description of the measured flow rate at the outflow can be achieved. The variable tank approach is, however, not used frequently because it adds equations to the model and slows down the computations numerically. In some cases, such as the modelling of MBR, this approach is recommended in order to deal with hydraulic shocks while maintaining a constant flux. If the shocks are ignored, unrealistically high fluxes need to be allowed through the membrane.

11.4.2 Modelling of mixing

A proper description of the mixing in tanks can be done with the aid of the advection-dispersion equation:

$$\frac{\partial C(x,t)}{\partial t} + U(x,t)\frac{\partial C(x,t)}{\partial x} - D\frac{\partial^2 C(x,t)}{\partial x^2} = \rho \tag{11.22}$$

where C is the concentration (g/m^3)

U is the velocity (m/s)

ρ is the rate or conversion term (biological sources and sinks) (g/m^3/s)

As this partial differential equation (PDE) requires specific solvers that are computationally slow, another approach based on a discretisation of the space may be used. This results in the so-called tanks-in-series (TIS) model

$$\frac{dV_k C_k}{dt} = Q_k(C_{k-1} - C_k) + \rho_k \text{ with } k = 1, \ldots, N \tag{11.23}$$

where N is the number of TIS

V_k is the settling velocity (m/s)

Several approaches exist to determine the number of TIS to be used:

- Empirical approach:

$$N = \frac{7.4}{WH} LQ_{in}(1 + r) \tag{11.24}$$

  where W, H, L respectively represent the dimension of the tank, i.e. width, height, length, (m)

  Q_{in} represents the incoming flow rate (m^3/s)

  r is the recycle ratio.

  This is obviously a simple and cheap method of determining the number of TIS. However, the method is less accurate compared to more complex approaches. It can also be observed that the number changes drastically during wet weather flow when Q_{in} can increase up to 3 times.
- Tracer test: An alternative approach is to perform a so-called tracer study by injecting a pulse or step of an inert tracer at the reactor inlet. The inert nature of the tracer is important as the substance should not be adsorbed by the sludge and should not be degraded. The time series of recovered tracer is then measured at the outlet of the reactor under study. Models with different numbers of TIS can then be constructed and made to fit the observed data. This option is obviously more time-consuming and expensive, but is more accurate. Note that this also does not account for any wet weather flow, unless the tracer test is repeated with different inflow rate conditions.

Typical model-based responses to tracer injection are given in Levenspiel (1999). The pulse injection case is reproduced in Figure 11.20. It should be noted that any recycling should be switched off when performing a tracer test to allow for the simple interpretation of the tracer test data.

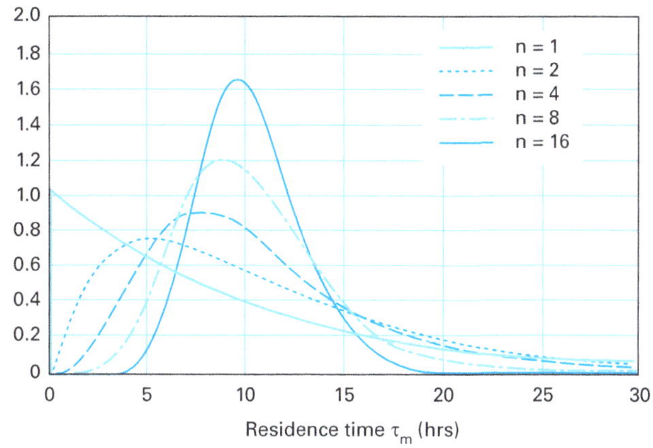

Figure 11.20 Illustration of the effect of a different number of CSTRs in series to describe the mixing behaviour; the *y*-axis represents the residence time distribution (RTD), τ_m represents the dimensionless time based on the mean residence time in all *N* tanks

From Figure 11.20 it can be observed that using one tank gives rise to the typical exponential decay model response of a CSTR. Increasing *N* results in the build up of a peak, which becomes higher in absolute value and whose occurrence is shifted further in time. This is the typical behaviour of a complete plug-flow system. When *N* reaches infinity, a perfect plug flow response is approximated, that is, all the tracer appears at once at the outlet.

An example of a tracer test performed at full-scale is shown in Figure 11.21. The plant consists of 3 parallel lanes, where one lane itself consists of 2 parallel lanes. The question is: how to model this plant from a hydraulic point of view. A tracer test can be performed, where a pulse of inert tracer is introduced in the influent splitter (large arrow) and the recovered tracer is measured at the outlet of the 3 lanes (Erlenmeyers).

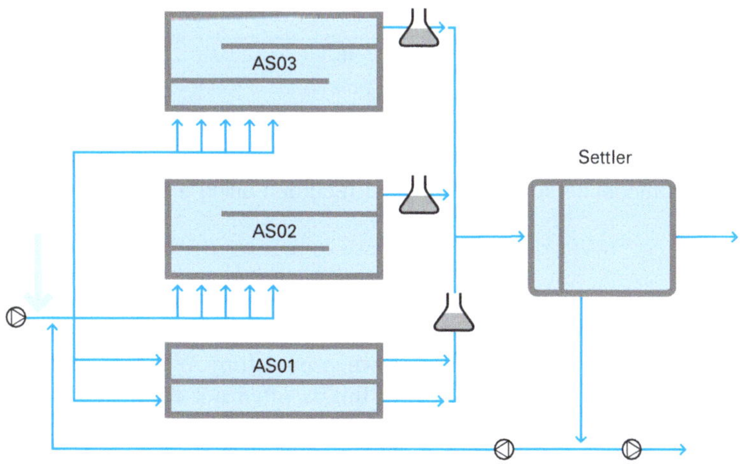

Figure 11.21 Illustration of the setup of a tracer study: point of tracer addition (large arrow) and sampling points (Erlenmeyers)

The results of the recovered tracer concentration versus the time of occurrence in lane AS01 with different TIS models are shown in Figure 11.22. It can be observed that using one TIS is not sufficient to model the mixing behaviour. However, when using two TIS with different volumes, the performance of the model is satisfactory. Similar results are obtained for the other 2 lanes. Note that different volume reactors are often required to fit the data.

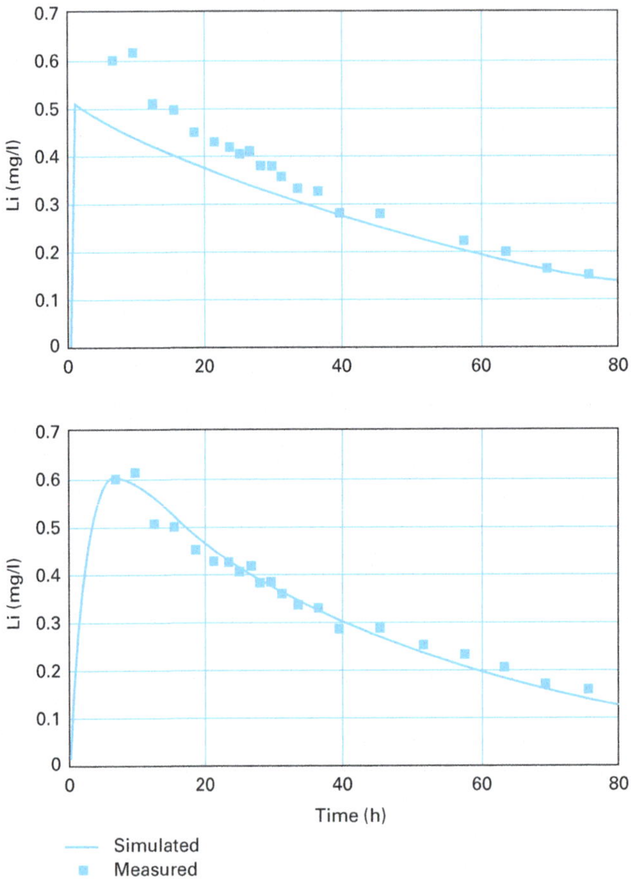

Figure 11.22 Illustration of model fits to RTDs in lane 1 recorded using a tracer test: using 1 CSTR (top) and 2 CSTRs in series (bottom)

The configuration that can then be used in a simulator platform is given in Figure 11.23. As can be seen, all lanes can be modelled using two TIS; a total of 8 CSTRs is needed to model the system.

Flow splitting works also deserve some attention when modelling WWTP. Splitting works are typically modelled as ideal flow splitters. However, in reality this is often not the case, and this can have a huge impact on the calibration part of a WWTP modelling exercise. One elegant way to investigate an influent splitting work is to investigate the sludge concentrations that occur in the different lanes. Theoretically, these should be the same when an equal loading is provided. The respective sludge concentrations in the different lanes for the previous example are given in Figure 11.24. In the case of an equal flow distribution, a line through the origin with slope 45° should be found. It can clearly be observed that this is the case for lanes 1 and 3. However, lane 2 exhibits a much larger sludge concentration, suggesting that a higher sludge load is going to that lane. Closer investigation of the influent splitting works reveals shortcut flows of both the influent and the return activated sludge from the secondary settlers. Accounting for this results in the slightly modified model configuration (Figure 11.23 right) including a more detailed influent splitting works description.

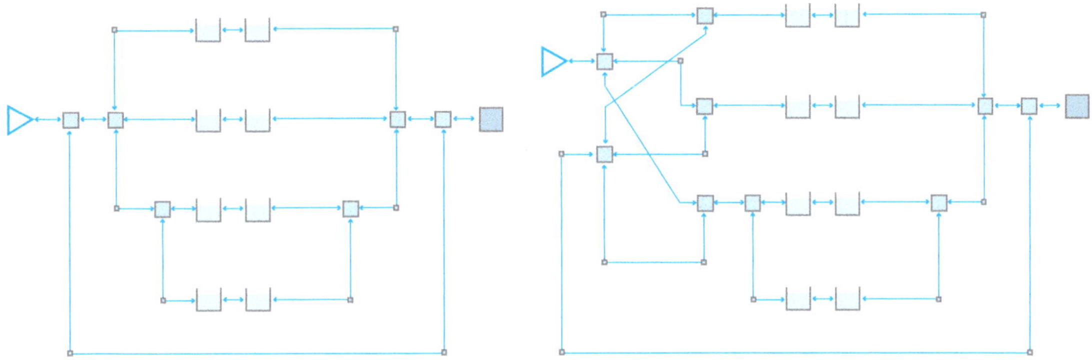

Figure 11.23 Illustration of how to account for flaws in a flow splitter (for explanation, see text)

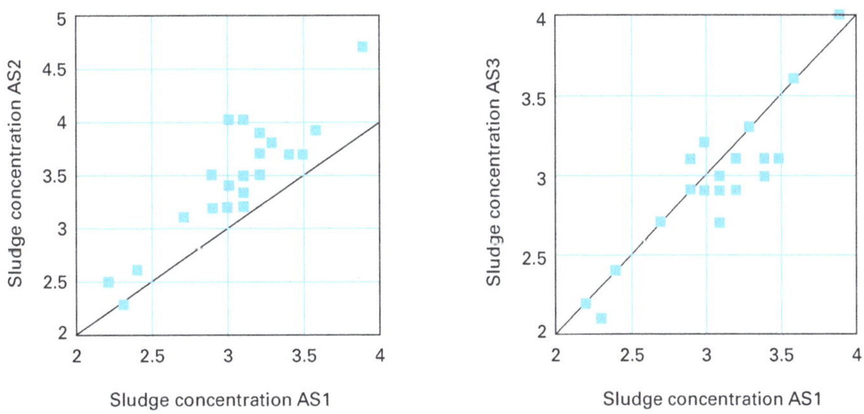

Figure 11.24 Illustration of differences in sludge concentration in different lanes caused by an improperly functioning flow splitting works

Another instructive example of a tracer test is shown in Figure 11.25 where samples are taken at different locations along a reactor. This clearly illustrates the ability of the TIS approach to model the mixing behaviour.

With regard to setting up a tracer experiment, some recommendations can be formulated:

- Choice of tracer: the tracer should be a non-biodegradable, non-adsorbing compound. Often used tracer materials are a Lithium salt and rhodamine.
- Injection of the tracer: the tracer should be injected as close as possible to the entrance of the reactor under study

- Data collection: samples should be taken as close as possible to the exit of the reactor under study. Samples should be collected at least during three times the hydraulic residence time. During this period about 20–50 samples should be taken.

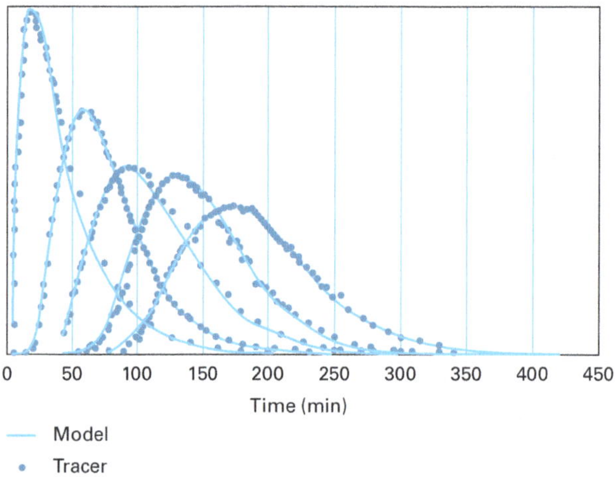

Figure 11.25 Conductivity measurements at different locations along a long reactor and a TIS-model describing the mixing properties of the reactor

11.4.3 Biological process modelling

Modelling of the biological processes in activated sludge started back in the 1970s and 1980s. The first milestone was the Activated Sludge Model no. 1 – ASM1 (Henze *et al.*, 1987). This was published as a Scientific and Technical Report from IAWQ and was innovative in respect to:

- nomenclature: all soluble components were addressed by a capital 'S', whereas particulate material carried the capital 'X'. Recently, an extension of this standardized nomenclature has been presented (Corominas and Rieger, 2010)
- the goal of the model and, hence, its focus was on a good prediction of sludge production, oxygen consumption and nitrogen removal
- the COD based modelling: as mentioned above, when using COD, it is possible to close the mass balance, which is not possible for BOD, for example.
- the use of a Gujer matrix for the compact representation of a set of ordinary differential equations; an example of a Gujer matrix is given in Figure 11.26, which is composed of the following parts:
    - the left column lists all *j* processes accounted for in the model
    - the top row lists all the *i* different components taking part in the processes
    - the right most column lists the process rates for the respective processes in the left column
    - the core part of the matrix represents the stoichiometric coefficients.
    - the right bottom cell indicates the kinetic (occurring in the right column) and the stoichiometric (occurring in the matrix core cells) parameters
    - the bottom row indicates the units that the different components are expressed in.

From the Gujer matrix one can derive the conversion term of the mass balances for all components by summing all the products of the stoichiometric coefficient v_{ij} of process j and the respective process rate ρ_j (only the process in which the component i takes part).

Components i	1: S_O	2: S_S	3: X_H	Process rate equation ρ_j
List of processes j				
Aerobic growth	$-\frac{1}{Y_H}+1$	$-\frac{1}{Y_H}$	+1	$\mu_H^{max} \cdot \frac{S_S}{K_S+S_S} \cdot X_H$
Lysis		+1	-1	$b_H \cdot X_H$
Observed transformation rate r_i	$r_i=\sum_j \nu_{j,i} \cdot \rho_j$ $[M_iL^{-3}T^{-1}]$			
Definition of stoichiometric parameters: Y_H Heterotrophic yield coefficient $[M_H M_S^{-1}]$	Dissolved oxygen (O_2)	Dissolved organic substrate (COD)	Heterotrophic biomass (COD)	Definition of kinetic parameters: μ_H^{max} Maximum specific growth rate $[T^{-1}]$; K_S Saturation coefficient for substrate $[M_{COD}L^{-3}]$; b_H rate constant for decay $[T^{-1}]$

Figure 11.26 Illustration of a Gujer matrix

From Figure 11.26, the following processes can be deduced to take place:

- Heterotrophic aerobic growth on readily biodegradable substrate (Ss)

$$\frac{1}{Y}S_S + \frac{1-Y}{Y}S_O \xrightarrow{\check{\mu}\frac{S}{K_S+S}X_B} X_B \tag{11.25}$$

where Y is the yield coefficient, i.e. the amount of biomass produced per unit substrate S_S
The equation is normalised to produce 1 mole of biomass

- Biomass decay

$$X_B + S_O \xrightarrow{bX_B} \tag{11.26}$$

The equation expresses the amount of dissolved oxygen that is required for endogenous respiration (i.e. respiration in the absence of an exogenous substrate). Note that we do not need to know what the biomass decays into.

When all the compounds are expressed in consistent units, all horizontal sums of the stoichiometric rows should equal 0, which otherwise provides a verification of mass leaks. This is called the continuity check. It can easily be verified that continuity is maintained for both processes and, hence, the entire model (note that oxygen is expressed as negative COD).

Finally, the mass balances for the 3 components can be written down easily using the Petersen matrix:

- Biomass X_B

$$\frac{dX_B}{dt} = \frac{Q}{V}(X_{B,in} - X_B) + 1\frac{\check{\mu}S_S}{K_S + S_S}X_B - 1bX_B \tag{11.27}$$

where μ is the growth rate (1/s)

The first term on the right hand side is the transport term (not derivable from the Petersen matrix). Term 2 and 3 are the respective contributions of the processes 1 and 2 to the conversion rate of X_B. The stoichiometric coefficients are shown explicitly for clarity.

- Readily degradable substrate S_S

$$\frac{dS_S}{dt} = \frac{Q}{V}(S_{S,in} - S_S) - \frac{1}{Y}\frac{\breve{\mu}S_S}{K_S + S_S X_B} \tag{11.28}$$

There is only one process consuming S_S.

- Dissolved oxygen S_O

$$\frac{dS_O}{dt} = \frac{Q}{V}(S_{O,in} - S_O) - \frac{1-Y}{Y}\frac{\breve{\mu}S_S}{K_S + S_S}X_B - 1bX_B \tag{11.29}$$

Two processes are consuming dissolved oxygen. As $S_{O,in}$ is typically zero, a term for aeration should be added. This process is, however, modelled separately and not included in the Gujer Matrix.

Aeration can be modelled in different ways. An often used model is one based on the oxygen transfer coefficient k_La. This coefficient is the product of k_L, the oxygen mass transfer coefficient from gas to water, and a, the surface area of the gas-water interface. It is therefore a measure for the amount of O_2 that is transferred from the gas phase to the liquid phase, where it becomes available for the biomass. Different aeration systems are used. Surface aerators bring the liquid in close contact with the headspace air (e.g. by splashing droplets into the air). Fine bubble diffusers are installed at the bottom of a tank and blow in compressed air as bubbles. For both systems the following reaeration model can be used (and is added as a term in the S_O mass balance of the ASM type models) expressed as Specific Oxygen Transfer Rate ($SOTR$ – g O_2/l/d):

$$SOTR = k_La(S_{O,sat} - S_O) \tag{11.30}$$

The other parameter used in this model, $S_{O,sat}$ is the saturated dissolved oxygen concentration in water. The reaeration rate is therefore dependent on the driving force being the difference between the actual S_O and the saturated dissolved oxygen concentration and the oxygen mass transfer. The saturated dissolved oxygen concentration is heavily dependent on temperature, whereas k_La depends on sludge concentration.

The different COD components considered in ASM1 are given in Figure 11.27.

As can be observed, the total COD is broken down into 3 main fractions: biodegradable, non-biodegradable and active biomass. The former two fractions are further split into soluble and particulate fractions yielding

- soluble (readily) biodegradable COD: S_S
- particulate (slowly) biodegradable COD: X_S
- soluble inert COD: S_S
- particulate inert COD: X_S

The biomass COD is split up into a heterotrophic (X_{BH} or X_H) and an autotrophic portion (X_{BA} or X_A), both governing specific processes as explained above.

The flow of COD in ASM1 is shown in Figure 11.28.

As can be observed, a readily biodegradable substrate S_S is used as a substrate in the aerobic growth of the heterotrophic biomass X_H. The latter decays and produces a fraction that is slowly biodegradable (X_S) and a fraction that is inert, which accumulates in the system. X_S, originating from biomass decay and also present in the influent can be hydrolysed back to S_S. Hydrolysis is the process where biomass produces enzymes to break down large molecules that are too large for them to take up through their cell membrane into small molecules that the membrane can accommodate. In the top group of processes in Figure 11.28 are the processes governing the nitrogen removal. Ammonium (S_{NH}) is used along with dissolved oxygen (S_O) to produce autotrophic biomass (X_A) and nitrate (S_{NO}). X_A further decays partially into particulate inerts (X_I) and particulate biodegradable COD (X_S).

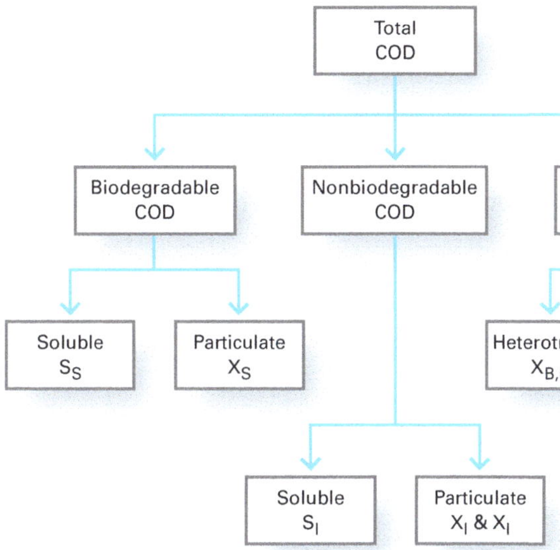

Figure 11.27 COD components of ASM1

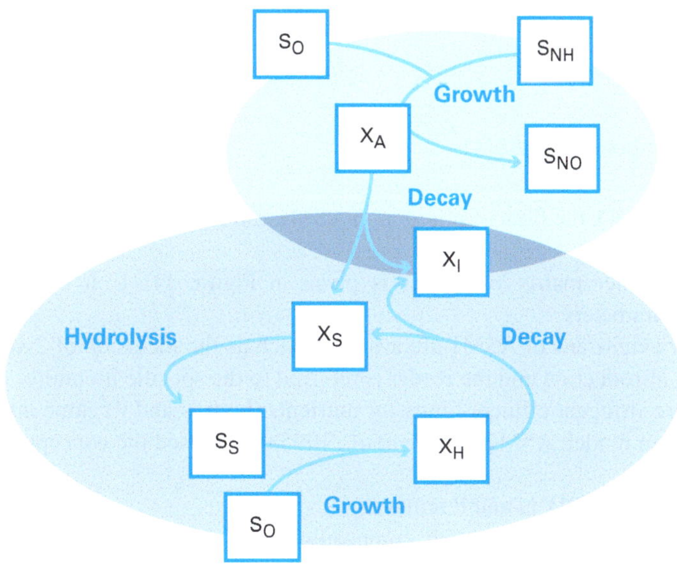

Figure 11.28 Flow of COD in ASM1

The different *N* components considered in ASM1 are given in Figure 11.29. Total nitrogen is the sum of nitrate (S_{NO}) and total Kjeldahl-N (TKN). The latter is further broken down into ammonium S_{NH}, organically bound N and active biomass N (X_{NB}), (the latter not being an actual component in ASM1, but merely defined through a N-fraction of biomass (i.e. 8%)). The organically bound N is partitioned further into soluble and particulate fractions which both have their biodegradable (S_{ND} and X_{ND}) and non-biodegradable (S_{NI} and X_{NI}) fractions. Again, the latter two are not actual components, but rather fractions.

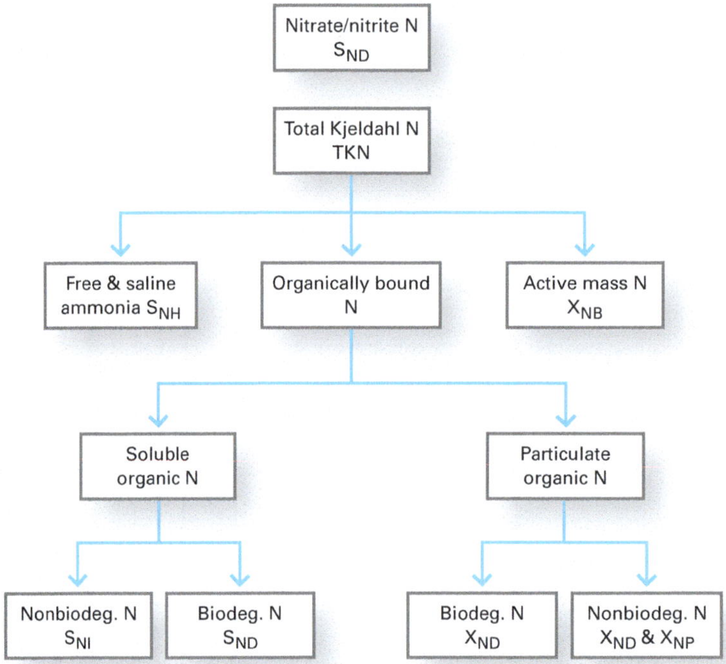

Figure 11.29 Nitrogen components of ASM1

In a similar way as for COD, the flow of nitrogen in ASM1 can be summarised in a self-explanatory graph; see Figure 11.30.

Finally, the complete Gujer matrix of ASM1 is given in Figure 11.31. It consists of 8 processes, 13 components and has 19 parameters.

In the literature many extensions of ASM1 are available such as the inclusion of 2-step nitrification. This is beyond the scope of this introduction and the reader is referred to the specific literature.

During the 1990s more stringent effluent limits for nutrients (both N and P) came into force. This led to the development of a follow-up model: ASM2 (Henze *et al.*, 1995a). It reused the concepts of ASM1 and its major assets are

- combined processes for COD, N and P removal
- a cell internal structure using an 'average' composition
- a TSS state variable in place of X
- chemical precipitation

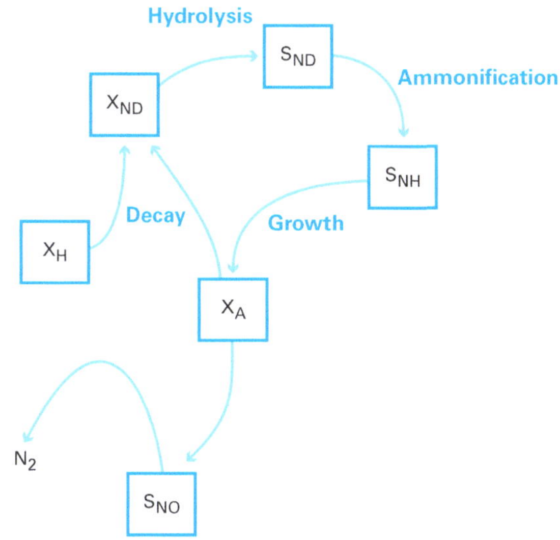

Figure 11.30 Flow of N in ASM1

	i	1 S_I	2 S_S	3 X_I	4 X_S	5 $X_{B,H}$	6 $X_{B,A}$	7 X_U	8 S_O	*Process rate*
j	**Process**									
1	Aerobic growth heterotrophs		$-\frac{1}{Y_H}$			1			$-\frac{1-Y_H}{Y_H}$	$\mu_{mH}\left(\frac{S_S}{K_S+S_S}\right)\left(\frac{S_O}{K_{OH}+S_O}\right)X_{B,H}$
2	Anoxic growth heterotrophs		$-\frac{1}{Y_H}$			1				$\mu_{mH}\left(\frac{S_S}{K_S+S_S}\right)\left(\frac{K_{OH}}{K_{OH}+S_O}\right)\left(\frac{S_{NO}}{K_{NO}+S_{NO}}\right)\eta_g X_{B,H}$
3	Aerobic growth autotrophs						1		$-\frac{4.57-Y_A}{Y_A}$	$\mu_{mA}\left(\frac{S_{NH}}{K_{NH}+S_{NH}}\right)\left(\frac{S_O}{K_{OA}+S_O}\right)X_{B,A}$
4	Decay heterotrophs				$1-f_P$	-1		f_P		$b_H X_{B,H}$
5	Decay autotrophs				$1-f_P$		-1	f_P		$b_A X_{B,A}$
6	Ammonification									$k_a S_{ND} X_{B,H}$
7	Hydrolysis organic compounds		1		-1					$k_H \frac{X_S/X_{B,H}}{K_X+X_S/X_{B,H}}\left[\left(\frac{S_O}{K_{OH}+S_O}\right)+\eta_h\left(\frac{K_{OH}}{K_{OH}+S_O}\right)\left(\frac{S_{NO}}{K_{NO}+S_{NO}}\right)\right]X_{B,H}$
8	Hydrolysis organic N									$\rho_7 (X_{ND}/X_S)$
		Unbiodegradable soluble matter (g COD m^{-3})	Readily biodegradable substrate (g COD m^{-3})	Particulate unbiodegradable matter	Enmeshed slowly biodegradable substrate (g COD m^{-3})	Active heterotrophic biomass	Active autotrophic biomass	Unbiodegradable particulates from cell	Oxygen (g (-COD) m^{-3})	

Figure 11.31 The Gujer Matrix for Activated Sludge Model No. 1

It is beyond the scope of this introduction to describe ASM2 in detail. Some specifications are given instead. The Gujer Matrix of ASM2 contains 19 processes and 19 components and not all variables are determined upfront. Continuity is guaranteed for COD, N, P and charge. ASM2d is a minor extension of the original ASM2 including anoxic P-uptake and PAO growth which was not included in ASM2, but yet observed in practice (Henze *et al.*, 1999). Therefore, ASM2d is the most frequently used model when a system including bio-P removal is to be modelled.

ASM3 was introduced in 2000 (Henze *et al.*, 2000a). This is merely a rewrite of ASM1 correcting some deficiencies observed during 10 years of using ASM1 in practice. Two major changes in concept were made:

- Endogenous respiration instead of death-regeneration
- Shift from hydrolysis to storage → cell internal components

Continuity is guaranteed for COD, N and charge.

A summary of the family of ASM models and their applicability is given in Table 11.7.

Table 11.7 Overview of the specific features and applicability of the ASM family of models

Model	Used for
ASM1	COD and N removal, death-regeneration
ASM2	COD, N and P removal
ASM2d	Extension of ASM2 with anoxic P-uptake
ASM3	COD and N removal, endogenous respiration; metabolic storage

Nowadays ASM1 is still the most widely used model in practice since it has proven its robustness and has a rather limited number of parameters compared to ASM2 and ASM3. More details of all models can be found in Henze *et al.*, 2000a and Henze *et al.*, 2008.

Next to the ASM models, a dedicated model for anaerobic digestion has been developed (Anaerobic Digestion Model Nr 1, ADM1). This is not discussed here, but details can be found in Batstone *et al.* (2002).

11.4.4 Settler modelling

There are several incentives for modeling primary and secondary settling processes. A review of different settling models can be found in Ekama *et al.* (1997).

For primary clarifiers, a good prediction of the load (and COD/N ratio) leaving the primary clarifier and entering the activated sludge reactors is of great importance as there is an optimum for efficient N-removal.

For secondary settlers incentives are:

- a sludge balance between biological tanks and clarifiers
- the dynamics of sludge motion between biological tanks and settler
- effluent quality (effluent suspended solids – ESS)
- settler performance: sludge blanket height (SBH)
- the introduction of control systems for polymer dosage in case of bad settling
- sludge production (thickening)
- settler design: placement of baffles

Settler models are classified according to their spatial detail:

- 0D (point settler)
- 1D (homogeneous in x- and y-direction)

- 2D (homogeneous in y-direction)
- 3D (non-homogeneous)

Besides their spatial description, settler models use a settling velocity model. This is discussed below. In some cases also processes can be included. Examples are the inclusion of hydrolysis in primary settler models and the inclusion of denitrification in secondary settler models to predict the amount of N_2 production which could generate rising sludge (N_2-bubbles take sludge flocs to the surface of the clarifier). In what follows, some of the important settling models are discussed in more detail.

0D-models or point settlers are ideal separators without volume. A point settler can be regarded as an ideal flow splitter, thereby conserving mass. The model is given by

$$\begin{aligned} Q_f X_f f_{ns} &= Q_{eff} X_{eff} \\ Q_f X_f (1 - f_{ns}) &= Q_u X_u \end{aligned} \tag{11.31}$$

where Q_f and X_f respectively represent the feed flow (m^3/s) and feed concentration (g/m^3) to the settler,
f_{ns} is the non-settleable fraction.
Q_{eff} and X_{eff} are the respective effluent flow rate (m^3/s) and effluent suspended solids (g/m^3)
Q_u and X_u are the respective underflow flow rate (m^3/s) and underflow sludge concentration (g/m^3)

The only degree of freedom or parameter is the non-settleable fraction, which can in some models be an increasing function of flow rate to mimic the effect of turbulence in the settler which affects the ESS. The major advantage is its simplicity and it can therefore be used if none of the above incentives is part of the modeling exercise. However, the major problem of this model is that it has no residence time. Some fixes have been proposed to correct this problem:

- point settler with volume mimicking the sludge retention using a number of CSTRs with the following equation

$$\frac{dM_X}{dt} = Q_f X_f - Q_{eff} X_{eff} - Q_u X_u \tag{11.32}$$

- point settler sequenced by a volume representing the sludge blanket 1D-models discretise the 1-D PDE

$$\frac{\partial X}{\partial t} = \frac{\partial}{\partial z}\left(D\frac{\partial X}{\partial z}\right) - \frac{\partial}{\partial z}(V_s X) - RX \tag{11.33}$$

using a finite difference method for the first derivative:

$$\frac{\partial X}{\partial z} = \frac{(X(j) - X(j+1))}{h}.$$

The latter is multiplied by the sludge settling velocity V_s, which is discussed below. This assumes that X is uniform in the horizontal plane and that no vertical dispersion occurs ($D = 0$). Moreover, R is often ignored assuming no biological reactions. The discretisation of the 1D-model is illustrated in Figure 11.32. 10 layers are chosen, where layer 6 is used as the influent layer. The resemblance of the discretised model to the actual solution of the 1D-PDE is illustrated in Figure 11.33. The left figure shows the fit of a discretised 10-layer model to an experimental dataset. The right figure shows the PDE-solution for different values of the dispersion coefficient D. It can be observed that both model outputs are similar. It seems that the dispersion coefficient is inversely related to the number of layers in the model. The figures also illustrate what type of information is provided by a 1D settler model: effluent and underflow sludge concentrations, sludge blanket height and total sludge mass.

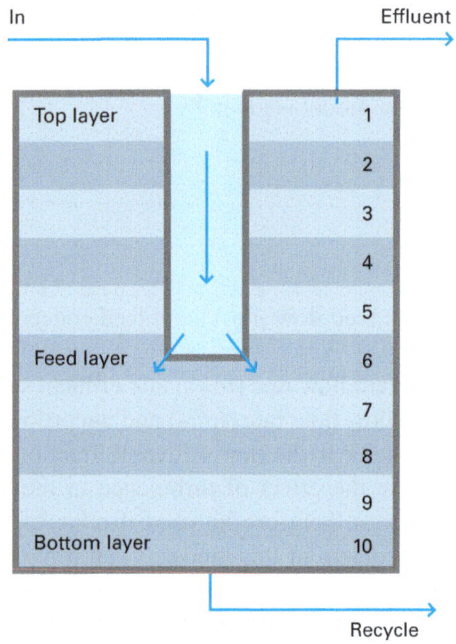

Figure 11.32 Illustration of the discretisation of the 1D settler model

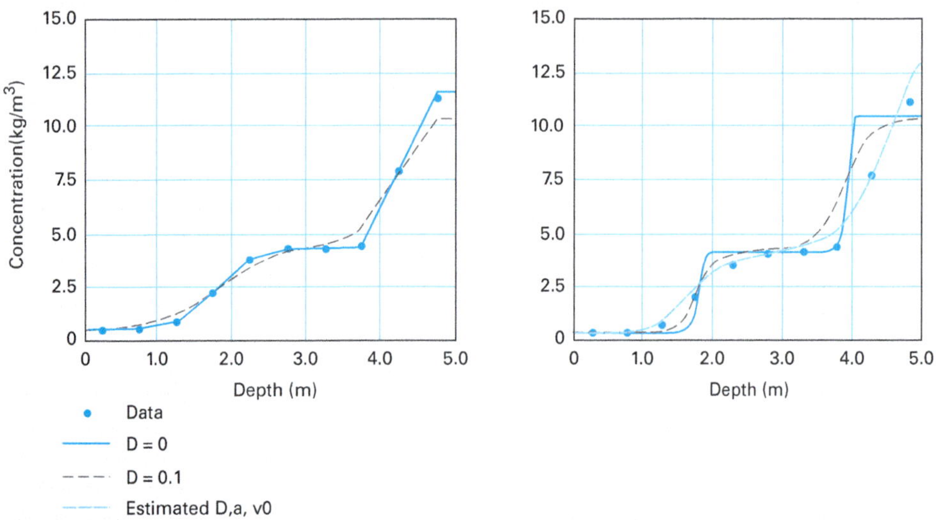

Figure 11.33 1D-settler performance of the 10-layer discretised model (left) and the PDE solution for different values of D (right)

2D and 3D models solve the Navier-Stokes equations for fluid dynamics. These are discussed in Section 5.2.1. A typical output of a computational fluid dynamics (CFD) clarifier model is given in Figure 11.34. It contains detailed 2D information concerning velocity and solids concentration, and clearly shows the sludge blanket and the concentration gradient.

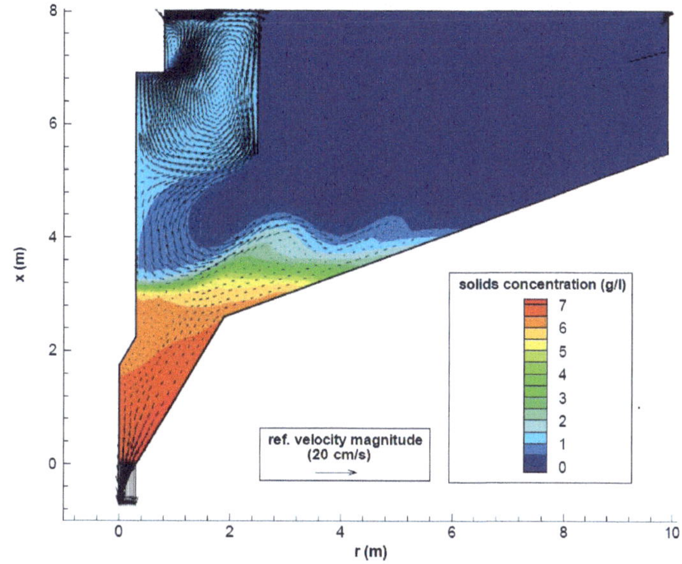

Figure 11.34 Illustration of a typical 2D-settler model output using Computational Fluid Dynamics

Next to the spatial detail, settling models need an expression for the settling velocity V_s. The latter is a sludge property and is typically measured using a batch settling test. The latter uses a 1 L cylinder where the sludge is allowed to settle and the supernatant-sludge interface is recorded versus time. This results in batch settling curves as illustrated in Figure 11.35.

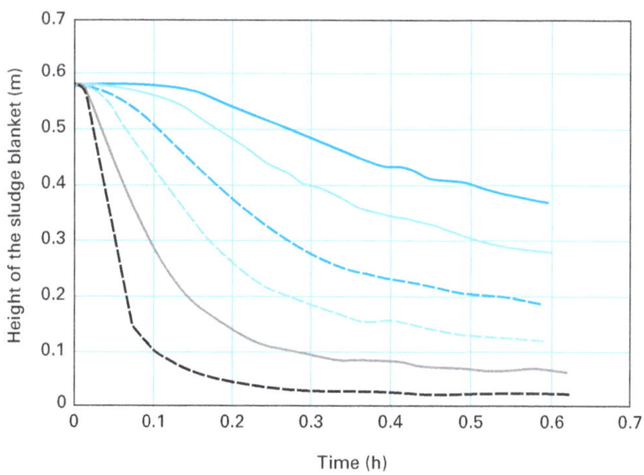

Figure 11.35 Batch settling curves for different initial sludge concentrations (from bottom to top: 3; 5; 7.7; 9.7; 12.7 and 15.6 g/L).

The curve consists of a couple of phases. After an initial transient phase, there is a linear part c, which is called the hindered settling velocity. This means that the sludge is not in contact, but the flocs can sense one another during settling. This linear part is used as a measure for the settling velocity V_S. After a while, compression settling kicks in, that is, the sludge flocs are actually in contact with one another and feel a resistance to further settlement. The way the curve then behaves depends on the compression properties of the sludge. It is clear that the batch settling curves change shape when the initial sludge concentration is altered. Higher concentrations result in less steep slopes, indicating slower settling. The different linear slopes are then typically plotted against the sludge concentration.

Several models have been proposed to fit v_S versus X datasets. The most widely used are:

- Vesilind (1968):

$$v_s = ke^{-nX} \tag{11.34}$$

- Takacs *et al.* (1991):

$$v_{sj} = v_0 e^{-r_h(X_j{'''}X_{\min})} - v_0 e^{-r_p(X_j - X_{\min})}$$

$$\text{with } 0 \leq v_{sj} \leq v_{00} \tag{11.35}$$

Both models are illustrated in Figure 11.36. The latter model is actually an extension of the former, correcting for the ridiculously large settling velocities predicted by the Vesilind equation at small sludge concentrations.

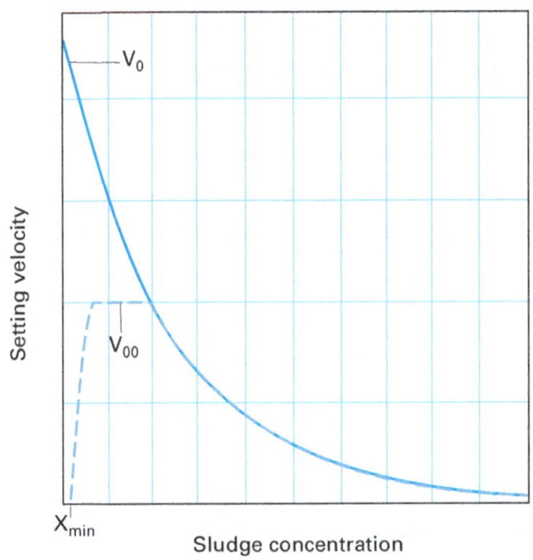

Figure 11.36 Illustration of the Vesilind equation (black) and the Takacs (blue) settling models

The spatial model combined with the sludge settling velocity models make up the typically used settler models and can predict the fate of particulate material in clarifiers.

Possible extensions

Next to the fate of particulates, there is also a need to deal with soluble materials because clarifiers typically have a large volume and therefore cause the solubles to have a residence time which causes a delay in the outflow of these components to the effluent. There are three options to deal with this:

- The solubles residence time is not modeled as such: this can be used when the time delay of the solubles in the effluent is of no interest
- The solubles residence time is modeled as a CSTR, which is able to mimic the delay to the effluent by introducing a minimal number of equations (i.e. one equation per soluble component, or 6 additional equations in the case of ASM1)
- A mass balance is written down for every soluble component in each layer of the model. This is the most detailed way of handling the problem, but introduces more equations to the system (i.e. 9 additional equations per soluble component, or 54 additional equations in the case of ASM1)

The effect of using a different number of layers for soluble components is illustrated for ammonium in Figure 11.37 for the three cases above.

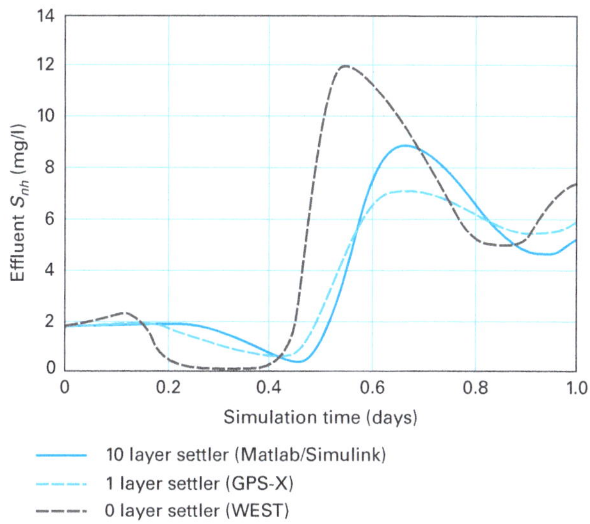

Figure 11.37 Illustration of the impact of assuming different layers to represent the soluble components in a settler model

Also, specific treatment is necessary for the particulate material. The reason for this is that in ASM several particulate components are present, whereas in the settler model only the TSS concentration is used. There are again three ways of dealing with the problem:

- A step response for a particulate component is defined and used to construct the dynamics of the settler underflow for that particulate component
- A lump-delump approach can be used to lump (i.e. sum) the different particulate components. During this lumping, the fractions of the different components are computed and stored. At the underflow, the different components are reconstructed from the same fractions depending on the predicted underflow sludge concentration.
- A propagation approach can be used in which for each layer a mass balance is written down for each particulate component. Again, this is the best way of dealing with the problem, albeit again adding more equations.

In both cases, the most sophisticated implementation is needed when modelling biological reactions in what are often called **reactive settler models**. These models predict denitrification activity which produces nitrogen gas production rates that can be coupled with loss of sludge through flotation. This process is illustrated in Figure 11.38.

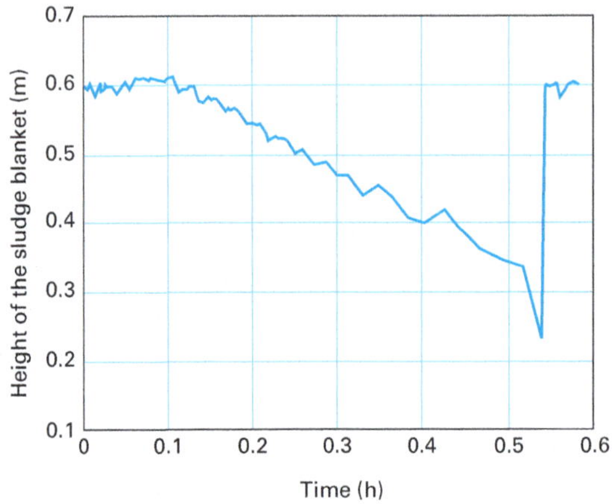

Figure 11.38 Illustration of the denitrification impact on sludge blanket height prediction in a reactive settler model

A second example of reactive settlers is the modeling of hydrolysis and ammonification processes in primary settlers.

A final model extension is for a polymer dosage. The idea is that the settling parameters are dependent on the polymer concentration. The effect of the polymer concentration on the parameters is modelled to deduce the clarifier performance. This can be achieved by measuring the batch settling curves after the addition of different concentrations of the polymer, and repeating the measurements for different sludge concentrations. An illustration of polymer addition at a fixed sludge concentration is shown in Figure 11.39. As can be observed, the settling velocity (i.e. slope of the linear part of the batch settling curve) increases as the polymer dosage increases.

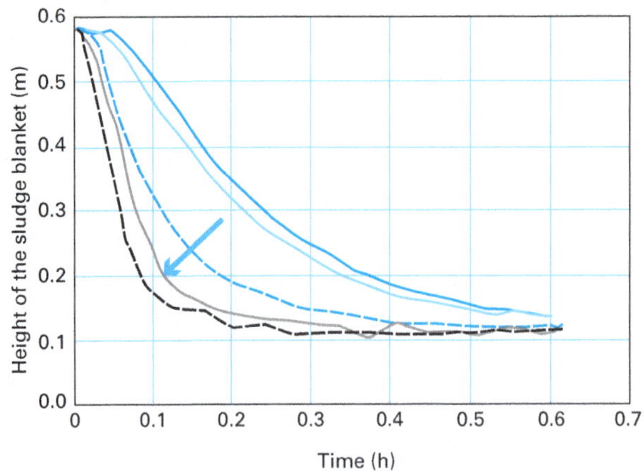

Figure 11.39 Illustration of the effect of polymer addition on batch settling curves at one sludge concentration

When this exercise is repeated for different sludge concentrations, a three dimensional graph is obtained as shown in Figure 11.40. In order to incorporate these results into a settler model, an empirical relationship can be sought between the settling velocity, the sludge concentration and the polymer addition. For the case shown, this was found to be:

$$V_s = 10.59e^{-X/(1.54P+2.5)} \tag{11.36}$$

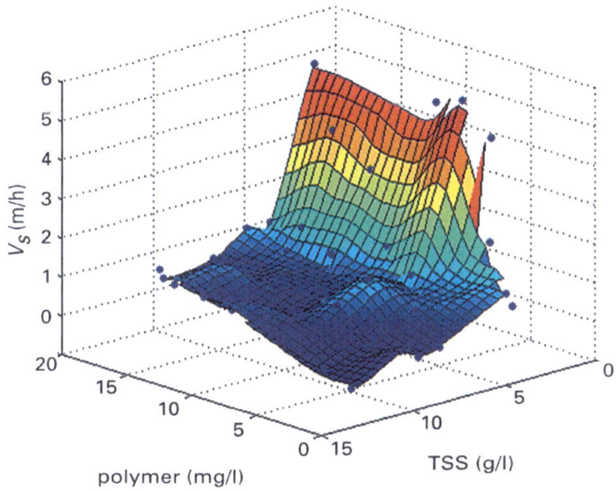

Figure 11.40 Three dimensional effect of polymer addition on batch settling curves at different sludge concentrations

11.4.5 Membrane filtration modelling

The detail with which the filtration step is modelled depends on the goal of the study. If there is no interest in the filtration behaviour of the plant, a perfect settler model can be used with a non-settleable fraction of zero (i.e. complete retention of solids). In this case, nothing can be said about the filtration efficiency or the dynamic variation of the transmembrane pressure (TMP). If the latter is needed, a more sophisticated model should be used. However, research is currently ongoing to develop these models. Most of them are based on Darcy's law. This gives the relationship between the pressure drop over the membrane and the permeate viscosity, filtration flux and the several contributions to the filtration resistance:

$$\Delta P(t) = \eta_p J_G (R_m + R_{irr} + R_c) \tag{11.37}$$

where R_m is the membrane resistance (1/m)

R_{irr} the irreversible fouling resistance (1/m)

R_c the filter cake resistance (1/m)

The former, also referred to as the clean water resistance, is easily determined by filtering clean water over a virgin membrane and logging the pressure drop versus the flux. The second resistance is due to pore blocking and is related to the reduction in the available membrane surface for active filtration, which can be modeled using a simple mass balance:

$$\begin{aligned} R_{irr}(t) &= R_m\left(\frac{A_0}{A(t)} - 1\right) \\ \frac{dA(t)}{dt} &= -\sigma_{COD} C_b(t) Q(t) \end{aligned} \tag{11.38}$$

where A_0 is the clean membrane surface (m^2)

$A(t)$ is the dynamic evolution of the membrane surface (m^2)

σ_{COD} is the fouling propensity of the sludge water COD (m^2/g)

C_b is the bulk water COD concentration (g/m^3)

Q is the filtration flow rate (m^3/s)

The contribution of the cake layer to the filtration resistance is based in the filter cake mass and its permeability. Also here, a mass balance can be used for the filter cake mass dynamics, which in its turn is dependent on the fraction of the bulk COD that is able to reach the filter cake. This can be empirically related to the critical flux and the shear:

$$
\begin{aligned}
R_c(t) &= \alpha \frac{w(t)}{A_0} \\
\frac{dw(t)}{dt} &= C_d(t)Q(t) \\
C_d(t) &= C_b(t)\left[J_L(t) \big/ J_m \right]^n
\end{aligned}
\tag{11.39}
$$

where α is filter cake permeability,

A_0 is the clean membrane surface (m^2)

R_c is the filter cake resistance (1/m)

Q is the infiltration flow rate (m^3/s)

$w(t)$ is the dynamic filter cake mass (kg)

C_d is the COD fraction that is able to reach the membrane

J_L is the current operational flux ($l/m^2/h$)

J_m is the critical flux above which all particulate material deposits on the membrane ($l/m^2/h$)

n is a fitting parameter.

An illustration of the model to predict cyclic dynamic transmembrane pressure between two backwash events is shown in Figure 11.41. This illustrates that the TMP-increase between two backwash events gradually increases in slope, indicating irreversible fouling. Once a certain maximum TMP is reached, a chemical cleaning is required. The slope itself depends on the cake filtration.

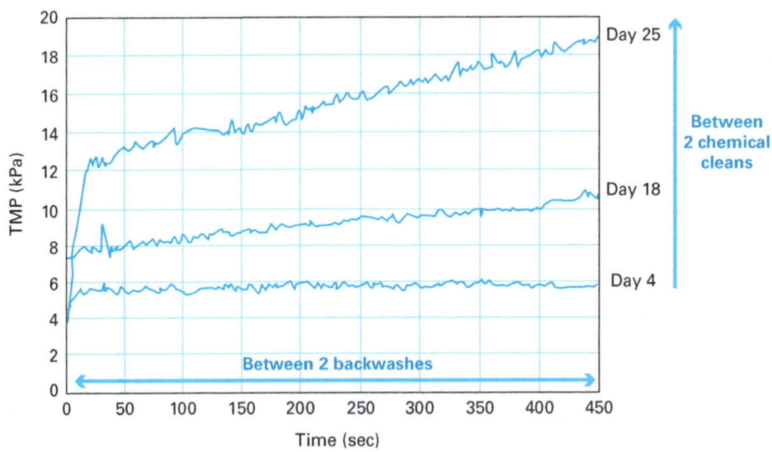

Figure 11.41 Illustration of short-term cyclic dynamic TMP (after Jiang, 2007)

Finally, Figure 11.42 illustrates long-term dynamic TMP. This illustrates the evolution of the irreversible fouling.

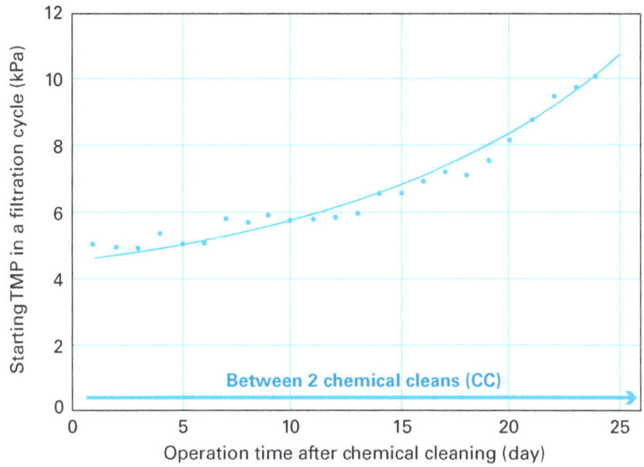

Figure 11.42 Illustration of long-term dynamic TMP (after Jiang, 2007)

11.5 SIMULATION SOFTWARE

A wastewater treatment simulator is software that allows the modeller to simulate a wastewater treatment configuration. A detailed overview of simulators can be found in Olsson and Newell (1999) and Copp (2002). General purpose simulators can be distinguished from specific wastewater treatment simulators. The former exhibit a high degree of flexibility, but the modeller has to supply the models that are to be used. Hence, the use of a specific software product depends on the skills of the modeller. However, every modeller should at least know what the model he or she is using contains. Blindly using models without knowing their content is dangerous, or quoting Rieger *et al.* (2010): 'a powerful tool is a more dangerous tool: it places more responsibility on the craftsmen who forge it, and requires a well trained operator for its efficient use'.

Examples of specific commercial wastewater treatment simulators are (in alphabetical order):

- www.wrcplc.co.uk/software
- WEST (www.mostforwater.com)

11.6 OUTLOOK

Wastewater technology is moving forward in many directions. It focuses on increasing the performance of the existing systems in many classical facets including resource consumption (energy, chemicals, construction materials, …) and environmental impact (COD, nitrogen, phosphorus, …). Increasingly, the use of improved operation schemes involving automatic control is supplementing improved design approaches to reach these objectives. For both approaches (operation and design) models have turned out to be key tools and there has been widespread acceptance of them in practice during the last decade.

The wastewater engineer is however, faced with increasing challenges: he needs to be aware of a multitude of design options (configurations) and operation modes. He must be able to discuss with control engineers and think along with them to increase process performance and he must be able to use advanced (dynamic) modelling tools to support the design and operation scheme he proposes.

But there is more to come and in some instances the challenge has already materialised. For instance, the horizon of the wastewater engineer is receding further and further away. Plant-wide modelling becomes a standard and requires the engineer to evaluate all aspects of a treatment plant, in particular the sludge treatment train (thickening, dewatering, digestion, …) and the return liquors (sidestreams) that affect the water line of the treatment plant. He must now be able to install sidestream treatment systems (Anammox, Babe, Demon, OLAND, Sharon,…) that are efficient on their own to deal with, for instance, the nitrogen load coming from digesters. Modelling challenges are such that the engineer must either be able to master a supermodel which follows all the relevant components throughout the plant or he must be able to set up interfaces between more traditional models (e.g. ASM1 and ADM1) so that the material flows between water and sludge line are properly described.

There are also developments in wastewater technology due to new environmental concerns. Attention is increasingly being given to micropollutants, in particular to such chemicals as personal care products, heavy metals, endocrine disrupting compounds, pharmaceuticals, etc. This is leading to the development of a wide range of new treatment technologies involving ozonation, activated carbon adsorption, membrane filtration, etc. Corresponding models are starting to be developed, but much basic research is still required to master these new technologies. In any case, the challenge is for the wastewater engineer to get to grips with the technology toolbox that is being developed.

Another important development is the rapidly increasing time span over which a treatment plant is being evaluated in detail. Awareness of climate change is being introduced in the evaluation of treatment plants, not only in terms of how the plant is exposed to the effects of climate change (e.g. through more extreme rain events) but also to what extent the plant is contributing to climate change through its resource use (energy, chemicals, …) and directly to greenhouse gas emissions (methane, nitrous oxide, …). The first models that evaluate greenhouse gas emissions and perform life cycle assessments to evaluate the overall impacts of treatment systems are being applied; these can be expected to become mainstream in years to come.

Finally, we may be at a brink of a paradigm shift in wastewater treatment: Sustainability demands that we acknowledge wastewater as a renewable resource from which water, materials (e.g., nutrients, bioplastics), and energy can be recovered. By shifting away from today's paradigm, which focuses on what must be removed from wastewater, to a new paradigm focusing on what can be recovered, treatment systems may begin to be described as resource recovery systems – a conceptual transformation that could allow the perceived impact of wastewater on communities to become a positive contribution.

11.7 CASE STUDY: UPGRADING LARGE WASTEWATER TREATMENT PLANTS: USE OF MODELLING AS A DECISION-MAKING TOOL IN SARAJEVO (BOZNIA AND HERCEGOVINA)

11.7.1 Introduction

Since the introduction of activated sludge models (Henze *et al.*, 1987 and Henze *et al.*, 2000a), mathematical modelling has become an irreplaceable tool for the upgrade, retrofit, troubleshooting and optimization of wastewater treatment plants (Brdjanovic *et al.*, 2000; Pinzon *et al.*, 2007; Brdjanovic *et al.*, 2007 and Henze and Comeau, 2008). This case study describes the use of mathematical models in a decision-making process concerning the reconstruction of a wastewater treatment plant (WWTP) in the city of Sarajevo, the capital of Bosnia and Herzegovina.

11.7.2 Modelling of Sarajevo sewage system

The Sarajevo sewage system includes a sanitary sewage system and a rainwater drainage system (the total length of sewer pipes in both systems is about 1,000 km with diameters up to 1.80 m). Although in theory the systems are separate, they suffers from cross connections, excessive infiltration and backflow from the river. The WWTP (with a maximum flow of 4.3 m^3/s and a designed capacity of 600,000 PE) was constructed in 1984 as a conventional activated sludge system for the removal of suspended solids (SS) and organic matter (COD) with anaerobic sludge digestion and biogas reuse for energy production. The plant was operational from 1984 until 1992. Severe problems were occasionally experienced because of an excessive accumulation of sand, grit and gravel at the inlet to the plant. Therefore, the plant operation was often interrupted, usually during and/or after periods of heavy rain or melting snow (Peric *et al.*, 1991). Several mitigation studies (Peric *et al.*, 1991 and JICA, 1999), etc have been carried out so far and all recommend the construction of a large sand trap in front of the existing inlet. However, the inlet is only the place where the problem is manifested; it is not necessarily the location of its origin. A recent model-based sediment transport study (DHI, 2009), using the DHI software package MIKE URBAN©, showed that indeed it may well be feasible to build several smaller sand traps within the sewer networks as an alternative to the centralized solution from the '90s.

The WWTP has been out of operation since 1992 as a result of devastation during the recent war (1992–1995). Untreated sewage has been discharging to the nearby river for almost two decades. The local government is currently planning to rehabilitate both a part of the network (addressing the sediment transport issue too) and the WWTP, under the framework of a substantial reconstruction programme to be financed by a loan from the World Bank. Four main questions arose during the decision-making process regarding the WWTP, namely: (*i*) should the plant be reconstructed in its original state as in 1984?, (*ii*) should it be upgraded by nutrient-removal (nitrogen and phosphorus) facilities to satisfy more stringent effluent requirements (EU standards) in the future?, (*iii*) what is the most appropriate technological process to be applied?, and (*iv*) what are the possible extent and scope of the reconstruction, given the available financial resources?

These questions were addressed (ASM Design, 2009) by the application of ASM models incorporated in the simulation software BioWin© of EnviroSim Associates Ltd; see Figure 11.43. In total, a dozen technological and engineering options (Figure 11.44) were considered, which took into account the original design, the present and future composition and quantity of sewage (up to 2030), the choice of wastewater treatment technology (an existing anoxic-oxic A/O system, or a future upgrade to an anaerobic-anoxic-oxic A2O/UCT system or oxidation ditch/Carousel© system), the choice of an aeration system (existing surface aeration by turbines, or an alternative with fine bubbles aeration by air diffusers), and the amount and composition of the sludge (and later biogas) generated in the process.

The model simulation showed that for all the scenarios which involve the adoption of EU standards for nutrient removal, the existing plant will not be able to achieve these standards. Therefore even for minimal improvement (e.g. nitrification), a large additional volume and aeration capacity is required. In order to reach such standards the sludge age needs to be extended from the designed 2–3 days to 10–15 days and the activated sludge section needs to be 3–5 times the current volume. Furthermore, 150% of the current secondary settling volume will be required. Extension by one, and later another, secondary settling tank will be necessary. The study showed that by the reconstruction of the primary treatment units, a reduction of an equivalent load of 320,000 PE will be achieved. Reconstruction of digestion to a full anaerobic digestion and biogas production as in the original design will also be necessary. Operating the existing aeration tanks with a diffused bubble aeration system will result in BOD, COD and SS concentrations in the effluent of below 25, 125 and 35 mg/L, respectively. It is possible to extend the existing A/O plant by an additional tank volume of 48.000 m^3 and retrofit it into an A2O (UCT) process to reach a nutrient level in the effluent of 10 mgN/L and 1 mgP/L (current EU standards for plants larger than 100,000 PE).

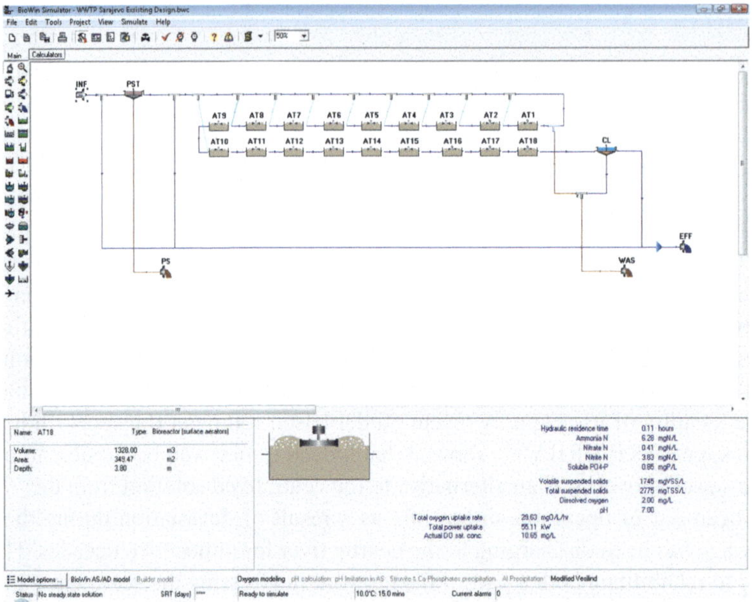

Figure 11.43 Hydraulic plant scheme in the BioWin simulator

Simulation Scenario Overview						
No.	**Description**	**Influent pre-treatment**	**Biological Tanks**	**Configuration**	**Aeration Type**	**Removal performance**
0	Existing reactors for BOD removal - installed surface aerator capacity, settled influent	Settled	Existing	Plug-flow	Turbines	BOD
1	Existing reactors for BOD removal - new surface aerator capacity, settled influent	Settled	Existing	Plug-flow	Turbines	BOD
2	Existing reactors for BOD removal - bubble aerators, settled influent	Settled	Existing	Plug-flow	Bubble	BOD
3	Existing reactors for BOD removal - bubble aerators, raw influent	Raw	Existing	Plug-flow	Bubble	BOD
4	Nitrifying Extended Volume Plug flow - bubble aerators, settled influent	Settled	Extended	Plug-flow	Bubble	BOD/Nitrifying
5	Nitrifying Extended volume Plug flow - turbines + bubble aeration, settled influent	Settled	Extended	Plug-flow	Turbines/Bubble	BOD/Nitrifying
6	**Denitrification, Extended Volume Plug flow - bubble aerators, settled influent**	**Settled**	**Extended**	**Plug-flow**	**Bubble**	**BOD/NIT/Denitrification**
7	Extended Volume Plug Flow for BOD removal - bubble aeration, raw influent	Raw	Extended	Plug-flow	Bubble	BOD
7a	Extended Volume Plug Flow for BOD removal - bubble aeration, raw influent	Raw	Extended	Plug Flow	Bubble	BOD
7b	Extended Volume Plug Flow for BOD removal - bubble aeration, raw influent	Raw	Extended	MLSS recirculation	Bubble	BOD
7c	Extended Volume Carrousel for BOD removal - bubble aeration, raw influent	Raw	Extended	Carousel	Bubble	BOD
8	Nitrifying Carousel type activated sludge system - bubble aeration, raw influent	Raw	Extended	Carousel	Bubble	BOD/Nitrifying
9	Denitrifying Carousel AO process - raw influent, bubble aeration	Raw	Extended	Carousel/AO	Bubble	BOD/NIT/Denitrification
10	**Denitrifying Plug flow AO process - raw influent, bubble aeration (EU standard)**	**Raw**	**Extended**	**Plug Flow/AO**	**Bubble**	**BOD/NIT/Denitrification**
11	A2O Plug flow process - raw influent, bubble aeration (EU standard)	Raw	Extended	Plug flow/A2O	Bubble	BOD/NIT/DEN/Bio-P
12	A2O Carrousel process - raw influent, bubble aeration (EU standard)	Raw	Extended	Carrousel/A2O	Bubble	BOD/NIT/DEN/Bio-P

Highlighted scenarios are best performing for settled and raw influent

Figure 11.44 Overview of scenarios examined in this model-based study

The study concludes that: (*i*) by reconstruction back to its original state and the replacement of surface aeration by a diffused air system the plant will be able to reach the current local standards regarding suspended solids and organic matter; (*ii*) in the case of the requirement for nutrient removal the plant will need substantial upgrade in terms of volume and aeration capacity; (*iii*) the existing A/O system is considered appropriate for current effluent quality standards, however the most appropriate technological process to be applied including nutrient removal is the modified UCT process (A2O); and (*iv*) based on the results of this study it will be possible to determine the scope of the reconstruction to match the available financial resources, given that the costs of preferred options are known.

The study clearly shows the usefulness of mathematical modelling in a decision-making process regarding investments in urban infrastructure in which a compromise often has to be sought between the prioritization of investments and the environmental benefits, as is the case in this study.

11.7.3 Acknowledgement

This article was prepared by Sebastian C. F. Meijer (ASM Design, Alexandar Numankade 205, 3572KX Utrecht, The Netherlands) and Damir Brdjanovic (Department of Urban Water and Sanitation, UNESCO-IHE Institute for Water Education. Westvest 7, 2611 AX Delft, The Netherlands and Department of Biotechnology, Delft University of Technology. Julianalaan 67, 2628BC Delft, The Netherlands).

11.8 KEYWORDS

By the end of the chapter you should have encountered the following terms; ensure that you are familiar with them!

Activated sludge
Aerobic
Anaerobic
Anammox
ASM1, ASM2, ASM3
Biofilm
Biofilter
Biological Oxygen Demand
Biomass
Chemical Oxygen Demand
Denitrification
Dilution
Dry weather flow
Eutrophication
Exfiltration
First flush
Floc
Gujer matrix
Infiltration
Membrane
Micro-organisms
Micropollutants
Nitrogen
Nutrient
Phosphorus
Polymer
Primary treatment
Secondary treatment
Settleable solids
Sludge blanket
Stoichiometric coefficient
Substrate
Suspended Solids
Tertiary treatment
Total Organic Carbon
Total Suspended Solids
Tracer
Wastewater
Wet weather flow

11.9 QUESTIONS

What are the primary characteristics of domestic wastewater?
Describe the three main forms of treatment of wastewater
What is the Gujer matrix, and how is it applied?
How are treatment processes modelled mathematically?
What is membrane filtration, and how does it work?

11.10 REFERENCES

ASM Design (2009). *Model Design Study WWTP Sarajevo; Model-Based Process Technology Selection for Rehabilitation of WWTP*. ASM Design, Sarajevo.

Batstone D., Keller J., Angelidaki I., Kalyuzhnyi S., Pavlostathis S. G., Rozzi A., Saunders W., Siegrist H. and Vaviliri V. (2002). *Anaerobic Digestion Model No. 1 (ADM1)*. IWA Task Group for Mathematical Modelling of Anaerobic Digestion Processes. IWA Publishing, London.

Brdjanovic D., Muthaiwala M., Moussa M. S., Army G. and van Loosdrecht M. C. (2007). Use of plant-wide modeling for optimization and upgrade of a tropical wastewater treatment plant in a developing country. *Water Sci. Technol.*, **56**(7), 21–31.

Brdjanovic D., van Loosdrecht M. C. M., Versteeg P., Hooijmans C. M., Alaerts G. J. and Heijnen J. J. (2000). Modelling COD, N and P removal in a full-scale WWTP Haarlem Waarderpolder. *Water Res.*, **34**, 846–858.

CEC (1991). *Directive Concerning Urban Waste Water Treatment*. 91/271/EEC. CEC, Brussels.

CEC (2000). *Directive Concerning Water Framework – Integrated River Basin Management for Europe*. 2000/60/EEC. CEC, Brussels.

Copp J. (2002). *The Cost Simulation Benchmark: Description and Simulator Manual*. Office for Official Publications of the European Community, Luxemburg, p. 154.

Corominas L. and Rieger L. (2010). New framework for standardized notation in wastewater treatment modelling. In: *Proceedings of the 2nd IWA-WEF Wastewater Treatment Modelling Conference*, Mont-Sainte-Anne, 1–3 June, Quebec, Canada.

De Clercq B., Coen F., Vanderhaegen B. and Vanrolleghem P. A. (1999). Calibrating simple models for mixing and flow propagation in waste water treatment plants. *Water Sci. Technol.*, **39**(4), 61–69.

DHI (2009). *Preparation of Investment Analysis and Modelling Study for the Sewer Collection Network in Sarajevo, Bosnia and Herzegovina: Model-Based Analysis of Sarajevo's Sewer System*. DHI, Horsholm.

Dochain D. and Vanrolleghem P. (2001). *Dynamical Modelling and Estimation in Wastewater Treatment Processes*. International Water Association, London.

Ekama G., Barnard J., Gunthert F. W., McCorquodale J. A., Parker O. S. and Wahlberg E. J. (1997). *Secondary Settling Tanks: Theory, Modelling, Design and Operation*. Scientific and Technical Report No. 6. IAWQ, London.

Henze M., Grady C. J., Gujer W., Marais G. V. R. and Matsuo T. (2000a). *Implementation of the Activated Sludge Model, Activated Sludge Models ASM1, ASM2, ASM2d and ASM3*. IWA Publishing, London.

Henze M., Grady C. P. L. J., Gujer W., Marais G. V. R. and Matsuo T. (1987). *Activated Sludge Model No. 1*. IAWQ Scientific and Technical Report No. 1. IAWQ, London.

Henze M., Grady Jr. C., Gujer W., Marais G. V. R. and Matsuo T. (1987). A general model for single-sludge wastewater treatment systems. *Water Res.*, **21**(5), 505–515.

Henze M., Gujer W., Gujer W., Marais G. V. R. and Matsuo T. (1995a). *Activated Sludge Model No. 2*. IAWQ, London.

Henze M., Gujer W., Mino T. and van Loosdrecht M. C. (1999). Activated Sludge Model No. 2d. *Wat. Sci. Technol.*, **39**(1), 165–182.

Henze M., Gujer W., Mino T. and van Loosdrecht M. C. (2000b). *Activated Sludge Models ASM1, ASM2, ASM2d and ASM3, IWA Task Group on Mathematical Modelling for Design and Operation of Biological Wastewater Treatment*. IWA Scientific and Technical Report, IWA, London.

Henze M., van Loosdrecht M. C. M., Ekama G. A. and Brdjanovic D. (2008). *Biological Wastewater Treatment: Principles, Modeling, and Design*. IWA Publishing, London.

Hug T., Benedetti L., Hall E. R., Johnson B. R., Morgaroth E., Nopens I., Rieger L., Shaw A. and Vanrolleghem P. A. (2009). Wastewater treatment models in teaching and training: the mismatch between education and requirements for jobs. *Water Sci. Technol.*, **59**(4), 745–753.

Henze M. and Comeau Y. (2008). Wastewater characterization. In: M. Henze, M. C. van Loosdrecht, G. Ekama, D. Brdjanovic (eds.), Biological WasteWater Treatment: Principles, Modelling and Design (pp. 33–52), IWA, London.

Jiang T. (2007). *Characterization and Modelling of Soluble Microbial Products in Membrane Bioreactors*. PhD thesis, Ghent University, Ghent.

JICA (1999). *The Feasibility Study of the WasteWater Treatment Plant of Sarajevo City in Bosnia and Herzegovina*. Final Report. Tokyo Engineering Consultants Co., Ltd/Nihon Suido Consultants Co., Ltd, Tokyo.

Judd S. and Judd C. (2006). *The MBR Book: Principles and Applications of Membrane Bioreactors in Water and Wastewater Treatment*. Elsevier Science Ltd, Oxford.

Kampschreur M., Poldermans R., Kleerebezem R., van der Star W. R. L., Haarhuis R., Abrna W. R., Jetten M. S. M. and van Roosdrecht M. C. M. (2009). Emission of nitrous oxide and nitric oxide from a full-scale single-stage nitritation-anammox reactor. *Water Sci. Technol. J. Int. Assoc. Water Pollut. Res.*, **60**(12), 3211.

Levenspiel O. (1999). *Chemical Reaction Engineering*. Wiley, Hoboken, New Jersey.

Nopens I., Batstone D., Copp J. B., Jeppson O., Volcke E., Alex J. and Vanrolleghem P. A. (2009). An ASM/ADM model interface for dynamic plant-wide simulation. *Water Res.*, **43**(7), 1913–1923.

Olsson G. and Newell B. (1999). *Wastewater Treatment Systems: Modelling, Diagnosis and Control*. IWA Publishing, London.

Peric M., Brdjanovic D. and Razdorov P. (1991). Upgrade of the wastewater treatment plant in Sarajevo (in Bosnian). *Naša Vodoprivreda*, **5/6**, 41–51.

Pinzon A., Brdjanovic D., Moussa M. S., Lopez Vazquez G. M., Meijer S. C. F., van Straten H. H. A., Janssen A. J. H., Amy M. G. and van Loosdrecht M. C. M. (2007). Modelling oil refinery wastewater treatment plant. *Environ. Technol.*, **29**(11), 1273–1284.

Rieger L. and Takacs I. (2010). Editorial: Status and future of wastewater treatment modelling. *Water Sci. Technol.*, **61**(4), 821–823.

Takács I., Patry G. and Nolasco D. (1991). A dynamic model of the clarification-thickening process. *Water Res.*, **25**(10), 1263–1271.

Tchobanoglous G., Burton F. and Stensel H. D. (2003). *Metcalf & Eddy Wastewater Engineering: Treatment and Reuse*. McGraw-Hill, NY.

UNESCO (2009). *The 3rd United Nations World Water Development Report: Water in a Changing World (WWDR-3)*. UNESCO Publishing/Earthscan, Paris, London.

USEPA (2004). *EPA Water Subtopics*. EPA. http://www.epa.gov/ebtpages/water/html (accessed 17 October 2010).

Vanrolleghem P., Benedetti L. and Meirlaan J. (2005). Modelling and real-time control of the integrated urban wastewater system. *Environ. Modell. Software*, **20**(4), 427–442.

Vesilind P. A. (1968). Theoretical considerations: design of prototype thickeners from batch settling tests. *Water Sewage Works*, 115, 302–307.

Chapter 12

Management of Water Quality in Integrated Drainage Systems

12.1 INTRODUCTION

Ideally, urban drainage systems are designed to minimise the frequency of surface flooding and to deal with the sewage such that the discharge of pollutants via overflows and from treatment works to receiving waters is minimised. Pollutants originating in an urban area reach the receiving waters via combined sewer overflows, uncontrolled runoff from the urban area, and the effluent from treatment works. The various pollutants can seriously affect not only the chemical composition of the receiving water, which has repercussions for human and animal health, but can also contribute adversely to the biota species present in the water. In addition, sediments and gross solids, such as plastic bags, can be deposited in the receiving waters, smothering breeding grounds for fish and other biota and generally destroying the aesthetics of the environment. Often, deposited fine sediments carry attached pollutants that are available for suspension during the next significant storm. The impact of a heavily polluted combined sewer overflow can lead to a dramatic reduction in the dissolved oxygen in the water such that fish cannot survive and die. Another consequence in a lake could be the appropriate condition for the rapid growth of algae, leading to what is termed 'eutrophication'. This can be serious for the intake of water for supply purposes.

Obviously there is a need to minimise the build up of oxygen-demanding substances such as faecal matter or detritus, to reduce substances such as oils that physically hinder re-oxygenation at the water surface, to avoid toxic compounds including ammonia, pesticides and some industrial effluents, and to limit biological activity.

All freshwater lakes and rivers have the capacity to assimilate some degradable wastes, by stabilising them through the natural self-purification processes in the water. However, when pollutant loads exceed the self-purification capacity of the receiving water, aquatic life is endangered and the use of the water for supply and recreational purposes is restricted. In contrast, the assimilative capacity of the water for many toxic substances is very low. Additionally, water polluted by drainage discharges can create unpleasant odours and may lead to the build-up of water-borne diseases such as cholera and typhoid, which are a particular hazard to health in tropical regions.

Good drainage design seeks to ensure that its continuous and intermittent discharges are assimilated by the receiving water, and that in so doing the quality of the receiving water is optimised and the treatment costs of the wastewater are minimised.

12.2 IMPACT OF POLLUTANTS ON RECEIVING WATERS

Consider an overflow discharging suspended solids and BOD to a receiving water. Figure 12.1 shows what is likely to happen downstream in terms of dissolved oxygen (Figure 12.1a), nitrogen (Figure 12.1b), flora (Figure 12.1c) and fauna (Figure 12.1d). The important indicators are the plots showing dissolved oxygen (Figure 12.1a) and the clean water fauna (Figure 12.1d). It is apparent that the clean water fauna plots follow that for dissolved oxygen closely. Note that the BOD (Figure 12.1a) rises or stays constant after release despite some of the pollutant being digested and depleting the dissolved oxygen. There is a long time lag as the bacteria multiply while digesting the pollutant. The Suspended Solids (SS) settle relatively quickly and act as a store for pollutants. Note that SS can contain quantities of toxic heavy metals.

Figure 12.1b shows ammonium ions (NH_{+4}) oxidised to nitrates (NO_{-3}). Many aquatic organisms, particularly fish, have a very low tolerance to this chemical. The ammonium concentration rises if the conditions are anaerobic, declines once aerobic conditions return, and the ammonium ions are oxidised to nitrates.

Flora are affected by the increased quantities of nutrients (PO_4 and NO_3) resulting in eutrophication. The fauna show the clearest pattern of response, and as a consequence the predictability of this response has resulted in the development of the many Biological Indices of pollution. In heavily polluted conditions only a few species dominate. For example, Tubificid worms can exist in near anaerobic conditions and having few competitors, multiply prolifically. As the oxygen levels increase these organisms are succeeded by Chironomids (midge larvae) and so on until the water is clean and well-oxygenated; obviously, there is a wide diversity of species all competing with each other. Refer to (Rauch *et al.*, 1998) for further details about the impact of wastewater discharges in receiving waters and rivers in particular.

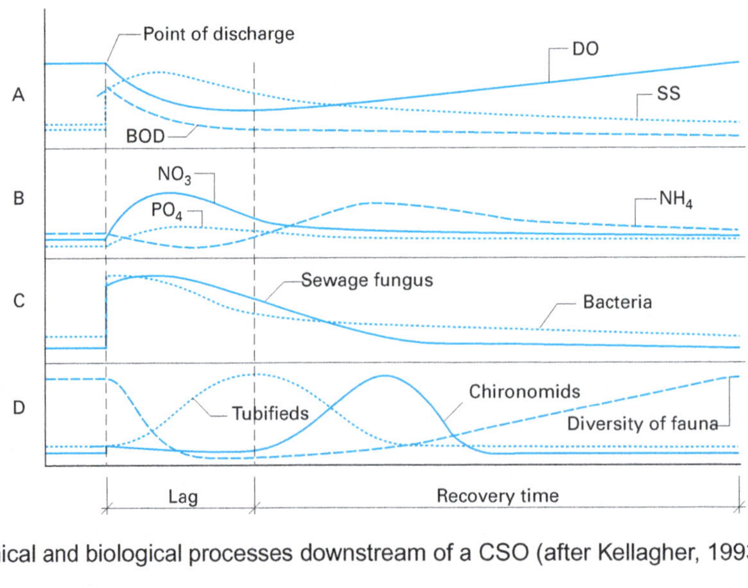

Figure 12.1 Chemical and biological processes downstream of a CSO (after Kellagher, 1993)

12.2.1 Oxygen demanding substances

Dissolved oxygen (DO) is the best indicator of the 'health' of a water source. A clean water source with little or no organic pollution will have an oxygen concentration of about 10 mg/l when in equilibrium with air in a temperate environment. This maximum concentration is highly temperature dependent and varies exponentially; see Table 12.1. Thus heated water with no impurities is an oxygen demanding substance.

Table 12.1 Maximum oxygen concentration

Temperature (C)	Dissolved oxygen (mg/l)
0	14.6
10	11.3
20	9.2
30	7.6

Oxygen is required by all the higher forms of life in a river and, in the absence of toxic impurities, there is a close correlation between DO and biodiversity. For example, most game fish die when the DO concentration falls below 4 mg/l. Oxygen is needed in the many natural treatment processes performed by micro-organisms that occur in a stream.

The quantity of oxygen required by micro-organisms to break down a given quantity of organic waste is termed the Biochemical Oxygen Demand (BOD) (expressed in mg/l). These organisms take time to fully digest and stabilise the waste, and the rate at which they do so depends on the temperature and the quantity of micro-organisms in the sample at the start.

BOD is measured by subjecting the sample to a test in the laboratory over a five day period at a constant temperature of 200°C. The test is carried out in the dark to prevent any oxygen production by photosynthetic organisms in the sample. The possible range of BOD is so large, between 2 mg/l in clean water to 50,000 mg/l in heavily polluted water, that heavily polluted samples are usually diluted. The test attempts to replicate conditions likely to occur due to the natural purification capacity of the water, and approximates the quantity of biodegradable material in the sample. As the digestion of oxidisable material in a sewer sample begins when it is taken, it is important that the sample is either refrigerated very quickly or the test is begun immediately. The 5 day period for the test is required to ensure that all of the oxidisable material is used up; this could be the case when there are substances in the sample that are toxic to the oxidising bacteria.

The Chemical Oxygen Demand (COD) test is quicker than the BOD test and measures the total oxygen demand in the presence of a powerful oxidising agent, such as ozone or a mixture of sulphuric acid and potassium dichromate; it is a measure of the total amount of oxygen required to stabilise the waste. COD has a higher value than BOD since the test material, such as a piece of wood, is oxidised and exerts a considerable demand, whilst in many circumstances the material may take several years (or for many materials, several thousand years) to fully oxidise. Therefore, in the example of a piece of wood, it is quite likely it will be either washed up on riverbank or swept to sea without exerting a significant oxygen demand on the receiving water.

It is important to note that the COD results need very careful examination, particularly where there are no BOD results, due to the likelihood of numerous substances being present, which exert a COD but will not exert a BOD. Wherever possible the BOD is therefore a preferable parameter. However, from experience, a relationship between BOD and COD can be determined for any site and used fairly reliably to provide guideline values for BOD from COD; see Butler and Davies (2000) and Table 12.3.

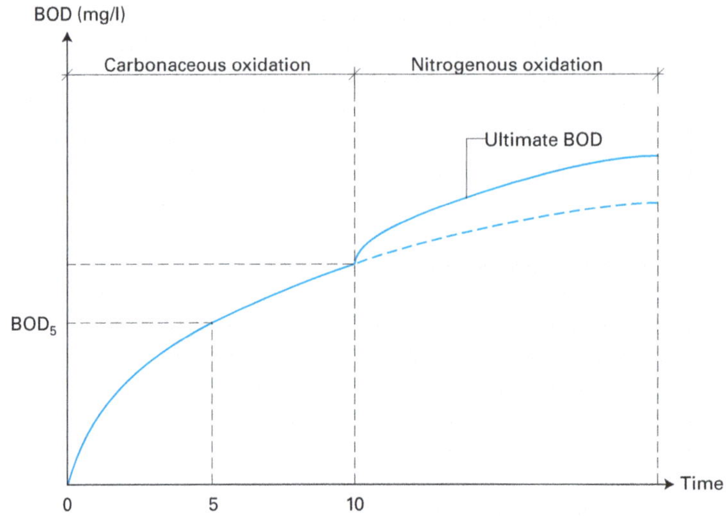

Figure 12.2 Typical BOD curve (after Kellagher, 1993)

Table 12.2 Typical Quality of Domestic Sewage

	Raw	Treated	Overflow
Suspended solids mg/l	300	5	200
BOD mg/l	367	10	300
Ammonia mg/l	39	7	30
COD mg/l	470	15	350

Table 12.3 Pollutant Concentrations in Urban Runoff

	Highway runoff	Residential areas	Commercial areas	Industrial areas
Suspended solids mg/l	28–1178	112–1104	230–1894	34–374
BOD5 mg/l	12–32	7–56	5–17	8–12
COD mg/l	128–171	37–120	74–160	40–70
Ammonia mg/l	0.02–2.1	0.3–3.3	0.03–5.1	0.2–1.2
Lead mg/l	0.15–2.9	0.09–0.44	0.1–0.4	0.6–1.2

12.2.2 Pollution dilution and oxygen sag

A complete understanding of the reaction kinetics, which governs the rate at which organic matter is stabilised, needs a good river water quality model. Such a model must be capable of replicating the self-purification ability of the river due to the reaeration of the water from the atmosphere through the water surface and at weirs and other structures; the latter increase the energy loss on the flow through turbulence, which is important in reducing pollutant concentrations. Normally, reaeration tends to balance the reduction in the oxygen levels in the river due to the demand of the biodegrading organics. Where the reaeration is insufficient this leads to the typical 'oxygen sag curves'. The rate of uptake of oxygen by the pollutants is dependent on the source of the pollutants, and the rate of reaeration is dependent on the characteristics of the river and the flow velocity. It is therefore necessary to calibrate these factors for different situations.

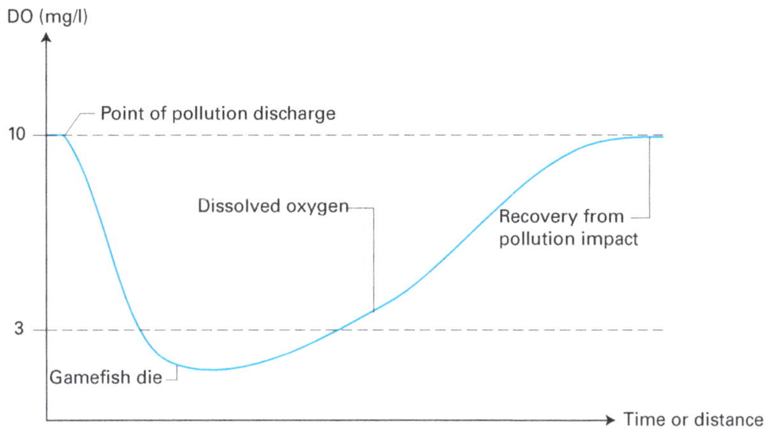

Figure 12.3 Dissolved oxygen concentration with distance (after Kellagher, 1993)

The processes of oxygen demand are made considerably more complex by the effect of photosynthesis in plants and algae in the river. These will produce oxygen by photosynthesis during daylight and will utilise oxygen for respiration during the night. Both the photosynthesis and respiration are temperature dependent and seasonal. They are also dependent on the characteristics of the river and on the background level of pollutants and nutrients, particularly nitrogen and phosphorus.

The discharges from sewage treatment works and storm overflows contain significant nutrient loads, and these alone may be sufficient to cause low oxygen levels. The change in oxygen levels between day and night can be of the same order as the oxygen sag caused by pollutant discharges. For continuous discharges the worst case is at night during summer when the respiration will be greatest and when the pollutants will be more concentrated because of low summer flows in the river. For intermittent discharges from storm overflows, the situation is considerably more complex because of the variation in pollutant load likely to be discharged at different times of the day.

12.2.3 First foul flush

Another factor making overflow discharges particularly problematic is that in non- continuous discharges there is often a phenomena, termed the 'first foul flush', whereby fine sediments deposited in the pipes during dry periods are eroded during a rainfall event and washed out of the system in the first phase of the overflow discharge; see Figure 12.4.

Since many of the most significant pollution events occur when the river has low flows and the quantity of spill is relatively small (due to the low dilution factor), the first foul flush may be a very important determinant in the overall pollution impact. Although rare, there are scenarios when systems may produce a pollutant peak in the middle or even near the end of the discharge. Full physically based models enable the engineer to determine the extent and timing of such peaks.

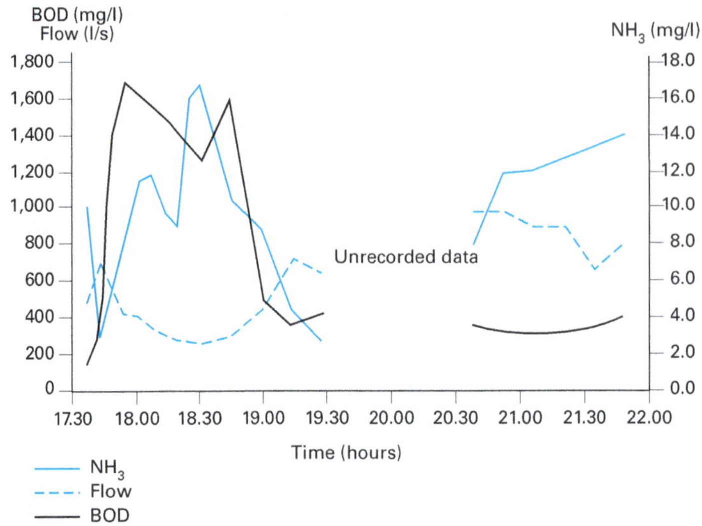

Figure 12.4 First foul flush (after Kellagher, 1993)

12.2.4 Bacteriological and pathogenic factors

The health hazards presented by faecal contamination of watercourses are serious, particularly due to CSOs. This leads to problems in developing countries with higher mortality rates. It is estimated that worldwide 5–10 percent of deaths are due to water related diseases. Therefore any untreated discharges of sewage to water courses should be avoided as far as possible.

A very large number of pathogenic organisms exist, each of which usually require a unique testing procedure that is generally very expensive. Also, many pathogens can present a significant risk to human health even in very small numbers, such as in some cases a factor of 10^{-9} times the number of Escherichia Coliform (E. coli). Incubation periods of over 24 hours are not uncommon. Unfortunately, there are no automatic monitoring techniques yet in commercial use. The detection of these pathogens relies heavily on indicator organisms present in faeces in far higher numbers than the pathogenic organisms. The most common faecal indicator is Escherichia Coliform, which has a fundamental test that is used throughout the world to test water samples for faecal contamination. Because of the severe problems in measuring micro-biological parameters involved in CSOs, most sophisticated methods of determining the quality of sewer water restrict themselves to more easily measured determinants such as BOD, COD and Suspended Solids.

12.2.5 Pollutants that hinder oxygenation at the surface

The most important chemicals in this group are oils, discharged to drainage systems by individuals, road accidents or illegal/accidental industrial discharges. Since oil floats on water and disperses rapidly into an extremely thin layer, a small quantity of oil discharged into a water body can prevent re-oxygenation at the surface and thus suffocate the organisms living there. Obviously the dispersal rate changes with oil viscosity, and the length of time the oil is a problem will partly depend on the surface of the receiving water as well. The importance of this group of chemicals is illustrated by the estimate that in as many as 40% of the closure of lowland river water supply intakes, the cause was oil contamination. Unfortunately there is little data on what proportion of these cases was caused by CSOs.

12.2.6 Toxic contaminants

There are three main sources of toxic contaminants that may be discharged from CSOs:

- industrial effluents: these could be anything from heavy metals to herbicides.
- surface washoff contaminants: including pesticides and herbicides. In many cases these contaminants make a larger contribution to the pollutant load than the domestic foul flow.
- substances produced naturally in the sewer: ammonia is almost certainly the most important, though nitrogen sulphide can also be significant.

Provided the particle size and density distribution of the potential contaminant is known, sophisticated models can model the flow of the pollutants in the network and a good river water quality model can model the effects once the contaminant has reached the waterway.

12.2.7 Discharges high in suspended solids

Discharges high in suspended solids invariably exert an oxygen demand. They also prevent light from penetrating the water and thus inhibit photosynthesis. Thirdly, if deposited, they present a reservoir of oxygen demanding particles that will prevent improvement of the water body and form an anaerobic layer on the bed, decreasing biodiversity. They also tend to 'concrete' the bed, thus being highly detrimental to particular locations, such as salmon spawning grounds. Thus even a relatively inert supply of suspended solids causes a variety of problems. If these suspended solids contain toxic substances, such as heavy metals, the problems are even more severe.

12.3 MODELLING EQUATIONS

The transport of pollutants through a sewerage network, the treatment works and the receiving water is a vital function of the whole drainage system, and therefore needs to be modelled. Unfortunately, the physics of pollutant transport is complex. Some pollutants are dissolved in the water and are therefore conveyed simultaneously with the water particles, though they are subject to molecular and turbulent diffusion. Some of these pollutants and others attach themselves to sediment particles, which are carried in suspension (the finer and lighter particles) or as a bed load (the coarser, denser particles). When the flow is sufficiently slow and lacks energy these particles can be deposited on the bed of the pipe and consolidated. Subsequently, if the flow is sufficiently large, sediment can be eroded. Because the sediment has different particle sizes and densities, distinct layers of sediment can be deposited on the bed, which can contribute physically to an armouring effect and chemically to processes that change the nature of the pollutants available. Given a large enough sediment deposit consolidated locally on the bed of a pipe then the flow can be affected through feedback from the deposit. These factors further complicate any calculation of the flow and the pollution transport.

Some simulation models address the process of deposition, but not layering or armouring. Also, the pollutants are generally regarded as being conservative, that is, they do not change their form over time. The complications of feedback from sediment deposits changing the conveyance of the pipe or channel are generally ignored.

The basic processes that the pollutant concentration is subject to in flowing water are advection and diffusion. The pollutant is carried (or advected) directly with the flow, just as if the pollutant particles were water particles. But within the flow the pollutant particles are moved around in a random manner with respect to the water by molecular diffusion (Brownian movement) and turbulence. There is also a dispersion introduced by the one dimensional averaging of the three dimensional structure of the flow in the pipe or channel. These three processes are represented by a diffusion effect in the governing equation. The resulting advection-diffusion equation complements the conservation of momentum and mass equations for the flow in the pipe or channel.

The chemical and biological processes that govern the changes in the pollutant concentration are introduced into the advection-diffusion equations as source terms, such as a decay term to represent the reduction of BOD by oxidation and a term for the re-aeration. Other terms would include photosynthesis and respiration and the erosion and deposition of sediment. Note that the advection-diffusion equation is required for each pollutant separately.

$$\frac{\partial C}{\partial t} + u\frac{\partial C}{\partial x} - D\frac{\partial^2 C}{\partial x^2} = -KC + R(C_s - C) \tag{12.1}$$

where K is the BOD rate coefficient
u is the velocity (m/s)
D is the longitudinal dispersion coefficient (m^2/s)
C is the concentration (kg/m^3)
C_s is the saturated DO concentration (kg/m^3)
R is the re-aeration coefficient

The re-aeration coefficient depends on the velocity and depth of flow and on the presence of weirs and other sources of turbulence.

The numerical solution of this equation is normally done by splitting the solution into two steps: the first dealing with the advection (hyperbolic) terms, and the other with the second order diffusion (parabolic) terms. Typically, the advection step will be solved using, say, Roe's scheme (see Priestley, 1989), and the diffusion step using Crank–Nicolson (1947). The source terms introduces the pollutant in the inflows to the 1D system, whether wastewater, rainfall-runoff or groundwater.

A full physically based water quality model for the sewerage system or the river is complex and requires careful use. Simpler models have been developed specifically for the analysis of pollutant impact on rivers, such as SIMPOL. This was developed originally for use with the first edition of the Urban Pollution Management (UPM) manual (FWR, 1998). Subsequently, SIMPOL 2 was released as a software package with the second edition, only to be updated to SIMPOL 3 with the third edition. SIMPOL 2 is based on Excel spreadsheets in which the sewerage system and the river are separately represented. The flow and pollutant inputs from the single treatment works and the upstream boundary of the river are defined as random distributions at a one hour time step, and the inputs from the CSO are defined by a very simple lumped catchment (compartmentalised) model that is designed to be calibrated against a detailed physically based model. Obviously, SIMPOL 2 has limitations in representing complex systems. SIMPOL 3 is a considerable improvement on SIMPOL 2, but it is proprietary to UK-WRc and is not on general release.

Two additional software tools, Simon and Stanks, have been developed by Montgomery-Watson-Harza (MWH). These two software tools were developed by MWH and have been widely used on studies for Yorkshire Water by a variety of consultants. The Stanks component deals with the inflows to the river system and the Simon component represents the river processes and carries out the assessment against the standards. The tools are used in conjunction with a detailed model of the sewerage system in InfoWorks CS. The inputs from CSOs are taken from the sewer model and are manipulated by Stanks. The flow inputs for the treatment works can be taken from the sewer model but Stanks then allows a statistical distribution of pollutant concentration to be associated with this. The river boundary conditions are generated as statistical distributions. Stormwater sewers can take flow from a sewer model and can use either pollutants from the same model or a statistical distribution of pollutants.

The model includes all of the necessary river processes. It does not automatically check that the lowest DO concentration has been located. The model is run for individual events from a time series. The model is similar to SIMPOL in requiring CSO discharges from upstream reaches to be accumulated and represented as discharging into reaches further downstream. The model is easy to use although there are some gaps in the documentation that have led to confusion. It does report compliance with the various standards.

The application of the hydraulic-pollutant concentration model can be to pollutant transport in the sewerage network, the coastal waters or the river. In the case of the receiving waters the model may be restricted to a critical reach of river with identified upstream and downstream boundary conditions. Ideally, the flow and pollutant concentrations at the upstream boundary should be modelled using a rainfall-runoff model, whereas the inputs from the CSOs are determined from a detailed sewerage model. In studying the impact on the river or coastal waters it is necessary to run the models for a time series of 10 or 15 years of rainfall. It is essential therefore that the model runs quickly. If the physically based model is too slow consideration should be given to replicating the performance of the physically based model with a data driven model. Generally the analysis of the results of the model for the receiving water should provide an assessment of the frequency and duration of the pollutant concentration when it exceeds certain threshold values that are laid down as standards. Care is needed when assessing compliance with standards and the audit trail should record the decision made and the conclusions reached.

12.4 URBAN POLLUTION MANAGEMENT PROCEDURE

Much of the pollution in rivers, lakes and coastal waters arises from non-point sources, such as agricultural activity and radical changes in land use including the destruction of forests. But a significant component of the pollution is generated from point sources, whether wastewater from domestic uses discharged directly to the receiving water through overflows or from wastewater treatment works, or industrial waste from food processing plants or industry. Each country is aware of the damage that is being done by such wastewater discharges, and many are putting in place legislation to control the situation, such as the Asset Management Planning process put in place by OFWAT, the UK regulator. In addition, the European Commission has issued the Waste Water Treatment Directive (CEC, 1991) in which member states are to limit the pollution arising from stormwater discharges. It was estimated in the 1980s that of the 25000 sources of intermittent discharges in the UK one third were in need of attention.

To avoid the complexity and difficulty of controlling the precise impact of a stormwater discharge on the aquatic environment, standards are prescribed to which the discharge should conform. Such standards are set according to the general ability of the receiving waters to accommodate the pollution. Separate environmental quality objectives (EQOs) are defined for each desired use of the receiving water, whether to protect the aquatic life in a river (which is sensitive to low DO or high un-ionised ammonia), bathing waters (with the risk of unacceptable pathogens) and the general amenity of a water body (including gross sewage solids from CSOs). Derived Intermittent Standards (DIS) are presented for BOD and total ammonia in the UPM manual (FWR, 1998).

The question therefore arises as to whether intermittent discharges from CSOs and continuous discharges from wastewater treatment works meet the EQOs or other prescribed standards. To answer this question it is necessary to assess the nature of the discharges, and possibly to determine the impact that the discharges have on the receiving waters depending on circumstances of the impact. This is where the UPM comes into its own: it is designed to provide a structured procedure to investigate the issue; see Figure 12.5.

The UPM procedure consists of the following major tasks:

12.4.1 Initial planning

Task A.1 Make a preliminary assessment of the problem

Description	Identify the primary problem and its nature
Input	Project objectives and scope, flow survey or historical event data, EQOs
Output	Audit documentation, nature of problem
Sub-tasks	Compare with other similar case-studies
Tools	DSS, database

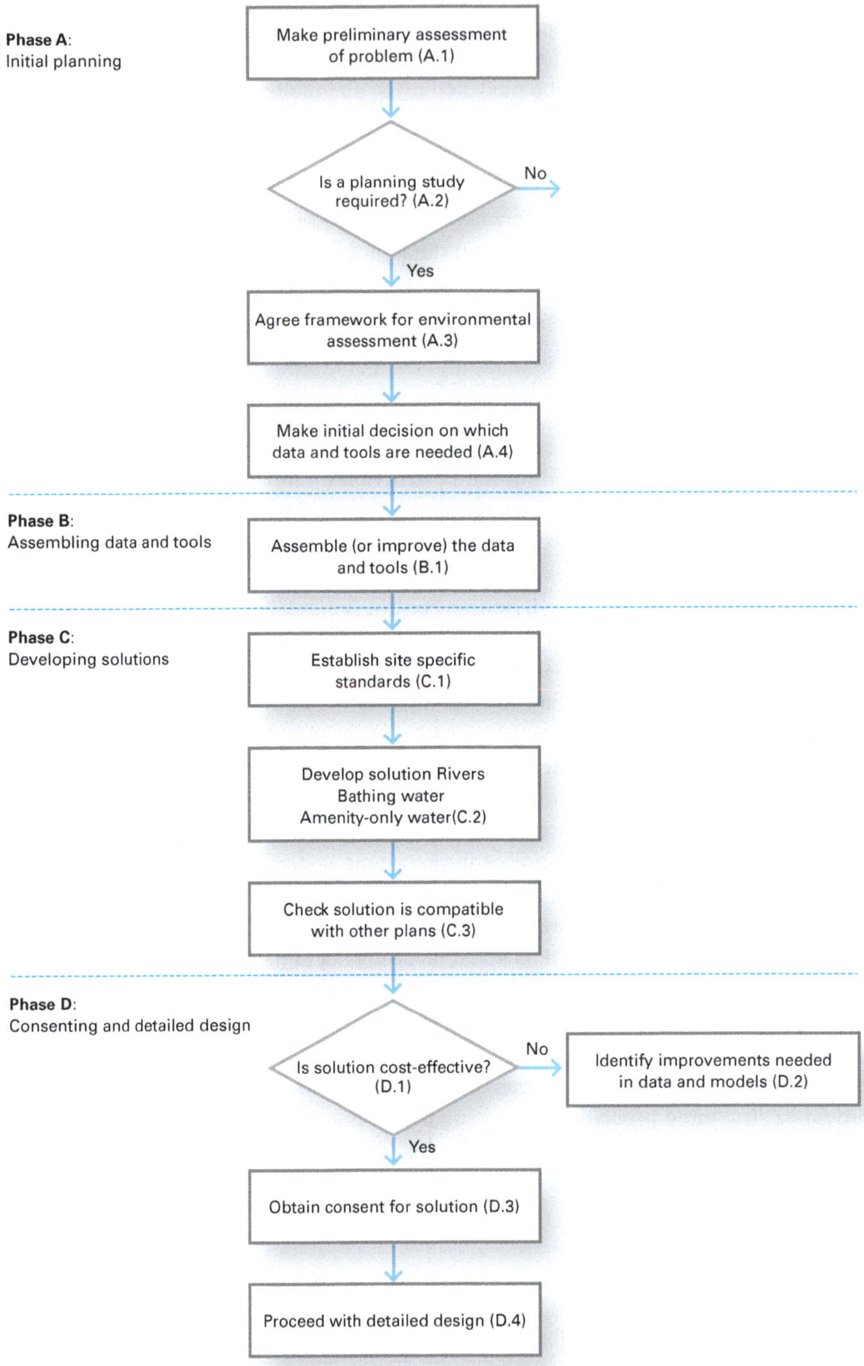

Figure 12.5 The UPM Procedure (FWR, 1998). Reproduced with kind permission of The Foundation for Water Research

UPM recommends that a simple approach based on an assessment of the discharge alone can be used if

1. The dilution of the foul dry weather flow in 5 percentile low river flow is greater than 8
2. There is no interaction with other discharges
3. The river characteristics give an allowable BOD of 15 mg/l or greater

Otherwise some form of impact assessment is required, such as for coastal waters. In this case, there is a need to consider the processes described in Figure 12.6.

UPM recommends that sewerage flow modelling is done according to Table 12.4.

The simple tank model could be SIMPOL.

Where the catchment is large and the sewage treatment works is thought to be significant in terms of its impact on the receiving waters then sewer quality modelling is recommended according to Table 12.5.

A STW model is used when there is a water quality problem downstream of the treatment works or the STW treatment works deteriorates significantly during wet weather flow and causes a problem in the river regardless of the situation with the CSO discharges.

Resource questions:

- Is a simplified approach justified?
- Is a SIMPOL model sufficient?
- Is a detailed STW model needed?

Task A.2 Is a planning study required?

Description	A decision is made on whether a detailed analysis of the various processes in the drainage system is required
Input	Preliminary project analysis, flow survey or historical event data, EQOs
Output	Audit documentation, decision on scope of ensuing study
Sub-tasks	Consult stakeholders
Tools	DSS

This task requires a positive or negative decision as to whether a detailed study is required or not.

Resource questions:

- Has enough research been done on the preliminary assessment to give confidence to the decision?

Task A.3 Agree the framework for the environmental assessment

Description	Establish the uses of receiving waters and identify the standards to apply
Input	Project objectives and scope, flow survey or historical event data, EQOs, stakeholder information
Output	Audit documentation, report on framework
Sub-tasks	Consult stakeholders, check Formula A for overflows, DO and un-ionised ammonia in rivers, spill frequency coliform count for batahing waters, and spill frequency and solids emission standard for amenity use
Tools	DSS

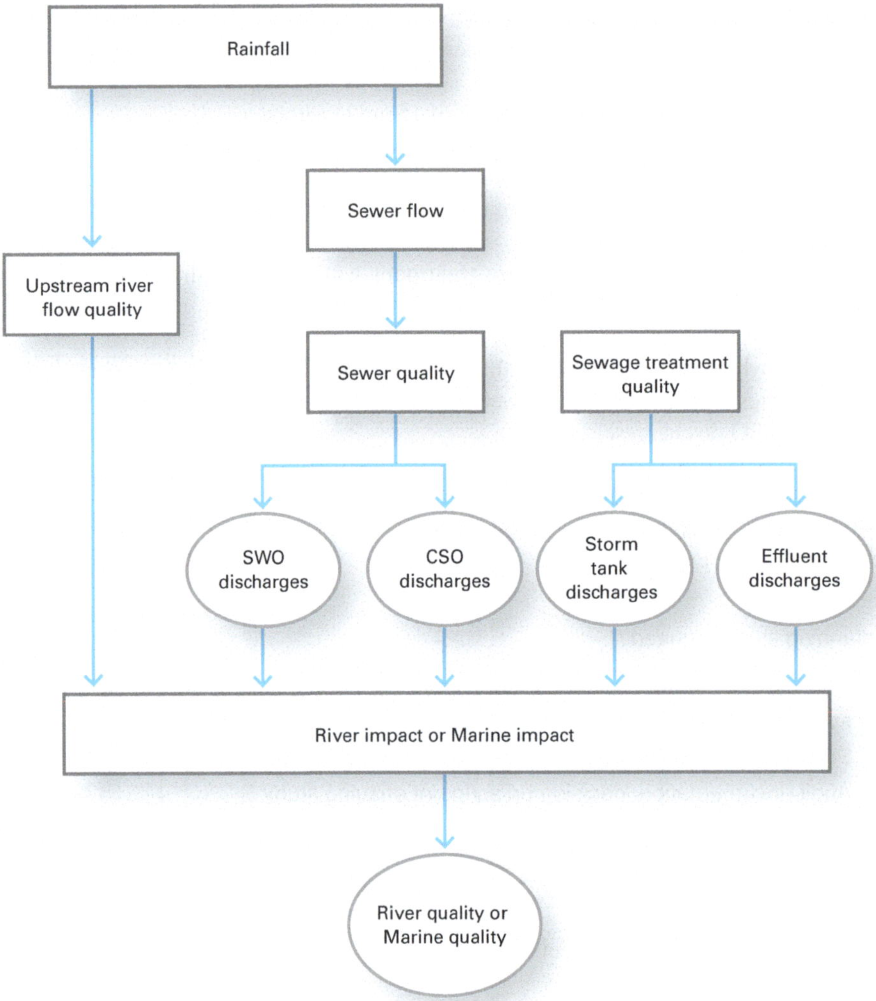

Figure 12.6 The main components of UPM method (FWR, 1998). Reproduced with kind permission of The Foundation for Water Research

Table 12.4 Application of water quality models (Reproduced with kind permission of The Foundation for Water Research)

Sewer Hydraulics	Urban catchment population		
	<5000	5000–20000	>20000
Straightforward	Simple tank	Simple tank	Detailed
Complex	Simple tank	Detailed	Detailed

Table 12.5 Application of STW model (Reproduced with kind permission of The Foundation for Water Research

Interaction with STW	Urban catchment population		
	<20000	>20000	
Not significant	Simple	Simple	Detailed
Significant	Simple	Detailed	Detailed

This task entails the review of the uses of the receiving waters, which should mean that consultations are held with the relevant stakeholders. In particular the environmental standards for protecting river aquatic life, bathing beaches and amenity use need to be agreed as these standards form the criteria against which the performance of the sewerage and sewage treatment works should be assessed.

Coastal discharges

Impact assessment for coastal waters, which are considered to be a 3 mile strip around the coast and specifically exclude estuaries, can be carried out using water quality tools. The type of assessment required is linked to the coastal waters category being considered. There are four categories of coastal waters where consideration is specifically given for obtaining Discharge Consent or are aspects to consider under the UWWTD. These are:

- Bathing Waters
- Non-designated bathing (other) waters
- Sensitive waters
- Shellfish waters

Bathing beaches

Bathing beaches are officially designated and registered. The UK interprets the conditions required as limiting the number of spills to 3 or fewer, where a spill is defined as being a volume that is greater than $50\,m^3$.

Fully separate systems are actually rarely fully separate. However even if they are, the amount of litter and faeces on the streets due to animals, will often mean that concentrations of coliforms etc will far exceed the allowable concentrations. This means in practice that stormwater outfalls have either to be picked up and discharged outside of the beach limits, or the outfalls need to be built sufficiently far out into the water to ensure sufficient dilution to ensure compliance is achieved.

What can be even more of a problem is that if a river discharges to the sea close to the beach and the normal currents drag the water along the bathing coast, compliance is virtually impossible due to background contamination of the river.

Shellfish waters and sensitive waters

The level of treatment required in the UWWTD (CEC, 1991) specifies that secondary treatment is required where STW discharge takes place in sensitive waters, which includes shellfish waters. In addition it is advised that no overflows from CSOs should take place into Shellfish Waters.

Resource question:

- Are the proposed standards communicated to everyone concerned with the project?

Task A.4 Make an initial choice of the data and tools needed

Description	Identify data and tools needed to support the decision making
Input	Project objectives and scope, flow survey or historical event data, available tools
Output	Audit documentation, assessment of data and tools
Sub-tasks	Review existing models
Tools	DSS, sewerage quantity and quality model, treatment works model, river water /coastal waters quantity and quality model

The planner has to decide whether a straightforward procedure with minimum data and simple modelling is adequate. If not, then the degree of complexity has to be agreed, with parsimony being an important factor in the decision making. The criteria for the decision in regard to discharges to rivers are identified under Task A.1 above. If the decision is to proceed with a full study then choices have to be made about the rainfall data, the sophistication of the sewerage model for both quantity and quality, the nature of the sewage treatment works model, the upstream flows and pollutant concentrations in the receiving river, and the nature of the river impact model for flow and quality or the marine impact model. UPM recommends the use of a spreadsheet based model SIMPOL (which is tank-based) for the sewerage and river impact models based on certain criteria dependent on urban catchment population, the characteristics of the river, and the performance of the sewage treatment works.

It is considered that there is only one approach to carrying out water quality modelling in terms of rainfall requirements. This is to use a time series (real or generated) which has all the real characteristics of both rainfall intensities and depths and also antecedent conditions.

The procedure to use depends on the type of standards which have been set. They may be set at various return periods with a different permitted pollutant load, or concentration at each return period. In the latter case, storms of each return period will have to be analysed. Alternatively the standards may be set at only one return period which is assumed to provide appropriate protection at all other return periods.

For each return period to be analysed, it is necessary to choose the antecedent conditions as detailed above. These should normally represent average conditions, although sensitivity testing should then be carried out to determine whether variations from the average cause substantial changes in the conditions.

Dry weather flow conditions should generally use average daily flows and pollutant loading, although in systems with flat gradients and a large build up of sediments during dry weather this may not be representative and sensitivity testing should be carried out for flow rates at various times of day.

The antecedent dry weather period before storms in the UK averages between 2 and 3 days depending on the region. A value of 3 days is recommended unless detailed analysis of rainfall records is carried out to justify an alternative value.

The choice of rainfall data is between synthetic design storms as in the Wallingford Procedure, an annual time series rainfall (see Rainey and Osborne, 1991 and Henderson, 1986), or a long rainfall time series as provided by STORMPAC™ with appropriate ranking and selection of events.

The sewerage network model has been discussed in detail in Chapters 5 and 10, including the water quality component. A range of commercial and public domain modelling software products are available.

Similarly, there is a range of modelling software products for sewage treatment works on the market; see Chapter 9.

Again, a number of commercial and public domain modelling software products are available for river impact. It is important to distinguish between long term and short term water quality response. In the first instance hydrological river models such as HYRROM or NAM can be used to deduce the upstream river flow and quality, and modelling software products like MIKE 11 and ISIS are needed for the short term impact on the river.

In each case, the models need to be calibrated on flow survey data and verified against a separate set of data. Ideally in the river discharge and concentrations should be measured upstream of any studied input from urban catchments, near the critical point in the river and some distance downstream of the critical point. The data should include flows and concentrations that enable travel times and dispersion coefficients to be defined robustly.

Although UPM does not emphasise sensitivity and uncertainty in the data and in the modelling, it is appropriate that these should be tested and evaluated for the resulting models.

Resource questions:

- Have conclusions been reached on the five major aspects: rainfall, sewerage flow, sewer water quality, treatment works, receiving waters?

Task B.1 Assemble or improve the data and tools

Description	Collate the data and models available, collect additional data, develop or refine the necessary models
Input	Flow survey or historical event data, necessary available models
Output	Audit documentation, more complete database, calibrated and verified models
Sub-tasks	Define and collect additional data, test existing models, develop or refine (including calibrating and verifying) the necessary models
Tools	DSS, modelling software, calibration tools

Once the data and modelling tools are available, they need to be assembled and outputs and inputs integrated. Generally it is assumed that there is no feedback from a model downstream to a model upstream; this implies that the river does not prevent discharges flowing freely from overflows or treatment works. This is obviously not the case on those occasions when the river water level is too high. If this is a critical case for the performance of the whole drainage system then action must be taken to resolve the situation through structural works or, say, the introduction of pumping.

Resource question:

- Is the data sufficiently complete to proceed with the modelling?

Task C.1 Establish site specific standards

Description	Establish that the solution for intermittent discharges to rivers or marine waters is adequate and cost-effective
Input	Project objectives and scope, flow survey or historical event data, local knowledge, sewerage model, sewage treatment model, river impact model
Output	Audit documentation, BOD standard
Sub-tasks	Estimate river BOD/ammonia concentrations, explore the implications for DO of additional storage in the sewerage network
Tools	DSS, StormPac™, SIMPOL, modelling software for sewerage, wastewater treatment and river impact

The appropriate models need to be used to assess where DO is a minimum in the receiving river downstream of the CSOs and/or treatment works, or where the regulatory requirement for bathing beaches is most likely to fail. UPM gives detailed procedures for carrying out both the river and marine impact modelling.

Resource question:

- Have the standards been agreed with the regulator?

Task C.2 Develop the solution

Description	Use the confirmed models to explore possible structural or improved RTC solutions to the sewerage system or treatment works
Input	Project objectives and scope, flow survey or historical event data, confirmed models
Output	Audit documentation, proposed solution
Sub-tasks	Assess satisfactory impact on receiving waters
Tools	DSS, STORMPAC, SIMPOL, models

The appropriate solution may require structural adjustments to the overflow settings, storage available in the sewerage system, introduction of real time control of storage facilities or the treatment works, extension of the treatment works, and so on. Refinement of control rules for RTC systems could also be an option. What is important is to put together an integrated solution that helps to meet the standards in the receiving waters; for details proposed by UPM see Figure 12.7.

Once verified, models of the drainage system (and receiving water) can be used to define the performance of the system compared to the required standards. We can subsequently investigate changes to the system to improve the performance.

There are several possible ways in which standards can be expressed in order to measure the system performance depending on whether the sewerage model is being run in conjunction with a river model and the type of river model that is being used. The most likely scenarios are:

1. Total spill volumes or total spill loads in a series of events are used to determine worst case scenarios.
2. The peak pollutant concentration in the river is identified from a series of events run through the sewer and river models by determining which event gives the largest pollutant concentration.
3. A range of rainfall events based upon intensity-duration-frequency curves are run for pollutant loading and washoff to determine peak pollution loads at the outfall. These are compared with permitted concentrations to check on compliance or determine the most critical storm(s).

The critical storm durations for these three types of analysis are likely to be different, and may also be different from the critical durations for considering the peak flow to the treatment works, or the maximum amount of flooding in the catchment.

The critical durations can also change if the drainage system is modified. Constructing new sewers tends to reduce the critical duration for overflow spill volumes. Putting in detention storage increases the critical duration for overflow spill volumes. Changes to the receiving water can also change the critical duration. For example removing an upstream discharge so that the river quality improves may change the critical duration for pollutant concentrations.

The choice of critical duration is therefore an iterative process and may change during the study.

Choice of antecedent conditions

When considering rainfall and spill frequency it is necessary to ensure that the antecedent conditions for the runoff model are appropriate so that the return period of spill is the same as that of the rainfall. Similarly, when modelling pollutants there is a wide range of antecedent conditions which must be considered.

The return period of pollutant load spilled from the overflow is a function the antecedent dry period and rainfall characteristics. The rainfall will not necessarily be an extreme event to be the 'worst' case.

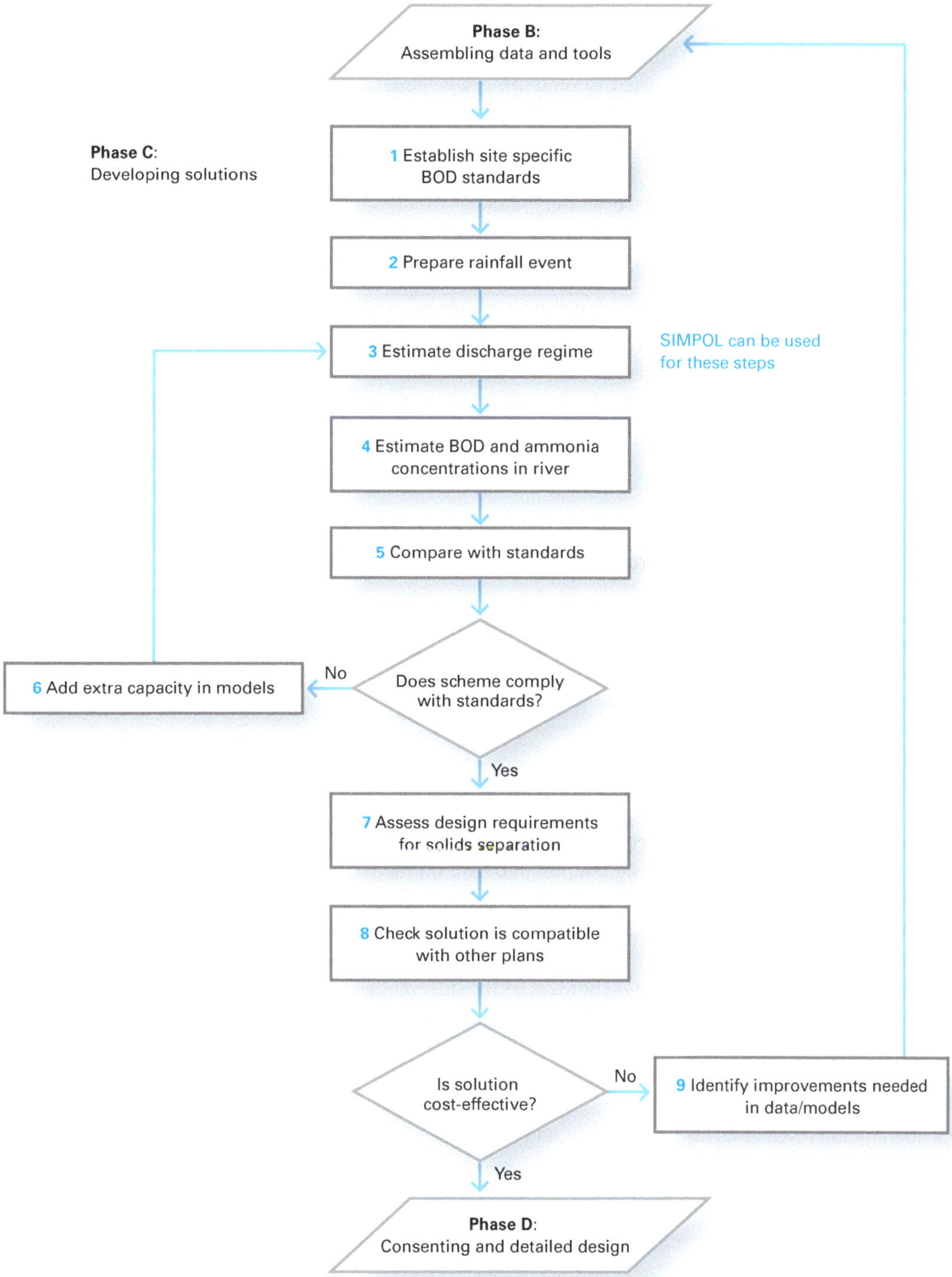

Figure 12.7 Flow chart for Phase C: developing solutions (rivers) (FWR, 1998). Reproduced with kind permission of The Foundation for Water Research

The duration of the dry period before the start of the event (the antecedent dry weather period) is one of the items in the sewerage water quality model. This is used to calculate the build up of pollutants on the catchment surface, and in sediment deposits in the sewers.

The rainfall data required for water quality modelling will generally represent a wider range of conditions than for hydraulic studies, usually with events with return periods of less than one year in terms of rainfall depth. It is very important to remember that storms which give the worst hydraulic conditions, for example the greatest overflow spill volume, may not be those which give the worst pollutant conditions considered either as spilled pollutant load, or as impact on the receiving water.

Task C.3 Check that the solution is compatible with other plans

Description	Examine the consistency of the solution with other initiatives and plans
Input	Solution, flood risk management plans, urban development plans, agricultural developments in the catchment upstream, water supply intakes downstream, etc
Output	Report on possible conflicts with other uses or aspects of the receiving waters management
Sub-tasks	Estimate risk of conflict
Tools	DSS

Resource question:

- Have the models undergone peer review?

Task D.1 Is the solution cost-effective?

Description	Compare the cost of the solution and the monetary benefit
Input	Solution, cost tables, benefit tables
Output	Cost-benefit report
Sub-tasks	Consult stakeholders
Tools	DSS

There can be a range of plans affecting the receiving waters, including the satisfaction of flood risk requirements, urban development increasing the possible load on the receiving waters, agricultural developments upstream in the river catchment, and so on. All of these other plans for the receiving waters need to be taken into account when finalising the present solution.

Resource questions:

- Have the appropriate champions/stakeholders of the different aspects been consulted?

Here the primary objective is to justify the benefit of the proposed solution in terms of its cost and benefits.

Resource question:

- Is the uncertainty in the cost and benefit data known?

Task D.2 Identify improvements needed to data and models

Description	Assess sensitivity and uncertainty of proposed solution to reveal what improvements may be needed to the data and models
Input	Solution
Output	Report on data and model improvements
Sub-tasks	Sensitivity and uncertainty analyses
Tools	DSS, sensitivity and uncertainty analysis tools

If the additional plans or the sensitivity or uncertainty in the proposed solution require a more robust analysis of the performance of the models, then some effort may still have to be put into refining the models.

Resource question:

- Has the uncertainty analysis shown which parameters and data need to be checked?

Task D.3 Obtain consent for the solution

Task D.4 Proceed with the detailed design

12.5 INTEGRATED MODELLING

Integrated modelling of different components of urban water cycle (Figure 12.8) has emerged as an important for several reasons. First, there is a growing need to manage the urban water cycle on a global basis. Second, a range of alternative technologies to process different aspects of the urban water cycle are becoming available. Third, advances in urban hydroinformatics enable different phases of the entire cycle to be modeled and to use such models to optimise each phase locally and in the global context. Initially, different components of urban water cycle have been modelled in a sequential way without the feedback between different models; see Figure 12.9. As an improvement to this approach, efforts have been made to incorporate feedback mechanism and to enable models to exchange information at each time step (this was defined as simultaneous or parallel modelling); see Figure 12.10.

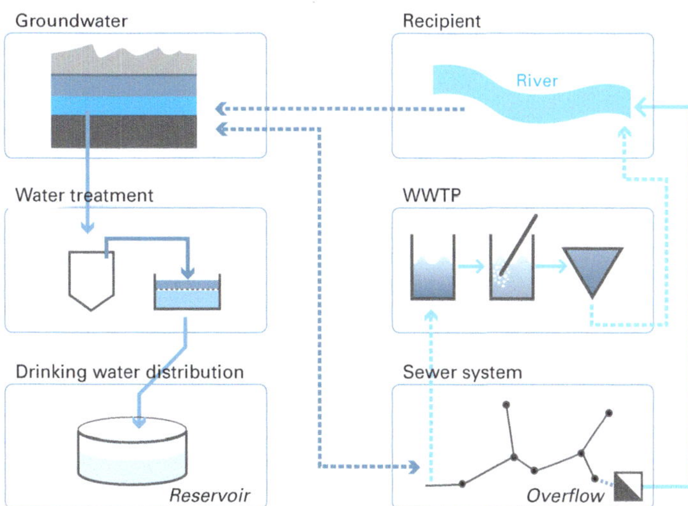

Figure 12.8 Schematisation of the urban water cycle and interactions between different components

The need for an integrated analysis has been discussed in various publications (see, for example, Rauch *et al.*, 1998; 2002; or Seyoum, 2009) and it has also been formulated within the EU Water Framework Directive 60/2000 (CEC, 2000) with a water-quality orientated view on the whole river basin system which calls for new ways of assessing its performance. Nowadays, there are attempts by commercial software companies to develop links between different detailed (i.e., refined) physically-based models (Moore *et al.*, 2004)

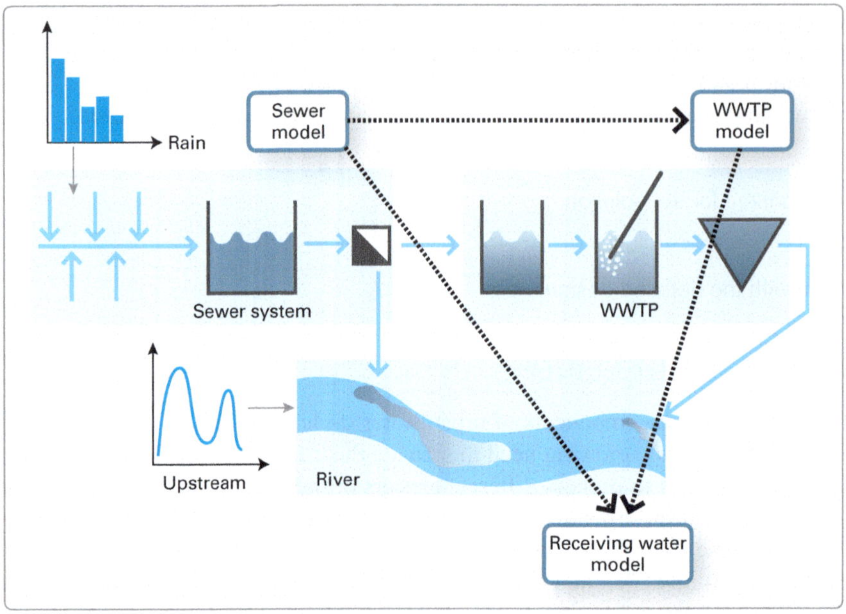

Figure 12.9 Schematisation of sequential modelling

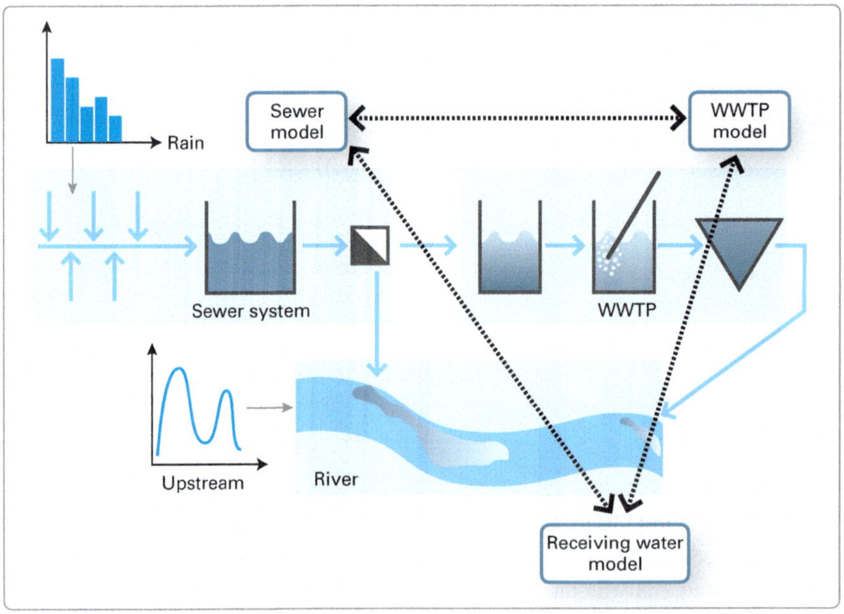

Figure 12.10 Schematisation of simultaneous (or parallel) modelling

but the effort to instantiate such models and link them together is still challenging. With the objective of facilitating the integration between different modelling systems, the EU funded the HarmonIT project, which was initiated to address the issues facing open modelling interface (OpenMI) protocols. Some of the issues addressed concerned questions of how to define, describe and transfer data explicitly (i.e. numerical data variables in space and time) between components on a time step basis, including the associated component access (Gijsbers, 2004). The HarmonIT project has turned OpenMI into a widely accepted 'standard' for model linkage in the water and environmental domain. To become an OpenMI linkable component, a model has to:

- be able to reveal information (what?, where?) to the outside world on the modelled variables it can provide, or which it is able to accept;
- submit to runtime control by an outside entity;
- be structured in a way that initialization is separate from computation, where boundary conditions are collected in the computation phase and not during initialization;
- be able to provide the values of the modelled variables for the requested points in time and space;
- be able to respond to a request, even when the component itself is time independent; if such a response requires data from another component, the component should be able to pass on the time as well in its own request.
- In case some values in the value set are missing, this should be flagged in the value set.
- In the exceptional case that an entire value set is unavailable, an exception needs to be thrown. Be aware that such an exception will stop the entire computation process and thus should be prevented whenever possible.

More information about HarmonIT project can be found at www.openmi.org and www.harmonit.org.

Many authors have reported the results of a water and contaminant balance analysis for an integrated urban water system, based primarily on simple monthly or annual averages using desktop or spreadsheet calculations, while a few utilise more detailed analysis tools such as simulation modelling software (Mitchell *et al.*, 2003). The simplified modelling systems include a simple generic urban water balance model based on empirical equations that has a variable time step, ranging from daily to annual with the primary purpose of providing preliminary screening tools for alternative water servicing options which are quasi-distributed daily time step models. Such models divide a study area into water system components (nodes) that are interconnected by drainage or supply links, or into clusters that comprise allotments, road corridors and open space. A sub-daily time step model called PURRS was developed to model the impact of local stormwater management devices on stormwater and water supply infrastructure.

Mitchell *et al.* (2003) developed a total urban water cycle modelling tool called UVQ. This is a daily time step, conceptual model, which represents an urban area in a quasi-distributed manner. UVQ includes representation of the climate, landuse and the existing infrastructure as these are the primary factors determining the water and contaminant balance of an urban area. Alternative methods for water supply, stormwater and wastewater service provisions are included, enabling the assessment of the impact of alternative water servicing approaches on the total water cycle. Three nested spatial scales are used in UVQ to describe the components of an urban area. The single allotment (unit block) represents a building and associated paved and pervious areas such as paths, driveways and gardens; the neighbourhood (cluster) consists of a number of identical unit blocks as well as roads and public open space; and the catchment represents the grouping of one or more clusters that may or may not have the same landuse. Conceptual routines are used to represent the rainfall-runoff process. Impervious surfaces are each represented as single stores that overflow when full. The concept of an effective impervious area is used to represent the proportion of impervious surface runoff that directly drains to the stormwater drainage system. Pervious area soil moisture stores are represented in one of two ways: a partial area saturation overland flow model or a two layer soil store model. The amount of evapotranspiration from pervious areas and of evaporation from impervious surfaces is calculated separately. The inflow of stormwater into the wastewater system is calculated as a proportion of the total surface runoff generated. The amount of leakage from distribution networks is assumed to be proportional to the daily bulk water use and the amount is directed into the groundwater store. All stormwater and wastewater stores, from unit block scale through to catchment scale, are represented as a straight side tank or reservoir. Overflow from the stores is simply the volume of inflow that exceeds the available storage capacity. Contaminants are modelled conservatively, with no conversion or degradation within the existing infrastructure. The performance of all treatment process from rain tanks through to catchment wastewater systems is specified in terms of the percentage removal of individual contaminant types or by setting output requirements.

CITY DRAIN is an open source software for integrated modelling of urban drainage systems and was developed by at the Institute of Environmental Engineering at the University of Innsbruck, Austria (Achleitner, 2006). CITY DRAIN adopts the Matlab/Simulink © environment, enabling a block wise modelling of different parts of an urban drainage system (catchment, sewer system, storage devises, receiving water, etc). Each block represents a system element (subsystem) with different underlying modelling approaches for hydraulics and mass transport. The open structure of the software allows the addition of the users own blocks and/or modified blocks (and underlying models) according to specific needs (Achleitner *et al.*, 2007). The computations in CITY DRAIN are based on a fixed discrete time steps approach in which each subsystem uses the same time increments, usually predetermined by the temporal resolution of the rain data used. Models implement for hydraulics and mass transport are formulated for discrete time steps Δt. A simple method to account for the initial loss is used. The volume of the basin (respectively the height) represents the volume of water retained due to initial losses. The volume of rain exceeding the basin's volume is considered to be the effective precipitation, contributing to the catchment surface flow. Permanent losses such as evapotranspiration can be either considered acting all the time or during dry weather only. In either case, the volume per time step to be evaporated is limited by the initial loss specified. The Muskingum method is used for flow routing in the catchment, sewers and river blocks. The system (catchment, channels and conduits) is considered as a whole. The model has separate blocks for handling combined and separate sewer systems. The model also has blocks for combined sewer overflow structures and pumping stations. Mass transport of pollutants is implemented for conservative matter/tracer substances. The model for the WWTP currently implemented is abstracted as a 'black box' accounting for cleaning efficiencies and maximum effluent qualities. Thus neither the flow nor the quality delivered to the treatment plant influences the level of treatment. The user is required to define cleaning efficiencies for the substances and the maximum effluent concentration. For reasons of simplicity a 'perfect' treatment plant in terms of emission standards is assumed. Currently an accurate process description incorporating an ASM 1 type process model is developed for implementation in CITY DRAIN (Achleitner *et al.*, 2007). The original CITY DRAIN (Achleitner *et al.*, 2007) is enhanced into the CITY WATER DRAIN system (Seyoum *et al.*, 2009) as part of the EU funded SWITCH project.

Another software product for modelling different aspects of the urban water cycle is KOSIM-WEST which represents a combination of the KOSIM hydrological catchment runoff and sewer transport model and WEST® (Solvi *et al.*, 2005). In the KOSIM model, the sewer system is represented by a number of reservoirs connected in series or in parallel. A conceptual rainfall–runoff model transforms the rainfall series into a flow series for the single sub-basin. Weirs and different types of storage basins may be considered in this package. The hydraulic calculations for these structures are based on the continuity equation, on maximum flow capacities and stage–discharge relations. Rainfall-runoff simulation in KOSIM distinguishes between impervious and pervious catchment areas. For impervious areas, wetting, depression (represented by a time-variant runoff coefficient) and evaporation losses are considered. Infiltration for permeable areas is simulated using Horton's approach, which has been adapted for use in long-term simulations. Flow routing in subcatchments is modelled by Nash cascades. Flow between the subcatchments is modelled by translation and addition of inflows from the subcatchments. From this modeling approach, it follows that backwater effects cannot be modelled by KOSIM. Structures such as pumps, overflows and different types of storage tanks can also be modelled. Surface flooding is not modelled explicitly by KOSIM; overflows occurring within subcatchments are assumed to be discharged into the river.

The Danish Hydraulic Institute (DHI) and Water Research Centre in the UK (WRc) developed an 'Integrated Catchment Simulator (ICS)' within a large EU-funded 'Technology Validation Project' (Clifforde, *et al.*, 1999). ICS is a graphical interface for setting up and running integrated models with feed forward /feed back of information. The present ICS version includes existing models for sewers (MOUSE), rivers (MIKE11), wastewater treatment plants (STOAT) and coastal areas. During the course of this project, these fairly complex constituent models were linked in various stages; first in a sequential way, and later in parallel.

SIMBA is a commercial toolkit for Integrated Urban Water Systems quality simulation and it is delivered by Institut für Automation und Kommunikation e.V. Magdeburg, Germany (http://www.ifak.de) (Freni *et al.*, 2003). It allows the holistic consideration of the entire urban wastewater system, including the sewer system, a wastewater treatment plant, the receiving water body as well as sludge treatment. Using information available from the plant, it is possible to set up simulation models for planning, design and optimization of plants within a reasonable time. SIMBA is an application-oriented tool. Without any knowledge of programming, and assisted by buttons, extensive libraries and menus, the user can build models and simulate plants within a short time following a three-step approach:

- selection of model blocks required
- connection of the blocks with lines according to plant configuration; parameterisation
- simulation, followed by evaluation of plots of pollutant concentrations and loads using the build-in Monitor feature

The aim of the block set SIMBA-sewer is the simulation of single dynamic events with interactions between the treatment plant and sewer system. Interaction between the treatment plant and sewer occurs, in case there is any kind of feedback present. Feedback can originate from hydraulic effects (backwater) or from control of the sewer system using information of the current state of the treatment plant. SIMBA-sewer is not a replacement for conventional sewer simulation systems but a special tool to investigate managed sewer systems and integrated systems. SIMBA-sewer does not describe the rainfall-runoff process. Information about the several influent components to a sewer system (wastewater, surface runoff, infiltration) as flow rates and concentrations must be provided by the user. This influent could be generated by conventional sewer simulation systems and provided as ASCII files for SIMBA sewer. The surface runoff models are not included in SIMBA sewer because usually no feedback exists between surface runoff, transport and storage in sewer systems. SIMBA sewer simulates:

- the transport of water and the load of dissolved and solid pollutants in sewer networks, through the application of a conceptual deposition/resuspension model
- the conversion processes, by means of depletion curves
- any kind of overflow and storage construction
- controllable inputs and controllers for sewer systems

Modelling of water transport is achieved by the diffusion wave approximation of the Saint Venant equation system – the hydrodynamic model includes storage, back water effects, change of flow direction, and flow separation in meshed sewer networks. SIMBA is able to simulate any closed or open kind of cross section profile, a change in roughness along the wetted perimeter, and pressurised flow in case of completely filled pipes.

Another integrated modelling package is SYNOPSIS ('software package for SYNchronous OPtimisation and SImulation of the urban wastewater System') which was developed to assist studies of urban wastewater systems following the need for an integrated perspective. The simulation package consists of three main simulation sub-programs for modelling water flow and quality processes in the urban drainage system, WWTP and river system. A number of auxiliary programs are also used for control, optimisation and file management. The urban drainage sub-model is based on the KOSIM program. The model allows for abstraction of wetting, depression and evaporation losses from rainfall–runoff. Subsequently, flow routing within a sub-catchment (incorporating surface and pipe storage effects) is accomplished using a cascade of linear reservoirs model. Flows between sub-catchments are routed by translation only. Pollutants are considered to originate from rainfall–runoff and domestic and industrial wastewater flows and to be completely mixed throughout the drainage system. No in-sewer processes are represented, neither sediment transport nor biochemical reactions, so any first foul flush cannot be predicted. However, the quality of the wastewater discharging at the overflows is specified with a greater fraction of rapidly biodegradable BOD (in the river model) than is given to the incoming WWTP effluent. Overflows and different types of storage tanks can be specified (if required) at the downstream end of each sub-catchment. The WWTP sub-model is based on a slight simplification of the IWA Activated Sludge Model No. 1. The model represents the main unit processes of a conventional activated sludge plant: storm tanks, primary clarifiers, activated sludge aeration tank and secondary clarifier. Quality parameters include SS, VSS, COD (in various fractions), NH_3-N and NO_3-N. These data are taken from the output of the urban drainage sub-model (Zacharof *et al.*, 2004). The river sub-model is based on the DUFLOW shell program. The river hydraulics is represented by solving the St. Venant equations using a four-point implicit Preissmann scheme. Pollutant mass transport is modelled using the advection–diffusion equation. Representation of river water quality is based on an extension of the classic Streeter–Phelps model, including BOD (readily and slowly biodegradable fractions), NH_3-N and DO as key variables. This allows the model to characterise a number of phenomena, including reaeration, deoxygenation, nitrification, sedimentation, sediment oxygen demand and photosynthesis. Discharges into the river originate from three sources: CSO spills, storm tank overflows and treatment plant effluent. Routines convert drainage system and WWTP model quality parameters to those required for the river model. River catchment runoff is modelled by a simplified time-series approach. The three sub-models are incorporated into one simulation program. The urban drainage system and WWTP are simulated in parallel. The river model is then run separately, as a sequential process. The programs are connected only at the necessary parts. This provides a hybrid of integrated simulation between pure sequential and parallel approaches. The simulation of the integrated system in this manner satisfies the requirements for investigating the relationship between overflows and receiving water quality (Zacharof *et al.*, 2004).

Also, another modelling system that has been developed is Cosmos, which is a conceptual simplified model for sewer system simulation. The model is conceptual both in the flow simulation part (i.e., the rainfall-runoff transformation and the subsequent propagation) and in the quality part (i.e., the build up and wash of solids). The system (catchment and channels) is considered as a whole (Calabro, 2001). The main goal of the model is to simulate the urban runoff quality. The model can be divided into four parts: the calculation of hydrological losses, the rainfall-runoff transformation and runoff propagation, the build up solids during dry weather and their wash off. To calculate hydrological losses the model considers primarily the loss due to retention in small reservoirs and after all others combined into a runoff coefficient. The other hydrological losses (for example, infiltration) are considered less important in the model mainly because urban catchments are mostly impervious and it was considered possible to take them in account as a whole using a runoff coefficient. To simulate the rainfall-runoff transformation of the linear reservoir method is used.

12.6 CASE STUDY 1: STRATEGIC AND OPERATIONAL MODELLING FOR MARINA RESERVOIR, SINGAPORE

12.6.1 Introduction

The Marina Reservoir (see Figure 12.11), formed by the construction of a 350-meter long barrage across the mouth of the Marina Channel, is fed by the largest and most urbanized catchment in Singapore. Together with two other reservoirs in the north-eastern part of Singapore (i.e. Punggol and Serangoon), the new reservoir increases the catchment area for water supply from half to two-thirds of the area of Singapore. More than 10% of Singapore's water demand is met with the conversion of Marina Bay into a fresh water reservoir, augmenting water supply from local catchments, and is one of the four 'taps' in Singapore's Four National Taps Strategy[1]. Two other benefits of the Marina Barrage are flood alleviation in low-lying areas of the city and the opportunity for a host of recreational possibilities, thereby enhancing Marina Bay for both the local population and tourists.

Figure 12.11 Artist impression of the future marina Reservoir (courtesy of URA)

The five main waterways draining into the Marina Reservoir (i.e. Singapore River, Stanford Canal, Rochor Canal, Kallang River and Geylang River) have been important for the development of many financial institutions and various commercial activities concentrated along their banks. As such, the waterways supported urban development with the majority of the population living and working along them. Industries such as boat repairs and goods processing plants flourished along these waterways (Hon, 1991). By the mid-1970s, the pollution problems due to the wastewater from the growing population and industries became more serious. The waterways degenerated into open sewers, with black and odoriferous discharges, which spread out into the surrounding seas from the mouth of the basin (Chou, 1998).

In 1977, a 10 year clean up programme for the Singapore River and the basin was initiated by the Singapore government. For many years this rehabilitated urban catchment in combination with the daily tidal flushing of Marina Bay did not reveal any significant water quality problems. However, with the construction of the freshwater reservoir in downtown Singapore and the intensification of activities both along the shore and in the open water, problems with water quality and its impact on different user functions are inevitable. The likely problems include:

- eutrophication, possibly leading to nuisance algae blooms (including toxic bluegreens), oxygen depletion and smell problems;
- silt, solids and organic matter runoff, leading to high turbidity (brownish water);
- bacterial contamination and related disease vectors;
- accumulation of micro-pollutants (heavy metals and organic substances) and accidental spills of chemicals;
- accumulation of diesel range and gasoline range organics;

[1] The other three National Taps are imported water, NEWater and desalinated water.

- litter accumulation; and
- mosquito breeding (when the fish stock is not yet developed).

It is perceived that water quality objectives will only be achieved by strict source control, aeration and a recirculation scheme with the Chestnut Avenue Waterworks producing water to drinking standards and having a capacity of 5 m^3/s. Furthermore, treated water will be redirected dynamically to critical regions of the reservoir by canals and secondary pumping stations.

12.6.2 The conversion process

From early 2008 till mid 2009 (Table 12.6), when a substantial part of the source control and mitigation programmes were still in the process of being implemented, the Marina Reservoir was operated as a well-flushed brackish water system with sufficiently high salinity to avoid any problems with bluegreens and mosquito breeding. During this period the reservoir was operated for flood prevention and recreational use, requiring water level control and well-controlled water quality.

Table 12.6 Transition from Marina Bay to a freshwater reservoir

	Marina Bay	Brackish reservoir	Transition to freshwater reservoir	Freshwater reservoir
Time Period	Till early 2008	Early 2008–mid 2009	mid 2009–early 2010	early 2010 onwards
Condition of Marina Bay/ Reservoir	Well-flushed estuary with no water quality problems	Well-flushed brackish reservoir with no significant water quality problems	Controlled conversion to freshwater system with possible water quality problems	Controlled freshwater reservoir with all measures implemented and with minimal problems

In 2009, the brackish water system was gradually converted to a freshwater reservoir. The duration of this conversion depended on rainfall and therefore on the monsoon season. Although the salinity concentration in the reservoir quickly dropped, it is taking a long time (up to the present) for the salinity to reach sufficiently low levels such that the water is suitable for treatment to potable water standard. It is this period of conversion from a brackish to a freshwater system that poses some water quality problems due to the expected degradation of living organisms in the Marina Reservoir.

12.6.3 Water quality modelling framework

A modelling framework (see Figure 12.13) has been set up to better understand the reservoir system under different salinity conditions, to guide the implementation of mitigation measures and to prepare for the future operational management of the reservoir. The framework consists of:

- Rainfall run-off from the sub-catchments of the Upper and Lower Peirce Reservoirs and Marina Reservoir: hydrological model HYMOS (Chowdhary *et al.*, 2004).
- The conveyance of water through the drains, canals and rivers: 1D hydraulic model SOBEK (Van Gils, 1998).
- Emissions from point and non-point sources (emission model) based on the latest land use information, stormwater sampling data and routine monitoring data for the canal and river systems, other available

measurements for specific source contributions (atmospheric deposition, sewers, industry, golf courses, construction sites, green zones, backyards, foodstalls, etc.) and literature data (characteristic emission data).

- Hydrodynamics, transport, water and sediment quality for the three reservoir systems: 3D water and sediment quality model Delft3D (Lesser *et al.*, 2004 and Los *et al.*, 2008).
- Real-time control (SOBEK RTC module) operation of the barrage (as integral part of the coupled 1D-3D model).

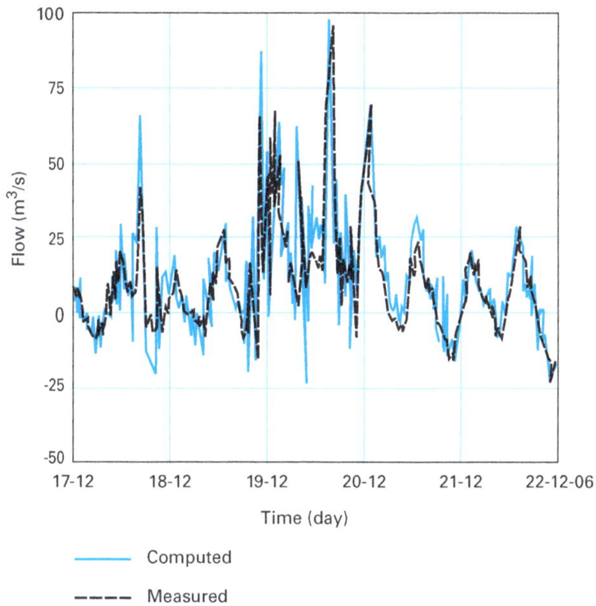

Figure 12.12 Example of hydraulic model results for the Singapore River

All of these modules were coupled to allow for an integrated analysis of the water quality management options (see Figure 12.13).

The 1D-3D modelling system has provided increased insight into the functioning of the interlinked systems through a description of the complex hydrodynamics (water balances, stratification and effects of aeration) and water quality parameters (resulting in among others detailed nutrient mass balances, algal blooms, species composition including occurrence of nuisance blooms, the occurrence of low oxygen concentrations, bacterial pollution and high turbidity).

Model simulations have shown that even after the implementation of selected source control (e.g. sewer rehabilitation and silt management) and mitigation measures such as aeration and recirculation, some water quality problems in the reservoir may still arise. Simulations also show that a combination of mitigation measures, in particular sewer rehabilitation, silt control, re-circulation and treatment and aeration, will help to significantly improve water quality in the Marina Reservoir.

i. Without mitigation measures: The model predicts that recreational water quality standards will not be met 20% of the time in Marina Bay, 38% of the time in Kallang Basin and 7% of the time in the Marina Channel. Marina Reservoir will be prone to algal blooms. Silty water is also expected about 10–15 times a year in Marina Bay and Marina Channel, and about 30 times a year in Kallang Basin.

ii. With comprehensive mitigation measures: The model predicts that the exceedance of the recreational water quality standards will be reduced to 10% of the time in Marina Basin, 30% of the time in Kallang Basin, and 3% of the time in the Marina Channel. They will also reduce the potential for algal scum formation, and improve the situation of silty water by 80–100%.

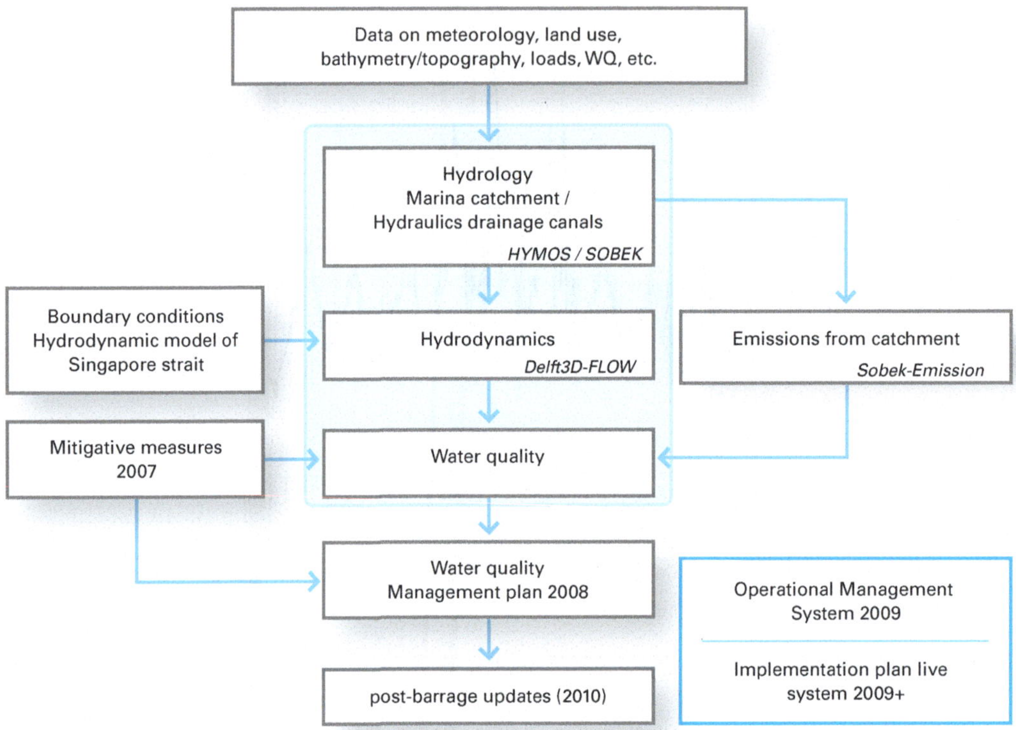

Figure 12.13 Schematized stepwise modelling approach indicating the phasing and timing of activities and the various models used. The models will be re-calibrated after the transition period

Given the complexity of the catchment, the water quality models predict that there would still be times when water quality in the Marina Reservoir may exceed target levels especially after heavy storms. Nevertheless, it is predicted that with all the programmes in place, water quality will be similar to that of Lower Seletar Reservoir, which is currently a popular spot for recreational activities.

The Operational Water Quality Management Plan aims to address and prioritise the most practical and effective strategic and operational options for the management of water quality in the Marina Reservoir. This means that in addition to the various measures to prevent deterioration of water quality in the reservoir, curative measures are needed to mitigate the problems. Although many alternative measures have been discussed in the literature, in practice, most of them do not work well in systems with a constant pollution load or do not combine well with the existing constraints (i.e. drinking water use, controlled water level, etc.).

12.6.4 Operational Management System

To address the future operational water quality management problems in the Marina Reservoir, an operational management system (OMS) to support day-to-day decision making based upon continuous rainfall forecasting (using numerical weather prediction modelling and radar data), online water quantity and quality monitoring information streams, and the 1D-3D water quality model forecasting (see Figure 12.14) has been developed. The first version of the OMS addresses the optimal tidal flushing and water level control scheme for brackish conditions that has been put in place for the period of 2008 to 2009, i.e. the period before the system was converted to a freshwater reservoir.

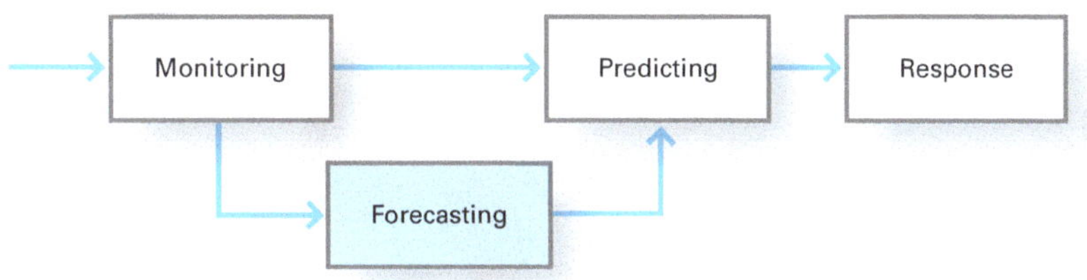

Figure 12.14 Set-up of the OMS

The general role of the OMS is to provide up-to-date information about present and expected water quality conditions in the coupled Marina, Upper and Lower Peirce Reservoir systems.

The on-line OMS output is available on individual PCs of selected users. The future 'live' OMS will be implemented in the control room of the Marina Barrage.

This OMS (see Figure 12.15) integrates real-time data from various sensors and external models, facilitates automated and interactive forecasting of both water quantity and water quality using a comprehensive modelling framework, and relates measurement and model data to predefined water quality thresholds. These functions of the OMS are illustrated in Figure 12.16.

The OMS is a client-server based application, where reservoir operators can view both measurement and model output using the client software. This client also facilitates interactive forecasting by the reservoir operators within the decision support framework. The OMS is set up using the Delft FEWS software package described in (Gijsbers *et al.*, 2008).

The modelling framework as implemented in the OMS includes the upstream catchment in 1D and the downstream reservoir in 3D. The 1D and 3D model components exchange information on a time step level using a methodology described in Twigt *et al.* (2009). The modelling framework takes both the effects of external forcing (like rainfall, wind and, heat flux) and reservoir operation (like barrage operation and aeration) into account.

Both the OMS and the modelling framework are continuously in development, to include additional modelling functionalities, additional monitoring data, and also other reservoirs in Singapore.

12.6.5 Organisational set-up and capacity building

The 1D-3D modelling framework and Operational Management System have been transferred to strategic and operational teams within PUB. Both teams work close together to manage the Marina Reservoir in the best possible way; see Figure 12.17.

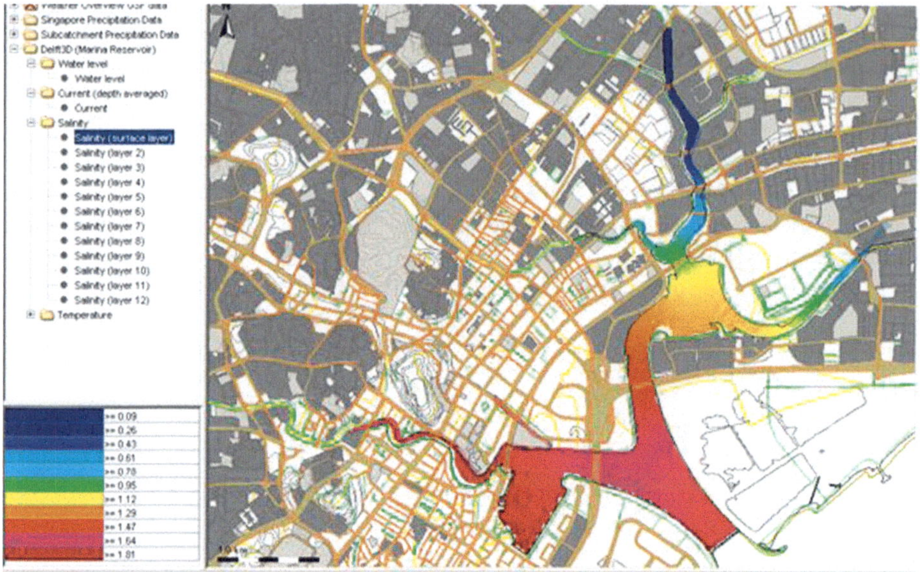

Figure 12.15 Snapshot of the OMS (salinity simulation)

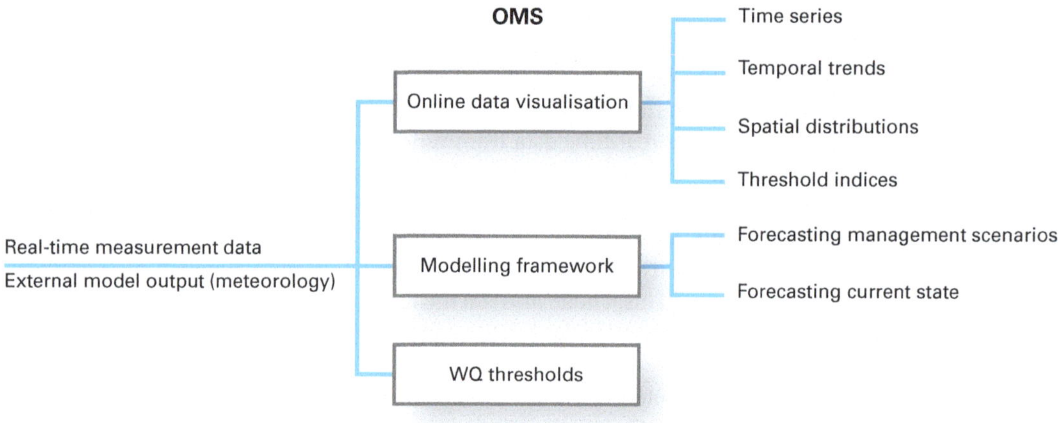

Figure 12.16 Main components of Operational Management System (OMS) for Marina Reservoir

12.6.6 Conclusions

The modelling study has shown that the programmes which PUB is currently implementing, especially for sewer rehabilitation, silt control, recirculation and aeration, will significantly improve the water quality in the Marina Reservoir. Nevertheless, the model study has revealed that there will still be periods when water quality is unlikely to meet target levels, especially after heavy storms. PUB will thus continue to engage the people, in both the public and private sectors to reduce the sources of pollution in the Marina Reservoir catchment, and to explore new and innovative measures through R&D and the ABC Waters Programme to mitigate their effects.

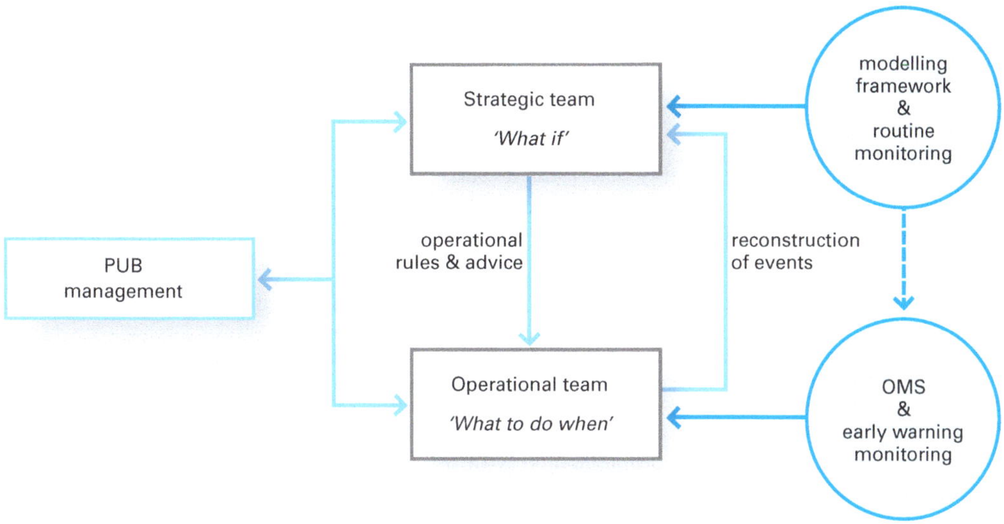

Figure 12.17 Organisational set-up of Operational and Strategic teams illustrating the specific activities and available means (tools and monitoring)

The modelling framework and OMS will significantly enhance strategic and operational management through structured approaches based upon the best information and knowledge available.

The stepwise modelling approach presented in this paper has led to a unique coupled 1D-3D modelling framework describing the water quality processes in the coupled Marina catchment-multiple reservoir system. This framework is implemented in the Operational Management System to support day-to-day operational management by making forecasts and recommendations on the mitigating or regulatory actions to take.

12.6.7 Acknowledgement

This case study was prepared by Tjitte Nauta (Deltares – Delft Hydraulics, PO Box 177, 2600 MH, Delft, The Netherlands; www.deltares.nl) and Kok Meng Tan and Hans Eikaas (Public Utilities Board, 40 Scotts Road #22-01; Environment Building Singapore 228231, Republic of Singapore; pub.gov.sg). The authors are grateful to PUB for allowing the presentation of the current state-of-the-art development of the modelling framework and operational management system considering the sensitivity of describing intermediate results. Whereas the 1D3D system and real-time monitoring and existing forecasting components are already integrated in an Operational Management System (OMS), the decision-support components are the subject of ongoing research under the framework of the Singapore-Delft-Water-Alliance (SDWA).

12.7 CASE STUDY 2: MODELLING THE INTERACTION BETWEEN DRAINAGE SYSTEM, WASTEWATER TREATMENT PLANT AND RECEIVER WATER AT PATTAYA BEACH

12.7.1 Introduction

In many places in South East Asia the seawater in the coastal zone is affected by the functioning of a nearby drainage system. Pattaya City in Thailand is one of such examples. The area changed too quickly from a small fishing village to a big city with high potential for tourism. The environment of Pattaya Beach, especially water quality, was seriously damaged, to a level that nobody was swimming on the beach due to high concentrations of Faecal Coliforms. Although there were no sewage treatment plants in Pattaya in the 1970s, the sea water in the coastal area of Pattaya was not yet polluted, primarily due to the small population and low number of visitors. However, since the early 1990's, with unbalanced infrastructure development and the booming number of visitors, water quality has considerably deteriorated. The result of many water quality surveys revealed high counts of the Coliforms bacteria in the coastal water of the Pattaya beach: The highest number of Coliforms bacteria found at the same sampling station along Pattaya beach increased from 6000 MPN/100 ml in 1976 to 11,000 MPN/100 ml, 118,000 MPN/100 ml and 240,000 MPN/100 ml in 1978, 1984 and 1986 respectively. It should be noted that the recommended value by the Thai EPA for Coliforms in sea water (bathing water) is set to be below 1000 MPN/100 ml.

In order to analyse the functioning of sewer system, its interaction with WWTP and receiving waters and to come up with improvement measures, an integrated modelling system, has been developed, including the 1D dynamic model for the drainage system and a 2D dynamic model for the dilution of water discharged and the sea water. The models have been complemented with the historical description of input and output water quality from Wastewater Treatment Plant.

Each part of the system has been individually treated by different models and then linked together by the Integrated Catchment Simulator (ICS) – a tool specially developed for an intuitive and efficient model integration. The MOUSE™ and add-on MOUSE TRAP™ were used to simulate the sewer system and the water quality in the sewer system. The WWTP is not considered important because of the limited time thus only described on historical data, the input and output records. For the receiving water, MIKE 21™ has been applied to simulate the dilution process of wastewater into receiver water and to evaluate the water quality of Pattaya beach.

12.7.2 Description of the study area

Pattaya, ranked as the second best travelling spot in Thailand, features a coastal lowland area, beach and islets. The Pattaya beach, especially, is the most attractive beach in the country. Being influenced by monsoon in summer and winter, the weather is warm throughout the year. Pattaya is located about 150 km to the Southeast of Bangkok. The total area of the city is 208.1 km^2, including land, water and islands. It consists of Tambon Na Klua (Coral Island, Krok Island, and Sak Island), Tambon Nong Prue, part of Tambon Huai Yai and Tamban Nong Pa Lai, 154.66 km^2 of water. It has a 15 km long beach.

The urban Pattaya catchment is divided into two main zones: Na Klua Zone and Pattaya Zone. The Na Klua Zone has been divided into about 46 sub-catchments, and the Pattaya Zone into 44 sub-catchments. Among these, there are some catchments with a high density of tourism with different water consumption standards.

In response to increased wastewater and storm water loads, the Public Work Department planned to improve the wastewater and drainage system. They managed to install large pipes at Kasemsuwan road to solve flood problems in Central Pattaya and South Pattaya. At the same time, the Pattaya City has improved the weirs by expanding them to be 6 m wide. The new wastewater treatment plant has been built and in operation since November, 2000.

The existing sewer system in Pattaya is a combined sewer system and serves 80% of the city area. The closed sewers are designed for a return period for 5 years, while a 10 years return period has been used for design of open canals. The system has two pumping stations at the end of the pipe system and seven weirs discharging a mixture of sewage and rainwater directly onto the beach (see Figure 12.19). Consequently, the Pattaya beach is seriously polluted, primarily due to domestic wastewater discharging from illegal settlements on the shoreline, hotels and restaurants together with the CSO during heavy rain.

Figure 12.18 A view of Pattaya Beach (Photograph: Ole Mark)

Figure 12.19 Combined sewer overflow onto the beach at Pattaya (Photograph: Ole Mark)

12.7.3 Modelling the hydraulic process and pollutant transport in the drainage system

The description of the flow in the drainage system of Pattaya is based on the hydrological and the 1D network hydrodynamic model. These models have been set-up based on the database from PCD (Pollution Control Department). The modelling of pollutant is carried out by using the advection-dispersion model, assuming the bacteria to behave as a dispersed, conservative pollutant. These assumptions are valid as the concentration time in the Pattaya's sewer system is rather short. It has been decided to use the Total Coliforms as the only indicator in the advection-dispersion model because the objective was to check the bathing water quality in the Pattaya beaches.

Two cases were analysed: before and after the WWTP has been built. Before the WWTP was built, the wastewater was discharged directly into the sea from several discharge points along the beach. After building the WWTP, half of the wastewater was pumped back to WWTP and treated there before discharging into the sea. At present, the WWTP has a capacity of 65,000 m^3/day and will be increased up to 137,500 m^3/day in next ten years.

In future, dry weather flow will be drained through block pipes laid along the coast and pumped back to the WWTP, then to be discharged to the small stream running to the sea. During the rain events when the rainfall exceeds 20 mm, the weirs will be opened to discharge a mixture of rain and wastewater to the sea.

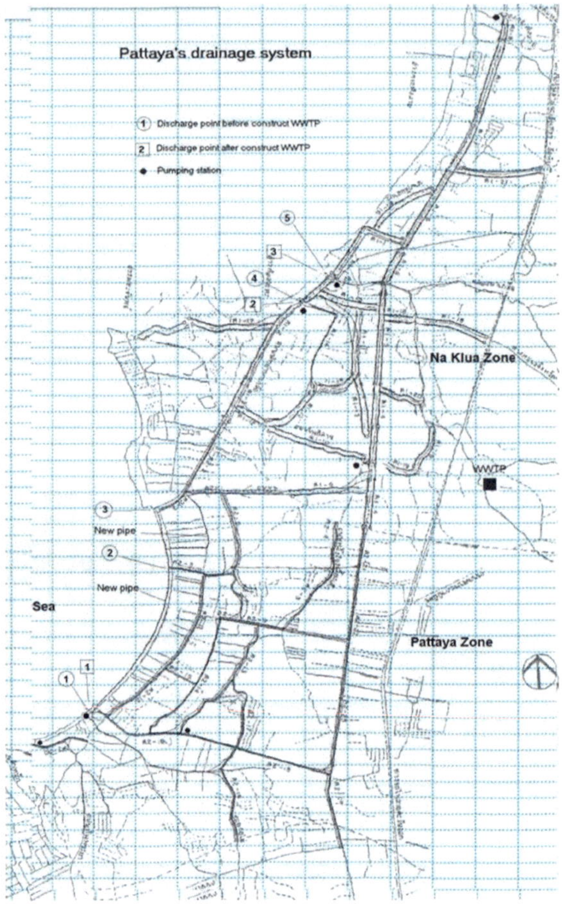

Figure 12.20 The layout of the pipe network of in the model of the Pattaya sewers

The standard water consumption for inhabitants in Pattaya is 185 l/person/day and for tourists it is 290 l/persons/day. For sub-catchments with the presence of tourists, the number of tourists is about 20% of the local population. Based on these data it is calculated that the standard of water consumption for these catchments is 203 l/person/day.

The rainfall data used as input to the hydrological model is a synthetic rainfall with a one year return period, developed on the basis of data from the weather station in the nearby Butchery. One-year return period has been found as the most suitable for the simulation of the commonly occurring situation.

There was a limited amount of data available to calibrate the model. For this purpose, a historical rainfall event featuring a 5-year return period was used. For the water quality part, the box rainfall of 20 mm total depth was used.

The results of the drainage network model were applied subsequently as input for the WWTP, which has been considered to be a 'black box'. The measured output from the WWTP was then used as along with the CSO's from the network model, as input for the 2D model of coastal waters.

12.7.4 Modelling of the receiving water along the Pattaya beach

The bathymetry file of the upper Gulf of Thailand was set up by digitising the analogue map then, the Nested Grid Hydrodynamic Module was used to build the model, with increasing level of detail as approaching to the zone of interest, i.e. the Pattaya beach.

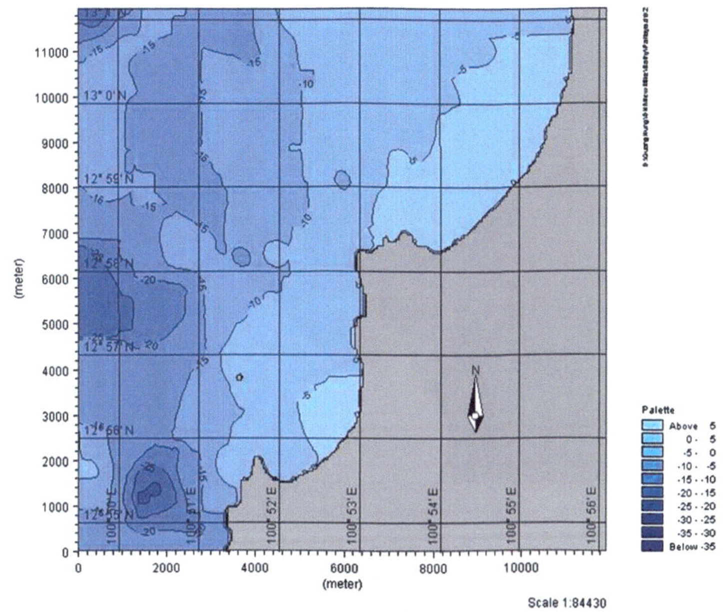

Figure 12.21 The bathymetry of Pattaya cover an area 12 × 12 km with grid spacing 50 × 50 m

The model was run for a one month period, one simulation in dry season (April) and the other in rainy season (July). The rainy season proved to be more relevant, and was selected for the analyses.

The simulation results show the changing of the concentration of Total Coliforms before and after the WWTP was built. Thus, the positive impact of WWTP on improving water quality can be seen clearly. The water quality can be considered suitable for swimming (the concentration of Total Coliforms is under 1000 MNP on the nearby coast). Figure 12.23 shows the Total Coliforms concentration after 3-day simulation without WWTP and with WWTP.

After the WWTP was built, the water quality at Pattaya beach was improved, the number of Total Coliforms range from 500 to 3,000 MPN/100 ml, comparing to its range from 2,000 to 9,000 MPN/100 ml before the WWTP was built. The WWTP removed a largea mount of bacteria, but after the rain, there are still some places where the concentration of Total Coliforms still can not meet the standards for bathing water. The outlet from WWTP discharges directly to the Naklua River, causing significant impacts within the Naklua zone. To improve the water quality in Pattaya beach, an analysis of long deep sea outfall was carried out, Figure 12.23.

12.7.5 Conclusions

The models were able to describe the interaction between three parts of the sewer system. The target of the study was to understand the interaction between drainage system, WWTP and receiver water (river and sea)

and to analyse several improvement scenarios. The work described here has proved all processes associated with the pollution emissions (i.e., sewer system transport capacity and CSOs, WWTP response and receiving water behaviour) must be taken into account in order to improve the bathing water quality. This can be certainly achieved by the application of an integrated modelling system.

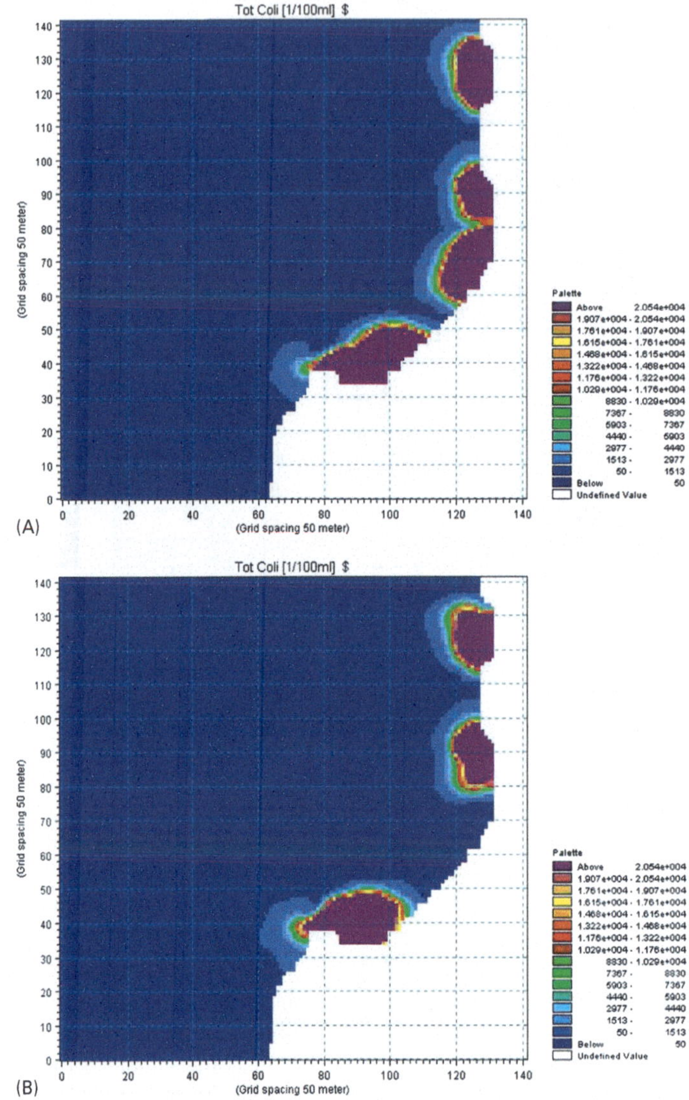

Figure 12.22 The result at the end of a 3-day simulation without WWTP (a) and with WWTP (b)

The advantages of integrated modelling applied here are:

- Increased understanding about the interaction between the sanitary system and the coastal water,
- A holistic approach to design and management of urban drainage systems,
- Ability to represent dynamic changes in the system during the simulation period,

- Integrated model does not required more work compared to the work involved with individual optimisation of each system, but it is certainly more difficult as it requires better planning and knowledge of all three systems,
- Integrated modelling is feasible and should be further explored.

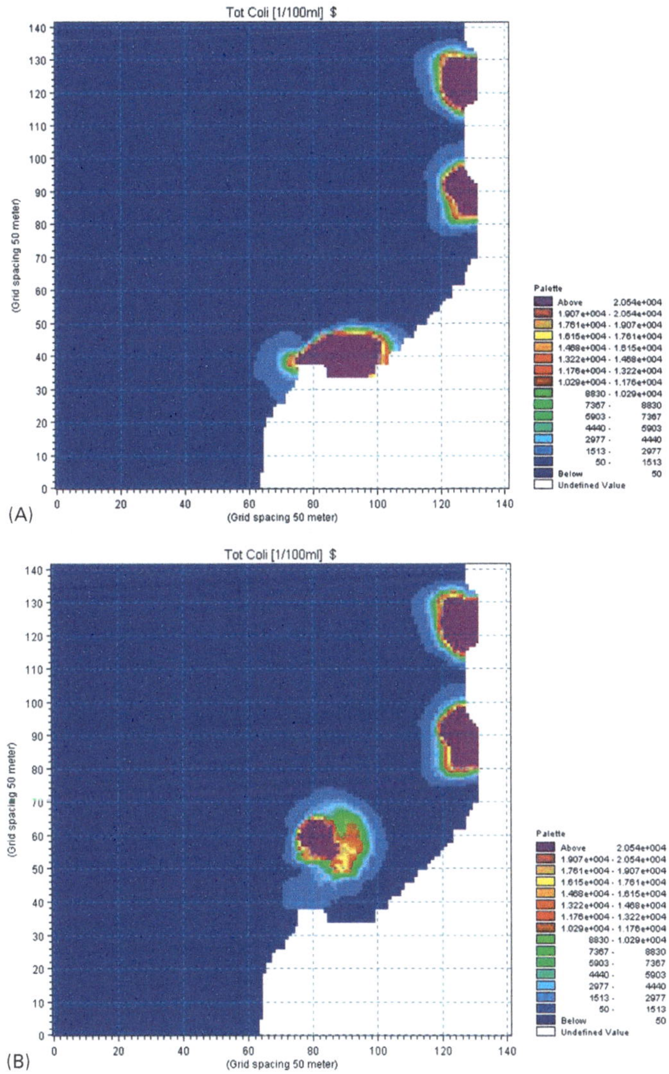

Figure 12.23 Total Coliforms concentrations after construction of the WWTP and deep sea outfall at Pattaya beach

12.7.6 Acknowledgement

This case study was prepared by O. Mark (DHI – Institute for Water and Environment, Agern Allé5 DK-2970, Horsholm, Denmark) and S. Weesakul (Water Engineering and Management Program, School of Engineering and Technology, Asian Institute of Technology, P.O. Box 4, Klong Luang, Pathumthani, 12120, Thailand).

12.8 KEYWORDS

By the end of the chapter you should have encountered the following terms; ensure that you are familiar with them!

Advection
AMP
Biological oxygen demand
Chemical oxygen demand
Combined sewer overflows
Derived intermittent standards
Diffusion
Dispersion
Dissolved oxygen
E. coli
Environmental quality objectives
Eutrophication
Fauna
First foul flush
Gross solids
Nitrates
Oxygen sag
Pathogens
Phosphate
Pollutants
Reaeration
Receiving waters
Sediment
Self-purification
Suspended solids
Treatment works
Urban Pollution Management

12.9 QUESTIONS

What are the primary impacts of nitrates on receiving waters?
Why are pathogens so important, and how can they be controlled?
What is the role of sediment transport in water quality management?
How is oxygen affected by pollutants in receiving waters?
What are the advantages of the UPM procedure?

12.10 REFERENCES

Achleitner S. (2006). *Modular Conceptual Modelling in Urban Drainage Development and Application of CITY DRAIN*. PhD dissertation, University of Innsbruk, Austria.

Achleitner S., Möderl M. and Rauch W. (2007). CITY DRAIN © – An open source approach for simulation of integrated urban drainage systems. *Environ. Modell. Software*, **22**(8), 1184–1195.

Butler D. and Davies J. (2000). *Urban Drainage*. Spon Press, London.

Calabro P. S. (2001). Cosmoss: conceptual simplified model for sewer system simulation. A new model for urban runoff quality. *Urban Water*, **3**, 33–42.

CEC (1991). *Directive Concerning Urban Waste Water Treatment*. 91/271/EEC. CEC, Brussels.

CEC (2000). *Directive Concerning Water Framework – Integrated River Basin Management for Europe*. 2000/60/EEC. CEC, Brussels.

Chou L. (1998). The cleaning of Singapore River and the Kallang Basin approaches, methods, investments and benefits. *Ocean Coast. Manag.*, **38**(2), 133–145.

Chowdhary H., Dijk M. v. and Ogiuk H. (2004). Implementation of hydrological data processing system in HYMOS in India. In: *6th International Conference on Hydroinformatics*, P. B. Liong (ed.). World Scientific Publishing Company, 21–24 June, Singapore.

Clifforde I. T., Tomicic B. and Mark O. (1999). Integrated wastewater management – A European vision for the future. In: *Proceedings of 8th International Conference on Urban Storm Drainage*. 30 August–2 September, Sydney, Australia.

Crank J. and Nicolson P. (1947). A practical method for numerical evaluation of solutions of partial differential equations of the heat conduction type. *Proc. Camb. Phil. Soc.*, **43**, 50–67.

Freni G., Milina J., Maglionico M. and DiFederico V. (2003). *State for the Art in Urban Drainage Modelling, CARE-S (EU Project "CARE-S Computer Aided REhabilitation of Sewer networks: Project Number EVK1-CT-2001-00167")*, IWA, London.

FWR (1998). *Urban Pollution Management Manual*. Foundation for Water Research, Marlow.

Gijsbers P. J. A. (2004). *The HarmonIT Document Series: Part C – The org.OpenMI.Standard Interface Specification* (v.0.99), www.opwnmi.org (accessed 4 October 2010).

Gijsbers P. J. A., Werner M. G. F. and Schellekeus J. (2008). Delft FEWS: A proven infrastructure to bring data, sensors and models together. In: *International Congress on Environmental Modelling and Software*, International Environmental Modelling and Software Society (iEMSs), 7–10 July, Barcelona.

Henderson R. (1986). Rainfall time series for sewer system modelling. Report WRC-ER-195E WRc Engineering, Swindon.

Hon J. (1991). *Tidal Fortunes: A Story of Change: The Singapore River and Kallang River*. Singapore Landmark Books, Singapore.

Kellagher R. (1993). *Sewer Modelling for Rehabilitation*. EU SPRINT Project Reports. HR Wallingford, Wallingford, UK.

Lesser G., Roelvink J., van Kester J. and Stelling G. S. (2004). Development and validation of a three-dimensional morphological model. *Coast. Eng.*, **51**(8–9), 883–915.

Los F., Villars M. and van der Tol N. W. M. (2008). A 3-dimensional primary production model (BLOOM/GEM) and its applications to the (southern) North Sea (coupled physical-chemical-ecological model). *J. Marine Syst.*, **74**(1–2), 259–294.

Mitchell V. G., Diaper C., Gray S. R. and Rahilly M. (2003). UVQ: modelling the movement of water and contaminants through the total urban water cycle. In: *International Hydrology and Water Resources Symposium*, 10–14 November, Wollongong, NSW.

Moore R., Tindall. and Fortune D. (2004). Update on the HamonIT project: The OpenMI standard for model linking. In: *Proceedings of the 6th International Conference on Hydroinformatics*, World Scientific Publishing Company, 21–24 June, Singapore.

Priestley A. (1989). Roe-type schemes for super-critical flows in rivers. *Numer. Anal. Rep.*, **13**, 89.

Rainey C., Osborne M. P. (1991). The classification of storms in the annual time series. HR Wallingford, UK.

Rauch W., Aalderink H., Krebs W. and Vanrolleghem P. A. (1998). Requirements for integrated wastewater models – driven by receiving water objectives. *Water Sci. Technol.*, **38**(11), 97–104.

Seyoum S. D., Vojinović Z., and Price R. K. (2009). Integrated modelling of urban wastewater systems. In: *8th International Conference on Urban Drainage Modelling*, 7–12 September, Tokyo, Japan.

Solvi A.-M., Benedetti L., Gille S. and Schlosseler P. (2005). Integrated urban catchment modelling for a sewer-treatment-river system. In: *10th International Conference on Urban Drainage*, 21–26 August, Copenhagen, Denmark.

Twigt D., De Goede E., Zijl F. and Schwanenberg B. (2009). Coupled 1D–3D hydrodynamic modelling, with application to the Pearl River Delta. *Ocean Dynam.*, **59**(6), 1077–1093.

Van Gils J. A. G. (1998). The Sobek "Process Editor" – a flexible tool for tailor-made quality modelling. In: *Proceedings of the 3rd Hydroinformatics Conference*, 24–26 August, Copenhagen, Denmark.

Zacharof A. I., Butler D., Schultze M. and Beck M. B. (2004). Screening for real-time control potential of urban wastewater systems. *J. Hydrol.*, **299**(3–4), 349–362.

Chapter 13

Urban Flood Risk Management

13.1 INTRODUCTION

Urban floods are nowadays even more frequent and more devastating than previously as societies succumb to pressure for development in areas more prone to flooding. A flood can be defined in terms of state variables that characterize its sources (e.g., rainfall, wind, waves, etc.), pathways (e.g., spilling manholes, overtopping river banks, structural flood protection structures, urban floodplain, overland flow-paths, etc.), receptors (e.g., property, people, environment, infrastructure, etc.), or a combination of these. Within the concept of *urban flood risk*, sources and pathways are related to the flood hazard, whereas receptors are related to the resilience of an urban area. The flood risk can be defined as the probability of the negative effects of a flood (or the flood hazard posed by a particular rainfall event) times the cost of damages, and it is a measure of the degree of exposure of the community and properties at risk to the flood hazard. Therefore, flood risk provides a quantification of hazards for different objects that are exposed to floods. Each object will have a different vulnerability or, in other words, a different potential to be damaged. The risk depends on three elements: hazard, vulnerability, and exposure. Exposure is often defined as the number of people and/or other elements at risk that can be affected by a particular event. If any of these three elements in risk increases or decreases, then the risk increases or decreases respectively. Living in a flood-prone area, a road parallel to a river bank that may be overtopped or a toxic chemical in the stormwater drain may present a hazard (or threat), meaning that these circumstances present the potential for harm. It is thus obvious that hazards are sometimes things that we can do little about. Usually, the process of defining flood hazards for an area involves the delineation of hazards (generally in terms of velocity and depth data obtained from flood modelling work) across the inundated area for different rainfall events. In terms of the risk that flood hazards pose, typically, it is quantified through damage-frequency analysis.

Changes in flooding are reflected in changes in the state variables over a range of time scales. Any phenomenon that may change the time-averaged state of the flooded system can be referred to as a driver

(Hall *et al.*, 2003). Therefore, flood risk results from a combination of different drivers, some of which are physical while some are socio-economic, cultural or institutional. Some of these drivers can be under the control of flood managers, such as the construction and operation of flood protection systems. Many other drivers, including increased values of house contents, deforestation, urbanisation (affecting the run-off patterns) or floodplain encroachment, are outside of the control of flood managers and fall within the policy or regulation domain. Some other drivers which are also outside direct control are climatic change and rainfall variability, population increase and migration, or economic growth. The distinction between different types of drivers is not clear, and often the risk of flooding results from a combination of controlled and uncontrolled natural and human factors.

While considering all drivers it is not surprising that the process of urbanisation itself might prove to have a much more dramatic impact than any other driver. The past century has witnessed a process of large scale urbanization (The World Bank, 2000). This process is still ongoing (often at a very high rate) for metropolitan areas, most notably in the deltas of Asia, Latin America and Africa. The relentless migration of people from rural areas to cities puts growing pressure on urban services, and especially on the management of emergencies and disasters. Certain sections of society are particularly vulnerable, especially the poor who have to resort to living in areas that are at high risk from natural disasters, such as flooding and landslides, and in shelters that readily collapse during earthquakes. City managers and the directors of the various emergency services are under pressure because of the apparent growing frequency of such disasters, and the difficulties in assessing risk and managing appropriate responses to emergencies. Furthermore, institutional capacity required for implementing prevention management, mitigation measures, and emergency response and relief is still inadequate in most cities in the world.

In Europe, the Floods Directive (2007/60/EC) (CEC, 2007) stresses the importance of mapping urban flood risks, and national initiatives argue for the need to identify flood risk and incorporate its assessment within the land use and town planning system (Pitt, 2008). Such assessments, together with the complexity of institutional and technical arrangements required to manage urban floods, represent a series of challenges for both developed and less developed parts of the world.

Nowadays, the challenges facing urban flood managers range from how to obtain reliable predictions of flood hazards to how to involve society meaningfully in the process of selecting the measures for risk reduction. This chapter provides an overview of the key issues that are related to the planning, development and implementation of sustainable solutions within the urban flood management framework.

13.2 URBAN FLOODS AND THEIR IMPACTS

As discussed earlier, the amount or volume of runoff (and the risk of flooding) generated by a rainfall event in an urban area, the rate of rise of water levels, depths and velocities depend on several factors. These include rainfall characteristics (intensity, duration, etc.), land characteristics (percentage of developed area, topography, slope, permeability, degree of moisture already present in the soil, etc.), surface vegetation characteristics (extent, type, etc.) and drainage system characteristics (conveyance, efficiency, capacity, etc.). Another factor that can increase the potential of urban floods is the lack of appropriate planning and development regulations and policies. Furthermore, the impact of a flood will depend on its type and mechanism. The five major types of urban floods can be distinguished:

- *Fluvial floods* which result from rivers breaching or overtopping flood defences and inundating urban areas;
- *Coastal floods* which result from tidal or storm surges in cities close to the coast or in deltas;
- *Pluvial floods* which are due to heavy rainfall directly on the urban area such that the runoff exceeds the capacity of the drainage systems;

- *Flash floods* which are caused by the rapid response of ephemeral streams to heavy rainfall, related, *inter alia* to steep slopes. Recent work on flash floods (Ruin *et al.*, 2008) and their rapid response during storms requires a reexamination of the central assumptions concerning the role of flood protection during flash floods and the importance of flash floods compared with other types of flooding.
- *Groundwater floods* which are caused by the groundwater table rising. This tends to occur in low-lying areas which are underlined by permeable rocks (aquifers) after much longer periods of sustained high rainfall. Compared to other types of floods (e.g., flash floods) flooding due to groundwater has much longer response time and it takes much longer to dissipate because groundwater moves much more slowly than surface water and will take time to flow away underground.

Certainly, there could be a case where two or more of the above mentioned flood types are combined.

In addition to thc distinction between different types of floods, the magnitude and severity of floods can be broadly classified into localised (minor) flooding, larger-scale (major) flooding and catastrophic (extreme) flooding. Such a classification is usually done on the basis of the population and number of structures affected. However, the flood risk perception may differ from one culture to another, from one country to another or even from one city to another. For example, in communities with a 'flood culture' (those that experience floods relatively frequently) a particular magnitude of flooding may be considered as acceptable whereas in some other cities the same flooding may not be acceptable at all. In all cases, public involvement and participation is essential for the definition of acceptable flood standards.

Flooding can produce widespread impacts in urban areas. Any type of industrial, commercial, or residential development located on a floodplain is vulnerable to flooding. In flood incidents, buildings can sometimes experience small damage which can be easily repaired and sometimes significant damage which is beyond repair. Household contents, furnishings and business inventories can be easily ruined if they are not removed to a safe location before the flood strikes. Very often urban floods cause extensive damage to public utilities, disruptions to the delivery of services, community life and even the economy. Loss of power and failure of communications due to floods are also very common. In some cases, drinking water and wastewater treatment facilities may be temporarily put out of operation. Other impacts such as erosion, sedimentation, land slides, relocation of people from homes, etc. are also experienced during certain flood events. In addition to these, people can face the threat of risks such as electric shocks, gas poisoning or even explosions, the outbreak of waterborne diseases, danger to traffic on roads, etc. Furthermore, severe floods can also cause psychological traumas or even loss of human lives due to incidents of physical instability and drowning.

Public perception of floods is rather subjective and it may differ between individuals. This is due to the fact that despite the negative effects of floods, there may also be some positive consequences. For example, recharge of groundwater, provision of water for crops, supply of fresh water for irrigation and domestic use, increased fishery resources, improved navigation transport, economic opportunities for provision of services to those affected by the flood water, etc. Overall however, the economic losses due to floods are much more significant than the benefits as they usually affect a much larger area than the area directly inundated and they may endure for a much longer time than the flood incident. This is not only due to the physical damage to buildings and infrastructure but also due to the loss of working hours as the working population may not be able to get to work. Also, depending on the service and products provided, some commercial businesses and industries may be more severely affected than others.

13.3 URBAN FLOOD MANAGEMENT PROCESS

The main aim of the urban flood management process is to develop, implement and maintain sustainable flood mitigation strategies that will enable safe occupation and use of urban land for present and future residents; see

Figure 13.1. In this process, emphasis is placed on the importance of developing flood management plans that address existing, future and continuing flood risk for the entire urban area, rather than an ad hoc or individual basis.

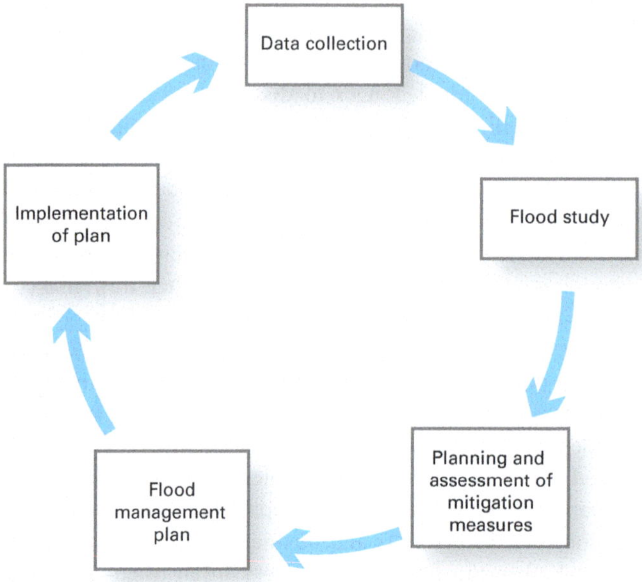

Figure 13.1 The flood management process

In general terms, solutions to flood management problems should be assessed in a broader context with the aim to produce positive outcomes by:

- reducing the social and financial costs resulting from flood-related risks;
- increasing the social, economic and ecological values of flood-prone areas.

The flood management process (as given in Figure 13.1) involves a number of discrete tasks and subtasks which can be described as follows:

- Formation of a consortium by involving drainage system planners, experts and all relevant stakeholders. This consortium is needed to oversee and facilitate the process of developing and implementing flood management plans for a given area.
- Data collection; Within the flood management process, the data collection task refers to the compilation of existing data and collection of additional data for system analysis, modelling and decision making. For a more detailed description of data collection see Chapter 4.
- Flood modelling study; In a typical flood modelling study, physically-based models are used to estimate the existing and probable future runoff rates and associated characteristics (depths of water, velocities, etc.) on the basis of rainfall characteristics, topography, hydrologic soil characteristics, land use pattern and morphology of the natural and artificial water courses. Once computed, the results from 1D or even 1D-2D coupled models are geo-referenced to the location of objects and structures and as such used to calculate potential damages and to produce various inundation maps.

- Planning and assessment of mitigation options;
- Public consultation and active stakeholder participation;
- Development and implementation of the flood management plan;

It should be noted that this process has a periodic character and as such it requires periodic review and updating of the flood management plan to incorporate new data (e.g., land use change, etc.) and to update the priorities as the flood mitigation works are progressively completed.

13.4 DELINEATION OF FLOOD HAZARDS

Flood hazards are typically delineated from numerical model results or from satellite images. For delineation of flood hazards from numerical model results specialist GIS mapping techniques are needed to map the results of 1D and 2D models onto a 2D map or 3D view. In this respect, 2D models (especially if they use the structured grid network) are much easier to handle since they have their own spatial framework which can be easily translated onto the 2D map, whereas 1D models do not have a 2D spatial framework and the development of such a framework needs to be created.

In terms of the satellite images, the procedure of delineation of flood hazards involves image projection, classification of flooded and non-flooded areas, depth and frequency estimation and production of hazard maps.

13.5 UNCERTAINTY

It is important to note that there will always be a degree of uncertainly within the process of flood risk assessment. The sources of uncertainty start from the measurement process (e.g., accuracy of the measuring device, limited number of raingauges to adequately capture spatial variability of rainfall, difficult conditions at the measuring location in a pipe, etc.). Uncertainties associated with our measurements will inevitably propagate further to the process of acquiring the knowledge (e.g., using the model to predict/extrapolate) and continue to the decision-making process (e.g., selection and design of a particular flood mitigation measure). It is then obvious that such uncertainties should be identified, if possible quantified and reported; or otherwise we can have little confidence in our decisions. However, the potential for quantifying uncertainty in urban flood modelling studies is often limited as it depends on the level of complexity of the modelling system used, amount and quality of data as well as the experience of the modeller.

13.6 EVALUATION OF IMPACTS OF URBAN FLOODS

Evaluation of impacts due to flooding in urban areas requires a detailed assessment of the potential risks that are possible with respect to a selected rainfall event (see Teng *et al.*, 2005 and Parkinson and Mark, 2005). Whichever approach is adopted to select design events, a historical worst-case scenario or the probability (or frequency) based approach, the associated flood damages are typically estimated for each of the selected rainfall events. Initially, the flood hazard across the urban area is determined in terms of flow depths and velocities and then transformed into associated tangible and intangible damages; see McConnell, 2000. The hazard-specific risk is usually determined as the product of the probability and the level of impact of the hazard occurring. Inevitably such calculations are data intensive; hence, the need for models to augment the data. As this analysis is used to identify property and population at different risk levels, it can be also used to identify emergency responses that may be required, including the need for temporary shelters and evacuation strategies.

Typically, several steps are required before we can identify risks and come up with the most appropriate measures for effective flood management. These steps connect from the data collection and modelling tasks and proceed towards the mapping, estimation of damages and evaluation of mitigation measures. Figure 13.2 provides a schematisation of this process.

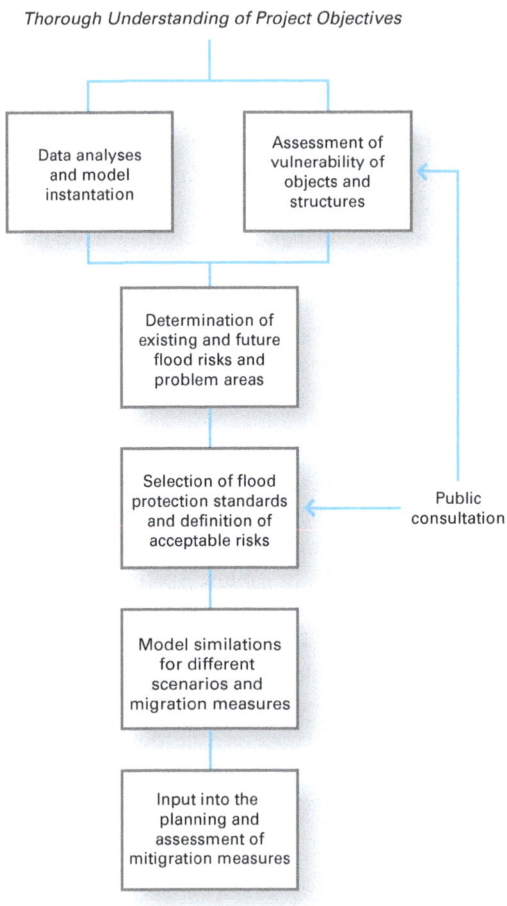

Figure 13.2 A schematisation of steps associated with effective flood management

Once the necessary model simulations are carried out, the model results are processed and incorporated into the GIS maps. In order to evaluate the risk to communities, properties and infrastructure effectively, it is important to estimate the distribution of hazards and the magnitudes of flood-related damages.

Estimation of storm water flood damages is normally required where major remedial works are proposed and as such it has the following objectives:

- To assess the relative benefits and costs of rehabilitation (i.e., remedial, mitigation) options for managing storm water in the catchment,
- To assess the cost of achieving performance standards and as a guide for the setting of affordable performance standards.

Flood damages can be broadly divided into *tangible* and *intangible* damages, Figure 13.3.

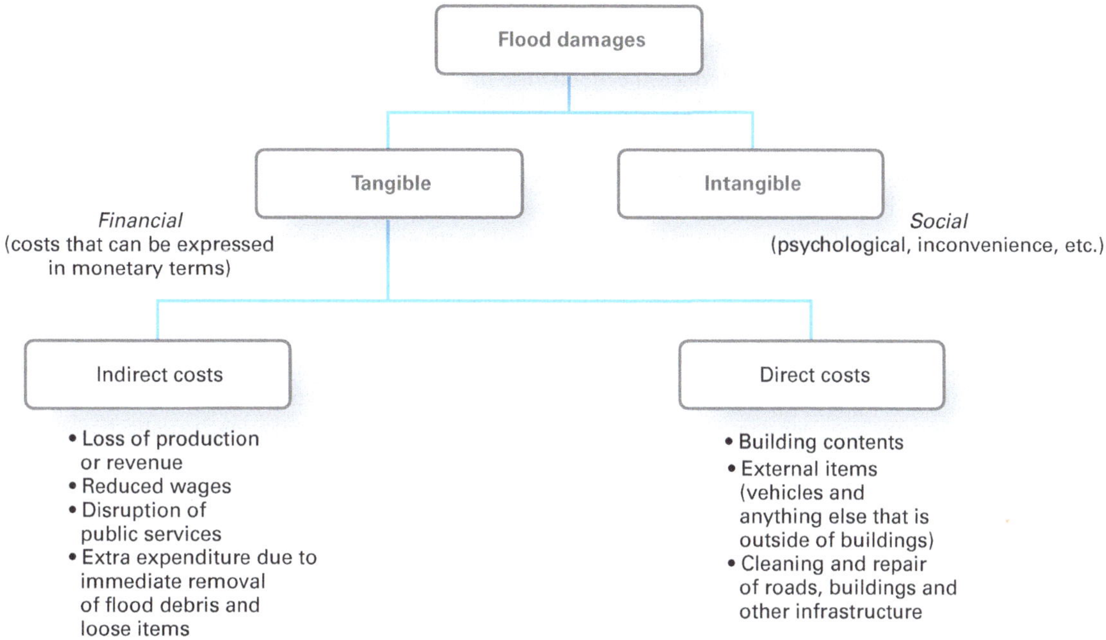

Figure 13.3 Classification of different flood damages

Those damages that can be estimated and expressed directly in monetary terms are referred to as *tangible* damages (e.g., damages to properties, infrastructure, etc.); see Penning-Rowsell and Chatterton, 1977 and Kanchanarat, 1989. Damages that are difficult to identify in monetary terms are referred to as *intangible* damages (e.g., loss of social values, loss of life, anxiety, etc.); see Lekuthai and Vongvisessomjai (2001). Tangible damages may be classified further into *direct* and *indirect* damages. *Direct* damages are those that have occurred from a direct interaction with the flood (e.g., property and infrastructure damages). The extent and magnitude of direct damages that can be incurred in a flood event greatly depend on the *vulnerability* of objects and structures exposed (e.g., depth damage curves for different buildings, etc.). Vulnerability here refers to the characteristics that limit a household, society, economy, or even an ecosystems' capacity to anticipate, manage, resist or recover from the impact of a flood event.

Indirect damages include reduced economic activity and individual financial hardship, as well as adverse impacts on the social comfort of a community. Indirect damages also take into account disruptive impacts such as loss of trading time and loss of market demand for products. The methodologies for estimating these categories of flood damages may differ from country to country, but essentially, they all tend to address and quantify various aspects of possible damages; see for example Tang *et al.*, 1992 and Nascimento *et al.*, 2007.

The purpose of the flood damage estimation work is to provide technically sound, accurate and up to date flood modelling and flood mapping, and flood risk assessment information for the purpose of storm water management, asset and development planning and risk management for flood-prone areas. Furthermore, the final output from flood damage analysis work is not only useful for comparing the effectiveness of different flood alleviation measures but also for the development of disaster management plans and for decision-making at the local, regional, and national levels.

13.7 FLOOD MITIGATION MEASURES

Within the broad flood management framework, three distinctive modes of management can be defined: pre-flood, during flood and post-flood modes. With the knowledge of the inherent risks for the various flood hazards there is a range of possible mitigation measures that can be applied to each of the three modes. The pre-flood management mode aims to reduce the flood risk in the long term, whereas, the second and third modes aim to enable timely issuing of flood warnings and undertaking of emergency response and recovery activities.

Typically, the flood management measures are often divided into *structural* and *non-structural*. Another classification that can be found is *conventional (i.e., piped)* and *natural (i.e., sustainable or ecological)*. It is important to note that a single category of measures is not sufficient on its own: what is needed is an integrated approach that designs and balances different measures.

Structural measures can be defined as physical interventions to the physical conditions of a drainage system. They involve building facilities such as channel/pipe networks, flood embankments, retaining walls or underground or above ground storages or areas (such as parks), or designing interconnected, designated flow paths over the urban surface (usually along roads and between certain structures following the natural ground topography). We can construct such measures to various levels of protection, usually based on: 1) national standards for flood protection (e.g., 10-year ARI event); 2) the optimum level of costs and benefits based on a cost-benefit analysis; or 3) meeting established levels of risk acceptable to the affected community. A useful source of information on the design and construction of urban drainage facilities is CIRIA which provides access to wide and rich variety of information (http://www.ciria.org); see also Balmforth *et al.*, 2006.

It is important to note that all structural elements of drainage systems have their own hydraulic capacity limits. At some point in their operational life time they will reach their limits and eventually cause flooding. This is due to the fact that they are designed to handle flows/volumes up to a certain standard (e.g., 10 year or even 100 year return period) and storms beyond that will exceed their limits. It is therefore important to accept that the flooding is inevitable in the urban environment. To compensate for the limitations associated with structural measures it is important to consider the use of appropriate non-structural measures in order to reduce these risks. Non-structural measures can be defined as interventions which are based on mechanisms that influence human behaviour (e.g., regulatory, financial, legal, etc.). Similar to structural measures, there is also a range of non-structural measures which may be considered: introduction of a legislation that can enforce planning criteria, changing land use, especially restricting construction on natural flood plains (or in flood affected areas) introducing flood insurance schemes where the national economy permits, and implementation of real-time forecasting and warning systems.

The so-called natural (or sustainable, ecological) measures are also referred to as 'best management practices' (BMPs in US), 'sustainable drainage systems' (SUDS in UK) or 'water sensitive urban design' (Australia). The main aim of these measures is to enable better management of runoff (in terms of both quantity and quality) at source by preserving natural drainage patterns and emulating the natural hydrological cycle. Examples are: runoff reuse, soakaways, infiltration trenches, swales, green roofs, permeable paving, ponds, wetlands, etc.

13.8 FLOOD FORECASTING AND WARNING SYSTEMS

The purpose of real-time forecasting and warning systems is to enable better preparation activities in advance of the flood event (e.g., lowering the water level in a reservoir or a pond, activation of pumps, evacuation of residents, traffic regulation, road closure, etc.) and enhancing the preparatory lead time. This measure aims to reduce the loss of life and property in vulnerable areas, especially in the inner floodplains. Typically, a combination of real-time measurement/communication technologies and modelling tools are used to forecast rainfall and other hydrometeorological factors. A well proven modelling system is required to make reliable forecasts of flows and flood levels along the floodplain. In order to make the warnings effective for wide dissemination, the forecasts have to be translated to readily understandable forms, and related to the population and infrastructure by linking model outputs to the GIS flood maps. Conceptually, the construct of such a system would consist of three distinctive units: *data acquisition and processing*, *model simulations and results mapping*, and *information dissemination*.

In the first part, telemetry (i.e., data acquisition) and pre-processing are concerned with the use of meteorological data and simulation of a hydraulic model to predict an upcoming flood given the required lead time. With the development of weather radar such as Next Generation Radar (NEXRAD) and Local Area Weather Radar (LAWR), it is possible to make very accurate forecasts of the upcoming rainfall events with a lead time of up to 3 hours. In certain cases (e.g., in the case of flash floods), the short response time of flooding between peak rainfall and runoff makes the introduction of warning systems in urban areas more difficult. This may necessitate the forecasting of rainfall some hours or even days ahead. Many countries are nowadays making increasing use of ensemble rainfall forecasts produced by global circulation of the atmosphere and associated local area models. These can provide forecasts up to 10 days ahead. Remote sensing data or weather radar can supplement such forecasts by generating reliable information up to 3 hours ahead.

In the second part, either the model results are obtained or the pre-prepared library of model results is searched for appropriate information (e.g., risk maps, safe areas notification, evacuation routes, traffic control etc.).

Finally, the third part is concerned with information dissemination. At this stage, all necessary real-time information applicable not only to flood managers but also to the public is crucial for the success of such system. Technologies from the fields of wireless inter-linkage and spatial IT such as GPS, GSM, UMTS, 3G, WAP and MGS are opening the new possibilities for flood management and dissemination of warnings using real-time data service and high-speed data capture. Furthermore, XML-based Common Alerting Protocol (CAP) data format technologies have been developed for the purpose of exchanging public warnings and emergencies between alerting technologies. Such technologies based on mobile communications, spatial systems, mobile positioning, mobile terminals, embedded software, mobile database, and application server architectures are aimed to achieve an efficient and dynamic information service to reach individual users at different times and at different places. Further to the warnings issued by the flood warning system, a set of specific actions (e.g., control and diversion of traffic by the police or fire brigade, evacuation of children from schools, etc.), as defined within the flood hazard emergency response management plan, would be triggered.

Schematisation of such real-time flood warning system is given in Figure 13.4. The system schematised in Figure 13.4 would need to be able to release warnings in advance of the storms by the use of different communication means, and primarily it would include very valuable real-time-modelled flood risk forecasts. Such information can provide substantial enhancements and make possible informed decision making in adopting proper measures towards disaster preparedness, mitigation, control, planning and management.

13.9 REAL-TIME CONTROL SYSTEMS

Traditional solutions to the urban drainage problems are based on the concept of providing sufficient capacity for a given rainfall event (i.e., design event). With this approach, the dynamics of the system and the lack of flexibility in operating the system are often neglected. For example, a pumping facility is activated on the basis of a predefined water level at the wet well, independently of the situation elsewhere in the system. This could lead to the situation where the capacity of the system is exceeded locally whereas elsewhere in the system some storage capacity is still available. Therefore, the system will only be fully utilised when it is loaded with the design load and for all other loadings the system will perform sub-optimally, in which case a certain amount of storage capacity will never be utilised. In such situations, it is always beneficial to investigate the benefits of applying real-time control facilities before upgrading the system.

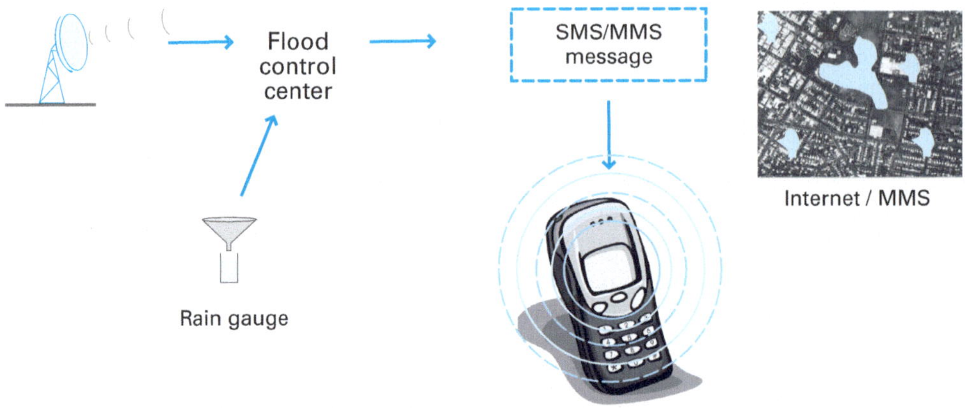

Figure 13.4 Schematisation of a real-time flood control center: Alerts are automatically sent in case of flood risks

A structure of real-time control systems that aims to achieve higher efficiency of the existing drainage system at relatively low implementation/operational costs can have different complexity. Generally, there are three different levels of control: *local*, *regional* (or *process*) and *central* (or *systems*) control. Any control system comprises the following elements: *sensor* (its purpose is to monitor the systems state and variables), *flow regulator* (its purpose is to control or maintain the process; examples are: pumps, gates, sluices, movable weirs, valves, orifices), *controller* (its purpose is to control the flow regulator; examples are: two point controller – on/off, the Proportional Integral Differential – PID controller, predictive controllers, etc.), and *communication system* (its purpose is to link the sensors, controllers and regulators in the system); see Figure 13.5. It is often used together with the SCADA (see also Chapter 4) system which is positioned on top of a real-time control system to control a process that is external to the SCADA system. Furthermore, in order to send and receive the data remotely, the entire system can be telemetered. Using the telemetry, system engineer or operator can obtain the sensor measurement information remotely in real time.

An efficient control strategy can reduce (or optimise) substantially the amount and frequency of CSO spills, flooding from a critical manhole or pump running time. In addition, it may lead to substantial cost savings as the number and storage capacities of the reservoirs required to keep spilling (from either overflow structure or a manhole) below a certain limit depends on the efficiency of the applied control strategy.

13.10 THE PRACTICE OF URBAN FLOOD DISASTER RISK MANAGEMENT

In general terms, natural disaster can be defined as an event that overwhelms the capacity of a household, community, city, business or a utility to resist or recover from the impacts without external assistance. Within the same context, and as described in the opening section of this chapter, the hazard is defined as a natural or other phenomenon with potential to bring harm. Usually, there are three types of events that can cause a disaster:

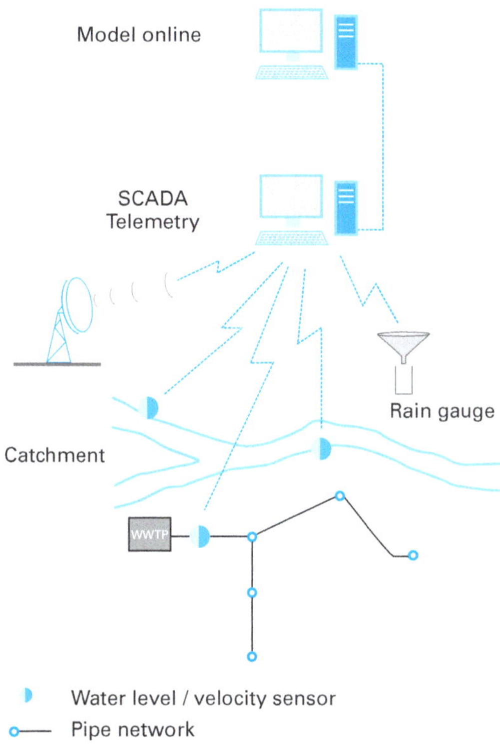

Figure 13.5 Schematics of a real-time control system for a drainage network

- Low frequency – high magnitude events (usually 'declared disasters') with long return period and large losses;
- High frequency – low magnitude events (often considered 'normal'/'routine' threats) that may occur daily/weekly/monthly but with relatively small impacts;
- High frequency – high magnitude events.

Disaster management usually refers to a *suite of measures* intended to strengthen management of risks and the consequences of disasters to reduce losses. These measures are broadly divided into:

- *Prevention*; It refers to measures designed to provide permanent protection or reduce the intensity of a hazardous event so it does not become a disaster (e.g., relocation of households from a flood-prone area, etc.).
- *Mitigation*; It refers to the measures taken well in advance of a hazard alert to minimise vulnerability of communities/households to a known/expected threat (e.g., construction of a detention pond, etc.).
- *Preparedness*; It refers to advance measures taken to predict, respond to and manage a hazard event, or in other words, measures that prepare people to react appropriately before, during and after it (e.g., dissemination of early flood warning information, etc.).
- *Response/relief*; This refers to measures taken to alleviate immediate hardship and meet basic needs for shelter, water, sanitation, health care as well as search, rescue and protection of those affected.
- *Recovery*; It refers to process undertaken by a disaster-affected community to fully restore itself to its pre-disaster level of functioning and which enables it to become even more disaster-resilient (e.g., flood-proofing essential community buildings, schools and clinics, etc.).

When speaking within the context of disaster risk management, disaster risk refers to the probability that a household, community, city, province or country is unable to anticipate, resist or recover from the losses sustained from a hazard or other threat without external assistance. Within the same context, disaster risk management refers to the increase in the likelihood that a household, community or a city will be able to anticipate, resist or recover from the losses sustained from a hazard or other threat without external assistance.

Risk management is a fundamental activity that leads to the evaluation of schemes for reducing but not necessarily eliminating the overall risk, as in many cases risk cannot be entirely eliminated. It includes assessing the potential for a hazard to occur and a vulnerability analysis to provide an understanding of the consequences should an event of a certain magnitude and frequency occur. Based on such work, various mitigation measures can be evaluated to assess their ability for reducing risk exposure. Once the preferred measures are identified, risk assessment of those measures needs to be carried out in order to develop effective disaster management plans and specific actions.

Risk identification comprises three complementary processes:

- *Hazard assessment;*
- *Vulnerability assessment;*
- *Coping capacity (or resilience);*

13.11 FLOOD RESILIENT COMMUNITIES

Even though evaluation of flood risks can be used as a measure to provide insight into the probability and consequences of flooding, more sophisticated measures of system performance can be adopted to provide insight into how other desirable characteristics of a system can be improved. A prime example is the concept of resilience, originating from ecology (Holling, 1979) where it is used to describe one of several stability properties of ecosystems. There is a growing recognition (Schlüter and Pahl-Wostl, 2007) that the dynamics of the natural and the social systems are closely related and dependent on each other. Twigg (2007) identified the key characteristics of resilient communities, as those with (a) capacity to absorb stress or destructive forces through resistance or adaptation (b) capacity to manage, or maintain certain basic functions and structures, during disastrous events and (c) capacity to recover or 'bounce back' after an event. By adopting the notion of resilience as an objective, it is accepted that no community can be completely safe from flooding, but the solution is to provide 'safe-fail' systems rather than the traditional engineering approach of attempting to provide 'fail-safe' solutions. It may be helpful to think of a disaster-resilient community as the safest possible community that we have the knowledge to design and build in a natural hazard context (Geis, 2000).

In the particular context of urban flooding, DEFRA (2008) has made a reference to what it considers to be the key challenges in going towards the flood-resilient communities. For example, DEFRA suggest that the key issues that need to be addressed are:

- Institutional barriers: Elected and unelected stakeholders from local government, the utility companies, the insurance industry, Environment Agencies, environmental consultants and flood victims need to determine and co-ordinate more effective working relationships so that a coherent voice and strategic guidance can be provided by an overriding agency or by allocating responsibility to an existing agency. Local authorities, because of their local knowledge and connections to the public, should be empowered to accept a leading role in this collaboration.
- Integration of flood management into the planning process (Long-term measures) – there is still a tendency for flood risk to be assessed and managed on a site by site basis thus inhibiting the potential for strategic solutions and involvement by the local community and key stakeholders. Better information and knowledge needs to be made available to planners to allow for the incorporation of flood measures into long terms urban development.
- New environmentally friendly measures and options (medium-term measures) should be available to control local surface water problems and tools for the planning and design should be developed and tested.

- Availability of data – The lack of robust data adversely affects flood risk management. Much data on past flood events is in the hands of private companies and there is no compulsion for them to make this data available to public bodies. Sharing data should be encouraged and robust local data could be collated.
- Informing the public – The public are often confused about who is responsible for urban flood risk management and are ill-informed about how best to protect their properties. Most would like a single contact covering all issues relating to flooding, including resilience measures.

Certainly, it is obvious that the lack of community participation in policy and decision making, and the absence of meaningful political commitment for urban flood management have seriously restricted medium- and long-term planning. Furthermore, action during disasters is hampered by the fact that the warnings issued by flood warning centers often do not reach the communities at risk due to various reasons (e.g., communication break-down or simply communication congestion are part of the problem, etc.). Therefore, part of the significant problems caused by flooding (or in other words, the low resilience of the communities with respect to flooding) is connected with communication, dissemination and preparedness and the potential of local communities to correctly interpret and make the most of the information and warnings provided by Flood Forecasting agencies.

Cities are complex systems as their characteristics such as emergence, self-similarity, self-organization and non-linear behavior of land use change with time. The complex and multidisciplinary nature of such environment requires a collective attention by planners, policy makers, flood engineers and decision-makers in order to develop strategies that will increase their flood resilience. Different cities in different parts of the world are designed under different circumstances and different objectives. For example, some Asian cities have drainage systems which are designed to deal with monsoon rain. Such systems sometimes contain certain structures which can appear as incredible huge to the European practitioners. In some cities in India streets are designed to convey the flood water with depths of up to 0.5 meters. Sidewalks or pavements in those cities are not only designed to walk on, but they are rather designed to act as islands to escape from floods.

The formulation and implementation of resilient strategies is not a straightforward task. Finding a proper balance between the interests of different stakeholders requires understanding the effects of combined strategies under different drivers. Uncertain future conditions related to, for example, changing political priorities, economic development, urbanisation, population growth and climate change, influence the outcome of different strategies, and thereby their importance. Although the decisions are inherently a political matter they certainly need sound technical foundation and the support by local stakeholders prior to their implementation. To facilitate the process of bridging the gap between the technical knowledge and standards set by stakeholders in the decision-making process it is necessary to combine a wide range of expertise and data into an integrated platform and to make it transparent and easy to use. To achieve this, an effective sociotechnical framework which integrates social, technical and economic concerns and facilitates stakeholders in making hard decisions is still a challenge.

13.12 CLIMATE CHANGE AND URBAN FLOOD MANAGEMENT

It is becoming more and more evident that the effects of global warming and climate change are causing an impact on urban areas. Even though these changes are likely to be due to a combination of natural and human activities it is believed that the main cause for global warming is the increase of concentration of greenhouse gases (GHG) (IPCC, 2001). Gaining an understanding of these phenomena and evolution of their potential future scenarios requires significant modelling efforts. Climate models are complex mathematical descriptions of the climate system and its interactions with different aspects (see for example May *et al.*, 2005). The key input into climate models is the level of emissions of important gases. The UN Intergovernmental Panel on Climate Change (IPCC) has defined 40 Special Report on Emissions Scenarios (SRES) scenarios which together encompass the current range of uncertainties of future GHG emissions arising from different drivers (IPCC, 2000); see Figure 13.6.

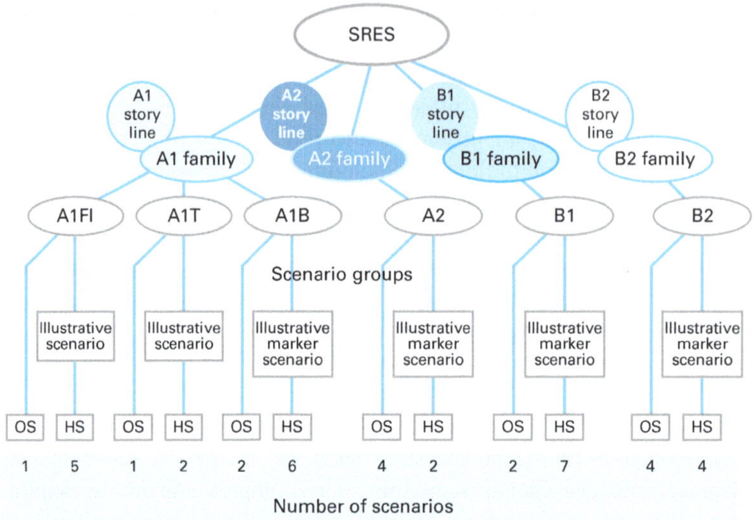

Figure 13.6 Schematic illustration of SRES scenarios, after IPCC, 2000

To assess impacts on urban areas climate forecasts need to be developed by using Global Circulation Models (GCMs). Current GCMs typically link dynamic models of the atmosphere and ocean so that the interactions between these two aspects of the climate system can be captured. Like any other model, these models are also adjusted (or calibrated) by tuning them until they can accurately predict past climate conditions. Downscaling climate data is a strategy for generating locally relevant data from Global Circulation Models (GCMs); see Figure 13.7. The typical strategy is to connect global scale predictions and regional dynamics to generate regionally specific forecasts. Downscaling can be done by applying various methods such as nesting, statistical regression, etc. Typically, local climate projections are derived from regional atmospheric climate model data which are derived from the global circulation model data (see for example Berggren *et al.*, 2007).

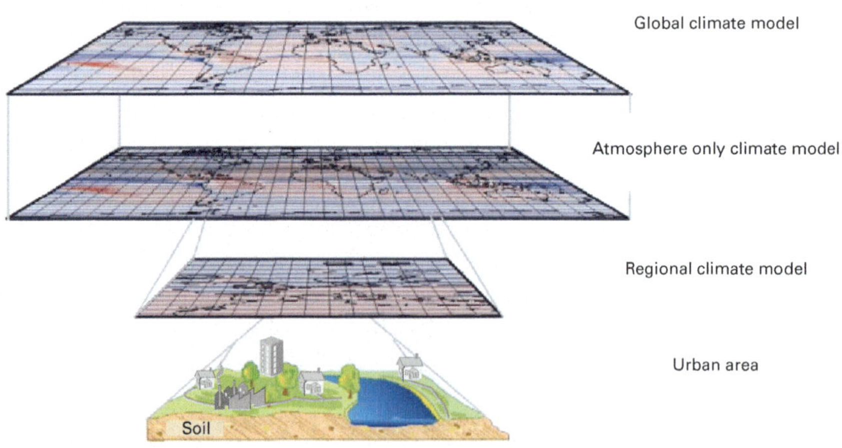

Figure 13.7 Downscaling of climate data

With the threats of climate change, increasing urban population and economic development great uncertainty exists about the necessary adaptation of the urban environment to flood risk. This requires development and employment of new measures (or perhaps extension of traditional measures which mainly have an impact on the pathway) that can enable management of the flood risk in a more flexible and holistic way. Such measures will need to have a positive impact on the source, the pathway, the receptor and the consequences of any flood event. There must be greater synergy and synchronisation between structural and non-structural flood mitigation and adaptation measures. Finally, a shift from pure technical to the sociotechnical environment that equally combines social and technical aspects is necessary in order to enable holistic assessment of different measures starting from planning and design, through their implementation and operation.

13.13 KEYWORDS

By the end of the chapter you should have encountered the following terms; ensure that you are familiar with them!

Different types of flooding
Flood hazards
Tangible damages
Digital terrain model
Flood mitigation measures
Intangible damages
Disaster management strategies
Flood resilience
SUDS

13.14 QUESTIONS

What are the different types of urban flooding?
What should be done in preparation for a flood disaster?
What procedures should be followed during a flood disaster?
How should a flood disaster be followed up?
What could be the potential solutions in response to potential climate change impacts?

13.15 REFERENCES

Balmforth D., Digman C., Kellagher R. and Butler D. (2006). *Designing for Exceedance in Urban Drainage – Good Practice.* Report C635. CIRIA, London.

Berggren K., Olofsson M., Viklander M. and Svensson G. (2007). *Tools for Measuring Climate Change Impacts on Urban Drainage Systems*. NOVATECH, 27 June–1 July, Lyon.

CEC (2007). *Directive Concerning EU Floods*. 2007/60/EEC. CEC. Luxemburg.

DEFRA (2008). *Improving Surface Water Drainage. Consultation to Accompany Proposals Set out in the Government's Water Strategy, Future Water*. Department for Environment, Food and Rural Affairs, London.

Geis D. (2000). By design: the disaster resistant and quality-of-life community. *Nat. Hazards Rev.*, **1**(3), 151–160.

Hall J., Evans E., Penning-Rowsell E. C., Sayers P. B., Thome C. R. and Saul A. J. (2003). Quantified scenarios analysis of drivers and impacts of changing flood risk in England and Wales: 2030–2100. *Global Environ. Change Part B: Environ. Hazards*, **5**(3–4), 51–65.

Holling C. (1979). Myths of ecological stability: resilience and the problem of failure. In: *Studies in Crisis Management*, pp. 97–109, Butterworth, Toronto.

IPCC (2000). *IPCC Special Report, Emission Scenarios, International Panel on Climate Change*. Contribution of Working Group III to the IPCC.

IPCC (2001). *Climate Change 2001, The Scientific Basis, International Panel on Climate Change*. Contribution of Working Group I to the Third Assessment Report of the IPCC.

Kanchanarat S. (1989). *Estimation of Flood Damage Function for Bangkok and Vicinity*. PhD thesis, AIT Bangkok.

Lekuthai A. and Vongvisessomjai S. (2001). Intangible flood damage quantification. *Water Resour. Manag.*, **15**(5), 343–362.

May W., Christensen B. O. and Christensen J. H. (2005). Potential future changes in heavy rainfall events in Europe simulated by HIRHAM regional climate model. In: *10th International Conference on Urban Drainage*, 21–26 August, Copenhagen, Denmark.

McConnell D. (2000). New directions in defining flood hazard and development control planning. In: *40th Annual Conference NSW Floodplain Management Authorities*, 9–12 May, Parramatta, Australia.

Nascimento N., Machado M. L., Baptista M. and Silva A. D. (2007). The assessment of damage caused by floods in the Brazilian context. *Urban Water J.*, **4**(3), 195–210.

Parkinson J. and Mark O. (2005). *Urban Stormwater Management in Developing Countries*. IWA Publishing, London.

Penning-Rowsell E. and Chatterton J. (1977). *The Benefits of Flood Alleviation: A Manual of Assessment Techniques*. Saxon House, Aldershot.

Pitt E. (2008). *Learning Lessons from the 2007 Floods*. Cabinet Office, London.

Ruin I., Creutin J.-D., Anquetin S. and Lutoff C. (2008). Human exposure to flash-floods – relation between flood parameter and human vulnerability during a storm of September 2002 in Southern France. *J. Hydrol.*, **36**, 199–213.

Saaty T. L. (1977). A scaling method for priorities in hierarchical structures. *J. Math. Psychol.*, **15**, 234–281.

Schlüter M. and Pahl-Wostl C. (2007). Mechanisms of resilience in common-pool resource management systems: an agent-based model of water use in a river basin. *Ecology Soc.*, **12**(2), 4.

Tang J., Vongvisessomjai S. and Sahasakmontri K. (1992). Estimation of flood damage cost for Bangkok. *Water Resour. Manag.*, **6**(1), 47–56.

Teng W. H., Hsu M. H., Wu C. H. and Chen A. S. (2006) Impact of flood disasters on Taiwan in the last quarter century. *Nat. Hazards*, **37**, 191–207.

The World Bank (2000). *The World Development Report 1999–2000*. The World Bank, Washington, DC.

Twigg J. (2007). *Characteristics of a Disaster-Resilient Community: A Guidance Note*. Benfield UCL Hazard Research Centre, UCL London.

Chapter 14

Management of Urban Water in Developing Countries

14.1 INTRODUCTION

Much has been written and said about the need for the provision of adequate water and wastewater services in urban areas of developing countries. Even though the targets and agendas have been set, the issues concerning such services will continue to be one of the biggest challenges for the 21st century. These issues will be particularly difficult to solve in those countries where a significant proportion of the urban population does not have access to basic water and wastewater/drainage services. Developing countries are often defined as nations with a lower level of material well being. An overview of countries with relation to different income groups is given in Figure 14.1. UN reports indicate that 75% of total world's population can be classified as poor (i.e., they earn less than 20% of world's income) and the majority are in Asia, Africa and Latin America and the Caribbean.

The need to address urban water services in developing countries becomes urgent as the migration to urban areas continues to be a threat, causing an increase in the pressure on the various infrastructure and services provided. As reported by the UN, nearly all of the urban growth projected in 21st century will occur in developing countries (Learner.org, 2010). In the 1950s only New York and Tokyo were large enough to qualify as megacities (i.e., cities with more than 10 million residents), but population experts project that six cities – São Paulo, Mexico City, Mumbai (formerly Bombay), Delhi, Dhaka, and Tokyo – will have exceeded the 20 million mark by 2015, with New York, Jakarta, and Lagos close behind (Figure 14.2).

According to UN reports, increasing population growth and rapid urbanisation are alarming signals that it is incumbent on the world community to address some of the most essential urban infrastructural aspects in the developing world in order to avoid catastrophic consequences (Figure 14.3).

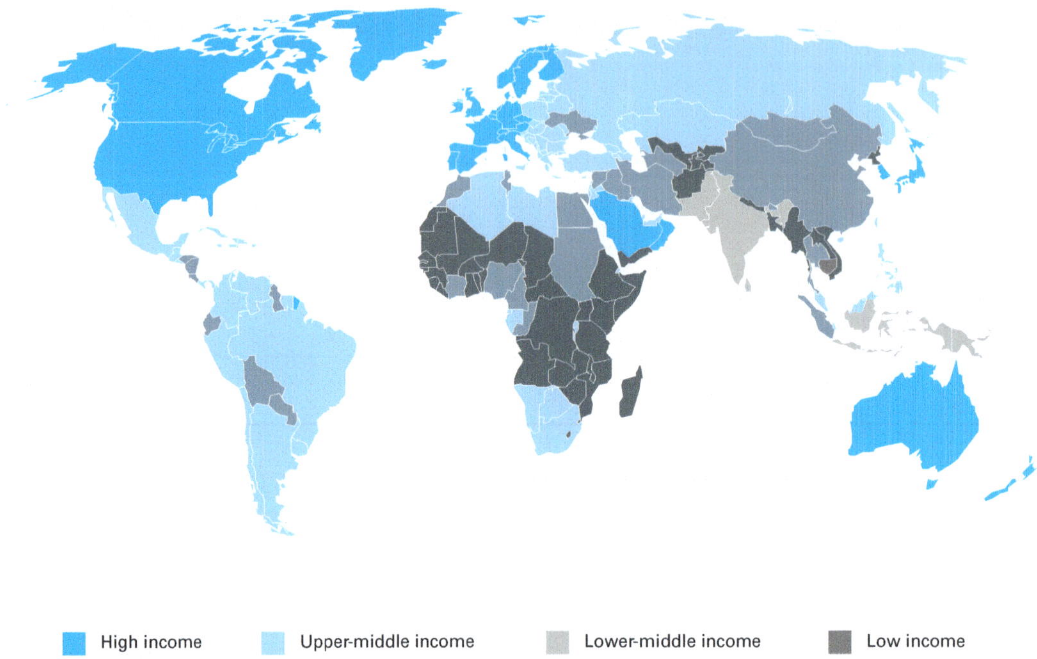

Figure 14.1 World Bank income groups (Source: commons.wikimedia.org)

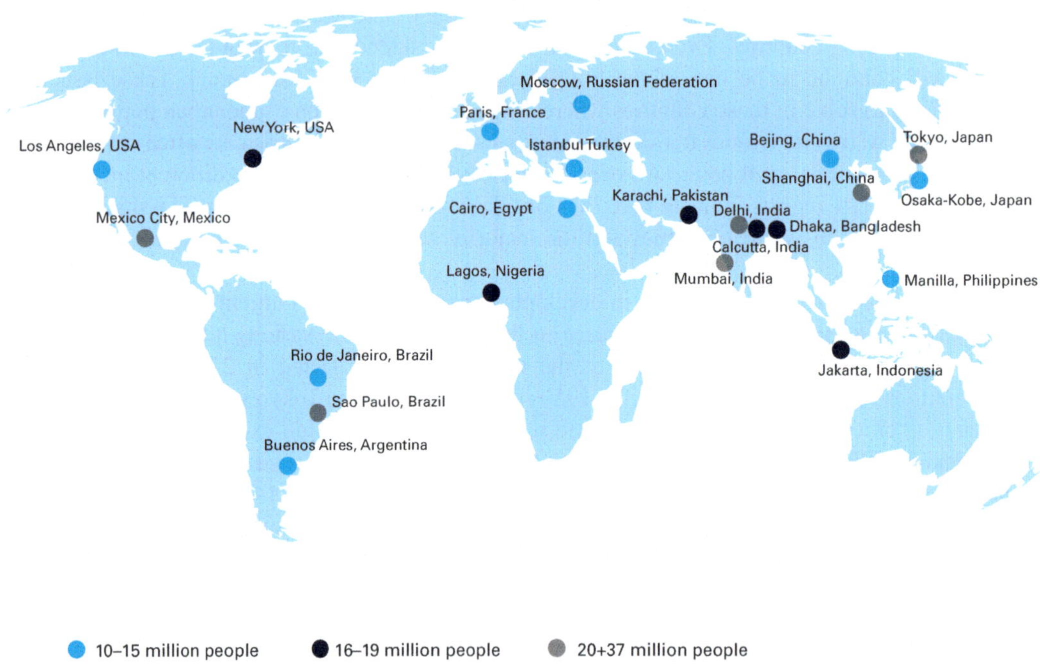

Figure 14.2 Cities with more than 10 million people by the year 2015 (Source: commons.wikimedia.org)

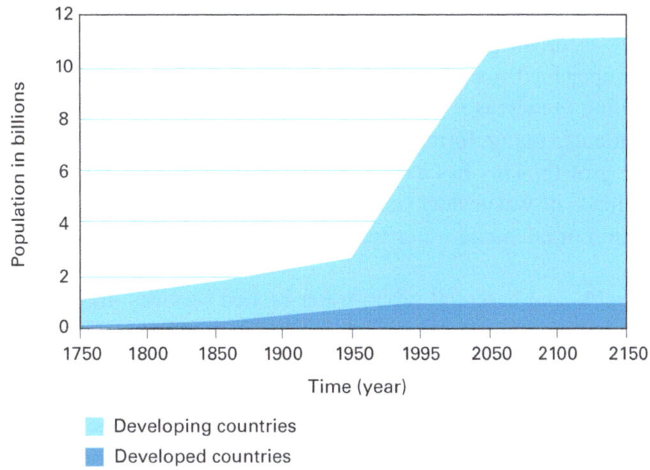

Figure 14.3 The worldwide trend of population growth (The Economist, 1998)

The significant difference in population growth figures between developed and developing countries (Figure 14.3) can be largely explained by the difference in educational levels and career prospects. Typically, higher education levels are linked to lower population growth rates. Also, many developing countries have large indigenous populations which have a very early marriage culture and lower educational levels.

In terms of the urbanisation process, human societies were originally predominantly rural-based. The urbanisation process has advanced dramatically since 1950 and recent urban population figures indicate an increase from 2.9 billion in 2000 to 5 billion in 2030 (UNPD, 2002). In terms of the provision of water and wastewater/drainage services in developing countries, the megacities have gained the greatest attention. They certainly do pose enormous challenges to urban water managers but overall they account for a relatively small percentage of the global population. It is observed that medium to small sized urban centres of developing countries are experiencing much of the urban growth and their trend is likely to continue or even accelerate.

Over the past decades, the economies of developing countries have not performed well. This situation has made its impact on the allocation of adequate resources to tackle some of the important water and wastewater related issues. Furthermore, the lack of proper planning and management practices, insufficient or even nonexistent financial resources and the poor condition of existing water/wastewater infrastructure have only aggravated the situation. Rapid urbanisation and population growth in urban areas have caused the limited amounts of water available to be transferred from agricultural sector to the urban water sector. In return, farmers who suddenly lost their livelihood from agricultural work are now forced to migrate to cities and look to their survival in the new environment. The poorly managed urbanisation process has also resulted in extensive air, water, land and noise pollution.

The presence of informal or squatter settlements (i.e., slum areas or shanty towns) in the cities of developing countries brings a special dimension into the perspective when dealing with provision of adequate water/wastewater/drainage services. The United Nations Human Settlements Programme (UNHABITAT) defines these areas as those that lack basic services (water, sanitation and electricity), have substandard housing, experience overcrowding, are unhealthy and situated in hazardous locations, with insecure tenure and social exclusion. As reported by the UN, almost 1 billion people, most of them in developing countries, live in informal settlements, and this figure is expected to double over the next 30 years. In some cities, these areas may account for up to 80% of urban population and they are always highly congested, leaving virtually no space for construction of proper facilities. As more people from rural areas migrate to cities, that densities of such settlements are continuing to increase.

Outside informal settlements, many formal settlements (especially, business areas) have experienced a sudden fast rate of vertical growth. This has caused at the same time an enormous demand for water supply and the generation of high loads of wastewater (at approximately 80% of the water supplied). It is evident that this has impeded the provision of adequate water/wastewater services which were in many cases inadequate to start with.

Some of the most vulnerable and essential urban-water-related services impacted by the population growth and urbanisation are: adequate supply of fresh water, continuous distribution of drinking water, drainage of excess stormwater runoff, impact on and conveyance of surface waters, management of groundwater, impacts on the receiving environment, and the collection and treatment of wastewater. In many cities in developing countries the water supply systems are inadequate and as such they entail considerable health risks. Those residents who rely on public systems often have to boil the water prior to drinking. Other people tend to use bottled water which is not always adequately purified due to the lack of appropriate standards and/or mechanisms that could enforce and ensure production of safe bottled water. In developing countries, particularly in tropical areas, there are many diseases that are associated with impure water (e.g., schistosomiasis, malaria, blind flue, dengue fever, etc.). Some diseases may be transmitted by insects, mosquitoes or flies and some other diseases may be contracted directly from the water (e.g., cholera, typhoid, dysentery, hepatitis, etc.). Overall, water-related diseases can be grouped into: waterborne diseases (which are mainly spread through contaminated water), water-washed diseases (mainly infections), water-based diseases and water-related insect vectors.

Figure 14.4 Lack of adequate water supply, India (Source: UNESCO-IHE)

Pollution of the environment by poor communities is inevitable. The key reasons are: inadequate support by the government, nonexistence of land ownership, inadequate water/wastewater/drainage facilities, rapid growth rate, outbreak of diseases and lack of education.

Some of the urban water related issues in developing countries are given in Table 14.1.

Table 14.1 Urban water related issues in developing countries

	Alternatives	Issues
Water resources for supply	• Surface water • Groundwater • River • Sea	• Inadequate quantities • Intermittent supply • Sewer discharges into fresh water resources
Potable water treatment	• Treatment plants • Desalination plants • Well supply and some treatment	• Inadequate quality • Intermittent quality • Public health
Potable water distribution	• Distribution networks	• Intermittent distribution • Leakage • Non-revenue water • Corrosion • Public health
Drainage and sewerage	• Condominial (simplified) sewerage • Latrines • Settled (i.e., small bore) systems • Conventional drainage and sewerage networks (open, piped) • Road as a drain	• Too frequent pollution due to uncontrolled discharges • Exfiltration • Public health • Flooding • Corrosion • Sedimentation
Wastewater treatment	• No treatment • Primary treatment (septic tanks, conventional, advanced) • Stabilisation pond systems, • Aerobic biofilm reactors, • Activated sludge systems, • Land disposal systems	• Environmental pollution • Public health • Environmental pollution
Receiving waters	• Watercourse • River • Lake • Sea	

Institutional arrangements in developing countries are also poor. Historically, the urban water sector has been divided between different government agencies such as health, water, and education. This has led to poor coordination. In terms of financing and management issues, urban water utilities are predominantly public organisations and as such available funds for operation and maintenance, as well as construction of new systems are often limited. There is also an issue associated with inadequate pricing of water services and inefficient billing. Furthermore, employees of these organisations are often overstaffed, not properly trained and inefficiently utilised. Since the salaries in public services are much lower from the private sector, more capable people tend to gravitate towards the private sector.

With the view on the above issues, integrated management of different phases of water in urban areas is of major concern in helping to meet the Millennium Development Goals (http://www.un.org/millenniumgoals) especially in providing an improved urban environment affecting Goal 1: Eradicating extreme poverty and hunger, Goal 4: reducing child mortality, and Goal 7: Ensure environmental sustainability (especially targets

9, 10 and 11). Proper management of overall urban water systems therefore requires that attention is given to the engineered and natural subsystems, which include both deterministic and stochastic phenomena, while meeting the demands and expectations of the populations concerned. Furthermore, the use of hydroinformatics technologies can provide the necessary means to achieve better planning and resilience of those communities that live under the threat of water-related disasters.

14.2 ASIA

In a study commissioned by the Manila-based Asian Development Bank Biswa and Seetharam (2008) highlighted that Asian developing countries will face an unprecedented water crisis in as little as a decade if they are unable to better manage their supplies of the precious resource. The same report suggests that policymakers in developing countries have so far paid insufficient attention to the way they supply water, manage wastewater and fight water pollution. Certainly, one contributing factor to water/wastewater/drainage management problems is massive and rapid urbanisation, which is seen in megacities such as Dhaka (Bangladesh), Jakarta (Indonesia) and Karachi (Pakistan). This has caused rapidly increasing demand for water supply and generation of high loads of wastewater. For example, no city in India has a full-day's water supply. Most cities supply water only for a few hours a day. Furthermore, wastewater collected from cities is often discharged to nearby rivers, lakes or oceans with little or no treatment, which heavily contaminates water bodies around the urban centres and is already causing health and environmental problems. Therefore, a major challenge facing Asian developing countries is how quickly and how efficiently current water and wastewater management practices and processes can be substantially improved. With the limited financial resources available for the construction and operation of water/wastewater management systems, and the lack of expertise needed to manage them at all levels (i.e., managers, planners, engineers, plant operators and technicians), the resolution of this problem will be a major challenge.

Figure 14.5 Slum area in Bangkok, Thailand (Photograph: Zoran Vojinović)

Another issue facing urban water managers in Asia is the continuous increase of environmental standards and international requirements for their compliance. Many countries have now accepted, or are in the process of accepting, more stringent environments standards which in turn require improved wastewater collection and treatment systems. Acceptance and introduction of new standards, legislation and policies is one matter but provision of appropriate mechanisms that could enforce and ensure compliance is a completely different matter. In India, only 30% of its entire wastewater is being treated, with the remainder flowing directly into rivers or groundwater. The lack of wastewater facilities in many urban areas is a serious health risk. Open defecation is widespread and it was estimated in 2002 by the World Health Organization that around 700,000 Indians die each year from diarrhoea.

Figure 14.6 The metal roofs of Dharavi (Mumbai, India), Asia's largest slum, stretch into the distance, melding with the Mumbai skyline. In a city of 19 million, this land is prized by developers. Residents are reluctant to move

Overall, the level of investment in water and sanitation in Asia is relatively high, but the local government institutions in charge of operating and maintaining the infrastructure are weak and lack the financial resources to carry out their functions, partly due to very low tariff levels. In particular, innovative approaches have been initiated to increase rural sanitation especially through a community-led total approach to sanitation that emphasizes demand for services and community action instead of supply-driven programme of latrine construction by the government. In rural water supply, the focus has also shifted away from supply-driven to demand-driven approaches.

In addition to the basic provision of water/wastewater services, some of the other effects that need to be considered seriously are impacts from global warming and climate change and increases in extreme flood events on the one hand and droughts on the other. For the Asian monsoon countries it will be a very difficult task to predict how the rainfall patterns will change through the time, and how to develop sustainable strategies that will make water available for various uses over the entire year, and between years, especially during prolonged drought periods.

14.3 LATIN AMERICA AND THE CARIBBEAN

The Latin America and the Caribbean (LAC) region is often regarded as a 'Land of Contrasts and Disparities' and contains some of the areas with largest year-to-year climate variability in the world. Such variability and the expected longer-term changes in climate are putting significant pressures on agricultural production, water-resources management and human health.

Most of the countries in this region have a high mean temperature throughout the year and a wet season that may continue for several months. The wet season in these countries is usually associated with heavy rainfalls, tropical storms and hurricanes. According to the US National Hurricane Centre, on average, ten tropical storms develop over Atlantic Ocean, Caribbean Sea, and Gulf of Mexico each year. On average again, six of these storms become hurricanes each year (http://www.nhc.noaa.gov). The main hazards associated with tropical storms and hurricanes are storm surges, high winds, heavy rainfalls, flash floods and tornadoes. These are by far the most dangerous nature-related threats for those regions prone to such impacts.

Urban environments in the LAC region are usually situated in low-lying areas, with little consideration for wastewater and drainage issues and as such are subject to frequent outbreaks of water-related diseases as well as flash flooding from surrounding hills, associated with extreme rainfall events such as direct thunderstorms. In many cities in the LAC region urban areas have grown on river floodplains. For example, the valley of Tiete River (Sao Paulo, Brazil) has been occupied since the beginning of the century (Tucci, 2000). Even though the government has made considerable investments to increase the river flow capacity and to reduce the flooding, still the effects of rapid development and urbanisation have limited benefits of these measures and caused a further increase in flood frequency.

In the Caribbean, many stormwater systems, whether open channels or closed conduits, are often short, entering the ocean as low or mid-order streams. The drainage channels are typically inadequate to convey excess rainfall-runoff due to the limited capacity, obstructions and the morphological blockages of the stream bed (Figure 14.7 and Figure 14.8).

Figure 14.7 Lack of basic water and sanitation facilities, India (Source: UNESCO-IHE Institute for Water Education)

Very often, the streams have poor retention to respond to frequent and unpredictable flash floods (Covich and McDowell, 1996 and Smith *et al.*, 2003). Furthermore, those areas close to coastal zones are very vulnerable to inundation due to high water levels resulting from storm surge and groundwater seepage. Since the large contributing areas are steep, often with limited vegetation coverage, almost any rainfall at all is likely to cause erosion and land slides (Figure 14.10 to Figure 14.13).

During the small rainfall events typical impacts include temporary disruption to transportation systems and other inconveniences related to community life. During the heavy rainfall events, especially those caused by tropical cyclones, the associated large-scale flooding can cause widespread damage to the residential and commercial areas and even the loss of human lives (Figure 14.10 to Figure 14.13). Therefore, the ability for authorities to address the diverse problems of flooding due to tropical cyclones represents a major challenge.

Figure 14.8 Example of a natural stream in one of the stormwater catchments on St Maarten, N.A. (Photograph: Zoran Vojinović)

Figure 14.9 Example of a natural stream in one of the stormwater catchments on St Maarten, N.A. (situation after the storm). (Photograph: Kurt Ruan)

Figure 14.10 Land slides after the flash flood (St Maarten N.A., August 2005). (Photograph: Kurt Ruan)

Figure 14.11 Land slides after the flash flood (St Maarten N.A., August 2005). (Photograph: Kurt Ruan)

Figure 14.12 After the flash flood (St Maarten N.A., August 2005). (Photograph: Kurt Ruan)

Figure 14.13 After the flash flood (St Maarten N.A., August 2005). (Photograph: Kurt Ruan)

Figure 14.14 On 23rd October 2005, tropical storm Alpha brought flash floods and mudslides to Haiti and the Dominican Republic

Figure 14.15 Earlier, on 22nd September 2004, tropical storm Jeanne had already brought deadly flash floods to Haiti which killed more than 700 people

In the Caribbean area only, an overview on the natural disasters between 1979 and 2001 is given in Table 14.2 (Source: OFDA/CRED International Disaster Database (EM-DAT) 2002. #USAID/Jamaica 2000, Hurricane Lenny Recovery in the Eastern Caribbean).

In recent years greater attention has been given to the destruction caused by major hurricanes, even though such events may be rare for a single territory. Far more frequent are periods of heavy precipitation, severe storms, and flash floods that cause deaths and serious damage to the economy – and for which very little international assistance is available. Especially in the LAC region, sudden and heavy rainfall can unexpectedly occur. Usually, official and public awareness of floods is heightened during the summer-time rainy season. However, flash floods during the dry season may also sometimes catch the population by surprise. Unfortunately, disaster mitigation in many urban areas of LAC countries is very often confined to after-event recovery measures (reactive management) and very little attention is given to a proper detection-forecasting-warning-response process (pro-active management).

Apart from the flood related issues in the LAC region, there are also issues associated with malnutrition, overpopulation, air pollution, poor or nonexistent sanitation, high disease rates, and a desperate need to improve the supply of potable water. Similar to developing countries in Asia and Africa, there is also a significant portion of informal settlements. It has been reported by the UN human settlements agency HABITAT that more than 130 million Latin Americans live in slums without property rights and with no access to basic services such as water, sanitation and electricity. For example, only a small percentage of metropolitan Buenos Aires is served by a sewage system. In Mexico City 95% of the industrial and domestic wastewater is untreated and as such it flows through the city in a network comprised of open channels.

Table 14.2 Main natural disasters in the Caribbean (1979–2001), source: OFDA/CRED International Disaster Database (EM-DAT) 2002. #USAID/Jamaica 2000, Hurricane Lenny Recovery in the Eastern Caribbean

Year	Country (Hazard type)	Persons affected	Damage US$
1979	Dominic (David and Frederick)	72,100	$45 million
1980	St. Lucia (Allen)	80,000	$88 million
1988	Dominican Republic (Flood)	1,191,150	/
1988	Haiti (Gilbert)	870,000	$92 million
1988	Jamaica (Gilbert)	810,000	$1 billion
1989	Montserrat (Hugo)	12,040	$240 million
1989	Antigua, St. Kitts/Nevis, Tortolla (Hugo)	33,790	$3.6 billion
1991	Jamaica (Flood)	551,340	$30 million
1992	Bahamas (Andrew)	1,700	$250 million
1993	Cuba (Storm)	149,775	$1 billion
1993	Cuba (Flood)	532,000	$140 million
1994	Haiti (Storm)	1,587,000	/
1995	St. Kitts, Nevis, Saint Maarten, Netherland Antilles (Luis)	2,000	$400 million
1995	US Virgin Islands, Saint Maarten, Netherland Antilles (Marilyn)	12,000	$2.5 billion
1998	Dominican Republic (Georges)	975,595	$2.2 billion
1999	Puerto Rico (heavy rain) , Virgin islands, Anguilla, Saint-Barthélemy, St.Maarten, (Lenny) http://www.nhc.noaa.gov/1999lenny.html	17 death casualties and many more affected	$330 million US territ. only
2000	Antigua/Barbuda, Dominica, Granada, St. Lucia (Jenny)	/	$270 million
2001	Cuba (Michelle)	5,900,012	$87 million

Figure 14.16 Drainage channel in Port-au-Prince, Haiti (Source: http://www.chfinternational.org/node/34094)

Figure 14.17 Slum areas in Peru (Source: http://ehp.niehs.nih.gov/docs/1994/102-12/focus1peru.JPG)

The growth of population is another issue. For example, Honduras' population is growing at a 3.18% rate, Nicaragua's at 3.36% and Guatemala's at 2.88% which represent the highest in Central and South America. The growth rates in Central America are high due to a number of factors: very early marriage, lack of education (most of these countries have large indigenous populations and there are vast differences in educational levels). Chile and Argentina, with their higher overall education levels and heavy European influence, have a considerably lower population growth rates compared to the rest of the region. Costa Rica's growth rate is also quite low because of a strong family planning programme which has managed to reduce the family size from six to four children.

In terms of the environmental issues, pollution from sewer systems results from the diffuse contribution from areas without wastewater facilities (i.e., from informal settlements) and from direct discharges in places where the existing wastewater collection systems do not lead to a treatment plant (which is the case of formal but poor settlements). It is often the case that the wastewater from households is linked directly to the stormwater drainage network either legally or illegally.

The levels of wastewater treatment vary throughout the region, even within the countries. The cities of Brazil show that there can be great variation in levels of treatment, with some large cities without any treatment (Belo Horizonte), others with low treatment levels (Porto Alegre with 15%, São Paulo with 25%) while Brasília stands out with 64% of its sewage treated (Silveira, 2001). These figures do not include treatment in septic tanks because in general such treatment is inadequate in practice. A similar situation can be found in other LAC countries where receiving water bodies still receive high loads of organic and industrial pollution.

14.4 AFRICA

The African population growth rate is the largest in the world. On most indicators on the provision of water, sanitation and human settlements, progress remains slowest in Africa. It has been reported that more than 300 million Africans lack access to safe water and 14 countries on the continent suffer from water scarcity. In many African countries water quantity available for supply is usually not sufficient to meet the demands and one of the most common methods of controlling demand is the use of intermittent supplies. A serious problem arising from intermittent supplies, which is generally ignored, is the associated high level of contamination. This occurs in networks where there are prolonged periods of interruption of supply due to negligible or zero pressures in the system. Furthermore, leaky sewer pipes, foul water bodies, and open drainage may be the primary sources for water distribution networks contamination. Also, septic tanks designed to treat domestic sewage may constitute significant sources of groundwater contamination. While contamination by leaking sewers has been reported as a serious potential problem in urbanised areas in Africa, the most serious impact from sewage waste has been reported in the densely populated, less affluent parts of the world, where drainage ditches, unlined open canals and rivers are frequently used to convey a large variety of urban waste. Lagos (Nigeria) is considered to have typical conditions of an African city (MEPP, 1997). In terms of the sanitation practice, 30% of its population use pit latrines, 53% use flush toilets, 12% have water closets and 1% use public toilets. Other means of sanitation include bushes and plastic bags.

The World Bank estimates that 60% of Ethiopia's overall disease burden is related to poor sanitation and hygiene, and that water insecurity reduces the country's GDP growth by 38%. The Africa Water Vision 2025 launched by a number of UN agencies and African regional bodies at the Second World Water Forum in 2000 identifies a number of natural and human threats to water security in Africa. It notes that the natural threats are primarily associated with Africa's extreme climate and rainfall variability, and are worsening as a result of climate change. Some of the impacts include growing water scarcity, shrinking of some water bodies, and desertification. The identified human-related threats include:

- inappropriate governance and institutional arrangements in managing national and transactional water basins;
- depletion of water resources through pollution, environmental degradation, and deforestation;
- failure to invest adequately in resource assessment, protection and development; and
- unsustainable financing of investments in water supply and sanitation.

Urbanisation in Africa is very much linked to poverty. In sub-Saharan Africa, nearly 75% of urban dwellers live in informal settlements (Figure 14.18 to Figure 14.20).

Figure 14.18 Informal settlements in Nairobi, Kenya (Source: http://www.nairaland.com/nigeria/topic-51356.768.html)

The problem of water supply and sanitation facilities for the urban poor in African countries is very complex. Typically, a majority of residents that live in informal settlements obtain their water with jerry cans at tap stands and transport these back to their home. Not only the access to water is poor, but also the per capita consumption of water is well below acceptable standards. In formal settlements of the same cities, the scarcity of water creates conditions for an intermittent supply system and this problem of inequity in distribution leads situations where some parts may be supplied a few hours in about 2 to 3 days. This exposes the pipes to a long period of zero pressure, a factor which increases the risk of contaminants intrusion into those pipes (Figure 14.21).

Figure 14.19 Informal settlements in Nairobi, Kenya (Source: http://www.nairaland.com/nigeria/topic-51356.768.html)

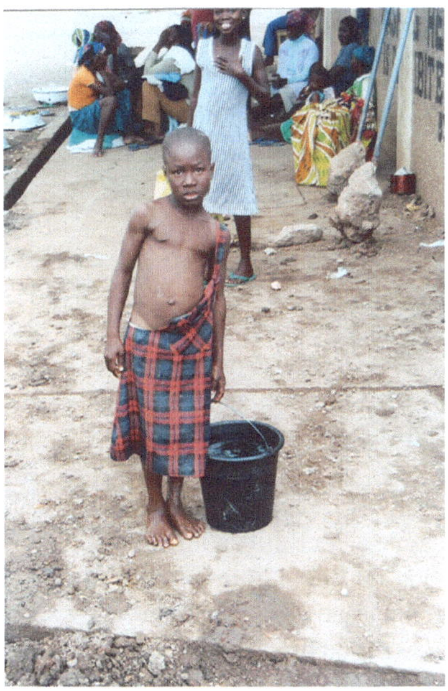

Figure 14.20 Lack of basic water and sanitation facilities, Ghana (Photograph: Nemanja Trifunovic)

In addition to poor water/wastewater/drainage services, there is also an issue with general public awareness and education. In African cities, sewer systems are often found to be blocked by leaves, stones and other material. Furthermore, some houses were built on the top of sewer lines, which have caused blockages and structural problems.

Figure 14.21 Water distribution pipes in an open channel in a slum area, Kigali, Rwanda (Source: UNESCO-IHE)

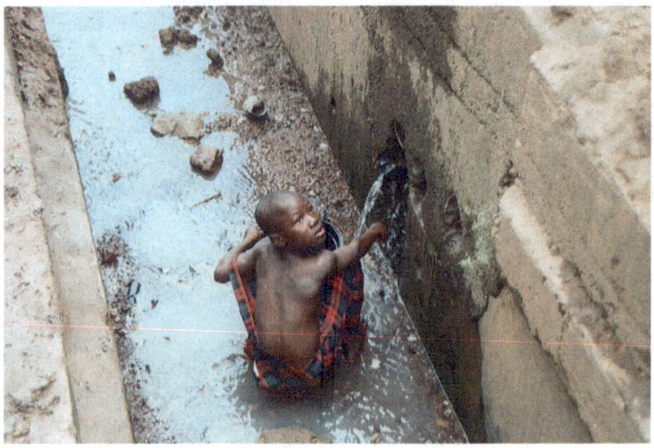

Figure 14.22 Lack of basic water and sanitation facilities in Ghana (Photograph: Nemanja Trifunovic)

14.5 TOWARDS BETTER PROVISION OF SERVICES

Provision of adequate urban water services in developing countries requires a series of different measures ranging from technical engineering, spatial (or town) planning, institutional and organisational as well as legal and economic/financial. Each of these measures has to be considered in an integrated view, by involving the stakeholders, so that the adopted measures can produce fruitful results. Some of these measures are:

- Water supply and distribution:
    - Rainwater harvesting;
    - Better groundwater management and utilisation (removal of pit latrines and waste dumps in the proximity of groundwater aquifer, small bores, etc.)
    - Minimisation of subsidence in urban areas due to groundwater extraction
    - Introduction of appropriate water quality standards and more stringent control of bottled water;
    - Efficient delivery of safer bottled water;
    - Desalination plants;
- Storm water flooding issues:
    - Use of both conventional as well as natural systems;
    - Better control of development and urbanisation activities;

    - Use of early warning systems;
- Wastewater issues:
    - Better control of population growth; Education changing people's attitudes, and provision of sufficient means to let women realise the number of children they want;
    - Use of ecological sanitation; Ecological sanitation, also known as *ecosan* or eco-san, is a sanitation process that uses human wastewater and sometimes immediately eliminates faecal pathogens from any still present wastewater (urine) at the source. The objectives are to offer economically and ecologically sustainable and culturally acceptable systems that aim to close the natural nutrient and water cycle.
    - Wastewater treatment and reuse;
    - Use of conventional (pipes, open channels and treatment facilities) and unconventional wastewater collection/treatment systems (e.g., floating macrophyte plants, duckweeds, sewerage led agricultural systems, etc.);
- Receiving water issues:
    - Reduction of pollution through better wastewater management practices;
    - Reduction in informal settlements on the banks of urban streams
    - Better management of solid waste
- Integrated approaches and strategies to achieve management of informal settlements. This issue requires better integration of these areas with other urban districts and it relies on better city level support.
- Stakeholder participation in the implementation of urban water management measures;
- Public education;
- Capacity building;
- Provision of adequate institutional and policy arrangements;

An overall lack of adequate finances will always represent an issue limiting the possibilities for improvements in the urban water sector of developing countries. Furthermore, the inability to achieve cost recovery due to the fact that informal settlements can constitute a considerable proportion of the total urban population is another issue. In this respect, a holistic optimisation and prioritisation of investments can significantly reduce the need for large-scale upgrading of individual systems. Also, the adoption of lower levels of service and the use of alternative construction materials (e.g., use of natural materials instead of concrete, etc.) will contribute to the overall reduction in the costs of improvement works. Certainly, the adoption of lower levels of service will lead towards a reduction in costs but they will require better practices such as source control, solid waste management, private sector participation as well community participation, otherwise such measures will increase costs in other respects (e.g., increase in damages due to more frequent floods, etc.). Community participation plays an important role, not only in the selection of improvement measures, but also to the successful functioning of most of these measures. For example, residents in a flood prone area need to be familiar with the capacity of their flood protection measures and with the related disaster management activities that need to be undertaken in case these measures fail.

14.6 CONCLUDING REMARKS

Proper management and provision of basic water/wastewater/drainage services in developing countries continues to be one of the biggest challenges for 21st century. Certainly, urbanisation is one of the first things that needs to be addressed in a controlled and sustainable fashion. At the same time, water scarcity represents a challenge for many cities in developing countries and as such it enforces thinking towards the collection and treatment of wastewater to such a level that it can be used as a new source of water.

Integrated approaches and strategies at different levels are required to achieve sustainable management of urban water systems.

Since climate change is likely to introduce high levels of risks and uncertainties, the management of water systems in urban areas of developing countries is going to be an exceedingly complex and difficult task for the forthcoming future. The increase in the frequency of extreme events such as droughts and floods demands more robust and flexible urban water infrastructures and management practices. Building flexibility and robustness into the design, construction and maintenance of urban water systems is certainly a way forward but will also mean higher financial costs, which may further strain the economies of these countries.

Certainly, some of the recent technologies such as the use of satellites and radars to monitor weather-related storm events and their movement, along with the high winds and rainfall intensities that they create, provide an invaluable means for managing floods and flood-related disasters. In this respect, the use of hydroinformatics technologies, such as numerical models and real-time flood-warning systems, can provide effective tools for minimising the impacts of these hazardous events, and as such they must be integrated in a wider disaster-risk reduction strategy, rather than being used only as 'stand alone' tools.

14.7 KEYWORDS

By the end of the chapter you should have encountered the following terms; ensure that you are familiar with them!

Capacity building
Floods
Integrated urban water management
Migration
Millennium Development Goals
Pollution
Population growth
Public education
Solid waste
Subsidence
Urbanization

14.8 QUESTIONS

What are the key issues concerning water supply to major cities in developing countries?
What are the key issues concerning wastewater and storm water collection in major cities in developing countries?
How do these issues differ from those in developed countries?

14.9 REFERENCES

Biswas A. (2008). Achieving water security for Asia. *Int. J. Water Resour. Dev.*, **24**(1), 145–176.
Covich A. P. and McDowell W. H. (1996). The stream community. In: *The Food Web of a Tropical Rain Forest*, D. P. Reagan and R. B. Waide (eds.). University of Chicago Press, Chicago, Illinois, pp. 433–460.
Learner.org (2010). *The Habitable Planet: Unit 5: Human Polpulation Dynamics // Section 6: Urbanization and Mega Cities*. Annenberg Media, www.learner.org (accessed 4 October 2010).
MEPP (1997). *Urban Services Improvement: A Willingness to Pay Study*. Ministry of Environment and Physical Planning, Lagos State Government, Lagos, Nigeria.
Smith G. C., Covich A. P. and Brasher A. M. D. (2003). An ecological perspective on the biodiversity of tropical island streams. *BioScience*, **53**, 1048–1051.
The Economist (1998). *Dirt Poor*. The Economist.
Tucci C. E. M. (2000). Urban drainage issues in developing countries. In: *Urban Drainage in Humid Tropics*, C. E. M. Tucci (ed.). UNESCO-IHP. I, Paris.
UNPD (2002). *World Urbanisation Prospects: The 2001 Revision*. ESA/P/WP.173. United Nations Population Division/United Nations Secretariat, New York.

Chapter 15
Future of Urban Water Management

15. 1 INTRODUCTION

This book has addressed a wide range of topics that demonstrate the complexity and challenges of urban water management and the considerable opportunities for improving the services provided. This is encouraging because the pressures on the urban water systems in cities around the world are increasing through migration, climate change, urban expansion, economic recession, changing attitudes towards quality of life, with repercussions for local and national politics and security. Like many other areas of human activity, urban water management has been revolutionized by burgeoning advances in information and communication technologies, affecting data management, the virtual world of modelling, support for decision making, and the networking of professionals and laypeople, giving all stakeholders the potential for contributions and 'ownership'. As reviewed in Chapter 4, data acquisition techniques are improving dramatically, and fortunately there are data archiving, analysis and retrieval tools in the form of GIS, digital terrain modelling and management information systems that enable the maximum amount of information to be extracted from survey, monitoring and remote sensing data. Chapter 5 sets out the mathematical basis of the computational simulation modelling that leads to the generation of the virtual world in which the modeller is king, and where experiments can be conducted in order to identify the optimal strategy or scenario for action in the physical world. An emphasis in the book is on the development of safe and reliable models whose results can be used with confidence, or where the uncertainty in the predictions is known and acceptable for negotiation. Part of the safety interest in modelling is supported by task based procedures whereby the engineer has computer based assistance in carrying out tasks in the right sequence and with the best available advice. In this way, the modeller can provide assurances to decision makers on the reliability of his results. Careful presentation of the model predictions can enable others become more interested in seeing how the various contributions fit together.

Chapters 8 to 13 explore information management in the context of water supply and distribution, wastewater collection and stormwater drainage, wastewater treatment and receiving water impact. The involvement of stakeholders is touched on but not obviously realised in terms of the management of information. In other words, much remains to be done in developing appropriate hydroinformatics tools to facilitate the generation, transfer and assimilation of information associated with urban water systems. Some of the key issues facing urban water management in developing countries are discussed in Chapter 14. It is further argued that integrated management of different phases of water in urban areas is of major concern in helping to meet the Millennium Development Goals where the use of hydroinformatics technologies can provide the necessary means to achieve better planning and resilience of those communities that live under the threat of water-related disasters.

Key areas for development are identified as follows.

15.2 DATA MANAGEMENT

Much has been achieved in developed countries such as the USA, Germany and UK in terms of improving facilities to measure rainfall and flows in conduits, especially in sewers with their harsh environment. Indeed, short term flow monitoring has been a fully commercial business opportunity in these countries since the early 1990s. With the advent of fine resolution remote sensing data for topography and land use, new modelling possibilities are being opened up, which in turn will increase pressure for better data on soil moisture and flooding extent in urban areas.

Many water companies and utility organisations have adopted the powerful facilities offered by advanced GIS tools. These provide them with better asset management procedures that are built on risk and the comparison and neighbourhood operations in the GIS. Closer links between GIS and optimisation tools should make available improved means of minimising costs of asset management and ensure that the assets are sustainable into the future. Advances in parallel computing will facilitate the optimisation processes, opening up more comprehensive forms of analysis.

Certainly, there is a need to enhance the use of quantitative, spatially distributed, urban water management information derived from remotely sensed satellite data. With advances in computer power, modelling and data processing tools we will see further expansion in the use of satellite data for applications such as model validation, flood hazard analysis, disaster risk assessment and warning, etc.

Much temporal data for water networks has to be collected directly in the underground conduits. The positioning of the monitors in a network is usually based on convenience of access rather than any other factor. But increasingly monitors can be placed almost anywhere. The problem then arises about how to maximise the information content from the monitors such that the data acquired more than adequately describes the performance of the system. This is to do with the information content in a given record. This can be defined using Shannon's definition of entropy. There are corresponding tools to define the mutual information content from time series for two or more monitors (Alfonso, 2010). Obviously the objective is to select, say, two monitors such that their mutual information is negligible (while each retains significant information in its own right).

Formal archiving of spatial and temporal data in GIS also opens up the opportunity to register the uncertainty associated with each data item. This uncertainty will be increasingly important as decision makers take it and risk into account in their analysis.

One of the most important applications of the growing amount of data collected in urban areas is better assessment of the risk of failure of each of the many components comprising the several urban water systems in a city, and the risk of failure in the performance of these systems. Such measures of risk enable better prioritization of surveys, maintenance and rehabilitation.

In terms of the urban flood management practice there is a growing need to design major systems in a way that they can complement the functioning of minor systems. Such developments will necessitate better (or more precise) terrain data collection, processing and modelling activities. In this respect, amongst other improvements, aspects that relate to the processing of LiDAR data will continue to be in the focus of engineers and scientists (Abdullah *et al.*, 2009).

15.3 MODELLING

The emphasis on data acquisition in urban water systems has lead quite rightly to an emphasis on direct measurement of particular parameters and variables, whether by 'on the ground' intervention or through remote sensing. Generally, measurements underground are very difficult to make, and are limited by access. But as with most water based systems, another way of acquiring data is through the conjunctive use of models with the direct data acquisition systems. A properly instantiated model can provide the estimate of data at points between the spatial and temporal measurements. In this sense the model can be regarded as an interpolation (and extrapolation) tool, filling out the knowledge of the temporal data. Similarly, the process of instantiating the model is likely to have provided a means of confirming the structural data for the model space. Models therefore constitute an important complement to the direct acquisition of data.

1D pipe network models are now in regular use around the world, but they still have shortcomings that have been compromised in order to make the models suitable for practical use. These shortcomings include the poor representation of the free surface-pressurized flow transition in closed conduits, the switch to the solution of the non-inertial equations by the implicit solvers for sub-critical flows with the Froude number approaching critical (or for super-critical flows), the assumption of a uniform fixed depth of deposited sediment along a pipe rather than a dynamically and spatially varying depth, and the poor representation of the pollutant processes in sewers. As the awareness of the potential of parallel processing using multi processor systems grows, it is likely that the computationally intensive solutions to these problems will become more appropriate. Their introduction depends partly on the pressure from the urban water sector for such solutions.

A major development in modelling during the first decade of the 21st century has been the development of robust and easy to use modelling software products for 2D flood flows in urban areas. Such flows can be generated directly from rainfall on the urban area, or can emerge from overloaded combined or storm drainage networks or can be due to over-bank flows from urban streams. An interesting further development is the possibility of having 2D catchment runoff models. These are generally supposed to be no more accurate in terms of the outflow from a sub-catchment to a storm drain than the traditional compartmentalised model based on the calibrated unit hydrograph, linear (or nonlinear) reservoir or time area diagram. The reason is that it has been difficult to acquire the topographic data for the urban (ground) surface and the horizontal resolution of the 2D model would have to be the order of a metre (or less) in order to accommodate the significant variations in the topography. With such fine resolution there is less difficulty in incorporating road kerbs and other detailed geometric features, but there remain problems in calibrating such a model, that is, in selecting appropriate roughness values and ensuring that other significant headlosses due to steps etc are formally included. Another problem is the considerable computing time needed with the large spatial extent (in terms of grid cells) and the small time steps required. Parallel processing is likely to make a valuable contribution to these problems. Also, the development of methodologies that can enable more efficient model applications at coarse spatial resolutions while retaining important information at the fine spatial resolution (i.e. storage and conveyance characteristics of the complex urban geometry) will continue to be of concern for researchers and practitioners alike (Vojinović *et al.*, 2010). The inclusion of meandering open channels (urban streams) in the 2D models has been innovatively addressed by some software suppliers (including MIKE21/MIKE11 and

SOBEK 1D/2D) through the integration of the 1D model for flow in the channels and the traditional 2D model for flow over the urban surface. Again, there are likely to be significant improvements in modelling these situations during the coming decade.

The 3D modelling of treatment works processes is already fairly well established in research but has yet to become standardised in the industry. This should happen though within the next ten years.

The complexities of physically based modelling in certain circumstances and the increasing availability of reliable data are indicating the potential for data driven modelling. This alternative paradigm opens up particular benefits in applications where the physics (or chemistry, biology, ecology) are poorly understood and imprecisely formulated. In such cases there are opportunities to introduce hybrid models that can capitalise on the advantages of each type of model.

Undoubtedly, a key aspect of modelling in the future will be the modelling of the integrated water cycle in urban areas, including rainfall modelling and forecasting, 2D rainfall-runoff, 2D above ground flood flows, below ground pipe network flows (distribution and collection), groundwater, urban streams and rivers, and river basin management. Furthermore, the application of 3D models will receive greater attention in the future. 3D models will help to increase our understanding of processes that relate to the interaction between above-ground surfaces and below-ground pipe network systems, flows in and out of gulley structures, overflow spills, interaction with receiving waters, etc. We are also likely to see further applications of higher dimensionality models for the calibration of simpler models.

In addition, there is the need to raise understanding of uncertainties and to improve the methods of addressing them with models that continue to be computationally expensive.

Furthermore, since there is a demand for more efficient design and operation in situations where urban water systems can be more flexible and adaptable, the use of models for real-time and near real-time applications will continue to attract the attention of researchers and industry practitioners.

Another modelling area which will see considerable growth in the coming decade is the use of intelligent agents. There are already a considerable number of applications in other areas than urban water management, and learning from their success we can expect to see new applications that improve the performance of water distribution and collection networks. One particular area where this approach is likely to help is in the planning and development of new urban areas, that is, urbanisation as described in Chapter 3. Here we can identify the link between the paths taken by water distribution trunk mains and land use. So given the likely land use of an extended urban area using an intelligent agent model (Barredo *et al.*, 2003) and the layout of major roads we can determine the likely route of the extended water main (Sanchez Torres *et al.*, 2010). Given the route taken by the water main we can estimate the pipe sizes for the main in terms of the demands in the new urban area and explore the likely layout for the local distribution pipes fed from the main. Once the water supply network of pipes is determined the construction and operation costs of the extended network can be considered. Another important advantage of this approach however, is the possibility of considering the nature of any rehabilitation of the existing network in order to meet the demands of the urbanised area. A goal of this approach is the long term sustainability of the urban water services to a given city.

A similar but different approach is needed for wastewater collection and storm water drainage. Here the topography is of major importance, unlike the case of the water distribution network. Sewerage systems are normally designed to operate under gravity to an adjacent water course, so it is important to ensure that the main trunk sewers start within the urban area and follow the steeper slopes while acknowledging that they are also preferably constructed along the major roads. Once the layout is determined then the pipe sizes can be calculated using simple design methods such as the Rational method. Again it will be important to look at the extension of existing drainage facilities into the new urban area in terms of any rehabilitation that may be required. Just as for water distribution, the aim is to make the drainage networks sustainable in the long term.

Whereas the hydraulic modelling has made and continues to make rapid progress, with such tools being put into the hands of engineers with less training in computational hydraulic modelling than previously, the trend will be for the software suppliers to provide remote access to products running on their servers, and this in turn will open up new business opportunities for consultants offering support services. This trend will open up access to the modelling software products to a wider audience and should assist the communications between modelling professionals and concerned stakeholders. The work of Tumwesigye *et al.* (2005) as well as that of Abbott *et al.* (2006) and Abbott (2007) are classic examples of how the instantiation of web-based urban water system models can be accomplished.

There is also the need for associated improvements in GIS mapping techniques (e.g., translation of 1D model results onto a 2D map) and better use of 3D visualisation, multimedia and game technologies for improved presentation of model results as well as for empowering stakeholders and enabling effective exchange of information. For effective communication to take place between all stakeholders it is important to translate the data and technical information into the form of knowledge and understanding that can be assimilated by the non-technical stakeholders if they are to be able to intervene meaningfully and responsibly in the decision-making process.

15.4 DECISION SUPPORT SYSTEMS

The failure of the expert system initiative of the 1980s has left open an object-oriented approach to decision support. Already we are seeing better use of engineering/modelling procedures to achieve safe and reliable objectives and applications. It is likely that software tools will become available to implement procedures, such as the Sewerage Rehabilitation Manual procedure, and to have an even more effective transfer of knowledge to inexperienced engineers and modellers. This will bring about better, more safe and reliable, models in the future with application to a range of important issues in urban water management. The trend towards implementing access to DSSs over Internet, such as to the latest SRM framework from WRc (2010), is set to escalate. The integration between urban water system models and economic and other models has received considerable attention over the past two decades and we are likely to see an increase in this trend in the future.

15.5 INSTITUTIONAL AND SOCIO-ECONOMIC ISSUES

Even though there is growing acknowledgement among urban water professionals that stakeholders should receive more attention from the utility organisations, and that they should be incorporated formally in the decision making processes, most urban water projects were and continue to be treated only as technical constructs with some minor 'social aspects' appended. This is certainly due to the lack of socio-technological expertise to weave technical and social aspects together into synergic interactions to provide a balance between technical and social values. This is recognised within hydroinformatics, but there is a singular lack of appropriate tools with which to facilitate their involvement. Experiments have been conducted by Jonoski (2000) among others. He used a tool called Aquavoice based on a sophisticated computational hydraulic model of a river flooding situation to provide non-professionals with the possibility of using the same modelling tool as the professionals but within the limits of understanding of the non-professionals. This involved the development of Aquavoice as a form of simplified driver of the computational model. With the tool the non-professional stakeholders could investigate for themselves the issues that were of most concern. With the fact that there is a lack of appropriate expertise and tools that can enable a meaningful stakeholder participation, the socio-technological application in the urban water sector plays a critical role in addressing some of the most crucial problems of our time, and these are perhaps the most important challenges facing urban hydroinformatics today.

In addition to the driver interface for complex simulation models there are opportunities with community communication tools such as Twitter® and Facebook®. These open up the means for conducting surveys, brainstorming, reporting events and circumstances, and accessing advice and support services. This is very much an open field, with a range of tools available but with the initiative and insights needed by the stakeholders so that they can take advantage of the technologies. Fortunately, the real beneficiaries of the new technologies and tools should be the developing countries, provided sufficient capacity is available in the relevant organisations.

There needs to be more research done on the role of the institutional framework in supporting water utilities and in involving the public. A continual stumbling-block is the shortage of suitable indicators of performance. Such indicators should be established internationally in different areas of urban water management so that better comparison can be made between the existence and performance of water systems in different parts of the world (UNESCO, 2009).

15.6 FUTURE OF URBAN HYDROINFORMATICS

We have reached the end of the book. Undoubtedly there remains much to be written, and it is hoped that future editions of this book will correct what is deficient and introduce new aspects and emphases. What is apparent is that there will be rapid changes occurring in urban hydroinformatics, and practitioners will need to keep under continual review the progress made in developing new technologies, instruments and insights.

15.7 KEYWORDS

By the end of the chapter you should have encountered the following terms; ensure that you are familiar with them!

data management	institutional framework	socio-economic issues
decision support systems	modelling	

15.8 QUESTIONS

What facilities for water management do you think will be centre stage in 10 years time?
How can newly urbanised areas by served efficiently by extending existing water services?

15.9 REFERENCES

Abbott M. B. (2007). Managing the inner world of infrastructure. *Proc. ICE Civ. Eng.*, 26–30.

Abbott M. B., Tumwesigye B. M. and Vojinović Z. (2006). The fifth generation of modelling in hydroinformatics. In: *7th International Conference on Hydroinformatics*, 4–8 September, Acropolis, Nice, France.

Abdullah A. F., Vojinović Z., Price R. K. and Rahman A. A. (2009). Lidar filtering algorithms and DTM generation for urban flood modelling applications: review of current algorithms and filters test. In: *8th International Conference on Urban Drainage Modelling*, 7–11 September, Tokyo, Japan.

Alfonso L. (2010). Optimization of water level monitoring network in polder systems using information theory. *Water Resour. Res* (In Press).

Barredo J., Kasanko M., McCormick N. and Lavalle C. (2003). Modeling dynamic spatial processes: simulation of urban future scenarios through cellular automata. *Landscape Urban Plann.*, **64**, 145–160.

Sanchez Torres, A., Vojinović Z. and Price R. K. (2010). Determining the route for a water main in a new urbanising area. (In Press).

Tumwesigye B. M., Vojinović Z., Jonoski A. and Abbott M. (2005). Towards the new business model in hydroinformatics: a case in urban drainage modelling. In: *10th International Conference on Urban Drainage*, 21–26 August, Copenhagen, Denmark.

UNESCO (2009). *The 3rd United Nations World Water Development Report: Water in a Changing World (WWDR-3)*. UNESCO Publishing/Earthscan, Paris, London.

WRc (2010). *Sewerage Risk Management: Managing the Risks of Service Failure*. http://srm.wrcplc.co.uk/ (accessed 4 October 2010).

Glossary

activated sludge	Sewage treatment process in which air or oxygen is forced into sewage liquor to develop a biological floc which reduces the organic content of the sewage.
advection	Conveyance of suspended or dissolved material at the speed of the fluid.
aeration	Assimilation of oxygen into the water at any free surface with the air.
aerobic	Conditions in which dissolved oxygen is present.
air entrainment	Trapping of air in the form of bubbles in the water.
anaerobic	Conditions in which free oxygen is not present.
ancillary	A structure within a sewerage system which is not a pipe or conduit. Ancillary structures include overflows, storage tanks and ponds, pumping stations, outfalls, sluice gates and flap valves.
aquifer	Water-bearing stratum within the earth's crust.
Archimedes screw pump	Screw device to lift sewage mechanically through a reasonable height.
artificial intelligence	Artificial replication of human intelligence.
artificial neural network	Mathematical algorithm to model the connection between two time series.
attribute	Feature of an object.
base flow	Sustained or dry-weather flows not directly generated by rainfall. It commonly constitutes flows generated by domestic and industrial discharge and also infiltration.
base performance	Performance of the model when it replicates the existing conditions in reality.
bathing beach	Beach that is physically safe for bathing.
bed load	Coarser sediment that is in transport next to the bed of the channel.
bed slope/gradient	Slope of the channel as defined by the bed (or the lowest points of successive cross-sections).
benching	Near horizontal surface adjacent to the channel in a manhole or inspection chamber, or a large sewer.
biochemical oxygen demand (BOD)	Mass concentration of dissolved oxygen consumed under specific conditions by the biological oxidation of organic and/or inorganic matter in water.
boundary condition	Mathematical condition at the boundary of the solution space for a differential equation.

boundary layer	Conceptual space adjacent to a physical boundary where the velocity gradient is large, and therefore the shear stress is significant.
Boussinesq coefficients	Coefficients that are approximately unity and are generated for the non-linear terms in the 1D or 2D Saint Venant equations when integrating the 3D Navier-Stokes equations.
branch (of a sewer network)	A number of pipes connected in series that form part of a sewerage system.
calibration	Allocation of values to parameters such that the physically based model best fits the observed field measurements.
catchment	Area of the land and the associated underground aquifer(s) draining to a specific point in a single watercourse.
catchment area	Area draining to an inlet, sewer or watercourse.
chamber	Part of a manhole or inspection pit manufactured as a single entity and intended to be joined with other similar units.
channel	Conveyor of flow in a well-defined path.
characteristic curve	A line in solution space for a hyperbolic equation along which a certain quantity is constant.
Chezy equation	Form of the conveyance in the friction slope proposed by Chezy.
coefficient	A constant parameter that multiplies a function of the dependent variables.
Colebrook-White equation	Form of the conveyance in the friction slope proposed by Colebrook and White.
collection system	System of conduits, generally underground pipes and associated structures, which receives and conveys sanitary waste water and/or storm water.
combined sewer overflow; storm water overflow	Device, on a combined or partially separate sewer system or at a sewerage treatment works, that relieves the system of excess flow and discharges to an adjacent surface water sewer or watercourse.
combined system	Sewer system designed to carry both wastewater and surface water in the same pipeline(s).
compartmentalised model	Model for lumped physical processes in an entity defined in geophysical space.
compressibility	Feature of a given mass of material to occupy a smaller (or greater) volume.
computational hydraulics	Computer based solutions of dynamic hydraulic equations applied to real situations in the natural environment.
computer program	Sequence of instructions written in a dedicated language for implementation on a computer to achieve a specific goal.
conceptual	As conceived by the mind.
concentration	Mass of a substance per unit volume.
conduit	Conveyor of fluid with an open or closed cross-section.
confirmation	Conclusion that an object is fit-for-purpose.
connection	General term used for the location at which one pipeline joins another pipeline or a manhole or inspection chamber.
consolidated sediment	Sediment which has become compressed through time onto the pipe or channel bed. In normal flow conditions this material is effectively non-erodible.
continuing losses	Precipitation that infiltrates the ground after the initial losses have wet the surfaces and filled the storages and water runs off the surfaces.
Courant number	Ratio of the grid speed to the wave speed.
critical depth	Flow depth where the Froude number is unity.
critical flow	Flow with Froude number of unity.
critical sewers	Sewers with the most significant consequences in the event of structural failure.

data driven	Modelling paradigm that is inductive from data.
decision support system	Modelling, data analysis and expert advice to support decision making.
deposition	Process of settling solid material from fluid suspension.
depression storage	Precipitation retained in surface depressions which does not contribute to runoff.
demand point	Node in a water distribution network at which water is delivered.
density	Mass per unit volume.
detention tank	Device for the temporary storage of sewage.
diameter	Longest distance across the interior section of a circular pipe.
digital terrain model	A digital map of the elevation of the earth's surface without additional surface features ('bare earth').
diffusion	Redistribution of material in the flow due to Brownian (molecular) movement and/or turbulent fluctuations.
dispersion	Redistribution of material within the flow.
dissipation	Loss of quantity outside the system.
dissolved	Material distributed in a liquid at molecular level.
dissolved oxygen	Oxygen distributed in a liquid at molecular level.
diversion	Structure changing the direction of some of the flow.
domestic wastewater	Wastewater discharged from kitchens, laundry rooms, lavatories, bathrooms, toilets and similar facilities.
drain	Pipeline, usually underground, designed to carry wastewater and/or surface water from a source to a sewer.
Drainage Area Plan (DAP)	A full assessment of a sewer system's performance and condition made by the responsible organisation, investigating hydraulic, operational, structural and environmental performance. It also proposes a strategy to achieve the desired levels of service.
dry weather flow	Rate of flow in a drain or sewer system in dry weather conditions.
e-coli	Pathogen Escherichia Coliform.
ecology	Pertaining to biological entities.
effluent	Wastewater or other liquid, partially or completely treated, or in its natural state, flowing out of a pipe, or treatment plant.
environment	The context of water in the atmosphere, on the land surface and below ground, and in the oceans.
erosion	Detachment and movement of soil or sedimentary deposits by the flow of water, such as over the ground surface or in a pipe or channel.
Environment Agency (EA)	An Executive Non-Departmental Public Body responsible for the protection and improvement of the environment, and the promotion of sustainable development. The EA plays a central role in delivering the Environmental policies of Central Government and the Welsh Assembly.
event (rainfall)	Single occurrence of a rainfall period before and after which there is a sufficient dry period to define its effect on the sewer system.
exfiltration	Escape of flow from a sewer into surrounding ground.
extraneous water	Unwanted flow in a drain or sewer system.
first flush	High concentrations of pollutants observed at the beginning of a flow event.
floatables	Buoyant solids such as plastic bags.
flooding	Condition where wastewater and/or surface water temporarily escapes from or cannot enter an urban stream, drain or sewer system and flows with significant depth over the surface, causing disruption to traffic and can enter buildings (see also surface flooding).

Flood Studies Report (FSR)	Provides techniques for design flood and rainfall estimation in the UK.
flow simulation	Modelling of flows in drain or sewer systems.
flow survey (sewers)	Collection of data on the hydraulic behaviour of the network at a series of selected pints over a period of up to several weeks. Ultrasonic flow gauges are usually used. Rain gauges are also used if flow response to rainfall is being collected.
foul water	Domestic or industrial waste water.
Force-fitting	Process of making arbitrary changes to a model to make it fit observed data. Should not be undertaken.
formula A	Formula for calculating the "carry on" flow at a storm water overflow as recommended by the Technical Committee on Storm Overflows and the Disposal of Storm Sewage, 1970.
friction coefficient	Coefficient used to calculate the energy gradient caused by friction.
friction slope	Energy gradient caused by friction.
Froude number	Ratio of the inertial to the gravitational forces.
fuzzy logic	Analysis where variables are given fuzzy values.
Geographic information system (GIS)	A mapping system to analyse and display geographically referenced information.
gradient	Ratio between the vertical and the horizontal projections of a pipe length.
gravity	Force attracting bodies towards the centre of the earth.
grid	Regular or irregular decomposition of the solution space.
grit	Coarse sediment with high densities.
gross solids	Solids, usually organic in nature, either floating, suspended or deposited which have a polluting effect on the receiving water. Often restricted to visible solids with one dimension greater than 25mm.
groundwater	Water present in the sub-surface strata.
gully	Structure to permit the entry of surface runoff into the sewer system. It is often fitted with a grating and a grit trap.
head loss	Energy loss at a geometric structure such as a contraction or expansion, or a weir due to turbulence.
hydraulic analysis	Assessment of the hydraulic behaviour of a system. Simulation hydraulic modelling of a sewerage network to determine its performance.
hydraulic capacity	Maximum flow a pipe of given dimensions, slope and roughness can carry (often quoted as pipe-full capacity which is a little less than the maximum capacity).
hydraulic gradient	The slope of the hydraulic grade line; the rate of change of pressure head; the ratio of the loss in the sum of pressure head and position head to the flow distance. For open channels it is the slope of the water surface, which may or may not be parallel to the invert. For closed conduits under pressure, it is the slope of the line joining the elevations to which the water would rise in pipes freely vented and under atmospheric pressure.
hydraulic jump	Transition between sub-critical and super-critical flow in an open channel.
hydraulic performance	The measure of the capacity of the system or part thereof.
hydraulic radius	The ratio of cross-sectional area of flow to the wetted perimeter of the sewer.
hydrogen sulphide	Poisonous and highly corrosive gas which dissolves in water to form sulphuric acid.
hydrograph	A graph showing, for a given point on a stream or conduit, the discharge, stage, velocity, available power, or other property of water with respect to time.

hydroinformatics The study of the flow of information and the generation and transfer of knowledge related to the dynamics of water as it manifests itself in the physical world.

hydrostatic pressure The pressure in a (flowing) body of water in which the assumption that the vertical acceleration of water particles in the flow is negligible compared to the acceleration due to gravity is valid.

hyetograph A graphical representation of the variation in rate of rainfall over time; specifically a graphical representation of average rainfall, rainfall excess rates, or volumes over specified areas during successive units of time during a storm.

hysteresis Where the water level can take more than one value for a given discharge such that one value water level is taken as the discharge increases and a second higher value is taken as the discharge decreases; see loop rating curve or transition between free surface and pressurised flow in a pipe.

impact Effect of a pollution event on the receiving water.

impermeable surface (impervious) Surface that resists the infiltration of water. Usually a measure of roof and road surfaces in simulation modelling.

implicit knowledge Knowledge that resides in the minds of individuals.

incompressible fluid The assumption that the volume of a certain mass of the fluid cannot be decreased or increased.

indicators Accepted or derived parameter which can be measured directly or determined from separate measurements, and which indicates trends in a particular referent.

industrial discharge Outflow from an industrial unit that varies enormously depending on the processes carried out in the factory.

infiltration The unintended ingress of ground water into a drainage system (also termed parasitary flow in some countries).

inflow Flow which enters the sewer; this can be generated by rainfall or an industrial discharge or other particular connection.

information That which resolves uncertainty, or explicatory meaning of data.

initial losses Precipitation that goes into wetting, filling depression storages and infiltrating before runoff begins over the surface.

inlet
(1) A connection between the catchment area and a drain or sewer for the admission of surface or storm water.
(2) A structure at the entrance end of a conduit.
(3) The upstream end of any structure through which water may flow.

inspection chamber Chamber with a removable cover constructed on a drain or sewer that provides access from surface level only, but does not permit entry of a person.

instantiation Combination of modelling software, structural data and dynamic boundary condition data with the verification of the integrity of the structural data and the calibration of model parameters.

Integrated Urban Drainage (IUD) Approach to planning or managing an urban drainage system which leads to an understanding of how different physical components interact.

intelligent agent Modelling of one or more independent entities that relate together and with the environment through simple rules.

inundation The covering of a land surface by water from a drainage system whose capacity is insufficient to convey the flow.

invert Lowest point of the internal surface of the barrel of a pipe or channel at any cross section [EN476:1997].

inverted siphon Length of gravity drain or sewer which is lower than upstream or downstream lengths to allow the pipeline to pass below an obstacle, and which consequently operates under pressure.

joint Location at which the ends of two adjacent pipe units are joined together longitudinally.

junction	The connection of two branches of a network.
kinematic viscosity	Coefficient relating the shear stress to the velocity gradient normal to the shear stress divided by the density.
kinematic wave	Disturbance in open channel flow that travels without change in peak discharge.
knowledge	Justified true belief.
knowledge management	Acquisition, analysis, archiving and application of knowledge to further business or organisational goals.
lateral	Any pipe connected to a sewer.
lateral inflow	Flow entering a channel along its length.
leakage	Potable water lost from a water distribution system through cracks in pipes and imperfect joints.
legislation	Laws applied to a specific subject.
linear reservoir	Conceptual reservoir that releases water such that the released discharge is proportional to the volume of water stored.
losses	Proportion of the precipitation that does not run off through the outfall.
Major drainage system	The above ground drainage systems. These would include watercourses and rivers which form the principal drainage pathways for catchments and the overland flow paths on river flood plains and the urban environment. These are broadly classified into two types: within channel flows or overland flow paths.
manhole	Chamber with a removable cover constructed on a drain or sewer to permit entry by personnel.
Manning equation	Form of the conveyance in the friction slope proposed by Manning.
Minor drainage system	The underground piped drainage systems which are typically sewers but could also be culverted watercourses or highway drains.
model	A series of mathematical equations in a computer developed and used with the aim of replicating the behaviour of a system.
model space	Geometric space potentially occupied by the water.
model trees	Data driven modelling technique whereby the output is related to the input through linear, piecewise and continuous functions.
monitoring	Procedure of measuring effluent characteristics such as flows or pollutants by means of instruments.
Monte Carlo analysis	Way of constructing uncertainty in model output corresponding to uncertainty in one or more input variables.
Navier-Stokes equations	Basic equations for laminar flow of a Newtonian fluid.
nearest neighbour	Data driven modelling technique whereby the output is related to the nearest neighbours in a certain space constructed out of the input variables.
network	A collection of connected nodes and links; manholes and pipes when referred to in the context of sewers.
Newtonian fluid	One in which the stress is proportional to the gradient of the velocity normal to the stress.
node	Manhole, inspection chamber, outfall, rodding eye or other significant intermediate point.
nutrients	Chemicals that provide food for bacteria and other biological entities.
object	Entity that can be recognised as a whole.
objective	Target of a project.
off-line tank	Detention tank which is off the normal path of flow in a network but which comes into operation during periods of high flows.
Office of Water Services (OFWAT)	The economic regulator for the Water Industry in England and Wales.

on-line tank A detention tank through which the flow of sewage is conveyed in a sewerage system.

orifice A constriction in a pipeline to control the rate of flow.

outfall Final length of pipeline from which sewerage is discharged to a treatment works or receiving water.

outfall flap Hinged flap that closes when the water level in the receiving water is higher than the water level in the outfall.

overflow The intentional or unintentional discharge of sewage to the environment prior to treatment.

partially separate system A sewer system, normally of two pipelines, where one pipeline carries wastewater together with a designed volume of surface water and the other pipeline carries the balance of the surface water.

peak discharge The maximum flow rate at a point in time at a specific location resulting from a given storm condition.

physically based Where the physical processes are identified and included in a model separately through mathematical algorithms.

pipeline Assembly of pipes, fittings, masonry units and joints between manholes or other structures.

pollutant Dissolved or particulate material washed into and through sewers. Pollutants when discharged into receiving waters cause an adverse environmental impact.

potable water Water treated for human consumption.

precipitation Rainfall, snow hail or any other form of water falling from the sky to the earth's surface.

Preissmann slot Conceptual (imaginary) slot inserted in the top of a pipe or other closed conduit such that the flow can always be regarded as having a free surface; first proposed by Preissmann.

pressure Isotropic force at a point in a fluid.

pressurised flow Flow in conduits that are full of the fluid.

primary treatment First level of treatment in a waste water treatment works, usually involving the mechanical removal of sediment and gross solids.

procedure Recognised sequence of tasks to achieve a specific objective.

public-private partnership Partnership between government and the private sector to manage a resource such as water.

pump Device to increase the pressure or head of the fluid between one point in a network and another.

pumping station Building, structures and equipment used to transfer sewage through a rising main or otherwise to raise the sewage.

rainfall intensity Depth of rain falling in unit time, ie volume of rain falling in unit time per unit area.

rainfall-runoff Precipitation that runs off from a catchment through the outfall.

real time control Manipulation of flows in a system through the operation of a moveable device according to certain rules and depending on measurements taken in the system.

receiving water Any body of water such as the sea, a river, stream or lake as well as an aquifer into which drain or sewer systems discharge.

rehabilitation All measures for restoring or upgrading the performance of existing drain and sewer systems.

relining Installation of sleeve on the inside of a pipe to strengthen the structural integrity of the pipe and to limit infiltration.

renewal Construction of a new sewer, on or off the line of an existing sewer, the basic function and capacity of the new sewer being similar to those of the old.

renovation Work incorporating all or part of the original fabric of the drain or sewer by means of which its current performance is improved.

repair Rectification of local damage.

replacement Construction of a new drain or sewer, on or off the line of an existing drain or sewer, the function of the new drain or sewer incorporating that of the old.

retention time	Time during which sewage is held within the pumping installation.
return period	The reciprocal of the average annual probability of exceedence of a specific flow value or event.
Reynolds number	Ratio between the inertial and viscous forces in the flow.
Reynolds stresses	Six additional stresses in the Navier-Stokes equations including turbulent fluctuations, which arise from the non-linear inertial terms in the momentum equations.
risk	Probability times the cost of damage.
roughness	Measure of the texture of the boundary surface generating frictional resistance of a pipe or channel under turbulent flow.
routing	Modelling the propagation of a disturbance in open channel flow.
runoff	Water from precipitation which flows off a surface to reach a drain, sewer or receiving water.
runoff coefficient	Factor dependent on the ground catchment, and by which the rain water quantity per unit of time is multiplied in order to indicate the flow expected to be carried to the drain or sewer system.
Saint Venant equations	Mathematical expressions of Newton's laws for gradually varying open channel flow in 1D or 2D channels.
sanitary sewer	Sewer that carries liquid and water-borne wastes from residences, commercial buildings, industrial plants, and institutions, together with relatively small quantities of ground, storm, and surface waters that are not admitted intentionally.
screen	Grid that does not permit objects greater than a certain size to pass through it.
SCS method	Soil Conservation Service method.
secondary treatment	Second level of treatment in a waste water treatment works, which aims to remove fine sediment and specific chemicals such as nutrients using chemical and biological additives.
sediment transport	The movement of solids transported in any way by a flowing liquid.
sedimentation	The process of deposition and consolidation of suspended material carried by water, waste water, or other liquids, by gravity.
self-cleansing	Ability of the flow in a drain or sewer to carry away solid particles, which would otherwise be deposited in the pipe.
self-purifying capacity	Ability of receiving waters to recover from pollution by natural processes.
sensitivity	Measure of the response of the model output to variations in an input variable or parameter.
separate system	Sewer system, normally of two pipelines, one carrying wastewater and the other surface water.
septic sewage	Anaerobic sewage containing sulfides.
sewage	Wastewater and/or surface water conveyed by a drain or sewer.
sewer	Pipeline or other construction, usually underground, designed to carry wastewater and/or surface water from more than one source.
sewer system	Network of pipelines and ancillary works which conveys wastewater and/or surface water from drains to a treatment works or other place of disposal.
sewerage	Alternative term for 'Drainage Collecting System'.
sewerage system	A complete system of pipes, pumps, basins, tanks, unit processes, and appurtenances for the collection, transporting, treating, and discharging of waste water.
soil type	Characterisation of soil into one of five classifications.
shear stress	Resistance to angular deformation of the fluid.
silt	Sediment (often soil) consisting of particles between 0.002 and 0.02 mm in equivalent diameter.
simulation	Representation of specific conditions during a specific period in a sewerage system, treatment works, river, etc., by means of a computer model.

simulation model	Representation of physical system and its time related behaviour by a computer model.
siphon	Closed conduit, a portion of which lies above the hydraulic grade line, resulting in a pressure less than atmospheric and requiring a vacuum within the conduit to start flow. A siphon utilises atmospheric pressure to effect or increase the flow of water through the conduit.
social costs	Costs incurred by society as a result of sewerage works and for which water service companies have no direct responsibility. These include unclaimed business losses due to road closures and the cost of extended journey times due to traffic diversions.
soffit	The highest point on the internal bore of a pipe (opposite: invert).
software	Set of instructions written in a specific language for a computer to execute.
software validation	Confirmation that a software product does what it says it does.
solid waste	Material that is thrown away or discarded by individuals and groups.
solution space	1D, 2D or 3D and time dimension space occupied by the solution of the algorithmic equations.
source control	Management of water generated at an urban surface by rainfall such that infiltration and retention storage is increased.
speed of sound in water	Speed at which sound travels as pressure waves in water assuming the water is slightly compressible.
spill event	Isolatable period when an overflow discharges to watercourse.
spill frequency	Number of spill events over a given period.
stability	Property of a numerical scheme for algorithmic equations such that small numerical perturbations are damped out.
stakeholder	Individuals and organisations who share an interest in, and responsibility for, solving community problems or the operation of a company.
standards	Accepted targets to be achieved in a water body or activity, which circumvents knowing in detail the consequences of the event.
storage	Retention or detention of water with a specific objective.
storm	Usually, an occurrence of an event of rainfall, snow or hail. Often used in connection with a meteorological phenomenon which is either unusual or of great magnitude, rate, or intensity.
stream network	Natural network of streams defined by gravity and the geometry of the catchment topography.
sub-critical flow	State of flow when the water velocity is less than the velocity of the small surface wave, water levels tending to be stable.
super-critical flow	State of flow when the water velocity is greater than the velocity of the small surface wave, violent fluctuations in water level being possible.
surcharge	Condition in which wastewater and/or surface water is held under pressure within a gravity drain or sewer system, but does not escape to the surface to cause flooding.
surface flooding	Condition where wastewater and/or surface water escapes from, or cannot enter, a drain or sewer system and either lies on the surface or enters buildings from the surface (see also flooding).
surface water	Water from precipitation, which has not seeped into the ground and which is discharged to the drain or sewer system directly from the ground or from exterior building surfaces.
suspended solids (SS)	Insoluble solids that either float on the surface of, or are in suspension in, water, waste water, or other liquids.
tacit knowledge	Like implicit knowledge, resides in the minds and experiences of individuals.
tank sewer	Oversized sewer which acts as a detention tank.
task	Activity to fulfil a certain goal.
tertiary treatment	Third level of treatment of waste water in a treatment works whereby pollutants after the first two levels are removed.

time-area diagram	Method of determining the rainfall-runoff from a catchment whereby surface areas are designated in terms of the time it takes for a water particle to travel from an area to the outfall.
time of concentration	Time taken for runoff to travel from the hydraulically most distant point of the catchment area to a defined point in the drain or sewer.
time series	Sequence of measurements of a particular variable made in increasing time and preferably at a fixed time interval.
total cost	Aggregate cost of a scheme over its design life, being the sum of the construction, operating and maintenance costs, all calculated at the same time base.
trade effluent	Wastewater discharge resulting wholly, or in part, from any industrial or commercial activity.
truncation error	Measure of the error remaining from a numerical scheme derived for a specific mathematical expression.
turbulence	Random fluctuations in laminar flow introduced by irregular activity at the boundaries for the flow.
unaccounted for water	Water that is removed deliberately and illegally from a water distribution network, and whose quantity is unknown.
uncertainty	Measure of the imprecision in the definition of the specific values of a measured variable.
unit hydrograph	Assumption that there is a standard time dependent response of a catchment to an instantaneous or unit impulse of rainfall, and that the runoff can be derived in terms of the summation of the responses for each unit of rainfall in an event.
UPM Manual	Urban Pollution Management Manual.
urban catchment wetness index (UCWI)	Measure of the wetness of an urban catchment at the start of a rainfall event, which is derived as a function of the daily rainfalls before the event.
urban drainage	Includes both man-made and natural elements of an urban environment.
urbanisation	The expansion of the urban area in time.
utility services	Services provided to customers and industry such as gas, electricity, telephone, cable TV and water.
validation	Confirmation that a modelling software product meets certain standards.
valve	Orifice device to control flow in a pipe under pressure.
velocity	Speed of a water particle in a particular direction at a point in a fluid.
verification	Process of proving that the hydrographs produced by the calibrated model match closely with the recorded observed information on flows and depths at various locations on the sewer network, which has not been used in the calibration.
verified model	Model that has been shown to produce acceptable predictions by comparison with independent field measurements.
viscosity	Coefficient relating the shear stress in a Newtonian fluid to the velocity gradient normal to the stress.
vorticity	Property of the flow at a point and defined as twice the angular velocity in a direction along the axis of the velocity such that a positive screw applies.
Wallingford Procedure	A design and analysis procedure for urban drainage networks. Produced by HR Wallingford and the Institute of Hydrology UK in 1981.
Wastewater Planning Users Group. (WaPUG)	CIWEM's Urban Drainage Group, with a long history of promoting best practice in the field of urban drainage.
WaPUG Code of Practice	Model building guidance document 1996.
washoff	Material eroded and conveyed from an urban surface into a drainage network by the rainfall-runoff.
wastewater	Water changed by use and discharged to a drain or sewer system.
water hammer	Pressure waves in a pipe network due to the elasticity of the pipes and/or the compressibility of the fluid.

water table	Surface within soil or rock strata at which ground water saturation occurs.
watercourse	A natural or artificial channel along which water flows.
water quality	Level of impurities or pollutants in water.
watershed	Area of the land surface and the associated underground aquifer(s) draining to a specific point in a river.
weir	Structure across a channel or pipe designed such that itonly allows flow over its crest.
weir crest level	The level of the crest of a weir.
wet weather flow	The response of the flow in a drainage network to a rainfall event.
wet well	Chamber forming a part of a sewage pumping station into which sewage discharges prior to pumping. It can include submersible pumping equipment and pipe work.
WQO	Water Quality Objective.

Index

IWA Publishing's authorised EU representative for General Product Safety Regulations is Diane D'Arras, 15 rue Duret, 75116 Paris, France, e-mail: safety@iwap.co.uk.

Printed and bound by CPI Group (UK) Ltd, Croydon, CR0 4YY
12/05/2026
02108869-0018